MW00814868

New Directions in Wireless
Communications Research

Vahid Tarokh
Editor

New Directions in Wireless Communications Research

 Springer

Editor
Vahid Tarokh
Harvard University
School of Engineering &
  Applied Sciences
33 Oxford St.
Cambridge MA 02138
USA
vahid@seas.harvard.edu

ISBN 978-1-4419-0672-4   e-ISBN 978-1-4419-0673-1
DOI 10.1007/978-1-4419-0673-1
Springer Dordrecht Heidelberg London New York

Library of Congress Control Number: 2009928505

© Springer Science+Business Media, LLC 2009
All rights reserved. This work may not be translated or copied in whole or in part without the written
permission of the publisher (Springer Science+Business Media, LLC, 233 Spring Street, New York,
NY 10013, USA), except for brief excerpts in connection with reviews or scholarly analysis. Use in
connection with any form of information storage and retrieval, electronic adaptation, computer
software, or by similar or dissimilar methodology now known or hereafter developed is forbidden.
The use in this publication of trade names, trademarks, service marks, and similar terms, even if
they are not identified as such, is not to be taken as an expression of opinion as to whether or not
they are subject to proprietary rights.

Printed on acid-free paper

Springer is part of Springer Science+Business Media (www.springer.com)

This book is dedicated to Dr. Vijay K. Bhargava of the University of British Columbia by his friends, colleagues, collaborators and former students, in deep respect and admiration for his over 30 years of extraordinary leadership and innovation in the field of communications research.

Preface

In recent years, there has been significant progress in the area of wireless system design. Wireless local area networks have had enormous success and this is expected to continue on into the foreseen future. Third-generation cellular systems have been successfully deployed and are witnessing growing consumer demand. Fourth-generation systems are being deployed in the near future. Other emerging systems include sensor networks, body area networks, wireless personal area networks and positioning systems.

The above commercial success has ignited enormous advances in research in wireless system design in the last two decades. As a result, new concepts have been invented with amazing speed. This makes it important to have books available that are close to ongoing research. This issue was first pointed out by a number of colleagues who encouraged me to edit this work based on contributions from some of the foremost researchers in the area of wireless communications.

The book addresses a selected number of important emerging topics in wireless system design. The authors were chosen from among leading international researchers in their area. No system design is possible without understanding the underlying channels, thus Chapter 1 focuses on this topic. Much of modern systems use orthogonal division frequency multiplexing (OFDM) as the underlying air interface. This is an old technology, but only recently has been put to massive commercial use in wireless local area networks and cellular systems. Chapter 2 discusses this important technology in great detail. Contention between multiple users, channel estimation, synchronization, and various other issues in system design requires sequences with low correlation properties and this is the topic of Chapter 3. Chapter 4 considers the scheduling problem, i.e., allocation of resources to multiple users in a wireless system, an issue of critical importance in a multiuser system. Chapter 5 considers the design of iterative receivers in order to achieve efficiencies near theoretical limits. Chapter 6 considers multiple-input multiple-output (MIMO) and multiuser multiple-input multiple-output (MU-MIMO) systems. These techniques promise significant increase in spectral efficiencies of wireless systems. Yet another promising way to increase spectral efficiency is by producing directional beam patterns and Chapter 7 considers the use of distributed nodes for this purpose.

This is a special co-operation between elements of a network. Other methods of co-operation are discussed in Chapter 8. Interference is the main limiting factor in reducing the spectral efficiency of wireless systems, thus management, avoidance, and rejection of interference are important topics discussed in Chapter 9. An interesting solution to avoid interference is by using intelligent devices that can sense the environment and avoid transmission in occupied space–time–frequency resources. These cognitive devices have witnessed a recent explosion of interest for transmission in TV band by secondary devices and other applications. This is the topic of Chapter 10. Chapters 11 and 12 study bidirectional-coded co-operation and their more general extensions known as network coding. Finally Chapters 13, 14, and 15 discuss modern fourth-generation standards (LTE and WiMax) and standard proposals (UMB) and the realization of some of the topics covered in previous chapters in practice.

The editor is grateful to Ian Blake for his assistance in compiling this volume.

Cambridge, MA   *Vahid Tarokh*
                                                                                 December, 2008

Contents

1 Measurement and Modeling of Wireless Channels 1
David G. Michelson and Saeed S. Ghassemzadeh
   1.1   Introduction 1
   1.2   A Brief History 3
   1.3   Characterization of Wireless Channels 4
   1.4   Development of New Channel Models 9
   1.5   Measurement of Wireless Channels 11
   1.6   Recent Advances in Channel Modeling 13
        1.6.1   Channel Models for Ultrawideband Wireless Systems ... 13
        1.6.2   Channel Models for MIMO-Based Wireless Systems 16
        1.6.3   Channel Models for Body Area Networks 18
        1.6.4   Channel Models for Short-Range Vehicular Networks ... 20
        1.6.5   Channel Models for 60 GHz and Terahertz Systems 22
   1.7   Conclusions 24
   References 25

2 OFDM: Principles and Challenges 29
Nicola Marchetti, Muhammad Imadur Rahman, Sanjay Kumar,
and Ramjee Prasad
   2.1   Introduction 29
   2.2   History and Development of OFDM 30
   2.3   The Benefit of Using Multi-carrier Transmission 31
   2.4   OFDM Transceiver Systems 34
   2.5   Analytical Model of OFDM System 35
        2.5.1   Transmitter 35
        2.5.2   Channel 37
        2.5.3   Receiver 38
        2.5.4   Sampling 41
   2.6   Advantages of OFDM System 42
        2.6.1   Combating ISI and Reducing ICI 42
        2.6.2   Spectral Efficiency 43

        2.6.3   Some Other Benefits of OFDM System   44
   2.7 Disadvantages of OFDM System   45
        2.7.1   Strict Synchronization Requirement   45
        2.7.2   Peak-to-Average Power Ratio (PAPR)   45
        2.7.3   Co-channel Interference in Cellular OFDM   46
   2.8 OFDM System Design Issues   46
        2.8.1   OFDM System Design Requirements   46
        2.8.2   OFDM System Design Parameters   47
   2.9 Multi-carrier Based Access Techniques   49
        2.9.1   Definition of Basic Schemes   49
   2.10 Single-Carrier vs Multi-carrier, TDE vs FDE   52
        2.10.1   Single-Carrier FDE   52
        2.10.2   Single-Carrier vs Multi-carrier, FDE vs TDE   54
        2.10.3   Analogies and Differences Between OFDM and SCFDE .   54
        2.10.4   Interoperability of SCFDE and OFDM   56
   2.11 OFDMA: An Example of Future Applications   58
   2.12 Conclusions   60
   References   61

3 **Recent Advances in Low-Correlation Sequences**   63
   Gagan Garg, Tor Helleseth, and P. Vijay Kumar
   3.1 Introduction   63
   3.2 Cyclic Hadamard Difference Sets   64
        3.2.1   Introduction   64
   3.3 The Merit Factor of Binary Sequences   71
        3.3.1   Introduction   71
   3.4 Low-Correlation QAM Sequences   76
        3.4.1   Preliminaries   77
        3.4.2   Quaternary Family $\mathbb{A}$ 78
        3.4.3   Canonical 16-QAM Family $\mathbb{CQ}$ 78
        3.4.4   Extensions and Improvements   80
        3.4.5   Example: Generation of a 16-QAM Sequence   83
   3.5 Low-Correlation Zone Sequences   84
   3.6 Additional Notes   86
        3.6.1   Merit Factor   86
        3.6.2   QAM Sequences   87
        3.6.3   Low-Correlation Zone Sequences   87
   3.7 Conclusions   88
   References   88

4 **Resource Allocation in Wireless Systems**   93
   Jon W. Mark and Lian Zhao
   4.1 Introduction   93
   4.2 System Model   95
   4.3 The Inverse of Γ_S   99
   4.4 Convergence of Power Distribution Law 100

        4.4.1   With Zero Disturbance 100
        4.4.2   With Nonzero Disturbance 102
        4.4.3   With Power Constraints 103
        4.4.4   Capacity Analysis 105
  4.5   Optimal Data Rate Allocation 106
        4.5.1   Assumptions 106
        4.5.2   Optimal Spreading Factor (OSF) Selection 107
        4.5.3   Rate Selection for GRP 107
  4.6   Joint Rate and Power Adaptation 108
        4.6.1   OSF-PC 108
        4.6.2   GRP-PC 109
  4.7   Numerical Results 111
  4.8   Conclusions 115
  References 116

5   **Iterative Receivers and Their Graphical Models** 119
    Ezio Biglieri
    5.1   Introduction 119
    5.2   MAP Symbol Detection 119
          5.2.1   Factor Graphs and the Sum–Product Algorithm ... 121
          5.2.2   The Basic Factorization 123
    5.3   Channel and Codes: A Menagerie of Factor Graphs 124
          5.3.1   Modeling the Channel 124
          5.3.2   Modeling the Code 126
    5.4   Equalization 127
    5.5   Multiuser Detection 130
    5.6   MIMO Detection 131
    5.7   Multilevel Coded Modulation 133
    5.8   Convergence of the Iterative Algorithm 133
    5.9   Conclusions 135
    References 136

6   **Fundamentals of Multi-user MIMO Communications** 139
    Luca Sanguinetti and H. Vincent Poor
    6.1   Introduction 139
    6.2   System Model 140
    6.3   Capacity 141
          6.3.1   Capacity Region of the Gaussian MIMO MAC 142
          6.3.2   Gaussian MIMO Broadcast Channel 150
    6.4   Open- and Closed-Loop Systems 158
          6.4.1   Open-Loop Systems 159
          6.4.2   Closed-Loop Systems 160
    6.5   System Design 160
          6.5.1   Receiver Design for Uplink Transmissions 160
          6.5.2   Transmitter Design for Downlink Transmissions . 161
    6.6   Limited Feedback Systems 167

6.6.1   Channel Quantization 167
6.6.2   Random Beamforming 168
6.6.3   Transceiver Optimization 169
6.7   Conclusions 169
References 170

7   Collaborative Beamforming 175
Hideki Ochiai and Hideki Imai
7.1   Introduction 175
7.2   System Model and Beam Patterns of Fixed Nodes 177
7.2.1   Array Factor and Beam Pattern 178
7.2.2   Beam Patterns of Linear Arrays 180
7.2.3   Beam Patterns of Circular Arrays 183
7.3   Collaborative Beamforming by Randomly Distributed Nodes 185
7.3.1   Definition 186
7.3.2   Average Beam Patterns 188
7.3.3   Distribution of Beam Patterns 190
7.3.4   Distribution of Maxima in Sidelobe 194
7.4   Conclusions 196
References 197

8   Cooperative Wireless Networks 199
Behnaam Aazhang, Chris B. Steger, Gareth B. Middleton,
and Brett Kaufman
8.1   Introduction 199
8.1.1   Overview 199
8.1.2   Physical Layer Cooperation 200
8.2   System Model 202
8.2.1   Wide Area Network 202
8.2.2   Multiple Flows and Flow Priority 203
8.2.3   Cooperative Building Blocks 204
8.3   Learning About the Environment: Network State Information 205
8.3.1   NSI Overhead Management 206
8.3.2   NSI Metric 206
8.4   Finding the Optimal Cooperative Path 207
8.4.1   Routing Cooperative Paths 207
8.4.2   Trellis Representation 208
8.4.3   Timing, Interference, and Duplexing Management 209
8.4.4   Traversal Algorithms 210
8.5   Network Discovery 210
8.5.1   Filling the Trellis: Gathering States, Edges, and NSI 211
8.5.2   Filling the Trellis: Metanodes 212
8.6   Conclusions 212
References 213

9   Interference Rejection and Management 217
Arun Batra, James R. Zeidler, John G. Proakis, and Laurence B. Milstein
  9.1   Introduction 217
  9.2   Self-Interference Among Cooperating Systems 218
      9.2.1   Interference Suppression to Enable Spectrum Sharing ... 218
      9.2.2   Effects of Interference on Channel State Estimation 220
  9.3   Interference Mitigation in Block-Modulated Multicarrier
      Systems 223
      9.3.1   Interference Mitigation in an Uncoded Multicarrier
          System 224
      9.3.2   Interference Mitigation in Coded Multicarrier Systems .. 233
      9.3.3   Doppler Sensitivity of OFDM in Mobile Applications ... 235
  9.4   Interference Suppression in Broadcast MIMO Systems 236
      9.4.1   Linear Precoding of the Transmitted Signals 237
      9.4.2   Nonlinear Precoding of the Transmitted Signals:
          The QR Decomposition 239
      9.4.3   Vector Precoding 244
      9.4.4   Lattice Reduction Method for Precoding 246
  9.5   Conclusions 247
  References 248

10   Cognitive Radio: From Theory to Practical Network Engineering .. 251
Ekram Hossain, Long Le, Natasha Devroye, and Mai Vu
  10.1   Introduction 251
  10.2   Information-Theoretic Limits of Cognitive Networks 253
      10.2.1   Cognitive Behavior: Interference Avoidance, Control,
           and Mitigation 253
      10.2.2   Information-Theoretic Basics....................... 254
      10.2.3   Interference Avoidance: Spectrum Interweave 255
      10.2.4   Interference Control: Spectrum Underlay 256
      10.2.5   Interference Mitigation: Spectrum Overlay 259
  10.3   Cognitive Sensing with Side Information 264
  10.4   Interference Analysis 266
      10.4.1   A Network with Beacons 267
      10.4.2   A Network with Primary Exclusive Regions 268
  10.5   Practical Cognitive Network Engineering: Interference Control
      Approach 269
      10.5.1   Single-Antenna Case.............................. 270
      10.5.2   Multiple Antenna Case 273
  10.6   Practical Cognitive Network Engineering: Interference
      Avoidance Approach 273
      10.6.1   Single-Hop Case 274
      10.6.2   Multi-hop Case 282
  10.7   Conclusions 283
  References 284

11 Coded Bidirectional Relaying in Wireless Networks 291
    Petar Popovski and Toshiaki Koike-Akino
    11.1   Introduction 291
    11.2   Preliminaries 293
    11.3   Two-Way Relaying with Decoding at the Relay 295
            11.3.1   The Uplink Phase 295
            11.3.2   The Broadcast Phase 296
            11.3.3   Improved Broadcast Strategies 297
            11.3.4   Numerical Illustration 300
    11.4   Two-Way Relaying Without Decoding at the Relay 302
            11.4.1   Amplify-and-Forward (AF) 302
            11.4.2   Denoise-and-Forward (DNF) 303
            11.4.3   Compress-and-Forward (CF) 305
            11.4.4   Numerical Illustration and Variations 306
    11.5   Achieving the Two-Way Rates with Structured Codes 307
            11.5.1   Parity-Check Codes for Binary Symmetric Channels 307
            11.5.2   Gaussian Channel 309
    11.6   Signaling Constellations for Finite Packet Lengths........... 312
            11.6.1   XOR Denoising 312
            11.6.2   Adaptive Denoising with Quintary Cardinality 313
            11.6.3   End-to-End Throughput Performance 314
    11.7   Conclusions 315
    References 316

**12 Minimum Cost Subgraph Algorithms for Static and Dynamic
    Multicasts with Network Coding** 317
    Fang Zhao, Muriel Médard, Desmond Lun, and Asuman Ozdaglar
    12.1   Introduction 317
    12.2   Problem Formulation 320
            12.2.1   Wireline Networks 320
            12.2.2   Wireless Networks 322
    12.3   Decentralized Min-cost Subgraph Algorithms for Static Multicast . 324
            12.3.1   Subgradient Method for Decentralized Subgraph
                     Optimization 325
            12.3.2   Convergence Rate Analysis 328
            12.3.3   Initialization and Primal Solution Recovery 334
            12.3.4   Simulation Results 335
    12.4   Min-cost Subgraph Algorithms for Dynamic Multicasts 340
            12.4.1   Nonrearrangeable Algorithm 340
            12.4.2   Rearrangeable Algorithms 342
            12.4.3   Simulation Results 345
    12.5   Conclusions 347
    References 348

13 Ultra Mobile Broadband (UMB) 351
Masoud Olfat
13.1 Introduction 351
13.2 UMB Overall Architecture 352
13.3 UMB Physical Layer 355
    13.3.1 Superframe Structure 356
    13.3.2 UMB FL Channelization 360
    13.3.3 Reverse Link in UMB 366
13.4 UMB MAC Layer 374
13.5 Other PHY/MAC-layer features in UMB 384
13.6 Conclusions 386
References 386

14 Mobile WiMAX 389
Masoud Olfat
14.1 Introduction 389
14.2 Standardization Process 390
    14.2.1 WiMAX Forum 391
14.3 WiMAX Network Architecture 393
    14.3.1 Network Reference Models 394
    14.3.2 ASN profiles 395
    14.3.3 Mobility Management 397
14.4 Physical Layer 399
    14.4.1 S-OFDMA Frame Structure 401
    14.4.2 Subchannel Permutation 402
    14.4.3 Frame Structure 406
    14.4.4 Channel Coding 407
    14.4.5 Multiple Antenna Modes in Mobile WiMAX 409
    14.4.6 Power Control and Link Adaptation 413
14.5 Medium Access Control Layer 416
    14.5.1 Quality of Service 420
    14.5.2 Power Saving Mode 421
    14.5.3 Multicast Broadcast Services 422
    14.5.4 Handoff 423
    14.5.5 Security and Authentication in WiMAX 425
14.6 WiMAX Performance 426
14.7 Future Work Toward IMT-Advanced 427
    14.7.1 Conclusions 428
References 429

**15 An Overview of 3GPP Long-Term Evolution Radio Access
Network** 431
Sassan Ahmadi
15.1 Introduction 431
    15.1.1 Chronology of 3GPP Air Interface Technology
        Development 432

        15.1.2   3GPP LTE System Requirements 433
15.2   Overall Network Architecture 434
15.3   LTE Protocol Structure..................................... 436
15.4   Overview of the LTE Physical Layer 438
        15.4.1   Multiple Access Schemes........................... 438
        15.4.2   Operating Frequencies and Bandwidths 439
        15.4.3   Frame Structure 442
        15.4.4   Physical Resource Blocks.......................... 443
        15.4.5   Modulation and Coding 444
        15.4.6   Physical Channel Processing 444
        15.4.7   Reference Signals 446
        15.4.8   Physical Control Channels 448
        15.4.9   Physical Random Access Channel................... 450
        15.4.10 Cell Search...................................... 452
        15.4.11 Link Adaptation................................. 452
        15.4.12 Multi-antenna Techniques in LTE 453
15.5   Overview of the LTE Layer 2 454
        15.5.1   Logical and Transport Channels 455
        15.5.2   ARQ and HARQ in LTE........................... 458
        15.5.3   Packet Data Convergence Sublayer (PDCP)............ 458
15.6   Radio Resource Control Functions (RRC).................... 458
15.7   Mobility Management and Handover in LTE 460
15.8   LTE Performance 462
15.9   Future Work Toward IMT-Advanced 462
15.10 Conclusions 464
References 464

Index 467

List of Contributors

Behnaam Aazhang
Center for Multimedia Communication, Department of Electrical and Computer Engineering, Rice University, e-mail: aaz@rice.edu

Sassan Ahmadi
Intel Corporation, Oregon, e-mail: sassan.ahmadi@intel.com

Arun Batra
Department of Electrical and Computer Engineering, University of California, San Diego, e-mail: abatra@ucsd.edu

Ezio Biglieri
Universitat Pompeu Fabra, Barcelona, Spain, e-mail: e.biglieri@ieee.org

Robert Calderbank
Princeton University, Princeton, NJ, e-mail: calderbk@princeton.edu

Natasha Devroye
Department of Electrical and Computer Engineering, University of Illinois at Chicago, e-mail: devroye@ece.uic.edu

Gagan Garg
Department of Computer Science and Automation, Indian Institute of Science, Bangalore, e-mail: gagan.garg@gmail.com

Saeed S. Ghassemzadeh
AT&T Labs - Research, Florham Park, NJ, e-mail: saeedg@research.att.com

Tor Helleseth
Department of Informatics, University of Bergen, Bergen, Norway, e-mail: torh@ii.uib.no

Ekram Hossain
Department of Electrical and Computer Engineering, University of Manitoba, e-mail: ekram@ee.umanitoba.ca

Stephen D. Howard
Defence Science and Technology Organisation, Edinburgh, Australia,
e-mail: Stephen.Howard@dsto.defence.gov.au

Hideki Imai
Chuo University, Bunkyo, Tokyo, e-mail: h-imai@elect.chuo-u.ac.jp

Brett Kaufman
Center for Multimedia Communication, Department of Electrical and Computer
Engineering, Rice University, e-mail: bkaufman@rice.edu

Toshiaki Koike-Akino
School of Engineering and Applied Sciences, Harvard University,
e-mail: koike@seas.harvard.edu

P. Vijay Kumar
Department of Computer Science and Automation, Indian Institute of Science,
Bangalore, e-mail: vijayk@usc.edu

Sanjay Kumar
Center for TeleInFrastruktur (CTIF), Aalborg University, Denmark,
e-mail: sk@es.aau.dk

Long Le
Department of Aeronautics and Astronautics, Massachusetts Institute of
Technology, e-mail: longble@mit.edu

Desmond Lun
The Broad Institute of MIT and Harvard, e-mail: dlun@broad.mit.edu

Nicola Marchetti
Center for TeleInFrastruktur (CTIF), Aalborg University, Denmark,
e-mail: nm@es.aau.dk

Jon W. Mark
Department of Electrical and Computer Engineering, University of Waterloo,
e-mail: jwmark@bbcr.uwaterloo.ca

Muriel Médard
Massachusetts Institute of Technology, e-mail: medard@mit.edu

David G. Michelson
University of British Columbia, e-mail: davem@ece.ubc.ca

Gareth B. Middleton
Center for Multimedia Communication, Department of Electrical and Computer
Engineering, Rice University, e-mail: gbmidd@rice.edu

Laurence B. Milstein
Department of Electrical and Computer Engineering, University of California,
San Diego, e-mail: milstein@ece.ucsd.edu

Hideki Ochiai
Yokohama National University, Hodogaya, Yokohama, e-mail: hideki@ynu.ac.jp

Masoud Olfat
Clearline, Herndon, VA, e-mail: masoud.olfat@clearline.com

Asuman Ozdaglar
Massachusetts Institute of Technology, e-mail: asuman@mit.edu

Vincent Poor
Department of Electrical Engineering, Princeton University,
e-mail: poor@ princeton.edu

Petar Popovski
Department of Electronic Systems, Aalborg University, Aalborg, Denmark
e-mail: petarp@es.aau.dk

Ramjee Prasad
Center for TeleInFrastruktur (CTIF), Aalborg University, Denmark,
e-mail: prasad@es.aau.dk

John G. Proakis
Department of Electrical and Computer Engineering, University of California,
San Diego, e-mail: jproakis@ucsd.edu

Muhammad Imadur Rahman
Ericsson Research, Kista, Sweden, e-mail: muhammad.imadur.rahman@
ericsson.com

Luca Sanguinetti
Dipartimento di Ingegneria dell'Informazione, Universita di Pisa, Pisa, Italy,
e-mail: luca.sanguinetti@iet.unipi.it

Songsri Sirianunpiboon
Defence Science and Technology Organisation, Edinburgh, Australia,
e-mail: Songsri.Sirianunpiboon@dsto.defence.gov.au

Chris Steger
Center for Multimedia Communication, Department of Electrical and Computer
Engineering, Rice University, e-mail: chrissteger@gmail.com

Mai Vu
Division of Engineering and Applied Sciences, Harvard University,
e-mail: maivu@deas.harvard.edu

James R. Zeidler
Department of Electrical and Computer Engineering, University of California,
San Diego, e-mail: zeidler@ece.ucsd.edu

Fang Zhao
Massachusetts Institute of Technology, e-mail: zhaof@mit.edu

Lian Zhao
Department of Electrical and Computer Engineering, Ryerson University, Toronto,
e-mail: lzhao@ee.ryerson.ca

Acronyms

2G	second generation
3G	third generation
3GPP	The Third Generation Partnership Project
3GPP2	The Third Generation Partnership Project 2
4G	fourth generation
AAA	accounting, authorization, authentication
AAS	advanced antenna system
AF	amplify and forward
AFD	average fade duration
AK	authentication key
AM	adaptive modulation
AMC	adaptive modulation and coding
ARQ	automatic repeat request
ASN	access service network
AWGN	additive white Gaussian noise
BAN	body area network
BC	broadcast channel
BCH	broadcast channel
BD	block diagonalization
BEP	bit error probability
BER	bit error rate
BF	beamforming
BLAST	Bell Labs Layered Space Time
BPA	belief propagation algorithm
BPSK	binary phase shift keying
BS	base station
BSC	binary symmetric channel
BTC	block turbo code
BWA	broadband wireless access
CC	convolutional code
CCE	control channel elements

CCH	common control channel
CCI	co-channel interference
CDMA	code division multiple access
CF	compress and forward
CID	connection ID
CIR	channel impulse response
CIR	carrier-to-interference ratio
CLPC	closed-loop power control
CNR	channel-to-noise ratio
CQI	channel quality indicator
CQICH	channel quality indicator channel
CR	cognitive radio
CSI	channel state information
CSN	connectivity service network
CTF	channel transfer function
CWG	Certification Working Group
DCD	downlink channel descriptor
DE	density evolution
DF	decode and forward
DFE	decision feedback equalizer
DL	downlink
DNF	denoise and forward
DPC	dirty paper coding
DRCH	distributed resource channel
DS	direct sequence
DS-CDMA	direct sequence code division multiple access
DSA	dynamic spectrum allocation
DSL	digital subscriber line
DSSS	direct sequence spread spectrum
EAP	extensible authentication protocol
ECC	error correction coding
ECM	EPS connection management
EIRP	effective isotropic radiated power
EM	expectation maximization
EPC	evolved packet care
EPS	evolved packet system
ERV	error vector magnitude
E-UTRAN	evolved UTRAN
EXIT	extrinsic information transfer
FA	foreign agent
FCH	frame control header
FDD	frequency division duplex
FDMA	frequency division multiple access
FDE	frequency domain equalization
FEC	forward error correction

FFT	fast Fourier transform
FH	frequency hopping
FHSS	frequency hopping spread spectrum
FL	forward link
FMT	filtered multitone
GBC	Gaussian broadcast channel
GMW	Gordon, Mills, and Welch (sequences)
GP	guard period
GRP	greedy rate packing
GW	gateway
GZF	greedy zero forcing
H-ARQ	hybrid ARQ
IC	interference cancellation
ICI	inter-carrier interference
IETF	Internet Engineering Task Force
IP	Internet Protocol
ISI	inter-symbol interference
ITS	intelligent transportation system
ITU	International Telecommunication Union
JDF	joint decode and forward
LAN	local area network
LBF	linear beam forming
LBS	location-based service
LC	linear combining
LDPC	low-density parity check
LLR	log-likelihood ratio
LTE	long-term evolution
MA	multiple access
MAC	medium access control
MAI	multiple access interference
MAN	metropolitan area network
MAP	maximum a posteriori
MBS	multicast broadcast services
MBSFN	multicast and broadcast single frequency network
MC	multi-carrier
MC-CDMA	multi-carrier code division multiple access
MCH	multicast channel
MFSK	M-ary frequency shift keying
MIMO	multiple input multiple output
ML	maximum likelihood
MLSE	maximum likelihood sequence estimation
MLSSE	maximum likelihood symbol-by-symbol estimation
MMSE	minimum mean square error
MMSEC	minimum mean square error combining
MPC	multipath component

MRC	maximal ratio combining
MS	mobile station
MT	multi-tone or mobile terminals
MU	multi-user
MuD	multi-user diversity
MUD	multi-user detection
NAP	network access provider
NBI	narrowband interference
NLOS	non-line of sight
NRM	network reference model
NSI	network state information
NWG	Network Working Group
OFDD	orthogonal frequency division duplexing
OFDM	orthogonal frequency division multiplexing
OFDMA	orthogonal frequency division multiple access
OSF	optimal spreading factor
OTA	over-the-air activation
PAM	pulse amplitude modulation
PAPR	peak-to-average power ratio
PC	power control
PCCH	paging control channel
PDCP	Packet Data Convergence Protocol
PDCCH	physical downlink control channel
PDN	packet data network
PEF	prediction error filter
PER	packet error rate = probability error rate
PHS	payload header suppression
PHY	physical layer
PMK	pairwise master key
PMP	point to multipoint
POMDP	partially observable Markov decision process
PSK	phase shift keying
PTS	pilot time slot
PUSC	partial usage subchannel
PUCCH	physical uplink control channel
QAM	quadrature amplitude modulation
QoS	quality of service
QPSK	quadrature phase shift keying
RACH	random access channel
RBF	random beamforming
RF	radio frequency
RLC	radio link control
RRM	radio resource management
RRA	radio resource assignment
RRC	radio resource control

RSSI	received signal strength indicator
RTD	round trip delay
RTTG	receive transmit transition gap
RWG	Regulatory Working Group
SA	security association
SAE	system architecture evolution
SAP	service access point
SC	single carrier
SDMA	space division multiple access
SDU	service data unit
SF	space frequency
SFBC	space frequency block code
SGW	service gateway
SIMO	single input multiple output
SINR	signal to interference+noise ratio
SIR	service to interference ratio
SM	spatial multiplexing
SNR	signal-to-noise ratio
SOFDMA	scalable OFDMA
SPA	sum product algorithm
SPWG	Service Provider Working Group
ST	space–time
STBC	space–time block code
STC	space–time coding
SS	spread spectrum
TCM	trellis-coded modulation
TCP	transmission control protocol
TD	transmit diversity
TDE	time domain equalization
TDMA	time division multiple access
TEK	traffic encryption key
TRAN	terrestrial radio access network
TTG	transmit/receive transition gap
TTI	transmission time intervals
TUSC	tile usage of subchannels
TWG	Technical Working Group
UCD	uplink channel descriptor
UL	uplink
ULP	ultra-low power
UMTS	universal mobile telecommunication systems
UTRAN	universal terrestrial radio access network
UWB	ultra-wide band
VBLAST	Vertical Bell Labs Layered Space–Time Architecture
VNA	vector network analyzer
VoIP	voice over IP

VSA	vector signal analyzer
WCDMA	wideband code division multiple access
W-OFDM	wideband orthogonal frequency division multiplexing
WAN	wide area network
WiMAX	worldwide interoperability for microwave access
WLAN	wireless local area network
WMAN	wireless metropolitan area network
ZF	zero forcing

Chapter 1
Measurement and Modeling
of Wireless Channels

David G. Michelson and Saeed S. Ghassemzadeh

1.1 Introduction

As wireless signals traverse the path from a transmitter to a receiver, they will be diffracted, scattered, and absorbed by the terrain, trees, buildings, vehicles, and people that comprise the propagation environment. In the process, the signal may be distorted or impaired in various ways. The presence of obstructions along the path may cause the signal to experience greater attenuation than it would under free space conditions. If the signal is scattered by obstacles located throughout the coverage area, replicas of the signal may take multiple paths from the transmitter to the receiver. Because the replicas will arrive at the receiver after different delays, the signal will experience *time dispersion*. Because the replicas will also arrive from different directions, the signal will experience *angular dispersion*. If either the scatterers or one of the terminals is in motion, rapid changes in the phase relationship between multipath components will cause the signal to fade randomly, perhaps deeply. Such variation in received signal strength over time is equivalent to *frequency dispersion*. The correlation between fading observed at the output of adjacent receiving antennas will depend upon the type and configuration of the antennas and the range of angles over which the incident signals arrive.

The objective of channel modeling is to capture our knowledge and understanding of the manner in which the propagation environment impairs and distorts wireless signals in a form useful in the design, test, and simulation of wireless communication systems. Designers and developers use such channel models to predict and compare the performance of wireless communication systems under realistic conditions

David G. Michelson
University of British Columbia, Vancouver, BC, Canada,
e-mail: davem@ece.ubc.ca

Saeed S. Ghassemzadeh
AT&T Labs - Research, Florham Park, NJ, USA,
e-mail: saeedg@research.att.com

V. Tarokh (ed.), *New Directions in Wireless Communications Research*,
DOI 10.1007/978-1-4419-0673-1_1,
© Springer Science+Business Media, LLC 2009

and to devise and evaluate methods for mitigating the impairments and distortions that degrade wireless signals. The importance of channel models in wireless system design has long been recognized. Indeed, some have proclaimed that

> Of all the research activities related to mobile radio that have taken place over the years, those involving characterisation and modeling of the radio propagation channel are among the most important and fundamental [1].

Channel models are the basis for the software simulators, channel emulators, and RF planning tools that are used during the design, implementation, testing, and deployment of wireless communication systems, as summarized in Fig. 1.1. They can also be used to precisely define the degree of impairment that a wireless system must be able to tolerate in order to (1) meet the requirements for certification by standards groups and/or (2) comply with contractual obligations.

Like any other mathematical model, a channel model is *an abstract, simplified, mathematical construct that describes a portion of reality.* In order to limit its complexity, a channel model must necessarily focus on those aspects of the channel that affect the performance of a system of interest and ignore the rest. As researchers develop more sophisticated signaling schemes in order to deliver faster, more reliable communications, it will be necessary to develop new channel models that capture the nature of the relevant impairments and their dependence on the environment. As systems are deployed in ever more demanding environments and, in some cases, in higher frequency bands, it will be necessary to extend existing models.

In this chapter, we review and summarize recent progress in measurement and modeling of wireless channels for mobile and personal communication systems and identify common issues. In Section 1.2, we present a brief history of the field. In Section 1.3, we review the approaches used to characterize wireless channels and propagation environments. In Section 1.4, we explore the process by which new channel models are developed. In Section 1.5, we consider the methods and approaches used to measure wireless channels. In Section 1.6, we review some of

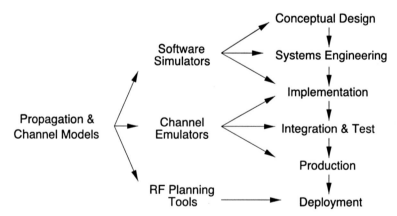

Fig. 1.1 The roles of channel models in the product development process

the key milestones achieved by the channel modeling community during the past decade. In Section 1.7, we conclude with some general remarks.

1.2 A Brief History

The need to understand and characterize wireless channels has been recognized since the earliest days of wireless communications. During the 1920s and 1930s, researchers began intensive studies of the ionosphere and its effect on wireless propagation at high frequencies. Such studies introduced methods such as swept-frequency channel sounding and concepts such as wide-sense stationary uncorrelated scattering (WSSUS) channels that are still used in various forms today [2]. The development of radar at the MIT Radiation Lab during the Second World War led to pioneering work concerning the effect of the troposphere, hydrometeors, and ground reflections on radiowave propagation at very high frequencies [3]. During the 1950s, the insights gained were used to plan and deploy the first long-distance point-to-point microwave systems. Many of these tools and techniques were later adapted for use in land mobile radio and cellular telephony deployments.

During the 1960s and early 1970s, several key discoveries ushered in the modern era of channel modeling for mobile and personal communication systems. Okumura et al. [4] revolutionized planning of mobile radio systems that operate over a broad range of frequencies between 100 MHz and 1 GHz by demonstrating that path loss in urban and suburban macrocell environments could be simply modeled in terms of the distance between the base station and remote terminal, the heights of their respective antennas above ground level, and the nature of the intervening terrain. Clarke [5] and others helped to transform mobile radio propagation from an empirical to an analytical science by showing that the U-shaped Doppler spectrum characteristic of signals received by mobile terminals could be modeled by a scenario in which incoming radio waves (1) propagate in the horizontal plane, (2) arrive over a uniform distribution of azimuthal angles, and (3) are received by an omnidirectional antenna.

Development of a rigorous treatment of linear time-varying wideband channels by Bello [6] provided essential tools and insights for analyzing wideband wireless channels that vary over time, e.g., due to changes in either the propagation environment or the location of the mobile terminal. The introduction of the spread spectrum cross-correlation technique for measuring the magnitude of the channel impulse response (CIR), i.e., the *power delay profile* (PDP), by Cox [7], made it possibly to routinely characterize time dispersion by wideband mobile channels. The channel measurement results later obtained by various research groups using this technique provided a solid foundation for the analysis and simulation of the first generation of digital mobile radio systems. By the early 1970s, it was clear that thorough characterization of the wireless channel was an essential first step in devising methods for achieving good link- and system-level performance in the presence of channel impairments and distortions. Many of these pioneering results were captured in a widely cited volume prepared by researchers at Bell Labs [8].

During the 1980s, the pace of development in mobile and personal communications increased dramatically as (1) business and private consumers expressed an unprecedented demand for wireless communications technology and (2) spectrum regulators opened up new spectrum, authorized new services, and set new goals for both performance and spectral efficiency. This prompted more intensive efforts to characterize the propagation environment and to develop the technologies required to realize next-generation systems. The European COST[1] 207 action concerning *Digital Land Mobile Communications* was conducted from 1984 to 1988 with a mandate to provide a firm technical foundation for development of GSM, the European standard for second-generation cellular telephony. Prior to COST 207, mobile radio propagation researchers and research groups tended to operate fairly independently and communicate mostly through conferences and journals. COST 207 brought together industry, government, and academic researchers from across Europe under a common umbrella and thereby encouraged more formal collaboration between channel modeling researchers. A key contributor to its success was the establishment of a mechanism by which those who would use the channel modeling results to evaluate alternative technology strategies and proposals could contribute to project goals and priorities.

The success of COST 207 set the stage for follow-on ventures, including COST 231 – *Digital Mobile Radio Towards Future Generation Systems* (1989–1996), COST 259 – *Wireless Flexible Personalised Communications* (1996–2000), COST 273 – *Towards Mobile Broadband Multimedia Networks* (2001–2005), and COST 2100 – *Pervasive Mobile and Ambient Wireless Communications* (2007–2009). The success of the COST actions encouraged similar collaborative channel modeling activities by standards groups. Such efforts yielded the fixed wireless channel models developed by the IEEE 802.16 Working Group on Broadband Wireless Access Standards, the MIMO (multi-input/multi-output) channel models developed for wireless LAN applications by IEEE 802.11's Task Group (TG) n, the ultrawideband (UWB) channel models developed by IEEE 802.15's TG3a and 4a, the 60 GHz channel models developed by IEEE 802.15's TG3c, the spatial channel models developed for wide area systems under the aegis of 3GPP (the Third Generation Partnership Project), and the enhanced wide area spatial channel models developed under the aegis of the Wireless World Research Forum's Wireless World Initiative New Radio (WINNER) project and the European Sixth Framework Programme.

1.3 Characterization of Wireless Channels

A channel model is a simplified representation of reality that captures those aspects of channel behavior that affect the performance of a particular class of wireless technologies. The fundamental principles of channel modeling for mobile and personal

---

[1] COST or *European Cooperation in the field of Scientific and Technical Research* is one of the longest running European programs that support cooperation among scientists and researchers across Europe

wireless communication systems that operate at frequencies of 800 MHz and above have been recounted in [8–11].

The ITU-R's IMT-2000 program has defined three basic propagation environments within which terrestrial mobile and personal communication systems are deployed. Picocells refer to indoor environments with transmitter–receiver separations of less than a few hundred meters. Microcells refer to outdoor environments in which both the base station and remote terminal antennas are placed below local rooftop level in the same street canyon (or adjoining side streets) with the remote terminal located at distances of up to 1 km away. In such cases, a *line of sight* (LOS) often exists between the base and remote. Macrocells refer to outdoor environments in which the base station antenna is placed well above local rooftop level while the remote terminal is placed well below local rooftop level at distances of up to several km from the base. In such cases, the link generally operates under *non-line-of-sight* (NLOS) conditions.

Path loss is the most fundamental measure of channel quality. In decibels, path loss, PL, is defined as

$$PL = P_t + G_t + G_r - P_r, \tag{1.1}$$

where P_t and P_r are the time-averaged power levels (in dBm) at the output of the transmitter and the input of the receiver, respectively, and G_t and G_r are the gains (in dBi) of the transmitting and receiving antennas. The relationship between path loss and the distance, d, between the transmitter and receiver generally follows a power-law relation and can be described by

$$PL(d) = PL_0 + n \cdot 10 \log_{10} \frac{d}{d_0} + X_\sigma, \tag{1.2}$$

where PL_0 is the value of path loss (in dB) at the reference distance d_0, n is the distance exponent, and X_σ is a zero-mean Gaussian random variable with standard deviation σ. The random variable X accounts for the location variability or *shadow fading* that is generally attributed to differences in the degree to which the path is obstructed at different points throughout the coverage area.

For systems with fractional bandwidths $\Delta f / f_0$ that are less than 20% where Δf is the occupied bandwidth of the signal and f_0 is the carrier frequency, path loss can generally be assumed to be constant over the band. For systems with large fractional bandwidths and/or which operate near a frequency where specific attenuation due to gaseous absorption changes rapidly, it may be necessary to model the frequency dependence of path loss as well. In such cases, it is reasonable to assume that the frequency and distance dependence of path loss are separable, yielding

$$PL(f, d) = PL(f)PL(d). \tag{1.3}$$

The relationship between path loss and frequency is generally found to follow a power-law relation that can be modeled by

$$\sqrt{PL(f)} \propto f^{-\kappa}, \tag{1.4}$$

where κ is the frequency exponent and $\kappa = 1$ in free space.

Signal Fading. Scattering by objects in the propagation environment causes multiple replicas of the received signal or *multipath components* (MPCs) to arrive via different paths. Small changes in the position of either the scatterers or either end of the wireless link will usually have only a small effect on the amplitudes of the *physical* MPCs that comprise a *resolvable* MPC.[2] However, the phase shifts between the physical MPCs may change significantly causing large changes in the strength of the resolvable MPC. If, over time, the signal follows a complex Gaussian distribution, the magnitude of the signal envelope, x, will follow a Rayleigh distribution,

$$p(x) = \frac{2x}{\Omega} \exp\left(-\frac{x^2}{\Omega^2}\right),$$ (1.5)

Ω is the average power in the signal. If the signal also has a fixed component, its magnitude will follow a Ricean distribution,

$$p(x) = \frac{2(K+1)x}{\Omega} \exp\left(-K - \frac{(K+1)x^2}{\Omega}\right) \cdot I_0\left(2\sqrt{\frac{K(K+1)}{\Omega}}\, x\right),$$ (1.6)

where K is the Ricean K-factor, $I_0(\cdot)$ is the zeroth-order modified Bessel function of the first kind, and Ω is the average power in the signal. For $K = 0$, the distribution reverts to Rayleigh. Otherwise, if neither distribution applies, others have been found to fit measured data, including (1) the Weibull distribution,

$$p(x) = \frac{\beta}{\Omega} x^{\beta-1} \exp\left(\frac{-x^\beta}{\Omega}\right),$$ (1.7)

where $\beta > 0$ is the Weibull fading parameter and Ω is the average power in the signal, and the distribution reverts to Rayleigh for $\beta = 2$, and (2) the Nakagami distribution,

$$p(x) = \frac{2}{\Gamma(m)} \left(\frac{m}{\Omega}\right)^m x^{2m-1} \exp\left(-\frac{m}{\Omega} x^2\right),$$ (1.8)

where $m \geq 1/2$ is the Nakagami m-factor and $\Gamma(m)$ is the Gamma function. Over a time interval during which the channel is stationary, knowledge of the form, scale, and shape of the fading distribution completely specifies the first-order statistics of the signal envelope.

Time-Varying Signals. The rate at which the amplitude and phase of a received signal varies over time is captured by the corresponding Doppler spectrum. The classic Doppler spectrum, expressed in baseband form by

$$R(f) = \frac{1}{\pi} \frac{1}{\sqrt{f_D^2 - f^2}}, \quad |f| \leq f_D,$$ (1.9)

---

[2] When the transmitted signal is a single carrier with constant frequency and amplitude, only one MPC can be resolved.

where f_D is the maximum Doppler frequency, is characteristic of signals received when the terminal at one end of the link is in motion [5]. In indoor or fixed wireless environments, the Doppler spectrum may take on other shapes, e.g., the peaky spectrum proposed in [12],

$$R(f) = \frac{2}{\pi^2}\sqrt{4f_D^2 - f^2}\ K\left(\frac{\sqrt{4f_D^2 - f^2}}{2f_D}\right), \qquad (1.10)$$

where $K(\cdot)$ is the complete elliptic integral. Computing the Doppler spectrum of a signal generally requires knowledge of both the amplitude and phase of the signal over time. The *average fade duration* (AFD) and the *level crossing rate* (LCR) offer an alternative method for capturing both the first- and second-order statistics of the signal envelope based upon amplitude-only received signal data.

Delay Spread or Time Dispersion. The data rate of a digital communication system is determined by the number of symbols that are sent per second and the number of bits that are represented by each symbol. As the symbol rate increases, time dispersion due to multipath scattering may cause time-delayed replicas of one symbol to be received within the time slot reserved for a subsequent symbol. The resulting *intersymbol interference* (ISI) may cause bit errors and will ultimately degrade the performance of the link. In spread-spectrum-based systems, however, use of multi-fingered rake receivers allows one to enhance the received signal using temporal diversity.

The time-varying impulse response of a wireless channel may be represented as the response of a tapped delay line filter with N taps and is given by

$$h(\tau, t) = \sum_{i=1}^{N} a_i(t)\delta(\tau - \tau_i), \qquad (1.11)$$

where a_i and τ_i are the complex (and time-varying) amplitude and the delay of the ith tap which corresponds to the ith resolvable MPC. The resolution of the taps (or the duration of the corresponding delay bins) is given by the inverse of the occupied bandwidth. A particular coefficient a_i may have both fixed and time-varying components and is completely described by the amplitude and phase distributions that define its first-order statistics and the Doppler spectrum that defines its second-order statistics. Past work has suggested that the arrival rate of the MPCs often follows a Poisson distribution.

For wideband systems that occupy bandwidths of several MHz or less, there are relatively few resolvable taps and the corresponding channel impulse response models are very simple. As the occupied bandwidth increases, the number of MPCs that can be resolved increases dramatically. As first observed by Saleh and Valenzuela [13], the resolvable MPCs may appear to form clusters leading to a channel impulse response of the form

$$h(t) = \sum_{\ell=1}^{L}\sum_{k=1}^{K} a_{k,\ell}\exp(j\phi_{k,\ell})\delta(t - T_\ell - \tau_{k,\ell}), \qquad (1.12)$$

where L is the number of clusters, K is the number of rays within each cluster, T_ℓ is the delay of the ℓth cluster, and $a_{k,\ell}$, $\phi_{k,\ell}$, and $\tau_{k,\ell}$ are the amplitude, phase, and delay of the kth tap within the ℓth cluster. While such clustering is very apparent in some environments, it is not as apparent in others. Some have found that an exponential decay multiplied by a noise-like variation with lognormal statistics provides an equally valid representation over a wide range of deployment scenarios within typical residential and commercial environments [20].

Linear Time-Varying Wideband Channels. The time-varying channel impulse response can be expressed in alternative forms. Because the wireless channel is linear and time variant (LTV), the simple Fourier transform pair that relates the LTI impulse response $h(t)$ and LTI frequency response $H(j\omega)$ of linear time-invariant systems must be replaced by a more complicated set as described in [6] and as given by

$$
\begin{array}{ccc}
h(\tau, t) & \begin{array}{c} \xleftarrow{F^{-1}} \\ \xrightarrow{F} \end{array} & S(\tau, \nu) \\[2ex]
F^{-1} \uparrow\downarrow \; F & & F^{-1} \uparrow\downarrow \; F, \\[2ex]
T(f, t) & \begin{array}{c} \xleftarrow{F^{-1}} \\ \xrightarrow{F} \end{array} & H(f, \nu)
\end{array}
\tag{1.13}
$$

where $S(\nu, \tau)$ is the Doppler-delay-spread function, $H(f, \nu)$ is the frequency-dependent Doppler spread function, $T(f, t)$ is the time-varying frequency response, t and f denote time and frequency while τ and ν denote delay and Doppler frequency, and F and F^{-1} denote the Fourier and inverse Fourier transforms, respectively. Thus, time dispersion is equivalent to frequency variation (or selectivity) and frequency dispersion is equivalent to time variation (or selectivity).

Angle of Arrival. If a system uses either a directional antenna or multiple antennas to achieve greater performance, one must account for the distribution of angles over which incoming MPCs arrive at the receiving antenna. In the case of directional antennas, the convolution of the *angle-of-arrival* (AoA) distribution with the free space antenna pattern gives the *effective* antenna pattern which determines the antenna's effectiveness in rejecting interfering signals from different directions. In the case of multiple antenna elements, the AoA distribution determines the mutual correlation between signal fading observed on adjacent elements.

The AoA distribution is characterized by its mean direction and angular extent. Three common distributions which have been found to fit the AoA distributions that are observed across the azimuth angle ϕ in macrocell and/or in indoor environments include the uniform distribution,

$$
p(\phi) = \frac{1}{2\pi},
\tag{1.14}
$$

the zero-mean Gaussian distribution,

$$
p(\phi) = \frac{1}{\sqrt{2\pi}\,\sigma_\phi} \exp\left(-\frac{\phi^2}{2\sigma_\phi^2} \right),
\tag{1.15}
$$

where σ_ϕ is the standard deviation of the distribution, and the zero-mean Laplace distribution,

$$p(\phi) = \frac{1}{\sqrt{2}\sigma_\phi} \exp\left(-\frac{|\sqrt{2}\phi|}{\sigma_\phi}\right), \qquad (1.16)$$

where σ_ϕ is once again the standard deviation of the distribution.

Spatial Channel Models. When both the angle of arrival and the time of arrival are known, one can identify the locations of individual scatterers. The result is referred to as a spatial channel model [14]. A method for extending the Saleh–Valenzuela (S-V) CIR model to the spatial domain is described in [15].

1.4 Development of New Channel Models

Development of a new channel model begins with discussion between the channel modeler and the wireless system designer/developer. First, they must agree upon which aspects of channel behavior are important and must be captured, and which can be ignored. If important aspects are neglected, the model will not be useful. If, however, too many aspects are considered, the resulting model could be overly complex and would likely require considerable additional effort to develop.

The channel modeler and the designer/developer must also agree upon the nature of the physical environment(s) to be considered and the manner in which the transmitting and receiving antennas will be deployed. This will often be captured in the form of *usage scenarios* that will describe, in broad terms, how devices that employ the technology will be used. They must also decide whether the model is to be broadly representative of the scenarios in which wireless devices based upon the technology are likely to be used, i.e., *site-general*, and the extent to which it must capture the manner in which the channel parameters depend upon the design parameters that describe the configuration of the link.

The nature and degree of the propagation impairments observed on a wireless channel will be affected by the gains, beamwidths, polarizations, and orientations of the transmitting and receiving antennas. If the width of the angle-of-arrival distribution of incident signals is narrower than or at least comparable to the beamwidth of the receiving antenna, then one can usually separate the distortions introduced by the wireless channel (which are captured by the channel model) from the distortions introduced by the antennas (which are captured by the antenna model). If the two sets of distortions cannot be easily separated, one often has little choice but to model them together. The combination of the wireless channel and the transmitting and receiving antennas is often referred to as the *radio channel.*

The nature and degree of the propagation impairments also depend upon many design parameters and environmental factors including the carrier frequency, the distance between the transmitting and receiving antennas, the relative heights of the antennas above ground level, the nature, height, and density of the scatterers in the environment and the nature of any obstructions that lie between the antennas. The decision to fix a design parameter or environmental factor, treat it as an independent variable or simply ignore it will depend upon (1) the extent to which the channel parameters are affected by that design parameter or environmental factor and (2) the likely range of values which the design parameter or environmental factor might take on in the usage scenario.

The channel modeler and the designer/developer must decide whether to develop the model by simulation, by measurement, or some combination. Although simulation-based methods such as ray tracing are potentially less expensive and time-consuming than measurement-based approaches, they are limited by the assumptions upon which they are based and the possibly tremendous amounts of detail regarding the type and location of the scatterers in a typical environment that one may need to supply to them. Measurement-based methods are widely used to characterize wireless channels because they can provide results that are (1) of immediate use to designers and developers and (2) useful in the validation of results obtained from simulation-based methods. The limitations of measurement-based approaches are described in the next section. Measurement- and simulation-based approaches to channel modeling are increasingly seen as complementary; many channel modeling studies employ both approaches.

Once the decision to collect channel measurement data has been made, whether as the primary basis for the channel model or to validate simulation results, the channel modeler must configure a suitable channel sounder. Alternative approaches are described in the next section. Important considerations include (1) whether the channel is static or time varying, (2) the nature of the antennas, including the manner in which the antenna pattern varies with frequency and, if applicable, the degree of mutual coupling between co-located antennas, (3) non-linearities in the transfer functions of active devices used in the instrument, especially if multi-carrier or other complex signals are used as stimulus signals, (4) the amount of phase noise in signals generated by oscillators in the system, (5) the size, weight, and transportability of the equipment, (6) the sensitivity of the equipment to the environment, especially temperature, and (7) cost.

The next step is to collect the required measurement data and reduce them, i.e., extract the channel parameters of interest. Often, measurement campaigns are conducted in two stages, as depicted in Fig. 1.2. Development runs are used to assess the performance of the channel sounder, identify potential models against which the measurement data can be reduced, and to provide an opportunity to fine-tune the instrument and the data collection protocol as required. Upon completion of the development runs, production runs are conducted in order to collect the vast amount of measurement data required to yield statistically reliable results. In order to ensure the consistency of the data set collected during production runs, changes to the equipment and/or the data collection protocol are strongly discouraged. The next step is to estimate the channel parameters and their marginal distributions, mutual correlation, relationship to environmental and design parameters, and so forth. The final step is to cast the results in the form of a model useful in the analysis, design, and simulation of wireless communication systems and verify that the model is consistent with the measurement data upon which it is based.

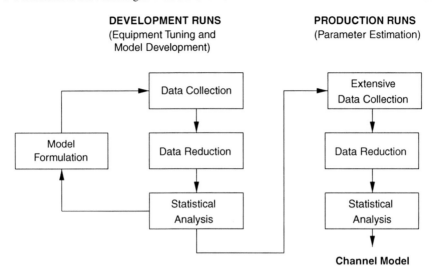

Fig. 1.2 Development of measurement-based channel models

1.5 Measurement of Wireless Channels

An instrument used to measure the response of a wireless channel is a stimulus-response test set commonly referred to as a *channel sounder*. Depending upon the aspects of the channel that are of interest, a channel sounder may take many different forms. A useful approach to classifying channel sounders distinguishes between instruments used to collect narrowband channel response data, wideband channel response data, and channel response data collected using multiple transmitting and/or receiving antennas.

Narrowband Channel Sounders. The simplest channel sounder consists of a source that transmits a single carrier and a narrowband receiver that measures the received signal strength at a remote location. Because single-carrier measurements only capture the channel response at a single frequency, the temporal resolution is effectively infinite and one cannot distinguish between replicas of the signal that arrive with different delays. In order to obtain useful estimates of the broadband path loss from single-carrier measurements, one must obtain the average values of the received signal strength based upon data collected by temporal and/or spatial sampling.

Wideband Channel Sounders. Transmitting a single carrier but sweeping it across numerous frequencies in quick succession and in synchronization with a tunable narrowband amplitude-only receiver allows one to measure the scalar frequency response of the channel. In cases where the position of the receiving antenna is fixed, such an approach allows one to suppress multipath fading by computing averages taken over frequency instead of location. Because this method does not permit one to measure the phase of the frequency response, one cannot estimate the channel impulse response by applying the inverse Fourier transform to the measured data.

A vector network analyzer (VNA) is a swept-frequency stimulus-response test set that measures the complex frequency response of a system by sweeping a single carrier across numerous frequencies in quick succession and comparing the amplitude and phase of the received signal at each frequency with those of the corresponding transmitted signal. Using a VNA to directly measure the complex frequency response of the channel allows one to recover the channel impulse response by applying the inverse Fourier transform to the measured response. While VNAs are frequently used to measure channel responses over short ranges, especially in indoor or enclosed spaces, their usefulness is limited by (1) the relatively slow rate at which individual frequency sweeps are collected and the resulting need to ensure that the channel is either static or, at the very least, changes over timescales far longer than the frequency sweep time, (2) the size and weight of most VNAs which generally renders them too large to be considered portable, and (3) the special effort required to synchronize the transmitter and receiver and provide amplitude and phase references over ranges greater than a few hundred meters.

The first practical method for measuring the wideband channel response over wide areas was reported by Cox [7] in the early 1970s. The technique involves transmitting a wide-band pseudo-random noise (PRN) signal and correlating the signal observed at the receiver with an identical PRN signal. The signal that appears at the output of the correlator is the complex impulse response of the channel. In modern versions of such channel sounders, the receiver is often a vector signal analyzer (VSA). A VSA is an RF tuner followed by a pair of high-speed analog-to-digital converters and a deep sample capture memory that allows one to record a complex time series that includes the in-phase and quadrature components of the received signal. The cross-correlation between the received signal and the PRN signal is then performed in software. The resolution of the channel sounder and the maximum resolvable delay is set by the chip duration and the length of the PRN sequence.

An alternative approach to broadband channel sounding involves transmitting a signal composed of multiple carriers with prescribed amplitude and relative phase, using the VSA to estimate the complex frequency spectrum, then taking the inverse Fourier transform of the complex envelope to recover the channel impulse response. In this case, the resolution of the channel sounder and the maximum resolvable delay is set by the number of carriers and the frequency interval between them. Whether one is transmitting pseudo-random noise or multiple carriers, one is effectively collecting frequency response data at all points simultaneously and thereby avoiding the blind time associated with the swept-frequency approach. VSA-based channel measurement schemes suffer from various limitations: (1) The maximum bandwidth that one can measure is determined by the maximum rate at which the analog-to-digital converters in the VSA's front-end can sample the received signal, (2) the maximum sample duration that one can measure is determined by the size of the deep memory used to store the received signal, and (3) the dynamic range and sensitivity of the receiver are determined by the resolution of the analog-to-digital converters in the VSA's front-end.

Multi-antenna Measurements. Use of multiple transmitting and receiving antennas in conjunction with suitable signal processing or data reduction techniques allows one to characterize aspects of the channel that depend upon the distribution of angles at which signals leave the transmitter and/or arrive at the receiver, e.g., the correlation between the fading signals observed at each antenna element in spatial diversity and MIMO transmission schemes. The simplest approach involves using fixed antennas at the transmitter and/or receiver. Mounting a single antenna on a mechanical positioner and moving it through a sequence of closely spaced points in space allows one to generate a virtual array and thereby characterize spatial correlation, temporal correlation, and angle-of-arrival distributions. Full characterization of MIMO channels involves estimating the angle of arrival distribution observed at the receiver for rays launched at each of the possible angles of departure from the transmitter. The result is referred to as a *double-directional channel model* [16].

Limitations of Measurement-Based Methods. Direct measurement of a wireless channel offers several advantages, as outlined in the previous section. However, the approach has several limitations: (1) It often takes considerable time and effort to collect and reduce a statistically significant amount of data. Moreover, redoing measurements if errors are found is equally time-consuming. (2) In many environments, the size, weight, and power consumption of the measurement equipment can be problematic. (3) If the width of the angle of arrival distribution exceeds the beamwidth of the antenna, then it will be difficult to separate distortions introduced by the channel from distortions introduced by the antenna. (4) It is generally difficult to accurately measure the phase difference between the transmitted and received signals unless steps are taken to substantially reduce frequency drift and phase noise, e.g., through a direct connection between the transmitter and receiver or use of precise frequency standards. (5) The expense and complexity of the measurement system increases rapidly as the measurement goals become more sophisticated. (6) Before applying the results, one must confirm that the measurement environment is sufficiently similar to the deployment environment. Nevertheless, the measurement-based approach has proven to be sufficiently useful and productive that it will likely remain the principal method for characterizing wireless channels for many years to come.

1.6 Recent Advances in Channel Modeling

In this last section, we review recent advances in channel modeling that have been motivated by the introduction of new signaling schemes (UWB and MIMO), new environments (body-centric communications and short-range vehicular environments), and/or new frequency bands (the 60 GHz and THz bands).

1.6.1 Channel Models for Ultrawideband Wireless Systems

In February 2002, the FCC released a report and order (R&O) that authorized the use of short-range wireless links that use at least 500 MHz of bandwidth between 3.1 and 10.6 GHz. The R&O limited the effective isotropic radiated power (EIRP) to −41.3 dBm/MHz or less in order to limit the interference that such ultrawideband (UWB) devices may cause to other services. Such a limit was deemed appropriate for UWB links of 10 m or less. From 2002 to 2007, the ITU-R's Task Group 1/8 under Study Group 1 assessed the compatibility between UWB devices and other radiocommunication services and prepared recommendations for UWB regulations. Since then, most jurisdictions around the world have authorized UWB devices although many allocations are considerably more restrictive than that authorized by the FCC.

The lead up to and release of the FCC R&O played a crucial role in stimulating interest in the potential of UWB wireless technology among both developers and researchers. In late 2001, the IEEE 802.15 Working Group on Wireless Personal Area Networks formed Study Group 3a (SG 3a) to assess the potential for developing a UWB-based physical layer (PHY) that could support data rates of hundreds of Mb/s over ranges of up to 10 m subject to the restrictions imposed by the FCC's emission mask. In late 2002, Study Group 4a (SG 4a) was formed with a mandate to assess the potential for developing a UWB-based PHY layer that would replace much of the functionality of ZigBee wireless technology but with lower power consumption and adding accurate real time location services (RTLS). SG 3a and 4a became full Task Groups in 2002 and 2004, respectively.

The FCC's decision to effectively restrict UWB data communication to the band from 3.1 to 10.6 GHz eliminated impulse radio from contention as the basis for a PHY layer for such systems and created a need for a fresh effort to characterize the UWB channel. One of the first tasks of the UWB SGs was to develop channel models suitable for fairly comparing alternative PHY and MAC layer proposals. Because IEEE 802.15.3a's main focus was on high-speed peripheral interconnect in residential and office environments, they produced four-channel models that correspond to LOS 0–4 m, NLOS (0–4 m), NLOS (4–10 m), and an extreme NLOS multipath channel based upon a 25 ns RMS delay spread [17]. Because IEEE 802.15.4a's main focus was on sensor networks, they took a different approach and produced eight channel models that correspond to residential, office, outdoor, and industrial environments [18]. They also produced a channel model for body area networks based upon UWB wireless technology, as will be described in more detail in a later section.

The results of the IEEE 802.15 channel modeling efforts reflect the notion that UWB channels differ from conventional wideband channels in several important respects: (1) UWB devices operate over a very wide frequency range, so both the distance and frequency dependence of path loss must be accounted for. (2) The extremely wide bandwidth occupied by a UWB signal allows the system to resolve extremely fine detail in the channel impulse response. Because the resolvable delay bins are so narrow, sparse channels with significant delays between resolvable MPCs often occur. The small-scale fading statistics are different from the wideband case because each resolvable MPC consists of fewer physical MPCs. (3) The frequency dependence of path loss distorts individual MPCs such that the time-varying impulse response given in (1.11) is now given by

$$h(t, \tau) = \sum_{i=1}^{N} a_i(t)\chi_i(t, \tau) * \delta(\tau - \tau_i), \qquad (1.17)$$

where $\chi_i(t, \tau)$ denotes the time-varying distortion of the ith echo due to frequency-selective interaction with the environment and $*$ denotes convolution. Because adjacent taps are influenced by a single physical MPC, the WSSUS assumption is no longer valid [19].

The density of scatterers varies greatly between environments. In most environments, the density of scatterers is low to moderate so clustering of MPCs can easily be observed. In other environments, the density of scatterers is so high that one cannot resolve individual clusters. IEEE 802.15.4a adopted two alternative models for the CIR. In the sparse, multi-cluster case, the modified Saleh–Valenzuela model applies and the shape of the corresponding power delay profile is given by the product of two exponential functions,

$$E\{|a_{k,\ell}|^2\} \propto \exp(-T_\ell/\Gamma) \cdot \exp(-\tau_{k,\ell}/\gamma), \qquad (1.18)$$

where Γ and γ are the cluster and ray decay constants, respectively, T_ℓ is the delay of the ℓth cluster, and $\tau_{k,\ell}$ is the delay of the kth ray within the ℓth cluster. As noted earlier, the cluster-based S-V model is only one of several options for modeling UWB channels. Other simpler models, including a single exponentially decaying cluster with lognormal variation, have been found to provide a representation with equal statistical validity over a wide range of deployment scenarios within typical residential and commercial environments [20].

In the dense, single-cluster case, the envelope of the PDP can be described as

$$E\{|a_{k,\ell}|^2\} \propto (1 - \chi \cdot \exp(-\tau_{k,\ell}/\gamma_{rise})) \cdot \exp(-\tau_{k,\ell}/\gamma_1), \qquad (1.19)$$

where χ denotes the attenuation of the first component, γ_{rise} describes how quickly the PDP rises to its maximum value, and γ_1 describes the decay after the maximum has been reached.

Although 802.15.4a found that the cluster arrival times are well described by a Poisson process, the inter-cluster arrival times are exponentially distributed, i.e.,

$$p(T_\ell|T_{\ell-1}) = \Lambda_\ell \exp\left(-\Lambda_\ell(T_\ell - T_{\ell-1})\right), \quad \ell > 0, \qquad (1.20)$$

where Λ_ℓ is the cluster arrival rate (assumed to be independent of ℓ). 802.15.4a models ray or resolvable MPC arrival times as a mixture of two Poisson processes where

$$p(\tau_{k,\ell}|\tau_{(k-1),\ell}) = \beta\lambda_1 \exp[-\lambda_1(\tau_{k,\ell} - \tau_{(k-1),\ell})]$$
$$+(1 - \beta)\lambda_2 \exp[-\lambda_2(\tau_{k,\ell} - \tau_{(k-1),\ell})], \quad k > 0, \quad (1.21)$$

β is the mixture probability and λ_1 and λ_2 are the ray arrival rates. In cases where the ray density is high and leads to a high MPC arrival rate, the CIR is represented by a tapped delay line model with regular tap spacings.

Details of the manner in which the propagation channel affects various UWB transmission schemes, including time-hopping impulse radio systems, direct sequence spread spectrum (DSSS) systems, orthogonal frequency multiplexing (OFDM) systems, and multiband systems are described in [19]. For example, the number of resolvable MPCs determines the number of fingers that a rake receiver will require in order to capture enough of the energy in the received signal. The range and coverage of UWB systems tend to degrade as the carrier frequency increases and the free space path loss and diffraction losses both increase.

1.6.2 Channel Models for MIMO-Based Wireless Systems

Consider a wireless communication system that uses only one transmission path to send data. Shannon's law gives the maximum capacity C_1 of the link in bits/s/Hz as

$$C_1 = \log_2(1 + \rho), \tag{1.22}$$

where ρ is the signal-to-noise ratio (SNR) at the receiver input. In practice, scattering by objects in the environment leads to multiple transmission paths between the transmitter and receiver. Such paths are often so closely spaced in angle that one cannot distinguish between them through simple beamforming. Instead, a more sophisticated approach is used which is based upon the use of *space–time* coding to distribute the data stream over the N_T transmitting antennas and recover the stream by suitably combining the signals received by the N_R receiving antennas [21]. Because the multiple transmission paths fade independently of each other, this approach also increases overall link reliability. Numerous methods for realizing MIMO-based systems in this manner have been proposed over the past decade [22].

The capacity C of a MIMO-based system with N_T transmitting antennas and N_R receiving antennas (in bits/s/Hz) is given by

$$C = \log_2 \left[\det \left(\mathbf{I}_{N_R} + \frac{\rho}{N_T} \mathbf{H} \mathbf{H}^* \right) \right], \tag{1.23}$$

where $(*)$ denotes the transpose conjugate, $\mathbf{H}$ is the $N_R \times N_T$ channel matrix and we have assumed that the N_T sources have equal power and are uncorrelated [21]. Further analysis suggests that the capacity of the system will reach its peak when the transmission paths experience uncorrelated Rayleigh fading. However, full appreciation of the strengths and limitations of these schemes requires that their performance be assessed in realistic propagation environments. A variety of analytical and model-based approaches have been proposed [23].

The most obvious way to characterize the MIMO wireless channel is to configure a channel sounder that directly characterizes $\mathbf{H}$. In a *true array* system, the channel sounder incorporates the coding and signal processing required to estimate all of the elements of $\mathbf{H}$ simultaneously. In a *switched array* system, the channel sounder is simplified by using high-speed switches to sequentially connect a single transmitter and a single receiver to all possible pairs of elements in the transmitting and receiving arrays in turn before the channel changes appreciably. Although such systems closely resemble practical MIMO-based systems and can accommodate time-varying channels, (1) the results are tied to specific antenna types and configurations, (2) it is difficult to separate the effects of mutual coupling between array elements from the correlation between transmission paths, and (3) the measurement system is relatively complex and expensive.

In a *virtual array* system, the channel sounder uses a single transmitter and receiver (e.g., a VNA) connected to single transmitting and receiving elements, respectively. Precision mechanical positioners move the transmitting and receiving elements to the points that define the virtual transmitting and receiving arrays. Such a system is much more versatile than the true or switched array systems because (1) it eliminates the effect of mutual coupling and (2) it allows an arbitrarily large number of points in the virtual array to be evaluated. Its major shortcoming is that it may take several minutes to translate the antenna elements to all of the points in the virtual array. Accordingly, the technique is mostly used in static indoor environments.

If the virtual receiving array is sampled finely enough and at enough points with the transmitting antenna fixed, it is possible to resolve the angle-of-arrival distribution at the receiving antenna using an AoA estimation algorithm such as ESPRIT or similar. The width of the AoA distribution is a strong indicator of the correlation between fading experienced at the output of adjacent antenna elements. If the frequency response of the channel is measured at each point, then the channel impulse response can also be estimated. Correlating the time-of-arrival associated with a given MPC with the corresponding angle of arrival allows one to estimate the spatial channel model that describes the propagation environment and to determine the extent to which scatterers form spatial clusters [14].

A *directional channel sounder* uses virtual arrays at both ends of the link to more fully account for the characteristics of the antennas and local scatterers. As above, the finite time required to sample the virtual array limits the approach to characterizing static channels. In essence, a directional channel sounder allows one to resolve the effect that a ray leaving the transmitter in a particular direction has on the time-of-arrival and angle-of-arrival distributions observed at the receiver. Directional channel models capture all of the information required to analyze a MIMO link and estimate its capacity [16].

Once the MIMO channel has been characterized, perhaps by a combination of simulation and measurement-based methods, designers and developers can begin using the channel models that describe environments of interest to design, test, and evaluate the performance of alternative antenna configurations, signaling schemes, and space–time codes. IEEE 802.11 TGn proposed a set of wideband MIMO channel models appropriate for comparing the performance of MIMO-based wireless LANs [24]. In the TGn channel models, each tap in the CIR is described by a channel matrix $\mathbf{H}$ which is resolved into a fixed LOS matrix $\mathbf{H}_F$ and a Rayleigh NLOS matrix $\mathbf{H}_v$,

$$\mathbf{H} = \sqrt{P}\left(\sqrt{\frac{K}{K+1}}\,\mathbf{H}_F + \sqrt{\frac{1}{K+1}}\,\mathbf{H}_v\right), \tag{1.24}$$

where K is the Ricean K-factor and P is the power contained in each tap. Because it is assumed that each tap contains a number of individual rays or physical MPCs, the complex Gaussian assumption can be justified. The correlation between antenna elements is determined by the *power angular spectrum* (PAS). Given the receive and transmit correlation matrices $\mathbf{R}_{tx}$ and $\mathbf{R}_{rx}$, respectively, $\mathbf{H}_v$ is given by

$$\mathbf{H}_v = \mathbf{R}_{rx}^{1/2} \, \mathbf{H}_{iid} \left(\mathbf{R}_{tx}^{1/2}\right)^{T} \tag{1.25}$$

where $\mathbf{H}_{iid}$ is a matrix of independent zero mean, unit variance, complex Gaussian random variables and the elements of $\mathbf{R}_{tx}$ and $\mathbf{R}_{rx}$ are the complex correlation coefficients between the ith and jth antennas in the transmitting or receiving array, respectively. An alternative approach uses the Kronecker product of the transmit and receive correlation matrices. TGn specified six models of this form that correspond to RMS delay spreads ranging from 0 to 150 ns.

3GPP and, later, the WINNER project, proposed a set of MIMO channel models appropriate for comparing the performance of MIMO-based systems used to provide wide area coverage in macrocell environments [25, 26]. For each of the scenarios considered, the WINNER project produced two types of channel models. The first is a generic model which captures the double-directional channel including the amplitude, phase, delay, angle-of-departure, angle-of-arrival and polarization of each ray in a manner which is independent of the details of the transmitting and receiving arrays. The second is a reduced-variability model which is suitable for calibration and comparison simulations. The results of the TGn, 3GPP and WINNER standards group activities and related COST actions concerning MIMO-based systems are summarized in [27].

1.6.3 Channel Models for Body Area Networks

Body area networks (BANs) are comprised of wireless links between ultra-low-power (ULP) wireless devices located in close proximity to the human body. Such devices may be implanted within the body (implanted nodes), attached to the skin, embedded within clothing, mounted on items attached to or carried by the body (body surface nodes), or located at distances up to 5 m away (external nodes). They may be used to (1) monitor the physiological condition of an individual for health care, athletic training, and workplace safety applications, (2) monitor environmental hazards in the vicinity of an individual in order to enhance workplace safety, (3) monitor and control the state of protective gear or safety equipment worn by individuals in hazardous environments, (4) communicate with other ULP devices in the immediate vicinity for personnel monitoring and authentication applications, (5) provide the individual with the means to command and control the ULP sensors and devices in his vicinity, e.g., through a wrist-, arm-, or chest-mounted control panel, and/or (6) relay signals from ULP wireless devices to distant networks, e.g., wireless LAN or cellular networks [28, 29].

Implementation of the wireless sensor nodes intended for use in body area networking applications presents special challenges. Not only must they be physically small in order to be unobtrusive, but they must operate from the same small battery for periods ranging from weeks to months at a time. Although simple transmitter–receiver pairs have been used to establish wireless connections in applications that involve just a pair of nodes, the effort required to provide them with full networking capabilities would be considerable. As a result, much interest has focused on devices based upon existing ULP wireless networking standards such as Bluetooth low energy (BLE) ZigBee, and IEEE 802.15.4a. At the same time, it is widely recognized that the wireless propagation environment in the vicinity of the human body is considerably different from the personal area and sensor network environments for which existing standards were developed. As a result, other techniques such as near-field techniques have been considered. In recognition of the growing interest in body area networking and the limitations of existing standards, IEEE 802.15 recently formed TG6 to develop a short-range wireless communication standard that has been optimized for this purpose.

Wireless signals may propagate from one sensor node to another via three types of paths: (1) through the body, (2) around the body, and (3) reflection or scattering from objects in the surrounding environment. Both electromagnetic field simulation studies and direct measurement have shown that propagation through the body is negligible at UHF frequencies and above. It is important to distinguish between direct transmission around the body and scattering from objects in the environment. Otherwise, link performance in open areas that have relatively few scatterers could be overestimated. In body area networking applications, antennas are located in close proximity to the body and their radiation characteristics are greatly affected. Determination of the extent to which antenna effects can be separated from propagation effects is an ongoing issue in body area channel modeling studies [30].

The first standardized model for body area networking environments was produced by IEEE 802.15.4a [18]. Additional details were reported in [31, 32]. It applies to UWB propagation between 3.1 and 10.6 GHz. The researchers characterized the body area channel using two alternative approaches: electromagnetic field simulation based upon the finite-difference time-domain (FDTD) approach and direct measurement using a VNA. Their major findings include the following: (1) the distance around the perimeter of the body is the correct measure of transmitter–receiver separation, (2) there are always two clusters of MPCs in the channel response – one due to direct transmission around the body and the second due to reflection from the ground, and (3) the small-scale fading statistics are best described by a lognormal distribution.

Their 802.15.4a BAN path loss model distinguishes between devices that are placed on the same side of the body and on opposite sides. The separation between clusters depends upon the position of the transmitting and receiving antennas with respect to each other and the ground. To incorporate this effect easily but without unduly complicating the model, they defined three scenarios corresponding to the transmitter placed on the front of the body and the receiver placed on the front, side, or back of the body. The distance ranges for those environments are 0.04–0.17 m, 0.17–0.38 m, and 0.38–0.64 m, respectively. Within each cluster, the very short transmission distances result in ray arrival times that are shorter than the delay resolution of the systems that they considered. Accordingly, they used a tapped delay line model to represent each cluster.

While several earlier efforts by others produced anecdotal results, IEEE 802.15.4a was the first to collect and reduce sufficient data to produce a preliminary model suitable for use in simulation. Their basic model is conservative; it does not include the effects of scattering from the environment which may be important if the receiver is otherwise well shadowed. However, they have suggested methods by which such scattering could be incorporated if required. The main purpose of the TG4a BAN channel models is to allow fair comparison of the performance of alternative PHY- and MAC-layer proposals. They are not intended to predict absolute measures of performance, nor do they address some important issues relevant to network layer issues. Thus, while TG4a's work represents a significant milestone in the characterization of UWB BAN channels, much additional measurement data are required and much additional work remains.

In January 2007, IEEE 802.15.6 formed a channel modeling committee and directed it to produce a new set of BAN channel models that will allow alternative PHY and MAC proposals to be fairly compared under its own standardization efforts. The committee presented its final report in November 2008 [33]. Their scenarios covered transmission between implanted, body surface, and external nodes. The scenarios involving implants were limited to the 402–405 MHz band. The scenarios involving nodes on the body surface include the 13.5 MHz, 5–50 MHz, 400 MHz, 600 MHz, 900 MHz, 2.4 GHz, and the 3.1–10.6 GHz bands. The scenarios involving external nodes were limited to the 900 MHz, 2.4 GHz, and the 3.1–10.6 GHz bands. The effect of body posture and body movement was included. Although the models formulated by TG6 represent a significant advance over those formulated by TG4a, they suffer from the same limitations: They are based on a limited amount of measurement data and are not suitable for predicting absolute performance. Once again, much additional work remains.

1.6.4 Channel Models for Short-Range Vehicular Networks

Intelligent transportation systems (ITS) are a suite of emerging technologies that will be used to make operation of land vehicles in urban centers or along transportation corridors safer and more efficient. A variety of wireless technologies have been proposed and/or evaluated for use in ITS applications including RFID technology, wide area cellular networks, and mobile satellite networks. Because much of the information that will be delivered and exchanged in ITS applications is time-sensitive and location-dependent, short-range vehicular networks have attracted particular interest in recent years.

Short-range vehicular networks comprise short-range wireless links between a vehicle (via an *onboard unit* (OBU)) and roadside units (also known as *roadside equipment* (RSE)) to form *vehicle-to-infrastructure* (V2I) networks, and between a vehicle and other vehicles in the immediate vicinity to form *vehicle-to-vehicle*

(V2V) networks. Anticipated applications of such networks include (1) enhancing traffic safety by providing warnings and alerts in real time, (2) easing traffic congestion by adaptively changing traffic rules, (3) providing location-dependent information to drivers, (4) aiding traffic regulation enforcement, (5) enabling electronic payments and toll collection, (6) assisting in direction and route optimization, (7) providing information concerning services for travelers, and (8) enabling automated highways.

Although interest in the potential for short-range vehicular networks to enable ITS applications dates back almost two decades, a major impediment to progress was the lack of a common, interoperable hardware platform that could be used in each of the envisioned roles. In the early 1990s, the ITS community proposed (1) that a standard for dedicated short range communications (DSRC) be developed in order to meet this need, (2) that such systems be deployed in or near the 5.8 GHz ISM band, and (3) that it support data rates of at least 1 Mb/s. Since the early 1990s, the European, Japanese, and American standards for DSRC have taken different paths. European and Japanese DSRC systems are single-carrier systems and are in active use, although mostly for electronic toll collection. In the United States, the DSRC standard is based upon IEEE 802.11p, a variant of the IEEE 802.11a OFDM-based standard that can operate in various licensed and license-exempt bands between 4.9 and 5.9 GHz and which incorporates enhancements to its MAC layer that are required for successful operation in mobility environments [34].

Usage models for short-range vehicular environments must account for four main features of the environment: (1) the nature of the link (V2V or V2I), (2) the speeds of the vehicle(s) at each end of the link, (3) the nature of surrounding environment, and (4) the density and speed of the vehicles that comprise the surrounding traffic. The number of combinations is large, so some discretion is required when selecting the subset to be characterized. Once the usage models have been identified, the characteristics of the link may be determined either by simulation using ray tracing combined with realistic models of objects in the environment, e.g., [35] or by direct measurement using a channel sounder that has been deployed in representative environments.

As with channel models for other environments, a short-range vehicular channel model must account for (1) variation of signal strength with distance, (2) variation of signal strength over time, and (3) time dispersion of the signal or, equivalently, the frequency selectivity of the channel. However, the vehicular environment is considerably more dynamic than other environments. First, at least one end of the link is a vehicle in motion. Second, many of the other vehicles that can obstruct or shadow that link are also in motion. Third, if the antennas used by the OBUs are placed below rooftop level, the vehicle itself will obstruct or shadow the link in certain directions. As a result, short-range vehicular channels are both time and frequency selective.

Measurements of the vehicular channel have recently been reported by several researchers, including [36, 37, 38]. Although differences between the scenarios considered make direct comparisons difficult, some general conclusions can be drawn.

First, unlike macrocell channels, which experience their longest delay spreads in open areas such as expressways or bridges where distant scatterers can make significant contributions to the response, short-range vehicular channels are influenced almost exclusively by local scatterers and experience their longest delay spread in street canyons under NLOS conditions whether they are formed by buildings in urban areas or by large trucks in the vicinity of the vehicle in highway environments. Second, shadow fading occurs much more rapidly in vehicular environments than in macrocell environments because the dominant obstructors are both smaller and closer to the vehicular terminal and are often in motion relative to the vehicle. Third, the Doppler spectrum frequently deviates from the classic U-shaped spectrum. This is likely due to the AoA distribution being extremely non-uniform.

Comparisons of the channel impulse responses experienced on vehicular and macrocell channels also show significant differences. First, taps in the channel impulse response persist for a much shorter time than in macrocell environments due to rapid changes in the configuration of the scatterers that contribute to the response. Finally, the amplitude distributions experienced on individual taps are frequently best described by a Nakagami distribution with an m-factor of less than 1, i.e., worse than Rayleigh. In any case, measurement-based modeling of vehicular channels is still at an early stage and standardized channel models have not yet been adopted by any of the major groups that are responsible for setting DSRC standards and certifying DSRC equipment.

1.6.5 Channel Models for 60 GHz and Terahertz Systems

In recent years, several groups have proposed that new wireless technologies capable of delivering data rates at 1 Gb/s and above be developed for deployment in the 60 GHz band. Such technologies would permit wireless replacement of very high speed short-range wired connections such as those based upon IEEE 802.3-2005 (Gigabit Ethernet) or IEEE 1394b-2002 (FireWire 800) [39]. Others with a longer view have proposed that new technologies capable of delivering data rates of 10 GB/s and above be developed for use in the unlicensed band between 300 GHz and 1 THz [40]. Proponents acknowledge that THz technology is still in its infancy and it will take at least a decade to deliver THz wireless devices to consumers.

Propagation at 60 GHz. At frequencies above 10 GHz, absorption due to atmospheric water vapor and oxygen play a significant role in determining the useful range of wireless links. In particular, wireless links that are deployed near the oxygen absorption line near 60 GHz experience losses of 10–15 dB/km beyond the usual free space and diffraction losses. While this precludes the use of 60 GHz systems for links longer than about 2 km, the losses are entirely manageable for (1) LOS links used to provide last mile connectivity in outdoor environments or (2) NLOS links used within a room in a home or office. Moreover, the rapid reduction in signal strength with distance is advantageous because it drastically reduces the interference caused by nearby systems in the same band and permits much denser deployment and a higher rate of frequency re-use than would otherwise be possible.

In recent years, spectrum regulators around the world have allocated a large amount of spectrum near 60 GHz for use by short-range wireless systems. In the United States and Canada, the band from 59 to 64 GHz has been allocated to license-exempt applications with a maximum output power of 27 dBm and an average power density that does not exceed 9 μW/cm^2, as measured 3 m from the radiating structure. In Japan, the band from 59 to 66 GHz has been allocated to license-exempt applications with a maximum output power of 10 dBm and a maximum effective isotropic radiated power of 57 dBm. Other jurisdictions, including Australia and Korea, have made similar allocations. It is widely expected that Europe and most remaining jurisdictions will soon follow. Compared to the regulatory hurdles which have plagued UWB outside the United States, the situation in the 60 GHz band is much more favorable [41].

Standards Activities at 60 GHz. Various standards groups are actively developing wireless technologies suitable for providing short-range multi Gb/s connectivity at 60 GHz. IEEE 802.15's Task Group 3c is developing a 60 GHz alternative physical layer for the high rate wireless personal area network (WPAN) developed by Task Group 3. In Europe, Ecma TC 48 is developing a similar standard. Various other groups are also developing technologies and/or proposing competing standards, including the WirelessHD consortium led by Broadcom, Intel, LG Electronics, Panasonic, NEC, Samsung, SiBEAM, Sony, and Toshiba. Although most groups have defined specific usage models in which the proposed systems are expected to operate at specified levels of performance, IEEE 802.15c plans the most ambitious coverage and is apparently the only group to sponsor a channel modeling committee.

The IEEE 802.15.3c channel modeling committee proposed channel models corresponding to LOS and NLOS links in residential, office, library, desktop environments (CM 1-8), and in a kiosk environment (CM-9). In the channel modeling committee's final report, they emphasize that their usage models are only representative of the many scenarios in which 60 GHz equipment might be deployed. The models that the committee has proposed are based upon measurement results that have been reported in the published literature, e.g., [42] and submitted directly to the committee [43].

The IEEE 802.15.3c 60 GHz channel models bear many resemblances to the IEEE 802.15.3a/4a 3.1–10.6 GHz UWB channel models. First, as in the UWB case, path loss depends upon both frequency and distance so their path loss models have been designed to capture both. Second, as in the UWB case, the occupied bandwidth of the signal is sufficiently wide that the channel impulse response is revealed with very fine resolution and MPCs are observed to arrive in clusters. Accordingly, the 60 GHz CIR model is also based upon the Saleh–Valenzuela model with extensions that capture certain unique aspects of the LOS component. The distribution of the cluster arrival and ray arrival times are described by a pair of Poisson processes. Analysis of measurement data has shown that both the cluster and ray amplitudes can be modeled by lognormal distributions.

Because the carrier frequency is so high, even walking speeds (1.5 m/s) can lead
to Doppler spreads of several hundred Hertz. Unlike IEEE 802.15.3a or 4a, the IEEE
802.15.3c channel model also captures the angular spread of the channel response in
the form of a power azimuth profile distribution. The distribution of the cluster mean
angle of arrival, conditioned on the AoA of the previous cluster, is uniform. The ray
AoAs within each cluster are modeled either by zero-mean Gaussian or zero-mean
Laplace distributions. The committee's task was made more difficult by the relative
lack of 60 GHz channel measurement data that has been reported in the literature,
and, in particular, the lack of measurement data that addresses the specific usage
models proposed by the committee. While the committee's standardized models
provide a useful basis against which alternative PHY or MAC layer proposals for use
in 60 GHz systems can be evaluated and compared, further measurement campaigns
are required in order to fill in key gaps.

Propagation in the THz Band. At frequencies between 300 GHz and 1 THz, atmospheric attenuation can reach hundreds of dB/km. At 300, 350, 410, 670, and
850 GHz, the atmospheric attenuation is sufficiently low, i.e., less than 50 dB/km,
to permit deployment of short-range links and the available bandwidth is approximately 50 GHz or greater. Only a few detailed studies of THz transmission in indoor
environments have been reported to date, e.g., [44]. Direct transmission will likely
perform best but is extremely susceptible to accidental and/or intermittent blockage.
As in the case of infrared wireless LANs, indirect transmission in which signals
reach the receiver via reflections from walls and ceilings may offer more consistent
performance. However, much work remains in order to determine the performance
that can be achieved in typical usage scenarios.

1.7 Conclusions

During the next decade, a new generation of wireless technologies will further improve the performance and reliability of wireless systems while increasing the range
of applications in which they can be used. Many of these systems will use new signaling schemes while operating in higher frequency bands and/or being deployed in
harsher environments than ever before. The degree to which these new technologies
will meet end user expectations will ultimately depend upon the accuracy and fi-
delity with which channel modelers characterize the impairments and distortions
that these systems will experience under realistic conditions. Measurement- and
simulation-based approaches to channel modeling are increasingly seen as complementary; many studies employ both. Recent progress by the channel modeling
community suggests that both the developers and end users of these new systems
will be well served.

Acknowledgments The authors thank Robert White, Arghavan Emami Forooshani, and Wadah
Muneer for their help in assembling the list of references.

References

1. J. D. Parsons, *The Mobile Radio Propagation Channel,* New York: Halsted Press, 1992, p. v.
2. A. H. Waynick, "The early history of ionospheric investigations in the United States," *Phil. Trans. R. Soc. Lond. A.,* vol. 280, no. 1293, pp. 11–25, 23 Oct. 1975.
3. D. E. Kerr, *Propagation of Short Radio Waves.* vol. 13 of the MIT Radiation Laboratory Series. New York: McGraw-Hill, 1951.
4. Y. Okumura *et al.,* "Field strength and its variability in VHF and UHF land-mobile radio service." *Rev. Elec. Commun. Lab.,* vol. 16, no. 9-10, pp. 825–873, 1968.
5. R. H. Clarke, "A statistical theory of mobile radio reception," *Bell Sys. Tech. J.,* vol. 47, pp. 957–1000, Jul.–Aug. 1968.
6. P. A. Bello, "Characterization of randomly time-variant linear channels," *IEEE Trans. Commun. Syst.,* vol. 11, no. 4, pp. 360–393, Dec. 1963.
7. D. C. Cox, "Delay Doppler characteristics of multipath propagation at 910 MHz in a suburban mobile radio environment," *IEEE Trans. Antennas Propag.,* vol. 20, no. 5, pp. 625–635, Sep. 1972.

Characterization of Wireless Channels

8. W. Jakes, Ed., *Microwave Mobile Communications,* New York: Wiley, 1974.
9. D. Greenwood and L. Hanzo, "Characterization of mobile radio channels," in *Mobile Radio Communications,* R. Steele, Ed., London: Pentech Press, pp. 92–185, 1992.
10. H. L. Bertoni, W. Honcharenko, L. R. Maciel and H. H. Xia, "UHF propagation prediction for wireless personal communication," *Proc. IEEE,* vol. 82, no. 9, pp. 1333–1359, Sep. 1994.
11. A. F. Molisch, *Wireless Communications.* New York: Wiley, 2005, pp. 43–170.
12. S. Thoen, L. Van der Perre and M. Engels, "Modeling the channel time-variance for fixed wireless communications," *IEEE Commun. Lett.,* vol. 6, no. 8, pp. 331–333, Aug. 2002.
13. A. A. M. Saleh and R. A. Valenzuela, "A statistical model for indoor multipath propagation," *IEEE J. Sel. Areas Commun.,* vol. 5, no. 1, pp. 128–137, Feb. 1987.
14. R. B. Ertel, P. Cardieri, K. W. Sowerby, T. S. Rappaport and J. H. Reed, "Overview of spatial channel models for antenna array communication systems," *IEEE Pers. Commun.,* vol. 5, no. 1, pp.10–22, Feb. 1998.
15. Q. H. Spencer, B. D. Jeffs, M. A. Jensen and A. L. Swindlehurst, "Modeling the statistical time and angle of arrival characteristics of an indoor multipath channel," *IEEE J. Sel. Areas Commun.,* vol. 18, no. 3, pp. 347–360, Mar. 2000.
16. M. Steinbauer, A. F. Molisch and E. Bonek, "The double-directional radio channel," *IEEE Antennas Propag. Mag.,* vol. 43, no. 4, pp. 51–63, Aug. 2001.

Ultrawideband Channel Models

17. A. F. Molisch, J. R. Foerster and M. Pendergrass, "Channel models for ultrawideband personal area networks," *IEEE Wireless Commun.,* vol. 10, no. 6, pp. 14–21, Dec. 2003.
18. A. F. Molisch, D. Cassioli, C. C. Chong, S. Emami, A. Fort, K. Balakrishnan, J. Karedal, J. Kunisch, H. G. Schantz, K. Siwiak and M. Z. Win, "A comprehensive standardized model for ultrawideband propagation channels," *IEEE Trans. Antennas Propag.,* vol. 54, no. 11, pp. 3151–3166, Nov. 2006.
19. A. F. Molisch, "Ultrawideband propagation channels – Theory, measurement, and modeling," *IEEE Trans. Veh. Technol.,* vol. 54, no. 5, pp. 1528–1545, Sep. 2005.
20. L. J. Greenstein, S. S. Ghassemzadeh, S. C. Hong and V. Tarokh, "Comparison study of UWB indoor channel models," *IEEE Trans. Wireless Commun.,* vol. 6, no. 1, pp. 128–135, Jan. 2007.

MIMO Channel Models

21. G. J. Foschini and M. J. Gans, "On limits of wireless communications in a fading environment when using multiple antennas," *Wireless Pers. Commun.* vol. 6, pp. 311–335, 1998.
22. D. Gesbert, M. Shafi, D. S. Shiu, P. J. Smith and A. Naguib, "From theory to practice: An overview of MIMO space-time coded wireless systems," *IEEE J. Sel. Areas Commun.*, vol. 21, no. 3, pp. 281–302, Apr. 2003.
23. M. A. Jensen and J. W. Wallace, "A review of antennas and propagation for MIMO wireless communications," *IEEE Trans. Antennas Propag.*, vol. 52, no. 11, pp. 2810–2824, Nov. 2004.
24. V. Erceg *et al.*, "TGn channel models," IEEE P802.11 Working Group for Wireless Local Area Networks, Doc. No. IEEE 802.11-03/940/r4, revised 10 May 2004.
25. D. S. Baum, J. Hansen, J. Salo, G. Del Galdo, M. Milojevic and P. Kyösti, "An interim channel model for beyond-3G systems," in *Proc. IEEE VTC 2005-Spring,* 30 May–1 Jun. 2005, pp. 3132–3136.
26. M. Narandžić, C. Schneider, R. Thomä, T. Jämsä, P. Kyösti, X. Zhao, "Comparison of SCM, SCME and WINNER channel models," in *Proc. IEEE VTC 2007-Spring,* 22–25 Apr. 2007, pp. 413–417.
27. P. Almers, E. Bonek, A. Burr, N. Czink, M. Debbah, V. degli-Esposti, H. Hofstetter, P. Kyösti, D. Laurenson, G. Matz, A. F. Molisch, C. Oestges and H. Özcelik, "Survey of channel and radio propagation models for wireless MIMO systems," *EURASIP J. Wireless Commun. Netw.* vol. 2007, p. 19, doi:10.1155/2007/19070.

Channel Models for Body Area Networks

28. A. Alomainy, Y. Hao, X. Hu, C. G. Parini and P. S. Hall, "UWB on-body radio propagation and system modelling for wireless body-centric networks," *IEE Proc. Commun.*, vol. 153, no. 1, pp. 107–114, Feb. 2006.
29. P. S. Hall and Y. Hao (Eds.), *Antennas and Propagation for Body-centric Communications.* Boston, MA : Artech House, 2006.
30. Y. Hao, P. S. Hall and K. Ito, (Eds.), *Special Issue on Antennas and Propagation on Body-Centric Wireless Communications, IEEE Trans. Antennas Propag.*, vol. 57, no. 4, Apr. 2009.
31. A. Fort, J. Ryckaert, C. Desset, P. De Donecker, P. Wambacq and L. Van Biesen, "Ultra-wideband channel model for communication around the human body," *IEEE J. Sel. Areas Commun.*, vol. 24, no. 4, pp. 927–933, Apr. 2006.
32. A. Fort, C. Desset, P. De Donecker, P. Wambacq and L. Van Biesen, "An ultra-wideband body area propagation channel model: From statistics to implementation," *IEEE Trans. Microw. Theory Tech.*, vol. 54, no. 4, pp. 1820–1826, Apr. 2006.
33. K. Y. Yazdandoost and K. Sayrafian-Pour, "Channel model for body area network," IEEE P802.15 Working Group for Wireless Personal Area Networks, IEEE P802.15-08-0780-02-0006, 12 Nov. 2008.

Channel Models for Vehicular Networks

34. J. Yin *et al.*, "Performance evaluation of safety applications over DSRC vehicular ad hoc networks," in *Proc. VANET 2004,* 1 Oct. 2004, pp. 1–9.
35. M. Toyota, R. K. Pokharel and O. Hashimoto, "Efficient multi-ray propagation model for DSRC EM environment on express highway," *Elec. Lett.*, vol. 40, no. 20, pp. 1278–1279, 30 Sep. 2004.
36. G. Acosta-Marum and M. A. Ingram, "Six time- and frequency-selective empirical channel models for vehicular wireless LANs," *IEEE Veh. Technol. Mag.*, vol. 2, no. 4, pp. 4–11, Dec. 2007.
37. I. Sen and D. W. Matolak, "Vehicle-vehicle channel models for the 5-GHz band," *IEEE Trans. Intell. Transp. Syst.*, vol. 9, no. 2, pp. 235–245, Jun. 2008.

38. I. Tan, W. Tang, K. Laberteaux and A. Bahai, "Measurement and analysis of wireless channel impairments in DSRC vehicular communications," in *Proc. IEEE ICC 2008*, 19–23 May 2008, pp. 4882–4888.

Channel Models for 60 GHz and Terahertz Systems

39. P. Smulders, "60 GHz radio: Prospects and future directions," in *Proc. 10th IEEE Symp. Commun. Veh. Technol.*, Benelux, Nov. 2003, pp. 1–8.
40. R. Piesiewicz, T. Kleine-Ostmann, N. Krumbholz, D. Mittleman, M. Koch, J. Schoebel and T. Kürner, "Short-range ultra-broadband terahertz communications: Concepts and perspectives," *IEEE Antennas Propag. Mag.*, vol. 49, no. 6, pp. 24–39, Dec. 2007.
41. C. Park and T. S. Rappaport, "Short-range wireless communications for next-generation networks: UWB, 60 GHz millimeter-wave WPAN and ZigBee," *IEEE Wireless Commun.*, vol. 14, no. 4, pp. 70–78, Aug. 2007.
42. T. Zwick, T. J. Beukema and H. Nam, "Wideband channel sounder with measurements and model for the 60 GHz indoor radio channel," *IEEE Trans. Veh. Technol.*, vol. 54, no. 4, pp. 1266–1277, Jul. 2005.
43. S. K. Yong, "TG3c channel modeling sub-committee final report," IEEE P802.15 Working Group for Wireless Personal Area Networks, Doc. No. IEEE 15-07-0584-01-003c, 13 Mar. 2007.
44. C. Jansen, R. Piesiewicz, D. Mittleman, T. Kürner and M. Koch, "The impact of reflections from stratified building materials on the wave propagation in future indoor terahertz communication systems," *IEEE Trans. Antennas Propag.*, vol. 56, no. 5, pp. 1413–1419, May 2008.

Chapter 2
OFDM: Principles and Challenges

Nicola Marchetti, Muhammad Imadur Rahman, Sanjay Kumar,
and Ramjee Prasad

2.1 Introduction

The nature of future wireless applications demands high data rates. Naturally dealing
with ever-unpredictable wireless channel at high data rate communications is not an
easy task. The idea of multi-carrier transmission has surfaced recently to be used for
combating the hostility of wireless channel and providing high data rate communications. OFDM is a special form of multi-carrier transmission where all the subcarriers are orthogonal to each other. OFDM promises a high user data rate transmission
capability at a reasonable complexity and precision.

At high data rates, the channel distortion to the data is very significant, and it is
somewhat impossible to recover the transmitted data with a simple receiver. A very
complex receiver structure is needed which makes use of computationally expensive
equalization and channel estimation algorithms to correctly estimate the channel, so
that the estimations can be used with the received data to recover the originally
transmitted data. OFDM can drastically simplify the equalization problem by turn-
ing the frequency-selective channel into a flat channel. A simple one-tap equalizer
is needed to estimate the channel and recover the data.

Future telecommunication systems must be spectrally efficient to support a number of high data rate users. OFDM uses the available spectrum very efficiently which
is very useful for multimedia communications. For all of the above reasons, OFDM
has already been accepted by many of the future generation systems [1].

Nicola Marchetti, Sanjay Kumar, and Ramjee Prasad
Center for TeleInFrastruktur (CTIF), Aalborg University, Denmark,
e-mail: {nm, in_sk, prasad}@es.aau.dk.

Muhammad Imadur Rahman
Ericsson Research, Kista, Sweden,
e-mail: muhammad.imadur.rahman@ericsson.com.

V. Tarokh (ed.), *New Directions in Wireless Communications Research,*
DOI 10.1007/978-1-4419-0673-1_2,
© Springer Science+Business Media, LLC 2009

2.2 History and Development of OFDM

Although OFDM has only recently been gaining interest from telecommunications industry, it has a long history of existence. It is reported that OFDM-based systems were in existence during the Second World War. OFDM had been used by the US military in several high-frequency military systems such as KINEPLEX, AN-DEFT, and KATHRYN [2]. KATHRYN used AN/GSC-10 variable rate data modem built for high-frequency radio. Up to 34 parallel low-rate channels using PSK modulation were generated by a frequency-multiplexed set of subchannels. Orthogonal frequency assignment was used with channel spacing of 82 Hz to provide guard time between successive signaling elements [3].

In December 1966, Robert W. Chang[1] outlined a theoretical way to transmit simultaneous data stream through linear band-limited channel without *inter-symbol interference* (ISI) and *inter-carrier interference* (ICI). Subsequently, he obtained the first US patent on OFDM in 1970 [4]. Around the same time, Saltzberg[2] performed an analysis of the performance of the OFDM system. Until this time, we needed a large number of subcarrier oscillators to perform parallel modulations and demodulations.

A major breakthrough in the history of OFDM came in 1971 when Weinstein and Ebert[3] used *discrete Fourier transform* (DFT) to perform baseband modulation and demodulation focusing on efficient processing. This eliminated the need for bank of subcarrier oscillators, thus paving the way for easier, more useful, and efficient implementation of the system.

All the proposals until this time used guard spaces in frequency domain and a raised cosine windowing in time domain to combat ISI and ICI. Another milestone for OFDM history was when Peled and Ruiz[4] introduced *cyclic prefix* (CP) or cyclic extension in 1980. This solved the problem of maintaining orthogonal characteristics of the transmitted signals at severe transmission conditions. The generic idea that they placed was to use cyclic extension of OFDM symbols instead of using empty guard spaces in frequency domain. This effectively turns the channel as performing cyclic convolution, which provides orthogonality over dispersive channels when CP is longer than the channel impulse response [2]. It is obvious that introducing CP causes loss of signal energy proportional to length of CP compared to symbol length, but, on the other hand, it facilitates a zero ICI advantage which pays off.

---

[1] Robert W. Chang, *Synthesis of Band-limited Orthogonal Signals for Multichannel Data Transmission*, The Bell Systems Technical Journal, December 1966.

[2] B. R. Saltzberg, *Performance of an Efficient Parallel Data Transmission System*, IEEE Transactions on Communications, COM-15 (6), pp. 805–811, December 1967.

[3] S. B. Weinstein, P. M. Ebert, *Data Transmission of Frequency Division Multiplexing Using the Discrete Frequency Transform*, IEEE Transactions on Communications, COM-19(5), pp. 623–634, October 1971.

[4] R. Peled, A. Ruiz, *Frequency Domain Data Transmission Using Reduced Computational Complexity Algorithms*, in Proceeding of the IEEE International Conference on Acoustics, Speech, and Signal Processing, ICASSP '80, pp. 964–967, Denver, USA, 1980.

By this time, inclusion of FFT and CP in OFDM system and substantial advancements in *digital signal processing* (DSP) technology made it an important part of telecommunications landscape. In the 1990s, OFDM was exploited for wideband data communications over mobile radio FM channels, *high-bit-rate digital subscriber lines* (HDSL at 1.6 Mbps), *asymmetric digital subscriber lines* (ADSL up to 6Mbps), and *very-high-speed digital subscriber lines* (VDSL at 100 Mbps).

Digital audio broadcasting (DAB) was the first commercial use of OFDM technology. Development of DAB started in 1987. By 1992, DAB was proposed and the standard was formulated in 1994. DAB services came to reality in 1995 in the United Kingdom and Sweden. The development of *digital video broadcasting* (DVB) was started in 1993. DVB along with *high-definition television* (HDTV) terrestrial broadcasting standard was published in 1995. At the dawn of the 20th century, several *wireless local area network* (WLAN) standards adopted OFDM on their physical layers. Development of European WLAN standard HiperLAN started in 1995. HiperLAN/2 was defined in June 1999 which adopts OFDM in physical layer. IEEE 802.11a has also adopted OFDM in its PHY layer.

Perhaps of even greater importance is the emergence of this technology as an enabler for future *4th generations* (4G) wireless systems, such as IMT-A. These systems, expected to emerge by the year 2015, promise to at last deliver on the wireless Nirvana of anywhere, anytime, anything communications. OFDM promises to gain prominence in this arena; therefore, it is expected to become the technology of choice in most wireless links.

2.3 The Benefit of Using Multi-carrier Transmission

Time dispersion represents a distortion of the signal that is manifested by the spreading of the modulation symbols in the time domain, also known as delay spread, and this is reflected by the ISI phenomenon. This is also reflected in frequency domain, by the inverse proportionality relation between coherence bandwidth and delay spread, i.e., the higher the delay spread, the lower the coherence bandwidth, and therefore the higher the channel frequency selectivity. For broadband multimedia communications the coherence bandwidth of the channel is always smaller than the modulation bandwidth. Thus in such conditions, the frequency selectivity effect cannot be avoided, which has a random pattern at any given time. This fading occurs when the channel introduces time dispersion and the delay spread is larger than the symbol period. Frequency-selective fading is difficult to compensate because the fading characteristics are random and may not be easily predictable. When there is no dispersion and the delay spread is less than the symbol period, the fading will be flat, thereby affecting all frequencies in the signal equally. Practically flat fading is easily estimated and compensated with a simple equalization [5, 6].

A single-carrier system suffers from ISI problem when the data rate is very high. According to previous discussions, we have seen that with a symbol duration T_{sym}, ISI occurs when $\tau_{max} > T_{sym}$. Multichannel transmission has surfaced to solve this problem. The idea is to increase the symbol duration and thus reduce the effect of ISI. Reducing the effect of ISI yields an easier equalization, which in turn means simpler reception techniques.

Wireless multimedia solutions require up to tens of Mbps for a reasonable QoS. If we consider single-carrier high-speed wireless data transmission, we see that the delay spread at such high data rates will definitely be greater than the symbol duration even considering the best-case outdoor scenario. Now, if we divide the high data rate channel over a number of subcarriers, then we have larger symbol duration in the subcarriers and the delay spread is much smaller than the symbol duration.

Figure 2.1 describes this very issue. Assuming that we have available bandwidth B of 1 MHz in a single-carrier approach, we transmit the data at symbol duration of 1 μs. Consider a typical outdoor scenario where the maximum delay spread can be as high as 10 μs, so at the worst-case scenario, at least 10 consecutive symbols will be affected by ISI due to the delay spread.

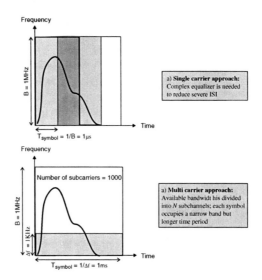

Fig. 2.1 Single carrier vs multi-carrier approach

In a single-carrier system, this situation is compensated by using equalization techniques. Using the estimates of channel impulse response, the equalizer multiplies complex conjugate of the estimated impulse response with the received data signal at the receiver. There are other well-known equalization algorithms available in the literature, such as adaptive equalization via LMS, RLS algorithms [7]. However, there are some practical computational difficulties in performing these equalization techniques at tens of Mbps with compact and low-cost hardware. It is worth mentioning here that compact and low-cost hardware devices do not necessarily function at very high data speed. In fact, equalization procedures take bulk of receiver resources, costing high computation power and thus overall service and hardware cost becomes high.

One way to achieve reasonable quality and solve the problems described above for broadband mobile communication is to use parallel transmission. In a crude sense, someone can say in principle that parallel transmission is just the summation of a number of single-carrier transmissions at the adjacent frequencies [8]. The difference is that the channels have lower data transmission rate than the original single-carrier system and the low rate streams are orthogonal to each other. If we consider a multi-carrier approach where we have N number of subcarriers, we can see that we can have $\frac{B}{N}$ Hz of bandwidth per sub-carrier. If $N = 1000$ and $B = 1$ MHz, then we have a subcarrier bandwidth Δf of 1 kHz. Thus, the symbol duration in a subcarrier will be increased to 1 ms (= $\frac{1}{1\,kHz}$). Here each symbol occupies a narrowband but a longer time period. This clearly shows that the delay spread of 1 ms will not have any ISI effect on the received symbols in the outdoor scenario mentioned above. So, we can say that the multi-carrier approach turns the channel to a flat fading channel and thus can easily be estimated.

Theoretically increasing the number of subcarriers should be able to give better performance in a sense that we will be able to handle larger delay spreads. But several typical implementation problems arise with a large number of subcarriers. When we have large numbers of subcarriers, we have to assign the subcarrier frequencies very close to each other, if the available bandwidth is not increased. We know that the receiver needs to synchronize itself to the carrier frequency very well, otherwise a comparatively small carrier frequency offset may cause a large frequency mismatch between neighboring subcarriers. When the subcarrier spacing is very small, the receiver synchronization components need to be very accurate, which is still not possible with low-cost RF hardware. Thus, a reasonable trade-off between the subcarrier spacing and the number of subcarriers must be achieved.

Table 2.1 describes how multi-carrier approach can convert the channel to flat fading channel from frequency-selective fading channel. We have considered a

Table 2.1 Comparison of single-carrier and multi-carrier approach in terms of channel frequency selectivity

Design parameters for outdoor channel	Required data rate		1 Mbps
	RMS delay spread, τ_{rms}		10 μs
	Channel coherence bandwidth, $B_c = \frac{1}{5\tau_{rms}}$		20 kHz
	Frequency selectivity condition		$\sigma > \frac{T_{sym}}{10}$
Single-carrier approach	Symbol duration, T_{sym}	1 μs	
	Frequency selectivity	$10\,\mu s > \frac{1\,\mu s}{10} \implies$ **YES**	
	ISI occurs as the channel is frequency selective		
Multi-carrier approach	Total number of subcarriers	128	
	Data rate per subcarrier	7.8125 kbps	
	Symbol duration per subcarrier	$T_{carr} = 128\,\mu s$	
	Frequency selectivity	$10\,\mu s > \frac{128\,\mu s}{10} \implies$ **NO**	
	ISI is reduced as flat fading occurs. CP completely removes the remaining ISI; and also inter-block interference is removed		

multi-carrier system with respect to a single-carrier system, where the system data rate requirement is 1 Mbps. When we use 128 subcarriers for multi-carrier system, we can see that the ISI problem is clearly solved. It is obvious that if we increase the number of subcarriers, the system will theoretically provide even better performance.

2.4 OFDM Transceiver Systems

A complete OFDM transceiver system is described in Fig. 2.2. In this model, *forward error control/correction* (FEC) coding and interleaving are added in the system to obtain the robustness needed to protect against burst errors. An OFDM system with addition of channel coding and interleaving is referred to as *coded OFDM* (COFDM).

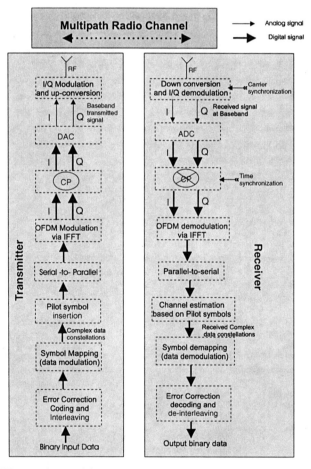

Fig. 2.2 OFDM transceiver model

In a digital domain, binary input data are collected and FEC coded with schemes such as convolutional codes. The coded bit stream is interleaved to obtain diversity gain. Afterward, a group of channel-coded bits are gathered together (1 for BPSK, 2 for QPSK, 4 for 16-QAM, etc.) and mapped to corresponding constellation points. At this point, the data are represented as complex numbers and they are in serial. Known pilot symbols mapped with known mapping schemes can be inserted at this moment. A serial to parallel converter is applied and the IFFT operation is performed on the parallel complex data. The transformed data are grouped together again, as per the number of required transmission subcarriers. Cyclic prefix is inserted in every block of data according to the system specification and the data are multiplexed in a serial fashion. At this point of time, the data are OFDM modulated and ready to be transmitted. A *digital-to-analog converter* (DAC) is used to transform the time-domain digital data to time-domain analog data. RF modulation is performed and the signal is up-converted to transmission frequency.

After the transmission of OFDM signal from the transmitter antenna, the signals go through all the anomaly and hostility of wireless channel. After receiving the signal, the receiver down-converts the signal and converts to digital domain using *analog-to-digital converter* (ADC). At the time of down-conversion of received signal, carrier frequency synchronization is performed. After ADC conversion, symbol timing synchronization is achieved. An FFT block is used to demodulate the OFDM signal. After that, channel estimation is performed using the demodulated pilots. Using the estimations, the complex received data are obtained which are demapped according to the transmission constellation diagram. At this moment, FEC decoding and de-interleaving are used to recover the originally transmitted bit stream.

2.5 Analytical Model of OFDM System

In this section, an analytical time-domain model of an OFDM transmitter and receiver, as well as a channel model, is derived.

2.5.1 Transmitter

The sth OFDM symbol is found using the sth subcarrier block, $\mathbf{X}_s[k]$. In practice, the OFDM signal is generated using an inverse DFT. In the following model, the transmitter is assumed to be ideal, i.e., sampling or filtering do not affect the signal at the transmitter side. Therefore, a continuous transmitter output signal may be constructed directly using a Fourier series representation within each OFDM symbol interval.

Each OFDM symbol contains N subcarriers, where N is an even number (frequently a power of 2). The OFDM symbol duration is T_u seconds, which must be a whole number of periods for each subcarrier. Defining the subcarrier spacing as $\Delta\omega$, the shortest duration that meets this requirement is written as

$$T_u = \frac{2\pi}{\Delta\omega} \Leftrightarrow \Delta\omega = \frac{2\pi}{T_u} = 2\pi\,\Delta f. \tag{2.1}$$

Using this relation, the spectrum of the Fourier series for the duration of the sth OFDM symbol is written as

$$\mathbf{X}_s(\omega) = \sum_{k=-N/2}^{N/2-1} \mathbf{X}_s[k]\delta_c(\omega - k\Delta\omega). \tag{2.2}$$

In order to provide the OFDM symbol in the time domain, the spectrum in (2.2) is inverse Fourier transformed and limited to a time interval of T_u. The time-domain signal, $\tilde{x}_s(t)$, is therefore written as

$$\tilde{x}_s(t) = \mathcal{F}\{\mathbf{X}_s(\omega)\}\,\varXi_{T_u}(t)$$

$$= \begin{cases} \frac{1}{\sqrt{T_u}} \sum_{k=-N/2}^{N/2-1} \mathbf{X}_s[k]e^{j\Delta\omega kt} & 0 \le t < T_u, \\ 0 & \text{otherwise} \end{cases} \tag{2.3}$$

$$\tag{2.4}$$

where $\varXi_{T_u}$ is a unity amplitude rectangular gate pulse of duration T_u. Following the frequency- to time-domain conversion, the signal is extended, and the cyclic prefix is added:

$$\tilde{x}'_s(t) = \begin{cases} \tilde{x}_s(t + T_u - T_g) & 0 \le t < T_g \\ \tilde{x}_s(t - T_g) & T_g < t < T_s \\ 0 & \text{otherwise} \end{cases}, \tag{2.5}$$

where T_g is the cyclic prefix duration and $T_s = T_u + T_g$ is the total OFDM symbol duration. It should be noted that (2.5) has the following property:

$$\tilde{x}'_s(t) = \tilde{x}'_s(t + T_u) \Leftrightarrow 0 \le t < T_g, \tag{2.6}$$

that is, a periodicity property within the interval $[0, T_g]$. The transmitted complex baseband signal, $\tilde{s}(t)$, is formed by concatenating all OFDM symbols in the time domain:

$$\tilde{s}(t) = \sum_{s=0}^{S-1} \tilde{x}'_s(t - sT_s). \tag{2.7}$$

This signal is finally upconverted to a carrier frequency and transmitted:

$$s(t) = \Re e\left\{\tilde{s}(t)e^{j2\pi f_c t}\right\}, \tag{2.8}$$

where $s(t)$ denotes the transmitted RF signal and f_c is the RF carrier frequency. For frequency hopping systems, the carrier frequency is changed at certain intervals. This is written as

$$f_c[s] = f_{c,0} + f_h[s],\qquad(2.9)$$

where $f_c[s]$ is the carrier frequency for the sth OFDM symbol, $f_{c,0}$ is the center frequency of the band, and $f_h[s]$ is the frequency deviation from the band center when transmitting the sth OFDM symbol. The period of $f_h[s]$ is χ, where χ is the hopping sequence period measured in whole OFDM symbols.

The transmitter model described in this section is illustrated in Fig. 2.3.

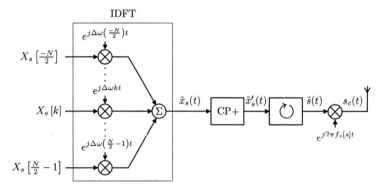

Fig. 2.3 Transmitter diagram for the OFDM analytical model given by (2.1)–(2.9). The subcarriers for the sth OFDM symbol each modulates a carrier; the carriers are separated by $\Delta\omega$. The resulting waveforms are then summed, and the CP is added. The symbol ⟳ represents the concatenation of the OFDM symbols given by (2.7). The resulting signal is then up-converted to a carrier frequency and transmitted

2.5.2 Channel

The channel is modeled as a time-domain complex-baseband transfer function, which may then be convolved with the transmitted signal to determine the signal at the receiver side. The channel baseband equivalent impulse response function for the uth user, $\tilde{h}_u(t)$ is defined as

$$\tilde{h}_u(\tau, t) = \sum_{l=0}^{L} h_{u,l}(t)\delta_c(\tau - \tau_l),\qquad(2.10)$$

where $h_{u,l}(t)$ is the complex gain of the lth multipath component for the uth user at time t. The channel is assumed to be static for the duration of one OFDM symbol, and the path gain coefficients for each path contribution are assumed to be uncorrelated. No assumption is made for the autocorrelation properties of each path, except in the case of frequency hopping systems. In such systems, the channel is assumed to be completely uncorrelated between two frequency hops, provided that the distance in frequency is sufficiently large.

As the channel is assumed to be static over each OFDM symbol, (2.10) is redefined as

$$\tilde{h}_{u,s}(t) = \sum_{l=0}^{L} h_{u,l}[s]\delta_c(t - \tau_l),$$ (2.11)

where

$$h_{u,l}[s] = h_{u,l}(t), \qquad sT_s \le t < (s+1)T_s.$$

The corresponding frequency-domain channel transfer function, $H_{u,s}$, can then be found using Fourier transformation:

$$H_{u,s}(\omega) = \mathcal{F}\left\{\tilde{h}_{u,s}(t)\right\}$$

$$= \int_{-\infty}^{\infty} \tilde{h}_{u,s}(t)e^{j\omega t}dt.$$ (2.12)

The time-domain channel model is illustrated in Fig. 2.4.

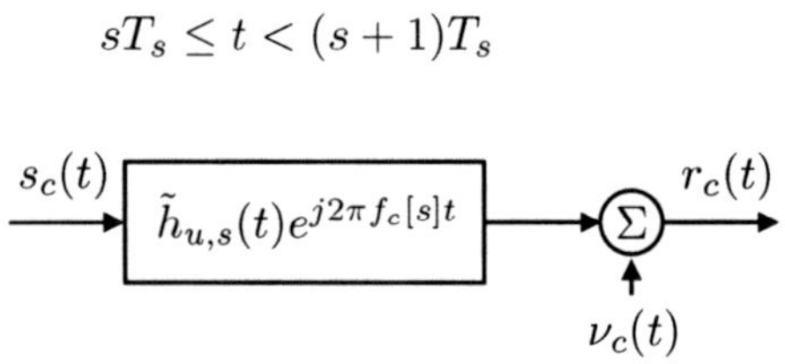

$$sT_s \le t < (s+1)T_s$$

Fig. 2.4 A diagram of the channel model given by (2.10)–(2.11). The transmitted signal passes through the channel, and noise is added

2.5.3 Receiver

The signal at the receiver side consists of multiple echoes of the transmitted signal, as well as thermal (white gaussian) noise and interference. The RF signal received by the uth user is written as

$$r(t) = \Re\left\{(\tilde{s}(t) * \tilde{h}_{u,s}(t))e^{j2\pi f_c[s]t}\right\} + v(t), \qquad sT_s \le t < (s+1)T_s, \quad (2.13)$$

where $v(t)$ is a real-valued, passband signal combining additive noise and interference. The receiver now has to recreate the transmitted signal. Aside from noise and multipath effects, other imperfections in the receiver may also affect this process:

- **Timing error**: In order to demodulate the signal, the receiver must establish the correct timing. This means that the receiver must estimate which time instant corresponds to $t = 0$ in the received signal (as seen from the transmitted signal point of view). As there are different uncertainties involved, a timing error of δt is assumed.

- **Frequency error**: Similarly, the local oscillator of the receiver may oscillate at an angular frequency that is different from the angular frequency of the incoming signal. This difference is denoted as $\delta \omega = 2\pi \delta f$.

The shifted timescale in the receiver is denoted as $t' = t - \delta t$. Furthermore, due to the angular frequency error $\delta \omega$, the down-converted signal spectrum is shifted in frequency. The down-converted signal is therefore written as

$$\tilde{r}(t) = (\tilde{s}(t') * \tilde{h}_{u,s}(t))e^{j\delta \omega t} + \tilde{v}(t'), \qquad sT_s \leq t < (s+1)T_s, \qquad (2.14)$$

where $\tilde{v}(t)$ is the complex envelope of the down-converted AWGN. The signal is divided into blocks each T_s-long, and the CP is removed from each of them. The sth received OFDM symbol block, $y'_s(t)$ is defined as

$$\tilde{y}'_s(t) = \tilde{r}(t' - sT_s), \qquad 0 \leq t < T_s. \qquad (2.15)$$

The signal block corresponding to $\tilde{x}_s(t)$, $\tilde{y}_s(t)$ is found by removing the CP from each $\tilde{y}'_s(t)$:

$$\tilde{y}_s(t) = \tilde{y}'_s(t + T_g), \qquad 0 \leq t < T_s - T_g, \qquad (2.16)$$

which can be rewritten as

$$
\begin{aligned}
\tilde{y}_s(t) &= \tilde{y}'_s(t + T_g), \qquad 0 \leq t \leq T_u \\
&= \tilde{r}(t' + T_g - sT_s) \\
&= (\tilde{s}(t' + T_g - sT_s) * \tilde{h}_{u,s}(t))e^{j\delta \omega t} + \tilde{v}(t' + T_g - sT_s) \\
&= (\tilde{x}'_s(t' + T_g) * \tilde{h}_{u,s}(t))e^{j\delta \omega t} + \tilde{v}_s(t') \\
&= (\tilde{x}_s(t') * \tilde{h}_{u,s}(t))e^{j\delta \omega t} + \tilde{v}_s(t'), \qquad (2.17)
\end{aligned}
$$

where $\tilde{v}_s(t')$ is the noise signal block of duration T_u corresponding to the sth OFDM symbol.

In order to recreate the transmitted subcarriers, N correlators are used, each one correlating the incoming signal with the kth subcarrier frequency over an OFDM symbol period:

$$Y_s[k] = \frac{1}{\sqrt{T_u}} \int_0^{T_u} \tilde{y}_s(t')e^{j\Delta \omega k t}\, dt. \qquad (2.18)$$

In order to determine the correlator output, (2.18) may be seen as taking the continuous Fourier transform of (2.17) multiplied by the rectangular pulse $\Xi_{T_u}(t)$ and evaluating it at the corresponding subcarrier frequency. Assuming that the timing error is low enough to avoid ISI

$$0 \leq \delta t < T_g - \max(\tau_l)$$

the continuous Fourier transform can be written as

$$
\begin{aligned}
Y_s(\omega) &= \mathcal{F}\left\{\tilde{y}_s(t)\Xi_{T_u}(t)\right\} \\
&= \mathcal{F}\left\{(\tilde{x}_s(t') * \tilde{h}_{u,s}(t))e^{j\delta\omega t} + \tilde{v}_s(t')\right\} * T_u e^{j\pi\frac{\omega}{\Delta\omega}}\operatorname{sinc}\left(\frac{\omega}{\Delta\omega}\right) \\
&= \mathcal{F}\left\{(\tilde{x}_s(t') * \tilde{h}_{u,s}(t))e^{j\delta\omega t}\right\} * T_u e^{j\pi\frac{\omega}{\Delta\omega}}\operatorname{sinc}\left(\frac{\omega}{\Delta\omega}\right) + N_s(\omega) \\
&= \mathcal{F}\left\{\tilde{x}_s(t') * \tilde{h}_{u,s}(t)\right\} * \delta_c(\omega - \delta\omega) * T_u e^{j\pi\frac{\omega}{\Delta\omega}}\operatorname{sinc}\left(\frac{\omega}{\Delta\omega}\right) + N_s(\omega) \\
&= e^{-j\omega\delta t}\mathcal{F}\left\{\tilde{x}_s(t) * \tilde{h}_{u,s}(t)\right\} * \delta_c(\omega - \delta\omega) * T_u e^{j\pi\frac{\omega}{\Delta\omega}}\operatorname{sinc}\left(\frac{\omega}{\Delta\omega}\right) + N_s(\omega) \\
&= e^{-j\omega(\delta t + \frac{\pi}{\Delta\omega})}\sum_{k'=N/2}^{N/2-1}\mathbf{X}_s[k']H_{u,s}(k'\Delta\omega)\operatorname{sinc}\left(\frac{\omega - k'\Delta\omega - \delta\omega}{\Delta\omega}\right) + N_s(\omega),
\end{aligned}
$$

$$(2.19)$$

where

$$
N_s(\omega) = \mathcal{F}\left\{\tilde{v}_s(t')\right\} * T_u e^{j\pi\frac{\omega}{\Delta\omega}}\operatorname{sinc}\left(\frac{\omega}{\Delta\omega}\right) \tag{2.20}
$$

is the Fourier transform of the AWGN contribution. The correlator output at the kth correlator is then found as

$$
\begin{aligned}
Y_s[k] &= Y_s(k\Delta\omega) \\
&= e^{-jk\Delta\omega(\delta t + \frac{\pi}{\Delta\omega})}\sum_{k'=N/2}^{N/2-1}\mathbf{X}_s[k']H_{u,s}(k'\Delta\omega)\operatorname{sinc}\left(\frac{k\Delta\omega - k'\Delta\omega - \delta\omega}{\Delta\omega}\right) \\
&\quad + N_s(k\Delta\omega).
\end{aligned}
$$

$$(2.21)$$

For zero frequency error, (2.21) reduces to

$$
Y_s[k] = e^{-jk\Delta\omega(\delta t + \frac{\pi}{\Delta\omega})}\mathbf{X}_s[k]H_{u,s}[k] + N_s[k], \quad \delta\omega = 0, \tag{2.22}
$$

where

$$
N_s[k] = N_s(k\Delta\omega), \tag{2.23}
$$

$$
H_{u,s}[k] = H_{u,s}(k\Delta\omega). \tag{2.24}
$$

From (2.21), it is seen that the kth correlator output, $Y_s[k]$, corresponds to the transmitted subcarrier, $\mathbf{X}_s[k]$, with AWGN, ICI, and a complex gain term (amplitude and phase shift) due to imperfect timing and channel effects. The analytical model for the receiver is illustrated in Fig. 2.5.

When estimating the channel, the constant phase rotation term and the channel transfer function would be estimated jointly (as the receiver cannot discern between the two). In the following, the timing delay phase shift is omitted for clarity. Defining the equalization factor for the kth subcarrier of the sth OFDM symbol and uth user as $Z_{u,s}[k]$, the subcarrier estimate is written as

$$\hat{X}_s[k] = Z_{u,s}[k]\,Y_s[k]$$
$$= Z_{u,s}[k]\,H_{u,s}[k]\,\mathbf{X}_s[k] + Z_{u,s}[k]\,N_s[k]. \qquad (2.25)$$

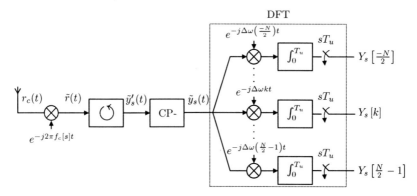

Fig. 2.5 Receiver diagram for the OFDM analytical model, given by (2.13)–(2.22). The received signal (suffering from multipath effects and AWGN) is converted down to baseband. The symbol ↺ represents the division of the received signal into blocks, given by (2.15). The CP is removed from each block, and the signal is then correlated with each subcarrier frequency, as shown by (2.18)

Assuming a zero-forcing, frequency-domain equalizer (as well as perfect channel estimation and zero frequency error), the corresponding equalizer gain is written as

$$Z_{u,s}[k] = \frac{1}{H_{u,s}[k]}$$

and (2.25) is rewritten as

$$\hat{X}_s[k] = \mathbf{X}_s[k] + \frac{N_s[k]}{H_{u,s}[k]}. \qquad (2.26)$$

It is observed that although this is an unbiased estimator for $\mathbf{X}_s[k]$, the signal-to-noise ratio decreases drastically for subcarriers in deep fades.

2.5.4 Sampling

Although the receiver may be modeled in the continuous time domain, an OFDM receiver uses discrete signal processing to obtain the estimate of the transmitted subcarriers.

When the received signal is modeled as a Dirac impulse train, i.e., an ideally sampled signal, (2.17) is instead written as

$$\tilde{y}_{s,d}(t) = \sum_{n=0}^{N-1} \tilde{y}_s[n]\delta_c(t - nT), \qquad (2.27)$$

where

$$T = \frac{T_u}{N} \qquad (2.28)$$

is the sample duration and

$$\tilde{y}_s[n] = \tilde{y}_s(nT), \quad n \in \{0, 1, ..., N-1\} \qquad (2.29)$$

is the discrete sequence corresponding to the sampled values of $\tilde{y}_s(t)$. When (2.27) is inserted into (2.18), the correlation becomes the discrete Fourier transform of the received signal. It can be shown, however, that (2.21)–(2.26) are still valid in the discrete-time case.

2.6 Advantages of OFDM System

2.6.1 Combating ISI and Reducing ICI

When signal passes through a time-dispersive channel, the orthogonality of the signal can be jeopardized. CP helps to maintain orthogonality between the sub-carriers. Before CP was invented, guard interval was proposed as the solution. Guard interval was defined by an empty space between two OFDM symbols, which serves as a buffer for the multipath reflection. The interval must be chosen as larger than the expected maximum delay spread, such that multipath reflection from one symbol would not interfere with another. In practice, the empty guard time introduces ICI, which is crosstalk between different subcarriers, meaning they are no longer orthogonal to each other [2]. A better solution was later found, that is, cyclic extension of OFDM symbol or CP, which is a copy of the last part of OFDM symbol, appended in front of the transmitted OFDM symbol [9] (see Fig. 2.6).

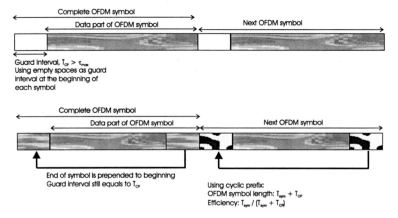

Fig. 2.6 Definition of cyclic prefix as the guard interval in OFDM systems

CP still occupies the same time interval as guard period, but it ensures that the delayed replicas of the OFDM symbols will always have a complete symbol within the FFT interval (often referred as FFT window); this makes the transmitted signal periodic. This periodicity plays a very significant role as this helps maintaining the orthogonality. The concept of being able to do this, and what it means, comes from the nature of IFFT/FFT process. When the IFFT is taken for a symbol period during OFDM modulation, the resulting time sample process is technically periodic. In a Fourier transform, all the resultant components of the original signal are orthogonal to each other. So, in short, by providing periodicity to the OFDM source signal, CP makes sure that subsequent subcarriers are orthogonal to each other.

At the receiver side, CP is removed before any processing starts. As long as the length of CP interval is larger than maximum expected delay spread τ_{max}, all reflections of previous symbols are removed and orthogonality is restored. The orthogonality is lost when the delay spread is larger than the length of CP interval. Inserting CP has its own cost, indeed we loose a part of signal energy since it carries no information. The loss is measured as

$$SNR_{loss\_CP} = -10\log_{10}\left(1 - \frac{T_{CP}}{T_{sym}}\right). \qquad (2.30)$$

Here, T_{CP} is the interval length of CP and T_{sym} is the OFDM symbol duration. It is understood that although we loose part of signal energy, the fact that we get zero ICI and ISI situation pay off the loss.

To conclude, CP gives twofold advantages, first occupying the guard interval, it removes the effect of ISI and by maintaining orthogonality it completely removes the ICI. The cost in terms of signal energy loss is not too significant.

2.6.2 Spectral Efficiency

Figure 2.7 illustrates the difference between conventional FDM and OFDM systems, where for $SC\ BW$ one means subcarrier bandwidth. In the case of OFDM, a better spectral efficiency is achieved by maintaining orthogonality between the subcarriers. When orthogonality is maintained between different subchannels during transmission, then it is possible to separate the signals very easily at the receiver side. Classical FDM ensures this by inserting guard bands between subchannels. These guard bands keep the subchannels far enough so that separation of different subchannels is possible. Naturally inserting guard bands results in inefficient use of spectral resources.

Orthogonality makes it possible in OFDM to arrange the subcarriers in such a way that the sidebands of the individual carriers overlap and still the signals are received at the receiver without being interfered by ICI. The receiver acts as a bank of demodulators, translating each subcarrier down to DC, with the resulting signal integrated over a symbol period to recover raw data. If the other subcarriers are all down-converted to the frequencies that, in the time domain, have a whole number

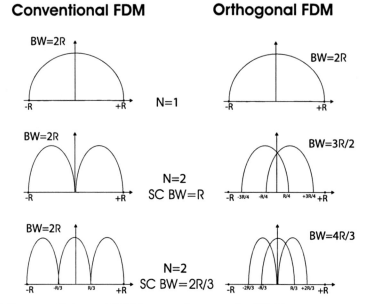

Fig. 2.7 Spectrum efficiency of OFDM compared to conventional FDM

of cycles in a symbol period T_{sym}, then the integration process results in zero contribution from all other carriers. Thus, the subcarriers are linearly independent (i.e., orthogonal) if the carrier spacing is a multiple of $\frac{1}{T_{sym}}$ [10].

2.6.3 Some Other Benefits of OFDM System

1. The beauty of OFDM lies in its simplicity. One trick of the trade that makes OFDM transmitters low cost is the ability to implement the mapping of bits to unique carriers via the use of IFFT [11].
2. Unlike CDMA, OFDM receiver collects signal energy in frequency domain, thus it is able to protect energy loss at frequency domain.
3. In a relatively slow time-varying channel, it is possible to significantly enhance the capacity by adapting the data rate per subcarrier according to SNR of that particular subcarrier [2].
4. OFDM is more resistant to frequency-selective fading than single-carrier systems.
5. The OFDM transmitter simplifies the channel effect, thus a simpler receiver structure is enough for recovering transmitted data. If we use coherent modulation schemes, then very simple channel estimation (and/or equalization) is needed, on the other hand, we need no channel estimator if differential modulation schemes are used.

6. The orthogonality preservation procedures in OFDM are much simpler compared to CDMA or TDMA techniques even in very severe multipath conditions.
7. It is possible to use maximum likelihood detection with reasonable complexity [12].
8. OFDM can be used for high-speed multimedia applications with low service cost.
9. OFDM can support dynamic packet access.
10. Single-frequency networks are possible in OFDM, which is especially attractive for broadcast applications.
11. Smart antennas can be integrated with OFDM. MIMO systems and space–time coding can be realized on OFDM and all the benefits of MIMO systems can be obtained easily. Adaptive modulation and tone/power allocation are also realizable on OFDM.

2.7 Disadvantages of OFDM System

2.7.1 Strict Synchronization Requirement

OFDM is highly sensitive to time and frequency synchronization errors, and especially at frequency synchronization errors, everything can go wrong [13]. Indeed, demodulation of an OFDM signal with an offset in the frequency can lead to a high bit error rate.

The source of frequency synchronization errors is two: first one being the difference between local oscillator frequencies in transmitter and receiver, second being relative motion between the transmitter and receiver that gives Doppler spread. Local oscillator frequencies at both transmitter and receiver must match as closely as they can. For higher number of subchannels, the matching should be even better. Motion of transmitter and receiver causes the other frequency error. So, OFDM may show significant performance degradation at high-speed moving vehicles [4].

To optimize the performance of an OFDM link, accurate synchronization is of prime importance. Synchronization needs to be done in three factors: symbol, carrier frequency, and sampling frequency synchronization. A good description of synchronization procedures is given in [14].

2.7.2 Peak-to-Average Power Ratio (PAPR)

Peak-to-average power ratio (PAPR) is proportional to the number of subcarriers used for OFDM systems. An OFDM system with large number of subcarriers will thus have a very large PAPR when the subcarriers add up coherently. Large PAPR of a system makes the implementation of digital-to-analog converter (DAC) and analog-to-digital converter (ADC) extremely difficult. The design of RF amplifier also becomes increasingly difficult as the PAPR increases.

The *clipping and windowing* technique reduces PAPR by non-linear distortion of the OFDM signal. It thus introduces self-interference as the maximum amplitude level is limited to a fixed level. It also increases the out-of-band radiation, but this is the simplest method to reduce the PAPR. To reduce the error rate, additional forward error correcting codes can be used in conjunction with the clipping and windowing method.

Another technique called *linear peak cancelation* can also be used to reduce the PAPR. In this method, time-shifted and time-scaled reference function is subtracted from the signal, such that each subtracted reference function reduces the peak power of at least one signal sample. By selecting an appropriate reference function with approximately the same bandwidth as the transmitted function, it can be assured that the peak power reduction does not cause out-of-band interference. One example of a suitable reference function is a *raised cosine window*. Detailed discussion about coding methods to reduce PAPR can be found in [2].

2.7.3 Co-channel Interference in Cellular OFDM

In cellular communication systems, CCI is combated by combining adaptive antenna techniques, such as sectorization, directive antenna, antenna arrays. Using OFDM in cellular systems will give rise to CCI. Similarly with the traditional techniques, with the aid of beam steering, it is possible to focus the base station's antenna beam on the served user, while attenuating the co-channel interferers.

2.8 OFDM System Design Issues

System design always needs a complete and comprehensive understanding and consideration of critical parameters. OFDM system design is of no exception, as it deals with some critical, and often conflicting parameters. Basic OFDM philosophy is to decrease data rate at the subcarriers, so that the symbol duration increases, thus the multipaths are effectively removed. This poses a challenging problem, as higher value for CP interval will give better result, but it will increase the loss of energy due to insertion of CP. Thus, a trade-off must be obtained for a reasonable design.

2.8.1 OFDM System Design Requirements

OFDM systems depend on four system requirements:

- **Available bandwidth:** Bandwidth is always the scarce resource, so the mother of the system design should be the available bandwidth for operation. The amount of bandwidth will play a significant role in determining number of subcarriers, because with a large bandwidth, we can easily fit in a large number of subcarriers with reasonable guard space.

- **Required bit rate:** The overall system should be able to support the data rate required by the users. For example, to support broadband wireless multimedia communication, the system should operate at more than 10 Mbps at least.
- **Tolerable delay spread:** Tolerable delay spread will depend on the user environment. Measurements show that indoor environment experiences maximum delay spread of few hundreds of ns at most, whereas outdoor environment can experience up to $10\,\mu s$. So the length of CP should be determined according to the tolerable delay spread.
- **Doppler values:** Users on a high-speed vehicle will experience higher Doppler shift, whereas pedestrians will experience smaller Doppler shift. These considerations must be taken into account.

2.8.2 OFDM System Design Parameters

The design parameters are derived according to the system requirements. Following are the design parameters for an OFDM system [2]:

- **Number of subcarriers:** Increasing number of subcarriers will reduce the data rate via each subcarrier, which will make sure that the relative amount of dispersion in time caused by multipath delay will be decreased (see Fig. 2.8). But when there are large numbers of subcarriers, the synchronization at the receiver side will be extremely difficult.
- **Guard time (CP interval) and symbol duration:** A good ratio between the CP interval and symbol duration should be found, so that all multipaths are resolved and not significant amount of energy is lost due to CP (see Fig. 2.9). As a thumb rule, the CP interval must be two to four times larger than the *root mean square* (RMS) delay spread. Symbol duration should be much larger than the guard time to minimize the loss of SNR, but within reasonable amount. It cannot be arbitrarily large, because larger symbol time means that more subcarriers can fit within the symbol time. More subcarriers increase the signal processing load at both the transmitter and receiver, increasing the cost and complexity of the resulting device [15].
- **Subcarrier spacing:** Subcarrier spacing must be kept at a level so that synchronization is achievable. This parameter will largely depend on available bandwidth and the required number of subchannels.
- **Modulation type per subcarrier:** This is trivial, because different modulation schemes will give different performances. Adaptive modulation and bit loading may be needed depending on the performance requirement. It is interesting to note that the performance of OFDM systems with differential modulation compares quite well with systems using non-differential and coherent demodulation

[16]. Furthermore, the computation complexity in the demodulation process is quite low for differential modulations.

- **FEC coding:** Choice of FEC code will play a vital role also. A suitable FEC coding will make sure that the channel is robust to all the random errors.

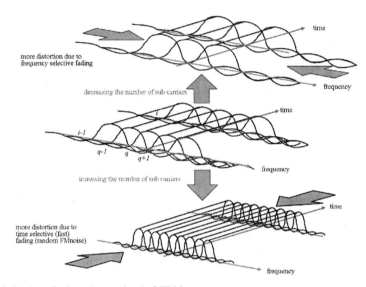

Fig. 2.8 Design of subcarrier spacing in OFDM systems

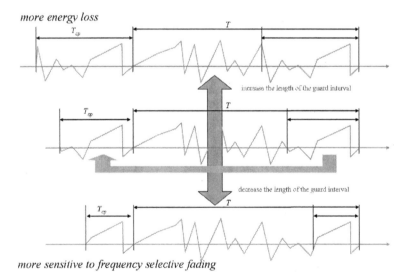

Fig. 2.9 Design of CP duration in OFDM systems

2.9 Multi-carrier Based Access Techniques

In this chapter, we present some of the main multiple access techniques that can be defined based on the original OFDM-type multi-carrier techniques.

2.9.1 Definition of Basic Schemes

It is imperative to understand the basic properties of three fundamental multi-carrier-based multiple access techniques, namely OFDMA, OFDM-TDMA, and OFDM-CDMA before embarking on studies related to probable access technique for 4G wireless communication systems, thus we here briefly summarize the basic properties of these three access schemes.

2.9.1.1 OFDM-TDMA

In OFDM-TDMA, a particular user is given all the subcarriers of the system for any specific OFDM symbol duration. Thus, the users are separated via time slots. All symbols allocated to all users are combined to form a OFDM-TDMA frame. The number of OFDM symbols per frame can be varied based on each user's requirement. Frequently, an error correcting code is applied to the data to compensate for the channel nulls experienced by several random bits. This scheme allows MS to reduce its power consumption, as the MS shall process only OFDM symbols which are dedicated to it. On the other hand, the data are sent to each user in bursts, thus degrading performance for delay-constrained systems [17].

Different OFDM symbols can be allocated to different users based on certain allocation conditions. Since the OFDM-TDMA concept allocates the whole band width to a single user, a reaction to different subcarrier attenuations could consist of leaving out highly distorted subcarriers [18]. The number of OFDM symbols per user in each frame can be adapted accordingly to support heterogeneous data rate requirements. An efficient multiple access scheme should grant a high flexibility when it comes to the allocation of time-bandwidth resources. On the one hand, the behavior of the frequency-selective radio channel should be taken into account, while on the other hand the user requirements for different and/or changing data rates have to be met [19]. For example, both for OFDMA and OFDM-TDMA, the usage of AMC on different subcarriers, as proposed in [20], may increase overall system throughput and help in further exploiting CSI.

2.9.1.2 OFDMA

In OFDMA, available subcarriers are distributed among all the users for transmission at any time instant. The subcarrier assignment is made for the user lifetime, or at least for a considerable time frame. The scheme was first proposed for CATV systems [21], and later adopted for wireless communication systems.

OFDMA can support a number of identical downstreams, or different user data rates [e.g., assigning a different number of subcarriers to each user]. Based on the subchannel condition, different baseband modulation schemes can be used for the individual subchannels, e.g., QPSK, 16-QAM, and 64-QAM. This is investigated in numerous papers and referred to as adaptive subcarrier, bit, and power allocation or QoS allocation [20, 22, 23, 24].

In OFDMA, frequency hopping, one form of spread spectrum, can be employed to provide security and resilience to inter-cell interference.

In OFDMA, the granularity of resource allocation is higher than that of OFDM-TDMA, i.e., the flexibility can be accomplished by suitably choosing the subcarriers associated with each user. Here, the fact that each user experiences a different radio channel can be exploited by allocating only "good" subcarriers with high SNR to each user. Furthermore, the number of subchannels for a specific user can be varied, according to the required data rate. Thus, multi-rate system can be achieved without increasing system complexity very much.

2.9.1.3 OFDM-CDMA

In OFDM-CDMA [25, 26], user data are spread over several subcarriers and/or OFDM symbols using spreading codes, and combined with signal from other users [27]. The idea of OFDM-CDMA can be attributed to several researchers working independently at almost the same time on hybrid access schemes combining the benefits of OFDM and CDMA. OFDM provides a simple method to overcome the ISI effect of the multipath frequency-selective wireless channel, while CDMA provides the frequency diversity and the multi-user access scheme. Different types of spreading codes have been investigated. Orthogonal codes are preferred in case of DL, since loss of orthogonality is not as severe in DL as it is in UL.

Several users transmit over the same subcarrier. In essence this implies frequency-domain spreading, rather than time-domain spreading, as it is conceived in a DS-CDMA system. The channel equalization can be highly simplified in DL, because of the one-tap channel equalization benefit offered by OFDM.

In OFDM-CDMA, the flexibility lies in the allocation of all available codes to the users, depending on the required data rates. As OFDM-CDMA is applied using coherent modulation, the necessary channel estimation provides information about the subcarrier attenuations; this information can be used when performing an equalization in the receiver [28].

2.9.1.4 Relative Comparison

As shown in the previous discussions, we can consider OFDM-TDMA as the most basic multiple-access scheme, while OFDMA scheme is an extension of OFDM-

TDMA, and in turn, OFDM-CDMA scheme as an extension of OFDMA. Going from OFDM-TDMA to OFDM-CDMA, we have increased the level of flexibility in multiple-access of the system, but at the same time increased the complexity (see Fig. 2.10). The OFDM-CDMA shall observe all requirements from OFDMA, plus its own requirements. And similarly, OFDMA must fulfill all requirements of OFDM-TDMA.

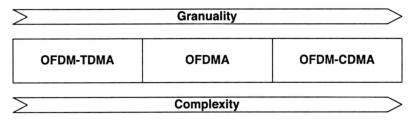

Fig. 2.10 Relative comparison of basic multi-carrier multiple-access techniques

Table 2.2 summarizes advantages and disadvantages of three basic multi-carrier multiple-access schemes.

Table 2.2 A summary of multiple access scheme

	Advantages	Disadvantages
OFDM-TDMA	Power savings (only receives own symbols) Simple resource allocation Easiest to implement	Relatively high latency Frequency-reuse factor ≥ 3 Lowest flexibility
OFDMA	Simple implementation Flexibility	Frequency-reuse factor ≥ 3
OFDM-CDMA	Spectral efficiency Frequency diversity MAI and inter-cell interference resistance Frequency-reuse factor $= 1$ Soft handover capability Highest flexibility	Requirement of power control Implementation complexity

OFDM-TDMA

OFDM-TDMA is simple to implement, but it may lack in highly delay-constraint system. For DL, basic OFDM-TDMA may not perform very well compared to other two schemes, but in UL, it may be very worthy. In UL, user time and frequency offset can cause real havoc in the system, and both OFDMA and OFDM-CDMA have to implement substantial procedure to combat the offsets, while OFDM-TDMA may be able to handle them quite easily. This is based on the fact that the entire bandwidth is allocated to a single user for several OFDM symbols, thus practically avoiding MAI.

OFDMA

The OFDMA scheme is distinguished by its simplicity, where the multi-access is obtained by allocating a fraction of subcarriers to different users. The benefit is that the receiver can be implemented in a relatively simple manner.

OFDMA is already in use in some standards, e.g., IEEE 802.16a, and can be used for both DL and UL. For the UL case, issues like synchronization are a major concern and are studied in several papers. Regarding the assignment of subcarriers, the literature provides no definitive answer whether it should be static/dynamic or contiguous/interleaved.

OFDM-CDMA

This scheme is potentially a very good scheme in DL due to its ability to exploit available frequency diversity, even coded-OFDMA can only make use of limited frequency diversity. It has been pointed that this scheme is vulnerable to near–far effect as a normal CDMA system is. Hence this scheme suits best mainly in an indoor DL scenario [29]. Now one is easily led to argue that in an indoor situation the coherence bandwidth is very large. In 5 GHz band, it ranges from 6 to 20 MHz. Thus to make use of its special advantage of providing frequency diversity, the system has to use a very wide band. Otherwise, even with a 20 MHz channel it will get as much frequency diversity as a coded interleaved OFDM system. In outdoor scenario, the loss of orthogonality due to severe channel coding may diminish the frequency diversity effect and introduce MAI to reduce the BER performance.

2.10 Single-Carrier vs Multi-carrier, TDE vs FDE

2.10.1 Single-Carrier FDE

A conventional anti-multipath approach, which was pioneered in voiceband telephone modems and has been applied in many other digital communications systems, is to transmit a single carrier, modulated by data using, for example, QAM, and to use an adaptive equalizer at the receiver to compensate for ISI [30]. Its main components are one or more transversal filters for which the number of adaptive tap coefficients is on the order of the number of data symbols spanned by the multipath. For tens of megasymbols per second and more than about 30–50 symbols ISI, the complexity and required digital processing speed become exorbitant, and this TDE approach becomes unattractive [31]. Therefore, for channels with severe delay spread, equalization in the frequency domain might be more convenient since the receiver complexity can be kept low. In fact, as for the OFDM, equalization is performed on a block of data at a time, and the operations on this block involve an efficient FFT operation and a simple channel inversion operation.

An SC system transmits a single carrier, modulated, for example, with QAM, at a high symbol rate. Linear FDE in an SC system is simply the frequency analog of what is done by a conventional time-domain equalizer. For channels with severe delay spread, SCFDE is computationally simpler than corresponding time-domain equalization for the same reason OFDM is simpler: because equalization is performed on a block of data at a time, and the operations on this block involve an efficient FFT operation and a simple channel inversion operation. Sari et al. [13, 32] pointed out that when combined with FFT processing and the use of a cyclic prefix, an SC system with FDE (SCFDE) has essentially the same performance and low complexity as an OFDM system. It is worth noting that a frequency domain receiver processing SC-modulated data shares a number of common signal processing functions with an OFDM receiver. In fact, as pointed out in Section 2.10.4, SC and OFDM modems can easily be configured to coexist, and significant advantages may be obtained through such coexistence.

Figure 2.11 shows conventional linear equalization, using a transversal filter with N tap coefficients, but with filtering done in the frequency domain. The block length N is usually chosen in the range of 64–2048 for both OFDM and SC-FDE systems.

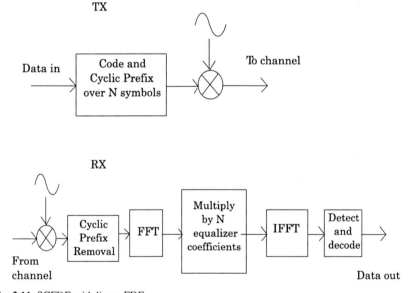

Fig. 2.11 SCFDE with linear FDE

A CP is appended to each block of N symbols, exactly as in OFDM. As an additional function, the CP can be combined with a training sequence for equalizer adaptation. An IFFT returns the equalized signal to the time domain prior to the detection of data symbols. Adaptation of the FDE's transfer function can be done with LMS, RLS, or LS minimization techniques, analogously to adaptation of time-domain equalizers [7, 33].

2.10.2 Single-Carrier vs Multi-carrier, FDE vs TDE

Some recent studies have clearly shown that the basic issue is not OFDM vs SC but rather frequency-domain equalization (FDE) vs time-domain equalization (TDE). FDE has several advantages over TDE in outdoor high-mobility propagation environments (usually with long tail channel impulse response).

The conventional approach to digital communications over dispersive channels is single-carrier transmission with time-domain equalization (TDE). TDE covers the simple linear equalizers, decision-feedback equalizers, as well as maximum-likelihood sequence estimation. These techniques have been in use for decades in digital microwave radio, and more recently in mobile radio systems. Although FDE was originally introduced in the late 1970s, it was not pursued and quickly disappeared from the literature [34].

Let us consider an FDE with N_{taps} taps. The FFT operator which forms the first stage of the equalizer gives N_{taps} signal samples denoted as $(Y_1, \ldots, Y_{N_{taps}})$. These samples are sent to a complex multiplier bank whose coefficients are denoted as $(F_1, \ldots, F_{N_{taps}})$. The coefficient values which minimize signal distortion are

$$F_n = \frac{H_n^*}{\mid H_n \mid^2} = \frac{1}{H_n}. \qquad (2.31)$$

Clearly, each coefficient is only a function of the channel frequency response at the corresponding frequency, and the equalizer is easily adapted to channel variations even if the number of taps is very large [34].

From the above discussions, SC-TDE is sufficient for channels with a small delay spread, because these channels can be equalized using a small number of taps. In contrast, SCFDE or OFDM are required on channels with a large delay spread, as these channels require a large number of taps and this leads to convergence and tracking problems with SC-TDE. Indeed, the normalized complexity of both OFDM and SCFDE is proportional to $\log(N_{taps})$, whereas the complexity of SC-TDE grows linearly with N_{taps}. For N_{taps} large, the complexity considerations clearly favor the use of frequency-domain techniques. In fact, these considerations indicate that the real problem is not OFDM vs SC, but instead FDE vs TDE [34].

2.10.3 Analogies and Differences Between OFDM and SCFDE

There is a strong analogy between OFDM and SCFDE. Analyzing the operation principle of OFDM, Sari et al. [13] noticed a striking resemblance to frequency-domain channel equalization for traditional single-carrier systems, a concept proposed more than three decades ago [35]. The motivation for frequency-domain equalization was due to the ability of this technique to accelerate the initial convergence of the equalizer coefficients. With a frequency-domain equalizer at the receiver, single-carrier systems can handle the same type of channel impulse responses as OFDM systems. In both cases, time/frequency and frequency/time transformations are made. The difference is that in OFDM systems, both channel equalization and receiver decisions are performed in the frequency domain, whereas in SCFDE systems the receiver decisions are made in the time domain, although channel equalization is performed in the frequency domain.

From a purely channel equalization capability standpoint, both systems are equivalent, assuming they use the same FFT block length. They have, however, an essential difference: *since the receiver decisions in uncoded OFDM are independently made on different carriers, those corresponding to carriers located in a region with a deep amplitude depression will be unreliable.* This problem does not exist for SCFDE, in fact *once the channel is equalized in the frequency domain, the signal is transformed back to the time domain, and the receiver decisions are based on the signal energy transmitted over the entire channel bandwidth.* In other words, the SNR value that dictates performance (assuming that residual ISI is negligible) corresponds to the average SNR of the channel. In fact, as noted in [13], the effect of the deep nulls in the channel frequency response is spread out over all symbols by the IFFT operation. Consequently, the performance degradation due to a deep notch in the signal spectrum remains small with respect to that suffered by OFDM.

The foregoing analysis indicates that with FDE, SC transmission is substantially superior to OFDM signaling. Without channel coding, OFDM is in fact not usable on fading channels, as deep notches in the transmitted signal spectrum lead to an irreducible BER. In order to work satisfactorily, *OFDM requires ECC with frequency-domain interleaving so as to scatter the signal samples falling in a spectral notch.* In this case, the interleaver uniformly distributes the low-SNR samples over the channel bandwidth. *In contrast, SCFDE can work without ECC.*

The main hardware difference between OFDM and SCFDE is that for SCFDE the transmitter's IFFT block is moved to the receiver. The complexities are the same. *Both OFDM and SCFDE can be enhanced by* adaptive modulation and *space diversity* [36].

The use of SC modulation and FDE by processing the FFT of the received signal has several attractive features:

- SC modulation has reduced PAPR requirements with respect to OFDM, thereby allowing the use of less costly power amplifiers;
- its performance with FDE is similar to that of OFDM, even for very long channel delay spread;
- frequency-domain receiver processing has a similar complexity reduction advantage with respect to that of OFDM: complexity is proportional to log of multipath spread;
- coding, while desirable, is not necessary for combating frequency selectivity, while it is needed in nonadaptive OFDM;
- SC modulation is a well-proven technology in many existing wireless and wireline applications, and its RF system linearity requirements are well known.

Comparable SCFDE and OFDM systems would have the same block length and CP lengths. The CP at the beginning of each block (Fig. 2.12), used in both SCFDE and OFDM systems, has two main functions:

- it prevents contamination of a block by ISI from the previous block;
- it makes the received block appear to be periodic of period N, which is essential to the proper functioning of the FFT operation.

If the first and the last N_{CP} symbols are identical unique word sequences of training symbols, the overhead fraction is $\frac{2N_{CP}}{N+2N_{CP}}$.

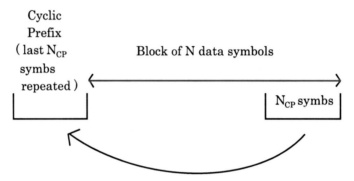

Fig. 2.12 Block processing in FDE

The following considerations provide a further support to the thesis of the similarity between SCFDE and OFDM. Precoding in OFDM disperses the energy of symbols over the channel bandwidth, i.e., it restores the frequency diversity broken by the IFFT operator. A common precoding matrix is the Walsh–Hadamard matrix which uniformly spreads the symbol energy across the channel bandwidth using orthogonal spreading sequences. Another precoding matrix that uniformly spreads the symbol energy over the channel bandwidth is the FFT matrix. This matrix cancels the IFFT matrix that generates the OFDM signal and the system reduces to SCFDE. From this discussion, it is clear that precoded OFDM mimics SCFDE. That is, *by precoding OFDM in order to restore frequency diversity, we get an SCFDE-type system* [34].

2.10.4 Interoperability of SCFDE and OFDM

Figure 2.13 shows block diagrams for OFDM and SC systems with linear FDE. It is evident that the two types of systems differ mainly in the placement of the IFFT operation: in OFDM it is placed at the transmitter to multiplex the data into parallel subcarriers; in SC it is placed at the receiver to convert FDE signals back into time-domain symbols. The signal processing complexities of these two systems

OFDM

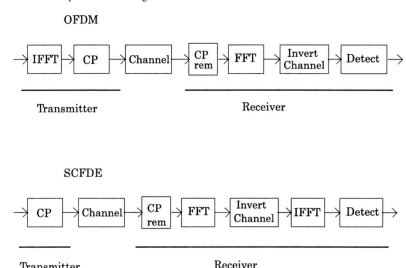

Fig. 2.13 OFDM and SCFDE signal processing similarities and differences

are essentially the same for equal FFT block lengths [31]. A dual-mode system, in which a software radio modem can be reconfigured to handle either SC or OFDM signals, could be implemented by switching the IFFT block between the transmitter and receiver at each end of the link, as suggested in Fig. 2.14. *There may actually be an advantage in operating a dual-mode system, wherein the base station uses an OFDM transmitter and an SC receiver, and the subscriber modem uses an SC transmitter and an OFDM receiver*, as illustrated in Fig. 2.15. This arrangement – OFDM in the downlink and SC in the uplink – has two potential advantages [31]:

- concentrating most of the signal processing complexity at the hub or base station. The hub has two IFFTs and one FFT, while the subscriber has just one FFT;

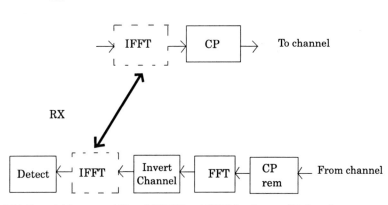

Fig. 2.14 Potential interoperability of SCFDE and OFDM: a "convertible" modem

- the subscriber transmitter is SC, and thus inherently more efficient in terms of power consumption due to the reduced power backoff requirements of the SC mode. This may reduce the cost of a subscriber's power amplifier.

HUB END   SUBSCRIBER END

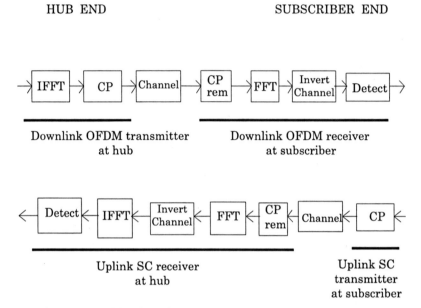

Fig. 2.15 Coexistence of SCFDE and OFDM: uplink/downlink asymmetry

2.11 OFDMA: An Example of Future Applications

OFDMA will be used as an access scheme in many high data rate and high-quality future services and applications. As an example, the ITU is currently working on specifying the system requirements toward next-generation mobile communication systems, called IMT-A. The IMT-A systems are expected to fulfill the requirements of so-called 4G systems and believed to be operational around year 2015. One of the key features for IMT-A is enhanced peak data rates to support advanced services and applications in the order of 100 Mbit/s for high-mobility and 1 Gbit/s for low-mobility conditions [37]. In order to support such high data rates the WRC 07 has identified the allocation of the following additional spectrum bands, 450–470 MHz, 698–862 MHz, 790–862 MHz, 2.3–2.4 GHz, and 3.4–3.6 GHz. Some of these bands support the channel bandwidth up to 100 MHz. It is envisioned that such high channel bandwidth in the new frequency bands will be optimized for small cell and low mobility scenarios such as dense urban and hot spot areas, especially indoor.

In order to provide high data rate and high-quality coverage in indoor local area, the solutions are emerging in the form of new devices and networks. These devices are expected to be deployed at very large scale in random manner to support the individual coverage in small area, accessing the spectrum from a common pool. In that situation the interference becomes the main concern. An efficient technique will be essential to support FSU to allow coexistence of such devices in the given area. FSU basically means allocation of spectrum by several devices over the same spectrum pool in flexible manner using the same radio access technology. OFDMA is considered as the most promising technique to facilitate FSU, because of its natural attribute of allocating the spectrum in flexible manner.

The following example available in [38] demonstrates this capability of OFDMA to allocate the spectrum in flexible manner. The example considers the deployment of four devices in an indoor small area. The chosen carrier frequency is 3.5 GHz with 100 MHz spectrum bandwidth, which is one of the newly allocated spectrum bands by WRC 07. TDD with perfect synchronization and equal uplink and downlink ratio is assumed. A transmission frame of 10 ms duration is considered. The indoor path loss propagation model and multipath channel model are based on the description presented in [39] which is proposed to ITU-R for evaluations of IMT-A systems. SISO antenna configuration is taken. The FSU target SINR threshold is defined and has to be overcome in order to allocate a spectrum unit, and allows coexistence of several devices in the small area, ranging from -10 to $20\,dB$ in steps of 5 dB [38]. The transmission characteristics of OFDMA for this scenario are listed in Table 2.3. Figure 2.16 demonstrates the flexibility of OFDMA in spectrum allocation at different FSU target SINR thresholds, presented together with the fixed spectrum allocation of 1, 1/2, and 1/4 frequency reuse schemes. The performance is measured in terms of average carried cell load against the average offered cell load. The impact of different FSU target SINR thresholds can be clearly seen in the figure. At very low SINR threshold (i.e. -10 and $-5\,dB$) the average carried cell load is very close to the offered load, which is very similar to reuse 1 case [38]. By lowering the threshold, higher amount of overlapping is allowed, resulting into higher average carried cell load. Instead, at very high SINR threshold the average carried cell load is lower compared to the offered load and becomes very close to 1/4 frequency reuse scheme [38], which is due to nearly orthogonal spectrum allocation. The carried cell load of 1/2 frequency reuse scheme lies between that of 1 and 1/4 frequency reuse schemes. Therefore, it is reflected that the OFDMA has the attribute to span over all the possibilities of fixed spectrum allocation schemes depending on the required SINR threshold. This allows coexistence of several devices in the small geographical

Table 2.3 Example of OFDMA characteristics used in local area indoor deployment toward next-generation mobile communication systems

Parameters	Settings
Carrier frequency	3.5 GHz
FFT size	2048
Number of subcarriers	1500
Sub carrier spacing	60 kHz
Transmission frame duration	10 ms
Slot duration	1 ms
Number of slots in one frame	10

area by dynamically and flexibly adapting the spectrum allocation. This feature of OFDMA provides self-adjusting and self-optimizing capabilities to the devices, and therefore it can be considered as a promising technology for the future applications.

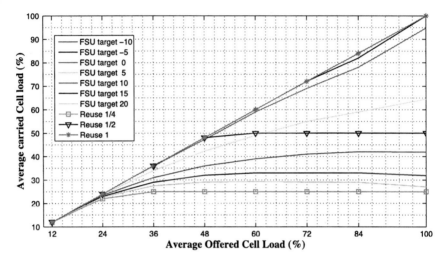

Fig. 2.16 Average carried cell load with different FSU target SINR thresholds using OFDMA [38] (©[2008] IEEE, Reprinted with permission, DySPAN 2008)

2.12 Conclusions

In this chapter, we have discussed the basics of multi-carrier transmission corresponding to current and future 4G wireless systems. It is evident that multi-carrier transmission is a very useful technique in combating the adverse effects of wireless channels and making use of the adversity to the benefit of the system. Modified single-carrier transmissions such as SC-FDE- and SC-FDMA-type transmission techniques also provide similar benefits. Thus, the transmission techniques mentioned in this chapter will be part and parcel of all (or most) immediate future wireless systems.

Section 2.1 emphasizes that outlights that future wireless applications will demand high data rate communication; this is not going to be an easy task, when dealing with the unpredictable wireless channel. The idea of multi-carrier transmission has emerged recently to be used for combating the hostile wireless channel and providing high data rate communications. OFDM is a special form of multi-carrier transmission where all the subcarriers are orthogonal, and it promises a high user data rate transmission capability at a reasonable complexity and precision.

Although OFDM has only recently been gaining interest from telecommunications industry, it has a long history of existence, and this is discussed in Section 2.2, where considerations about history and development of OFDM have been made. While Section 2.3 discusses spots the rationale behind using multi-carrier transmission, a general OFDM transceiver is described in Section 2.4. The analytical model for OFDM transceiver is presented in Section 2.5.

Advantages and disadvantages of this technique are discussed in Sections 2.6 and 2.7, respectively. Section 2.8 relates OFDM with a cellular system, considering system design requirements and parameters; multiple access techniques based on OFDM modulation are described in Section 2.9. Analogies, differences, and potential interoperability of OFDM and SCFDE are discussed in Section 2.10. Finally, Section 2.11 highlights an example of possible future application of OFDMA for spectrum sharing and FSU.

References

1. S. Hara and R. Prasad, *Multicarrier Techniques For 4G Mobile Communications*, Artech House Publishers, Norwood, MA, USA, 2003
2. R.V. Nee and R. Prasad, *OFDM for Wireless Multimedia Communications*, Artech House Publishers, Norwood, MA, USA, 2000.
3. W.Y. Zou and Y. Wu, *COFDM: An Overview*, IEEE Trans. Broadcast., vol. 41, no. 1, March 1995.
4. E.P. Lawrey, *Adaptive Techniques for Multiuser OFDM*, Ph.D. Thesis, James Cook University, Australia, December 2001.
5. W.C.Y. Lee, *Mobile Cellular Telecommunications Systems*, McGraw Hill Publications, New York, USA, December 1989.
6. J. G. Proakis, *Digital Communications*, McGraw-Hill, New York, USA, Fourth edition, August 2000.
7. S. Haykin, *Adaptive Filter Theory*, Prentice-Hall, Upper Saddle River, New Jersey, USA, Third edition, December 1996.
8. R. Prasad, *OFDM for Wireless Communications Systems*, Artech House Publishers, Norwood, MA, USA, September 2004.
9. D. Matic, *OFDM Synchronization and Wideband Power Measurements at 60 GHz for Future Wireless Broadband Multimedia Communications*, Ph.D. Thesis, Aalborg University, Denmark, September 2001.
10. U.S. Jha, *Low Complexity Resource Efficient OFDM Based Transceiver Design*, Ph.D. Thesis, Aalborg University, Denmark, September 2002.
11. C.R. Nassar et al., *Multi-carrier Technologies for Wireless Communication*, Kluwer Academic Publishers, Dordrecht, The Netherlands, 2002.
12. H. Rohling et al., *Broad-band OFDM Radio Transmission for Multimedia Applications*, Proc. IEEE, vol. 87, no. 10, pages 1778–1789, October 1999.
13. H. Sari et al., *Transmission Techniques for Digital Terrestrial TV Broadcasting*, IEEE Comm. Mag., vol. 33, no. 2, pages 100–109, February 1995.
14. Ove Edfors et al., *An Introduction to Orthogonal Frequency-Division Multiplexing*, Research Report, no. 1996:16, Division of Signal Processing, Lulea University of Technlogy, Sweden, September 1996.
15. T. Pollet et al., *BER Sensitivity of OFDM Systems to Carrier Frequency Offset and Wiener Phase Noise*, IEEE Trans. Comm., vol. 43, no. 2–4, pages 191–193, February–April 1995.
16. H. Rohling and T. May, *Comparison of PSK and DPSK Modulation in a Coded OFDM System*, Proc. IEEE VTC, pages 5–7, Phoenix, Arizona, USA, October 1997.
17. T. Bruns, *Performance Comparison of Multiuser OFDM Techniques*, EE360 Class Project, Stanford University, Spring 2001.
18. H. Rohling and R. Grunheid, *Performance of an OFDM-TDMA Mobile Communication System*, VTC, vol. 3, pages 1589–1593, April–May 1996.

19. H. Rohling and R. Grunheid, *Performance comparison of different multiple access schemes for the downlink of an OFDM communication system*, VTC, vol. 3, pages 1365–1369, May 1997.
20. C.Y. Wong, R.S. Cheng, K.B. Letaief and R.D. Murch, *Multiuser OFDM with Adaptive Sub-carrier, Bit, and Power Allocation*, IEEE J. Select. Areas Commun., Vol. 17, No. 10, pages 1747–1758, October 1999.
21. H. Sari and Y. Levy and G. Karam, *Orthogonal Frequency-Division Multiple Access for the Return Channel on CATV Networks*, IEEE ICT, Istanbul, April 1996.
22. D. Kivanc and H. Liu, *Subcarrier Allocation and Power Control for OFDMA*, IEEE Signals, Systems and Computers, Pacific Grove, USA, October 2000.
23. W. Rhee and J. M. Cioffi, *Increase in Capacity of Multiuser OFDM System Using Dynamic Subchannel Allocation*, IEEE VTC, Tokyo, Japan, May 2000.
24. C.Y. Wong, C.Y. Tsui, R.S. Cheng and K.B. Letaief, *A Real-time Subcarrier Allocation Scheme for Multiple Access Downlink OFDM Transmission*, IEEE VTC, Houston, USA, May 1999.
25. S. Hara and R. Prasad, *Overview of Multicarrier CDMA*, IEEE Comm. Mag., Vol. 35, No. 12, pages 126–133, December 1997.
26. R. Prasad, *Universal Wireless Personal Communications*, Artech House Publishers, Norwood, MA, USA, 1998.
27. N. Yee, J.P. Linnartz et al., *Multicarrier CDMA in Indoor Wireless Networks*, IEEE PIMRC, pages 109–113, Yokohama, Japan, September 1993.
28. T. Muller, H. Rohling and R. Grunheid, *Comparison of different detection algorithms for OFDM-CDMA in broadband Rayleigh fading*, VTC, vol. 2, pages 835–838, July 1995.
29. M. I. Rahman et al., *Comparison of Various Modulation and Access Schemes under Ideal Channel Conditions*, JADE Project Deliverable, D3.1[1], Aalborg University, Denmark, July 2004.
30. S.U.H. Qureshi, *Adaptive Equalization*, Proc. IEEE, September 1985.
31. F. Falconer, S.L. Ariyavisitakul, A. Benyamin-Seeyar and B. Eidson, *Frequency Domain Equalization for Single-Carrier Broadband Wireless Systems*, IEEE Comm. Mag., Vol. 40, No. 4, pages 58–66, April 2002.
32. H. Sari et al., *Frequency Domain Equalization of Mobile Radio and Terrestrial Broadcast Channels*, IEEE International Conference on Global Communications, pages 1–5, 1994.
33. M.V. Clark, *Adaptive Frequency-Domain Equalization and Diversity Combining for Broadband Wireless Communications*, IEEE JSAC, Vol. 16, No. 8, pages 1385–95, October 1998.
34. H. Sari, *Transmission and Multiple Access Techniques for Broadband Wireless Access Networks*, November 2005.
35. T. Walzman and M. Schwartz, *Automatic equalization using the discrete Fourier domain*, IEEE Trans. Inform. Theory, vol. 19, pages 59–68, January 1973.
36. A. Benyamin-Seeyar et al., Harris Corporation Inc., http://www.ieee802.org/16/tg3/contrib/802163c − 01_32.pdf, 2001.
37. T. Bruns, *Technical Specification Group Radio Access Network; Requirements for Further Advancements for E-UTRA (LTE- Advanced)*, 3GPP Technical Report 36.913, , V0.0.4, 2008.
38. S. Kumar et al., *Spectrum Load Balancing for Flexible Spectrum Usage in Local Area Deployment Scenario*, IEEE DySPAN, (© [2008], IEEE) 2008.
39. IST-Winner Consortium, *WINNER II Channel Models part I – Channel Models*, Technical Deliverable D1.1.2, IST-EU, September 2007.

Chapter 3
Recent Advances in Low-Correlation Sequences

Gagan Garg, Tor Helleseth, and P. Vijay Kumar

3.1 Introduction

This chapter aims to provide an overview of four topics of current interest relating to the design and analysis of sequences with low correlation. The topics in question are the discovery of new families of cyclic Hadamard difference sets over the past decade following a gap of almost 40 years, the recent realization of the existence of long sequences with larger merit factor than was previously suspected, the development of a theory of sequences possessing a low-correlation zone, and the recent novel construction of low-correlation sequences over the quadrature-amplitude-modulation (QAM) alphabet. While there has also been considerable recent interest on the topic of the design of sequences with low values of peak-to-average power ratio (PAPR), this topic has been addressed in-depth in the recent publication [1] by Litsyn.

Low-correlation sequences find application in signal synchronization, spread spectrum communications, signal identification in multiple access communication systems, multipath resolution, navigation, pulse compression, radar ranging,

Gagan Garg
Department of Computer Science and Automation,
Indian Institute of Science, Bangalore 560012, India,
e-mail: gagan.garg@gmail.com

Tor Helleseth
Department of Informatics, University of Bergen,
HIB, N-5020, Bergen, Norway,
e-mail: torh@ii.uib.no

P. Vijay Kumar
Department of Electrical Communication Engineering,
Indian Institute of Science, Bangalore 560012, India,
e-mail: vijayk@usc.edu

P. Vijay Kumar is on leave of absence from the Department of Electrical Engineering-Systems, University of Southern California, Los Angeles, CA 90089, USA.

V. Tarokh (ed.), *New Directions in Wireless Communications Research*,
DOI 10.1007/978-1-4419-0673-1_3,
© Springer Science+Business Media, LLC 2009

random number generation, and stream-cipher-based cryptography. Excellent introduction to low-correlation sequences and their applications can be found in Golomb [8], Golomb and Gong [2], Helleseth and Kumar [38], Sarwate and Pursley [3], and Fan and Darnell [4].

3.2 Cyclic Hadamard Difference Sets

3.2.1 Introduction

After a gap of 40 years following the uncovering of the family of Gordon–Mills–Welch difference sets in 1962 [5], some new constructions of cyclic difference sets with Hadamard parameters surfaced in the literature. This section will provide an overview of the new discoveries. Difference sets are closely related to binary sequences with two-level autocorrelation.

Definition 3.1. Let G be a group of order v. A (v, k, λ) cyclic difference set $\mathbb{D} = \{d_1, d_2, \ldots, d_k\}$ is a k-element subset of G, such that every $x \neq 0$ can be written as $d_i - d_j = x$ in the same number, λ, of ways as d_i and d_j run through $\mathbb{D}$. The design is said to be cyclic if the group G is cyclic.

We will consider difference sets in the additive cyclic group of integers $\mathbb{Z}_{2^m-1}$ modulo $v = 2^m - 1$, or equivalently in the cyclic multiplicative group of $\mathbb{F}_{2^m}$, the finite field with 2^m elements. The connection between cyclic difference sets and sequences with two-level autocorrelation is a consequence of the following result.

Lemma 3.1. *Let $\{s(t)\}$ be a binary sequence of length v that is the characteristic function of the difference set $\mathbb{D}$ (i.e., $s(t) = 1$ if $t \in \mathbb{D}$ and $s(t) = 0$ otherwise). Then the autocorrelation of $\{s(t)\}$ at shift τ satisfies*

$$\theta(\tau) = \sum_{t=0}^{v-1}(-1)^{s(t+\tau)-s(t)} = \begin{cases} v - 4(k - \lambda) & \text{if } \tau \neq 0 \ (mod \ v), \\ v & \text{if } \tau = 0 \ (mod \ v). \end{cases}$$

Conversely, any binary sequence satisfying a two-level autocorrelation function may be regarded as the characteristic function of a cyclic difference set.

Definition 3.2. Cyclic difference sets with parameters $(v = 4t - 1, k = 2t - 1, \lambda = t - 1)$ are referred to as *cyclic Hadamard difference sets*.

Note that $v - 4(k - \lambda) = -1$ for cyclic Hadamard difference sets which therefore are equivalent to binary sequences with two-level autocorrelation where the two correlation values are v in-phase and -1 out-of-phase. Among the most well-known classical Hadamard difference sets are the family of Singer difference sets. These correspond to the characteristic function of the complement (interchanging 0 and 1) of m-sequences, defined below.

Definition 3.3. A binary m-sequence $\{s(t)\}$ is a sequence of period $n = 2^m - 1$ that satisfies a linear recurrence relation of degree m of the form

$$\sum_{i=0}^{m} f_i s(t+i) = 0 \text{ for all } t \geq 0,$$

where $f_i \in \{0, 1\}$.

The trace map is a linear map from $\mathbb{F}_{2^m}$ to $\mathbb{F}_2$ defined by

$$Tr(x) = \sum_{i=0}^{m-1} x^{2^i}.$$

The trace function takes on the values 0 and 1 equally often as x runs through $\mathbb{F}_{2^m}$. The m-sequence can (after a suitable cyclic shift) be written as

$$s(t) = Tr(\alpha^t),$$

where α is a primitive element in $\mathbb{F}_{2^m}$ that is a zero of the characteristic polynomial $f(x) = \sum_{i=0}^{m} f_i x^i$ of the m-sequence. The m-sequence has 2^{m-1} ones and $2^{m-1} - 1$ zeros during a period. The autocorrelation of an m-sequence is therefore, for any $\tau \not\equiv 0 \pmod{2^m - 1}$, given by

$$A(\tau) = \sum_{t=0}^{2^m-2} (-1)^{s(t+\tau)-s(t)} = \sum_{t=0}^{2^m-2} (-1)^{Tr(\alpha^{t+\tau}-\alpha^t)}$$

$$= \sum_{x \in \mathbb{F}_{2^m}^*} (-1)^{Tr((\alpha^\tau-1)x)} = -1.$$

A difference set is said to have the Singer parameters if $(v, k, \lambda) = (2^m - 1, 2^{m-1} - 1, 2^{m-2} - 1)$ (or $(v, k, \lambda) = (2^m - 1, 2^{m-1}, 2^{m-2})$). Note that if $\mathbb{D}$ is a difference set with one parameter set, the complementary set $\mathbb{D}' = \mathbb{Z}_v \setminus \mathbb{D}$ is a difference set under the second set of parameters. Thus the two-valued autocorrelation of the m-sequence above implies that

$$\mathbb{D} = \{t \mid s(t) = 0\}$$

is a difference set with parameters $(2^m - 1, 2^{m-1} - 1, 2^{m-2} - 1)$, while the complementary set

$$\mathbb{D}' = \{t \mid s(t) = 1\}$$

yields a $(v, k, \lambda) = (2^m - 1, 2^{m-1}, 2^{m-2})$ difference set.

A second, well-known family of binary sequences with identical autocorrelation and also associated with difference sets with Singer parameters are the Gordon, Mills, and Welch (GMW) sequences [5, 6]. For k dividing m, let $Tr_k^m(x) = \sum_{i=0}^{m/k-1} x^{2^{ki}}$ be the trace from $\mathbb{F}_{2^m}$ to the subfield $\mathbb{F}_{2^k}$.

Theorem 3.1. *Let k, m be integers with $k \mid m$, $k \geq 1$. Let r, $1 \leq r \leq 2^k - 2$ satisfy $\gcd(r, 2^k - 1) = 1$. Let $a \in \mathbb{F}_{2^m}^*$ and let $\{s(t)\}$ be the GMW sequence of period $2^m - 1$ defined by*

$$s(t) = Tr_1^k\{[Tr_k^m(a\alpha^t)]^r\},$$

where α is a primitive element in $\mathbb{F}_{2^m}$. Then,
(i) $\{s(t)\}$ is balanced (i.e., 2^{m-1} ones and $2^{m-1} - 1$ zeros in a period).
(ii) The autocorrelation is two valued and is given by

$$\theta_s(\tau) = \begin{cases} -1 & \text{if } \tau \neq 0 \, (mod \, 2^m - 1), \\ 2^m - 1 & \text{if } \tau = 0 \, (mod \, 2^m - 1). \end{cases}$$

The known constructions for cyclic Hadamard difference sets can be classified according to the values of v. The Singer difference set and the GMW difference sets correspond to $v = 2^j - 1$ for some j. When $v = 4t - 1$ is a prime $p = 3 \,(\text{mod } 4)$ then the quadratic residues modulo p form a Hadamard difference set with parameters $(p, (p-1)/2, (p-3)/4)$. The binary sequence associated with this difference set is commonly referred to as the Legendre sequence. An additional construction due to Hall provides difference sets when the prime $v = 4t - 1$ is expressible as $v = 4x^2 + 27$. The remaining value for which a construction is available covers the case $v = p(p+2)$ where p and $p+2$ are both prime. These are the twin-prime difference sets. For more details, the reader is referred to Baumert [7] and Golomb [8].

Note that in all of the constructions above, the parameter v belongs to one of the following types: (i) $v = 3 \,(\text{mod } 4)$ is a prime, (ii) $v = p(p+2)$ where both p and $p+2$ are prime or else, (iii) $v = 2^j - 1$ for some $j > 1$. Based on extensive numerical evidence, Golomb has conjectured that a cyclic Hadamard difference set exists only if v is one of the above types.

Some 40 years after the appearance of the GMW construction, a variety of new and interesting cyclic Hadamard difference sets with Singer parameters were discovered in the late 1990s by several research groups. New difference sets were found through computer search by No et al. [9] that led them to postulate five new conjectures. We present the first of these conjectures, related to a Singer difference set (or equivalently a sequence with two-level autocorrelation) as an example. The sequences are expressed in terms of the trace function.

Conjecture 3.1. Let α be a primitive element of $\mathbb{F}_{2^m}$ and let $m = 2k + 1$ be odd. Let

$$b(t) = Tr(\alpha^t + \alpha^{(2^k+1)t} + \alpha^{(2^k+2^{k-1}+1)t}).$$

Then $\{b(t)\}$ is a binary sequence of period $2^m - 1$, with 2^{m-1} ones in a period, having a two-level autocorrelation, with an out-of-phase correlation value of -1.

This conjecture is a special instance of a more general conjecture of Dobbertin [10] that was proved subsequently by Dillon and Dobbertin [11]. A second conjecture formulated by No et al. [12] and given below does not involve the trace function.

Conjecture 3.2. Let $m = 3k \pm 1$, $d = 2^{2k} - 2^k + 1$, $N(z) = (z + 1)^d + z^d$, and $\mathcal{N} = Im(N)$. Define the following sequences:

(i) for m odd:
$$s(t) = \begin{cases} 1 \text{ if } \alpha^t \in \mathcal{N}, \\ 0 \text{ otherwise.} \end{cases}$$

(i) for m even:
$$s(t) = \begin{cases} 0 \text{ if } \alpha^t \in \mathcal{N}, \\ 1 \text{ otherwise.} \end{cases}$$

Then $\{b(t)\}$ is a binary sequence of period $2^m - 1$ with 2^{m-1} ones in a period, having a two-level autocorrelation, with an out-of-phase correlation value of -1.

This above conjecture was proved when m is odd by Dillon [13] and for m even by Dillon and Dobbertin [11]. An additional and important contribution of this conjecture was that it inspired a second conjecture by Dobbertin [10].

3.2.1.1 The Hadamard Transform

To describe the flavor behind the proofs of new difference sets with Singer parameters (or related sequences with two-valued autocorrelation) it is useful to recall some properties of the Hadamard transform.

The *Hadamard transform* $\hat{F}$ of a real-valued function F on $\mathbb{F}_{2^m}$ is defined by

$$\hat{F}(y) = \frac{1}{\sqrt{2^m}} \sum_{x \in \mathbb{F}_{2^m}} F(x)(-1)^{Tr(yx)}.$$

The inverse transformation is

$$F(x) = \frac{1}{\sqrt{2^m}} \sum_{y \in \mathbb{F}_{2^m}} \hat{F}(y)(-1)^{Tr(xy)}.$$

An important result is Parseval's identity implying

$$\sum_{x \in \mathbb{F}_{2^m}} F(x)G(x) = \sum_{y \in \mathbb{F}_{2^m}} \hat{F}(y)\hat{G}(y).$$

Let f be the characteristic function for a subset D of $\mathbb{F}_{2^m}^* = \mathbb{F}_{2^m} \setminus \{0\}$ (i.e., $f(x) = 1$ if $x \in D$ and $f(x) = 0$ otherwise). Define the corresponding binary sequence $\{s(t)\}$ such that $s(t) = f(\alpha^t)$, i.e.,

$$s(t) = \begin{cases} 1 \text{ if } \alpha^t \in D, \\ 0 \text{ otherwise.} \end{cases}$$

Let $\mathbb{D} = \{t \mid \alpha^t \in D\}$ and define $F(x) = (-1)^{f(x)}$. The autocorrelation of the sequence $\{s(t)\}$ for $\tau \neq 0 \pmod{2^m - 1}$ is

$$A(\tau) = \sum_{t=0}^{2^m-2} (-1)^{s(t+\tau)-s(t)} = \sum_{t=0}^{2^m-2} (-1)^{Tr(f(\alpha^{t+\tau})-f(\alpha^t))}$$

$$= -1 + \sum_{x\in\mathbb{F}_{2^m}} (-1)^{Tr(f(ax)-f(x))} = -1 + \sum_{x\in\mathbb{F}_{2^m}} F(ax)F(x),$$

where $a = \alpha^\tau$.

Then the condition for $\mathbb{D}$, with $|\mathbb{D}| = 2^{m-1}$, to be a cyclic difference set in $\mathbb{Z}_{2^m-1}$ with the Singer parameters $(v, k, \lambda) = (2^m - 1, 2^{m-1}, 2^{m-2})$ is that

$$\sum_{x\in\mathbb{F}_{2^m}} F(ax)F(x) = 0 \quad \text{for all } a \neq 1.$$

It follows from Parseval's identity that this is equivalent to

$$\sum_{y\in\mathbb{F}_{2^m}} \hat{F}(ay^t)\hat{F}(y^t) = 0 \quad \text{for all } a \neq 1$$

for $t = 1$ and hence true for any t relatively prime to $2^m - 1$.

Let $\gcd(k, 2^m - 1) = 1$, $s_k(x) = Tr(x^k)$, and $S_k(x) = (-1)^{Tr(x^k)}$, then

$$\hat{S}_k(y) = \sum_{x\in\mathbb{F}_{2^m}} (-1)^{Tr(x^k+yx)}$$

and for any $a \neq 1$, using Parseval's identity implies

$$\sum_{y\in\mathbb{F}_{2^m}} \hat{S}_k(ay)\hat{S}_k(y) = \sum_{x\in\mathbb{F}_{2^m}} S_k(ax)S_k(x) = \sum_{x\in\mathbb{F}_{2^m}} (-1)^{Tr((a^k-1)x^k)} = 0.$$

Thus to find new difference sets it is sufficient to find a set D with characteristic function f such that

$$\hat{F}(y) = \hat{S}_k(y^t) \quad \text{for all } y \in \mathbb{F}_{2^m}$$

where $\gcd(k, 2^m - 1) = 1$ and $\gcd(t, 2^m - 1) = 1$.

In 1998 Maschietti [14] discovered a surprising new family of cyclic Hadamard difference sets using the terminology of hyperovals.

Definition 3.4. A hyperoval in the two-dimensional projective geometry $PG(2, 2^m)$ is a set of $2^m + 2$ points no three on a line. Every hyperoval can be written in the form

$$D(f) = \{(1, t, f(t)) \mid t \in \mathbb{F}_{2^m}\} \cup \{(0, 1, 0)\} \cup \{(0, 0, 1)\},$$

where f is a permutation polynomial of degree $\leq 2^m - 2$ where $f(0) = 0$, $f(1) = 1$, and $f_s(x) = (f(x+s) + f(s))/x$, $f_s(0) = 0$ is also a permutation polynomial. If x^k is a monomial then $D(x^k)$ is called a monomial hyperoval.

An alternative characterization of monomial hyperovals is as follows.

Lemma 3.2. $D(x^k)$ *is a monomial hyperoval if and only if* $\gcd(k, 2^m - 1) = 1$ *and* $x^k + x + a = 0$ *has 0 or 2 solutions for all* $a \in \mathbb{F}_{2^m}$.

One can show that these conditions also imply that $\gcd(k - 1, 2^m - 1) = 1$. Monomial hyperovals are known to exist for the following values of k:

1. Singer: $k = 2^i$, $\gcd(i, m) = 1$.
2. Segre: $k = 6$, $m \geq 5$ odd.
3. Glynn I: $k = 2^{(m+1)/2} + 2^{(3m+1)/4}$ if $m = 1 \pmod 4$, $m \geq 7$
   or $k = 2^{(m+1)/2} + 2^{(m+1)/4}$ if $m = 3 \pmod 4$, $m \geq 7$.
4. Glynn II: $k = 3 \cdot 2^{(m+1)/2} + 4$.

Glynn has conjectured that this list exhausts the set of all monomial hyperovals. The difference sets of Maschietti based on monomial hyperovals are given in the following theorem. The proof given here is due to Dillon [13] since it is simpler than the original proof by Maschietti.

Theorem 3.2. *Let* $\gcd(k, 2^m - 1) = 1$ *and let* $x^k + x$ *be a two-to-one map on* $\mathbb{F}_{2^m}$. *Let* α *be a primitive element of* $\mathbb{F}_{2^m}$ *and let*

$$D = \mathbb{F}_{2^m} \setminus \{x^k + x \mid x \in \mathbb{F}_{2^m}\}.$$

Then $\mathbb{D} = \{t | \alpha^t \in D\}$ *is a cyclic difference set with Singer parameters* $(2^m - 1, 2^{m-1}, 2^{m-2})$.

Proof. Let $f(x)$ be the characteristic sequence of D. Then it is sufficient to prove that $\hat{F}(y) = \hat{S}_k\left(y^t\right)$ for some t where $\gcd(t, 2^m - 1) = 1$. Then direct calculations give

$$\hat{F}(y) = \frac{1}{\sqrt{2^m}} \sum_{x \in \mathbb{F}_{2^m}} (-1)^{f(x)+Tr(yx)}$$

$$= \frac{1}{\sqrt{2^m}} \left(\sum_{x \notin D} (-1)^{Tr(yx)} - \sum_{x \in D} (-1)^{Tr(yx)} \right)$$

$$= \frac{2}{\sqrt{2^m}} \sum_{x \notin D} (-1)^{Tr(yx)} = \frac{1}{\sqrt{2^m}} \sum_{x \in \mathbb{F}_{2^m}} (-1)^{Tr\left(y(x^k+x)\right)}$$

$$= \frac{1}{\sqrt{2^m}} \sum_{z \in \mathbb{F}_{2^m}} (-1)^{Tr\left(z^k+y^{\frac{k-1}{k}}z\right)} = \hat{S}_k\left(y^{\frac{k-1}{k}}\right),$$

where we use that $x^k + x$ is a two-to-one map and the substitution $x = y^{\frac{-1}{k}}z$. Note that $t = (k-1)/k$ modulo $2^m - 1$ is relatively prime to $2^m - 1$ since $\gcd(k, 2^m - 1) = \gcd(k - 1, 2^m - 1) = 1$ holds for monomial hyperovals. $\square$

Inspired by Conjecture 3.2 of No et al., Dobbertin [10] conjectured the following result, that was subsequently proved in the landmark paper by Dillon and Dobbertin [11].

Theorem 3.3. *Let $d = 2^{2k} - 2^k + 1$, $D(z) = (z+1)^d + z^d + 1$ where $gcd(k, m) = 1$ and $\mathbb{B}_k = Im(D)$. Define,*

$$s(t) = \begin{cases} 0 \ if \ \alpha^t \in \mathbb{B}_k, \\ 1 \ otherwise. \end{cases}$$

Then $\{b(t)\}$ is a binary sequence of period $2^m - 1$ with 2^{m-1} ones in a period, having a two-level autocorrelation, with an out-of-phase correlation value of -1.

Furthermore, Dobbertin [10] found an explicit trace description of the sequences given in Theorem 3.3 and applied this to determine the linear complexity of these sequences (or equivalently the 2-rank of the corresponding difference sets).

Lemma 3.3. *The linear complexity of the characteristic sequence of $L \setminus \mathbb{B}_k$ is given by $n(F_{k_1} - 1) + 1$, $k_1 = min(k', n - k')$ where $k'k = 1 \ (mod \ m)$ and F_i denotes the Fibonacci sequence (i.e., $F_0 = F_1 = 1$ and $F_{i+2} = F_i + F_{i+1}$).*

The linear complexity of the characteristic sequences of the sets above are therefore distinct for $k' < m/2$. Thus the corresponding difference sets are inequivalent and since $gcd(k, m) = 1$ the theorem gives $\phi(m)/2$ pairwise inequivalent difference sets.

Dobbertin [10] showed that several conjectured difference sets were explained by Theorem 3.3. For example all five conjectures by No et al. [9] were consequences of the proof of Dobbertin's conjecture. The special case of $k' = 2$ is equivalent to Conjecture 3.1 which therefore holds.

The case $k' = 3$ is related to Conjecture 3.2 but the 2-rank of the corresponding difference sets are different. Therefore Conjecture 3.2 by No et al. leads to inequivalent difference sets compared to the ones given by Dillon and Dobbertin in Theorem 3.3. Let $\mathcal{N}$ be the image of $N(x) = (x + 1)^d + x^d$, where $d = 2^{2k} - 2^k + 1$ and $3k = 1 \ (mod \ m)$. The connection is

$$\mathcal{N} = \begin{cases} \mathbb{B}_{1/3} + 1 & if \ m \ is \ even, \\ L \setminus (\mathbb{B}_{1/3} + 1) & if \ m \ is \ odd. \end{cases}$$

The proof of Conjecture 3.2 was proved for odd values of m by Dillon [13] and the proof of the m even case was proved by Dillon and Dobbertin [11].

It may also be interesting to observe that $k = 2$ in Theorem 3.3 corresponds to the Segre hyperoval $D(x^6)$.

The status at present is that all known difference with Singer parameters from sequences of period $n = 2^m - 1$ for $m \leq 10$ can be explained by existing constructions.

3.3 The Merit Factor of Binary Sequences

3.3.1 Introduction

Let $\{s(t)\}$ be a binary sequence of length n with elements $s(0), s(1), \ldots, s(n-1)$ where each $s(t)$ takes on the value $+1$ or -1. A classical problem is to study the aperiodic correlation of binary sequences.

Definition 3.5. The *aperiodic autocorrelation* of the binary sequence $\{s(t)\}$ at shift τ is given by

$$\rho_s(\tau) = \sum_{t=0}^{n-\tau-1} s(t)s(t+\tau) \quad \text{for } \tau = 0, 1, \ldots, n-1.$$

An important problem in digital communication that has been studied for more than 50 years is to construct sequences where the aperiodic autocorrelation coefficients are "collectively small" for all nonzero shifts in some suitable measure. Such sequences are of significant interest in several applications in synchronization, pulse compression, and radar. The problem of finding sequences with the best possible aperiodic correlation remains largely a challenging and unsolved problem. One possible measure is to require uniformly small autocorrelation coefficients. This motivates the definition of the following sequences, denoted Barker sequences.

Definition 3.6. A binary $\{-1, +1\}$ sequence $\{s(t)\}$ of length n is said to be a *Barker sequence* if the aperiodic autocorrelation values $\rho_s(\tau)$ satisfy

$$| \rho_s(\tau) | \leq 1$$

for all $0 \leq \tau \leq n-1$.

The Barker property of a sequence is preserved under the following transformations:

$$s(t) \rightarrow -s(t), \quad s(t) \rightarrow (-1)^t s(t), \text{ and } s(t) \rightarrow s(n-1-t).$$

Unfortunately only the following Barker sequences are known:
$n = 2 \quad ++$
$n = 3 \quad ++-$
$n = 4 \quad +++-$
$n = 5 \quad +++-+$
$n = 7 \quad +++--+-$
$n = 11 \ +++---+--+-$
$n = 13 \ +++++--++-+-+$
(where $+$ denotes $+1$ and $-$ denotes -1), as well as all equivalent sequences generated from these via the transformations defined above.

It was shown by Turyn and Storer [15] that there exist no further Barker sequences of odd length and further that if n is even, then $n = 0 \pmod{4}$ is necessary. There is overwhelming evidence that there are no Barker sequences of length > 13.

Conjecture 3.3. There are no Barker sequences of length $n > 13$.

It is not hard to see that the *periodic autocorrelation*

$$\theta_s(\tau) = \sum_{t=0}^{n-1} s(t)s(t + \tau),$$

where the correlation is taken over a full period of the sequence and indices are computed $\pmod{n}$, which satisfies

$$\theta_s(\tau) = n \pmod{4}.$$

Since for any sequence,

$$\theta_s(\tau) = \rho_s(\tau) + \rho_s(\tau - n),$$

it follows that

$$\theta_s(\tau) = 0, \quad \tau \neq 0$$

for any Barker sequence of length n, $n = 0 \pmod{4}$ and therefore such Barker sequences form cyclic difference sets.

Using a variety of known results on cyclic difference sets and other methods, the existence of Barker sequences of even length n, other than those defined above, has been ruled out for $1 \leq n \leq 1,898,884$ by Eliahou and Kervaire [16]. Additional nonexistence results may be found in Jedwab and Lloyd [17].

3.3.1.1 Sequences with High Merit Factor

Since there are so few Barker sequences a better measure of good aperiodic sequences might be appropriate. In 1972 Golay [18] introduced the *merit factor* of a sequence. The main problem is to find sequences with a high merit factor.

Definition 3.7. The *merit factor* F of a $\{-1, +1\}$ sequence $\{s(t)\}$ of period n is defined by

$$F = \frac{n^2}{2\sum_{\tau=1}^{n-1}\rho_s^2(\tau)}.$$

Note that collectively small values of the correlation coefficients (i.e., a small value of $\sum_{\tau=1}^{n-1}\rho_s^2(\tau)$) lead to a high merit factor. In the 1980s and 1990s researchers were able to construct arbitrarily long sequences with a provable asymptotic merit factor of 6. For a long time it was widely believed this was the maximum asymptotic value. In recent years new sequence constructions have led to an asymptotic merit factor larger than 6.34 achievable for very long sequences. However, some challenging problems still remain since there is yet no proof that the constructions give an asymptotic merit factor greater than 6.34. In this section we describe some of the highlights in the constructions of sequences with large merit factors.

It is interesting to observe that the merit factor problem is related to a problem in complex analysis where the problem has been studied independently using different techniques. Let $i = \sqrt{-1}$ and let

$$S(\omega) = \sum_{t=0}^{n-1} s(t)e^{i\omega t}, \quad 0 \le \omega \le 2\pi,$$

be the *Fourier transform* of $\{s(t)\}$. Straightforward calculations lead to

$$2 \sum_{\tau=1}^{n-1} \rho_s^2(\tau) = \frac{1}{2\pi} \int_0^{2\pi} [\mid S(\omega) \mid^2 - n]^2 d\omega.$$

Thus the merit factor may also be viewed as a measure of the deviation of the transform magnitude of $\{s(t)\}$ from the constant value $\sqrt{n}$. A reformulation of the expression above gives

$$n^2 \left(\frac{1}{F} + 1 \right) = 2 \sum_{\tau=1}^{n-1} \rho_s^2(\tau) + n^2 = \frac{1}{2\pi} \int_0^{2\pi} \mid S(\omega) \mid^4 d\omega,$$

where the right-hand side is the fourth power $(\mathcal{L}^4)^4$ of the $\mathcal{L}^4$ norm of the polynomial $\sum_{t=0}^{n-1} s(t)x^t$.

Thus finding sequences with a large merit factor is equivalent to finding polynomials with ± 1 coefficients with small $\mathcal{L}^4$ norm on the unit circle of the complex plane. This leads to difficult problems in complex analysis. Prior to Golay's definition of the merit factor in 1972, Littlewood and others studied the equivalent problem involving the $\mathcal{L}^4$ norm. For results on the problem in this formulation the reader is referred to Newman and Byrnes [19].

Let F_n denote the largest merit factor of any binary $\{-1, +1\}$ sequence of length n. Exhaustive computer searches carried out for $n \le 40$ have revealed that for $1 \le n \le 40$, $n \ne 11, 13$, $3.3 \le F_n \le 9.85$, and $F_{11} = 12.1$, $F_{13} = 14.08$. Note that F_{11} and F_{13} are achieved by Barker sequences. These are the two largest known merit factors and the only ones known possessing a merit factor ≥ 10.

From partial searches by Jensen, Jensen, and Høholdt [20], for lengths up to 117, the highest known merit factor is between 8 and 9.56. Currently F_n has been calculated for all $n \le 60$ by Mertens and Bauke [21] and for larger values of n in the range $61 \le n \le 271$ by Knauer [22].

The aperiodic autocorrelation coefficients $\rho_s(1), \rho_s(2), \ldots, \rho_s(n-1)$ of a randomly chosen sequence of length n are clearly dependent. In 1977 Golay [23] introduced a certain "ergodicity postulate," implying that the autocorrelation coefficients $\rho_s(\tau)$ for $\tau = 1, 2, \ldots, n-1$ can be treated as *independent* variables to find the correct asymptotic merit factor $\lim_{n\to\infty} F_n$. Golay established in [23, 24], using this controversial postulate, that

$$\lim_{n\to\infty} F_n = 12.32.$$

3.3.1.2 Construction of Sequence Families with High Merit Factor

There are some families of sequences for which the merit factor or the asymptotic merit factor has been explicitly determined. An early result by Newman and Byrnes [19] gives information about the expected value of the merit factor of a random sequence.

Theorem 3.4. *The mean value of $\frac{1}{F}$, taken over all sequences of length n, is $\frac{n-1}{n}$.*

Another useful observation by Jensen, Jensen, and Høholdt [20] is that for a family of sequences to have a nonzero asymptotic merit factor the sequences have to be asymptotically balanced.

Among the first sequence families found with merit factor >1 are the Rudin–Shapiro sequences which have an asymptotic merit factor of 3. These sequences were extended to another family with asymptotic merit factor 3 by Høholdt, et al. [25].

Theorem 3.5. *Let $x_0 = 1$ and $x_{2^i+j} = (-1)^{j+f(i)} x_{2^i-j-1}$ for $j = 0, 1, \ldots, 2^i - 1$, $i = 0, 1, \ldots, m-1$ where f is any mapping from the natural numbers into $\{0, 1\}$. Then the merit factor F is given by*

$$F = \frac{3}{1 - (\frac{-1}{2})^m}.$$

The Rudin–Shapiro sequence of length 2^m is obtained by letting $f(0) = f(2k-1) = 0$ and $f(2k) = 1$ for $k > 0$.

Another well-known family of sequences proven by Jensen et al. [20] to have an asymptotic merit factor of 3 is the family of m-sequences. We recall from the previous section that these are sequences of period $2^m - 1$ generated by a linear recursion associated with a characteristic polynomial that is a primitive polynomial of degree m.

Theorem 3.6. *The asymptotic merit factor of the family of m-sequences is 3.*

An "offset" sequence is one in which a fraction θ of the elements of a sequence of length n are chopped off at one end and appended to the other end, i.e., the sequence is a cyclic shift of the original sequence by $n\theta$ symbols. The asymptotic merit factor of several sequence families related to cyclic difference sets can be found in the papers by Høholdt and Jensen [26] and Jensen et al. [20]. In these papers, it is shown that

- among the class of all known cyclic difference sets, only the subclass of Hadamard difference sets gives rise to sequences with nonzero asymptotic merit factor.
- the asymptotic merit factor of m-sequences is equal to 3 and all offsets of the m-sequences have the same asymptotic merit factor.

The m-sequences are closely connected to Singer difference sets. Several of the best constructions of families of sequences leading to an asymptotic merit factor of 6 are also based on difference sets.

The Legendre symbol is defined by

$$\left(\frac{t}{n}\right) = \begin{cases} 1 & \text{if } t \text{ is a square (mod } n), \\ -1 & \text{if } t \text{ is a non-square (mod } n). \end{cases}$$

We will here use the convention that $\left(\frac{0}{n}\right) = 1$.

Definition 3.8. The *Legendre sequence* of period n, n a prime, is defined by $s(0) = 1$ and

$$s(t) = \left(\frac{t}{n}\right) \quad \text{for } t = 1, 2, \ldots, n - 1.$$

A natural extension of the Legendre sequences is the Jacobi sequences.

Definition 3.9. Let $p_1 < p_2 < \cdots < p_k$ be different primes and $n = p_1 p_2 \cdots p_k$. Then the *Jacobi sequence* of period n is defined by

$$s(t) = \left(\frac{t}{p_1}\right)\left(\frac{t}{p_2}\right) \cdots \left(\frac{t}{p_k}\right).$$

Jensen et al. [20] studied the modified Jacobi sequence of period $n = pq$.

Definition 3.10. The *modified Jacobi sequence* of period $n = pq$ is defined by

$$s(t) = \begin{cases} 1 & \text{for } t = 0 \ (\text{mod } q), \\ -1 & \text{for } t > 0 \text{ and } t = 0 \ (\text{mod } p), \\ \left(\frac{t}{p}\right)\left(\frac{t}{q}\right) & \text{otherwise.} \end{cases}$$

The special case when p and $q = p + 2$ are twin primes we obtain the twin prime sequences that correspond to the twin-prime difference sets encountered in the previous section.

Theorem 3.7. *The Legendre sequences and twin-prime set sequences, when offset by a fraction θ of their length, have asymptotic merit factor F satisfying*

$$\frac{1}{F} = \frac{2}{3} - 4 \mid \theta \mid + 8\theta^2, \quad \mid \theta \mid \leq \frac{1}{2} \tag{3.1}$$

so that F reaches a maximum value of 6 when $|\theta| = \frac{1}{4}$. Furthermore, if $\frac{(p+q)^5 \log^4 n}{n^3} \to 1$ as $n \to \infty$ the Jacobi sequences also have asymptotic merit factor satisfying (3.1).

Some generalizations to Jacobi sequences of period $n = p_1 p_2 \cdots p_k$ can be found in Borwein and Choi [27]. Golay [28] had earlier established (3.1) for Legendre sequences under the "ergodicity postulate."

In 1999, Parker [29] constructed two families of sequences of periods $2p$, where p is an odd prime and gave numerical evidence that each of the two families had an asymptotic merit factor of 6.

Thus the largest asymptotic merit factor proven for a family of sequences is 6. The excellent survey paper by Høholdt [30] in 1998 concluded that *we are fairly convinced that the maximal asymptotic value of the merit factor is 6 but we are far from having a proof of such a theorem.*

Since the appearance of this survey paper there has been significant and surprising progress related to the value of the largest possible asymptotic merit factor. In 1999, Kirilusha and Narayanaswamy observed that starting with a sequence from a family with asymptotic merit factor 6 and appending the initial part of the sequence to the sequence they were able to construct a sequence where the numerical experiments indicated a merit factor strictly greater than 6. Inspired by their work two independent research groups Borwein et al. [31] and Kristiansen and Parker [32] were able to construct families of sequences where the numerical evidence indicated an asymptotic merit factor greater than 6.34. Even though the numerical results are quite convincing, there is thus far no analytical proof that their infinite families lead to such a merit factor.

The construction of sequences with this record breaking merit factor is based on a Legendre sequence S and cyclic shifts of it by a fraction r of the period. The resulting sequence is denoted by S_r. Thereafter, one appends a fraction $t, 0 < t < 1$ of the initial portion of S_r, to S_r, i.e., the $\lfloor tn \rfloor$ initial bits in S_r are appended to S_r. Extensive numerical evidence suggests that for n large

- the merit factor of the new sequence is greater than 6.2 when $r = 1/4$ and $t \approx 0.03$,
- the merit factor of the new sequence is greater than 6.34 when $r \approx 0.22$ and $t \approx 0.06$.

A challenging and hard problem is to analytically establish the asymptotic merit factors of these new sequences. Of course even if this is possible the hard question still remains. What is the largest possible asymptotic merit factor $\lim_{n \to \infty} F_n$ of any binary sequence?

3.4 Low-Correlation QAM Sequences

Up until now the chapter has dealt exclusively with the correlation properties of a single sequence, either periodic or aperiodic. We now turn to the correlation properties of a family of sequences, relevant in CDMA applications.

Given a collection of complex-valued sequences $\{\{s_i(t)\} \mid i = 1, 2, \ldots, M\}$, we will use $\theta_{s_i, s_j}(\tau)$ (or more simply $\theta_{i,j}(\tau)$) to denote the correlation of sequences $s_i(t), s_j(t)$ at shift τ, i.e.,

$$\theta_{i,j}(\tau) = \sum_{t=0}^{2^r-2} s_i(t+\tau)s_j^*(t),$$

with the case $i = j$ corresponding to autocorrelation and $i \neq j$ corresponding to cross-correlation. The maximum correlation parameter $\theta_{\max}$ of the sequence set is defined by

$$\theta_{\max} = \max\left\{\, |\,\theta_{i,j}(\tau)\,| \ \big| \ \text{either } i \neq j \text{ or } \tau \neq 0 \right\},$$

and is frequently used as a measure of performance of a sequence family. In CDMA settings where phase modulation is used to carry data, low values of autocorrelation magnitude $|\,\theta_{i,i}(\tau)\,|$, $\tau \neq 0$ facilitate synchronization, while minimizing $\theta_{i,j}(\tau)$, $i \neq j$, all τ, helps minimize interference due to the presence of other users. Much of the literature on families of sequences with low correlation deals with either the $\{\pm 1\}$ binary (BPSK) or $\{\pm 1, \pm \iota\}$ quaternary (QPSK) alphabets. Low-correlation sequences over the quadrature amplitude modulation (QAM) constellation are of interest on account of their ability to carry a larger number of data bits per sequence period in comparison with the 1 or 2 bits associated with binary or quaternary phase-shift-keying modulation. Some recent results on the construction of low-correlation QAM sequences are presented below.

3.4.1 Preliminaries

An alternative description for the 16-QAM (see Fig. 3.1) constellation

$$\{a + \iota b \mid -3 \leq a, b \leq 3, \ a, b \text{ odd}\} \tag{3.2}$$

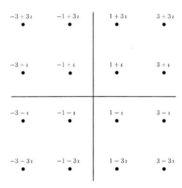

Fig. 3.1 16-QAM constellation

is given by [33–35]

$$\left\{\sqrt{2\iota}\left(\iota^a + 2\iota^b\right) \,\Big|\, a, b \in \mathbb{Z}_4\right\}, \tag{3.3}$$

where $\sqrt{2\iota}$ denotes $(1 + \iota)$. The latter representation suggests the use of quaternary sequences as building blocks in the construction of low-correlation sequences over the 16-QAM constellation. On account of its excellent correlation properties, a natural family to consider is quaternary Family $\mathbb{A}$.

3.4.2 Quaternary Family $\mathbb{A}$

This is a family of cyclically distinct sequences whose symbol alphabet is the set $\mathbb{Z}_4$ of integers modulo 4. As noted earlier, the characteristic polynomial of a binary m-sequence is a primitive polynomial. All the sequences belonging to Family $\mathbb{A}$ share a common characteristic polynomial and this polynomial may be generated by "lifting" a primitive polynomial over $\mathbb{Z}_2$ to a polynomial over $\mathbb{Z}_4$ as described below.

Let $f(x)$ be the (primitive) characteristic polynomial of a binary m-sequence of length $2^r - 1$ for some integer r. The coefficients of $f(x)$ are thus either 0 or 1. Next, regard $f(x)$ as a polynomial over $\mathbb{Z}_4$ and form the product $(-1)^r f(x) f(-x)$ (mod 4). The resultant polynomial is a polynomial in x^2 over $\mathbb{Z}_4$. Let the $\mathbb{Z}_4$ polynomial $g(x) = \sum_{i=0}^{r} g_i x^i$ of degree r be defined by setting $g(x^2) = (-1)^r f(x) f(-x)$ and consider the set of all quaternary sequences $\{a(t)\}$ having $g(x)$ as characteristic polynomial, i.e., sequences over $\mathbb{Z}_4$ satisfying the recursion $\sum_{i=0}^{r} g_i a(t + i) = 0$ for all t. It turns out that with the exception of the all-zero sequence, all the sequences generated in this way have period $N = 2^r - 1$. Thus, the recursion generates a collection, Family $\mathbb{A}$ of $2^r + 1$ cyclically distinct quaternary sequences [36–41]. A particular sequence within the family turns out to be in essence a binary sequence as all symbols are either 0 or 2; we discard this particular sequence in our discussion below and will henceforth regard Family $\mathbb{A}$ as having size 2^r.

If $\{u(t)\}$, $\{v(t)\}$ are two distinct sequences drawn from Family $\mathbb{A}$, then by the correlation of u, v we will mean correlation of the associated complex sequences $\{\iota^{u(t)}\}$, $\{\iota^{v(t)}\}$. In the case of Family $\mathbb{A}$, it is known [36,40] that $\theta_{\max} \leq \sqrt{N+1}+1$ where $N = 2^r - 1$ is the period of each sequence in the family. In the remainder of this section, we will make the approximation $\theta_{\max} \approx \sqrt{N}$.

3.4.3 Canonical 16-QAM Family $\mathbb{CQ}$

Let the 2^r sequences in Family $\mathbb{A}$ be partitioned into 2^{r-1} disjoint pairs $\{(\{a_i(t)\}, \{b_i(t)\}) \mid 1 \leq i \leq 2^{r-1}\}$. Each user in the CDMA scheme associated with canonical 16-QAM Family $\mathbb{CQ}$ is then assigned a unique pair $(\{a_i(t)\}, \{b_i(t)\})$ and uses the pair to construct the spreading code $s_i(t) = \sqrt{2\iota}(\iota^{a_i(t)} + 2\iota^{b_i(t)})$. The actual signal transmitted by the user is then given by $s_i(t, \kappa) = \sqrt{2\iota}(\iota^{a_i(t)+\kappa_a} + 2\iota^{b_i(t)+\kappa_b})$ where κ_a, κ_b lie in $\mathbb{Z}_4$ and represent the data-bearing portions of the signal, so that one period of the sequence carries 4 bits of data.

Canonical Family $\mathbb{CQ}$ [42] may thus be described as follows:

$$\mathbb{CQ} = \left\{ \left\{ s_i(t, \kappa) \,\middle|\, \kappa \in \mathbb{Z}_4^2 \right\} \,\middle|\, 1 \leq i \leq 2^{r-1} \right\}.$$

Since each sequence in Family $\mathbb{A}$ has period $N = 2^r - 1$, the same is true of the members of Family $\mathbb{CQ}$. The correlation between the spreading codes of two users takes on the form

$$\theta_{i,j}(\tau) \triangleq \sum_{t=0}^{N-1} s_i(t+\tau, 0) s_j(t)^* = 2 \sum_{t=0}^{N-1} \left[\iota^{a_i(t+\tau)} \quad \iota^{b_i(t+\tau)} \right] \begin{bmatrix} 1 \\ 2 \end{bmatrix} \begin{bmatrix} 1 & 2 \end{bmatrix} \begin{bmatrix} \iota^{-a_j(t)} \\ \iota^{-b_j(t)} \end{bmatrix}$$

$$= 2 \sum_{t=0}^{N-1} \left[\iota^{a_i(t+\tau)} \quad \iota^{b_i(t+\tau)} \right] \begin{bmatrix} 1 & 2 \\ 2 & 4 \end{bmatrix} \begin{bmatrix} \iota^{-a_j(t)} \\ \iota^{-b_j(t)} \end{bmatrix}.$$

If in this computation, we set $i = j$ and $\tau = 0$, we recover the energy of the sequence $s_i(t, 0)$. Since the energy of each sequence in Family $\mathbb{A}$ equals N and the correlation between any two sequences in Family $\mathbb{A}$ is of order $\sqrt{N}$, it follows that the energy of a sequence in Family $\mathbb{CQ}$ is well approximated by ignoring the contribution of the non-diagonal terms in the matrix $\begin{bmatrix} 1 & 2 \\ 2 & 4 \end{bmatrix}$, i.e.,

$$\| s_i(t, 0) \|^2 = \theta_{i,i}(0) \approx 10\, N.$$

When either $i \neq j$ or $\tau \neq 0$, we have $| \theta_{i,j}(\tau) | \lesssim 18 \sqrt{N}$. The calculations above can be verified to hold even in the presence of data modulation, i.e., even when the spreading codes $\{s_i(t, 0)\}$, $\{s_j(t, 0)\}$ are replaced by their data-bearing counterparts $\{s_i(t, \kappa)\}$, $\{s_j(t, \kappa')\}$.

Two comments are appropriate at this stage. First, to permit comparison of correlation values across sequence families, we need to scale the sequences in Family $\mathbb{CQ}$ by the factor $\frac{1}{\sqrt{10}}$ so that each sequence in the family, after normalization, will have energy equal to the sequence period N. Second, the definition of $\theta_{\max}$ given earlier assumed that data modulation was carried out by rotating the phase of the spreading code, i.e., via phase modulation. In the present instance, we are rotating instead the phases of the individual components. The appropriate definition of parameter $\theta_{\max}$ in this setting is given by

$$\theta_{\max} = \max\{|\theta_{(i,\kappa),(j,\kappa')}(\tau)| \mid \text{ either } i \neq j \text{ or } \tau \neq 0\}. \tag{3.4}$$

The difference here is that we do not include in the set (over which maximization is carried out), the correlations at shift $\tau = 0$ between different data modulations of the same spreading code. This is because in a receiver which employs a bank of $|\{(\kappa_a, \kappa_b) \mid \kappa_i \in \mathbb{Z}_4\}| = 16$ correlators, in which the different correlators carry out inner products with the different data-modulated versions of the same spreading code, large values of correlation values $\theta_{(i,\kappa),(j,\kappa')}(0)$ clearly do not correspond to interference nor will they negatively impact synchronization. Keeping these two points in mind, we arrive at the value $\theta_{\max} \lesssim 1.8\sqrt{N}$ for the maximum correlation parameter of Family $\mathbb{CQ}$.

A second parameter that governs CDMA performance is the minimum-squared Euclidean distance [42] between the pair $(\{s_i(t, \kappa)\}, \{s_i(t, \kappa')\})$ of sequences corresponding to different data modulations of the same spreading code. This is given by

$$d_{\min}^2 = \min_{1 \leq i \leq 2^{r-1},\ \kappa \neq \kappa' \in \mathbb{Z}_4} \left\{ d_E^2(s_i(t, \kappa), s_i(t, \kappa')) \right\},$$

which upon energy normalization turns out to be given by $d_{\min}^2 \approx 0.4N$. This compares very favorably with the value

$$d_{\min}^2 = \frac{\pi^2}{64} N \approx 0.15N,$$

for the case when 16-ary PSK data modulation is employed.

3.4.4 Extensions and Improvements

3.4.4.1 Extension to M^2-QAM

As in the case of the 16-QAM constellation, the M^2-QAM constellation

$$\{a + \iota b \mid -M \leq a, b \leq M,\ a, b \text{ odd}\} \tag{3.5}$$

has the alternative expression

$$\left\{ \sqrt{2\iota} \left(\sum_{k=0}^{m-1} 2^k \iota^{a_k} \right) \middle| a_k \in \mathbb{Z}_4 \right\}, \tag{3.6}$$

from which it is evident that a construction for a family of low-correlation M^2-QAM sequences, analogous to the construction of the 16-QAM Family $\mathbb{CQ}$ holds. It is not hard to deduce that the resultant M^2-QAM family is of common period $N = 2^r - 1$, family size $\frac{N}{\log_2 M}$, and is capable of transporting $2\log_2 M$ bits per sequence period. It turns out that the minimum-squared Euclidean distance for this family is given by

$$d_{\min}^2 \approx \frac{6}{M^2 - 1} N,$$

which is substantially larger than the value

$$d_{\min}^2 \approx \frac{4\pi^2}{M^4} N$$

that holds for the case when M^2-PSK data modulation is employed. The increase in size of the symbol alphabet, does however, cause the value of the maximum correlation parameter to rise to $\theta_{\max} \lesssim 3\sqrt{N}$ in the limit as $M \to \infty$.

3.4.4.2 Improvement Through Interleaving

It is possible to improve upon the correlation properties of the canonical Family $\mathbb{CQ}$ by interleaving a pair of sequences. While sequences in Family $\mathbb{CQ}$ have period $2^r - 1$ and take on the form

$$s_i(t, \kappa) = \sqrt{2\iota}\left(\iota^{a_i(t)+\kappa_a} + 2\iota^{b_i(t)+\kappa_b}\right),$$

all sequences in the interleaved Family $\mathbb{IQ}$ have period $N = 2(2^r - 1)$ and the corresponding sequence in the $\mathbb{IQ}$ family is specified in terms of the sequence pair:

$$p_i(t, \kappa) = \sqrt{2\iota}\left(\iota^{a_i(t)+\kappa_a} + 2\iota^{b_i(t)+\kappa_b}\right),$$

$$q_i(t, \kappa) = \iota\sqrt{2\iota}\left(\iota^{b_i(t)+\kappa_b} - 2\iota^{a_i(t)+\kappa_a}\right).$$

Given the sequence pair $(p_i(t, \kappa), q_i(t, \kappa))$, the transmitted sequence $s_i^{(I)}(t)$ in the interleaved Family $\mathbb{IQ}$ is obtained through interleaving via the Chinese Remainder Theorem (CRT), i.e., if

$$t_1 = t \quad (\text{mod } 2^r - 1) \quad \text{and} \quad t_2 = t \quad (\text{mod } 2), \tag{3.7}$$

then

$$s_i^{(I)}(t, \kappa) = p_i(t_1, \kappa) \quad \text{when } t_2 = 0 \quad (\text{mod } 2),$$

$$s_i^{(I)}(t, \kappa) = q_i(t_1, \kappa) \quad \text{when } t_2 = 1 \quad (\text{mod } 2).$$

This is shown below for the example when $r = 3$ so that $N = 7$:

$$\begin{bmatrix} p(0) & q(0) \\ p(1) & q(1) \\ p(2) & q(2) \\ p(3) & q(3) \\ p(4) & q(4) \\ p(5) & q(5) \\ p(6) & q(6) \end{bmatrix} = \begin{bmatrix} s^{(I)}(0) & s^{(I)}(7) \\ s^{(I)}(8) & s^{(I)}(1) \\ s^{(I)}(2) & s^{(I)}(9) \\ s^{(I)}(10) & s^{(I)}(3) \\ s^{(I)}(4) & s^{(I)}(11) \\ s^{(I)}(12) & s^{(I)}(5) \\ s^{(I)}(6) & s^{(I)}(13) \end{bmatrix}.$$

This ensures that the correlation of this two-dimensional sequence can be expressed as the sum of correlations of the constituent one-dimensional sequences. As an example, for even values of the time-shift parameter τ, i.e., for $\tau = (\tau_1, \tau_2)$ with

$\tau_1 = \tau \pmod{2^r - 1}$, $\tau_2 = \tau \pmod 2$, and $\tau_2 = 0 \pmod 2$, and in the absence of data modulation, we have

$$\theta_{i,j}(\tau) \triangleq \sum_{t=0}^{N-1} s_i(t+\tau, 0) s_j(t, 0)^*$$

$$= \sum_{t_1=0}^{2^r-2} p_i(t_1 + \tau_1, 0) p_j(t_1, 0)^* + \sum_{t_1=0}^{2^r-2} q_i(t_1 + \tau_1, 0) q_j(t_1, 0)^*$$

$$= 2 \sum_{t_1=0}^{2^r-2} \left[\iota^{a_i(t_1+\tau_1)} \; \iota^{b_i(t_1+\tau_1)} \right] \begin{bmatrix} 1 \\ 2 \end{bmatrix} [1 \; 2] \begin{bmatrix} \iota^{-a_j(t_1)} \\ \iota^{-b_j(t_1)} \end{bmatrix}$$

$$+ 2 \sum_{t_1=0}^{2^r-2} \left[\iota^{a_i(t_1+\tau_1)} \; \iota^{b_i(t_1+\tau_1)} \right] \begin{bmatrix} -2 \\ 1 \end{bmatrix} [-2 \; 1] \begin{bmatrix} \iota^{-a_j(t_1)} \\ \iota^{-b_j(t_1)} \end{bmatrix}$$

$$= 2 \sum_{t_1=0}^{2^r-2} \left[\iota^{a_i(t_1+\tau_1)} \; \iota^{b_i(t_1+\tau_1)} \right] \begin{bmatrix} 5 & 0 \\ 0 & 5 \end{bmatrix} \begin{bmatrix} \iota^{-a_j(t_1)} \\ \iota^{-b_j(t_1)} \end{bmatrix}.$$

We will get a similar set of equations for the case when τ is odd.

Interleaving in this manner (on account of partial cancelation) reduces the maximum normalized correlation parameter to $\theta_{\max} \lesssim 1.41\sqrt{N}$ for the case of 16-QAM and to $\theta_{\max} \lesssim 2.12\sqrt{N}$ for the case of M^2-QAM, with M large.

3.4.4.3 Other Attributes

Variable Data Rate Signaling. Family $\mathbb{CQ}$ supports variable-data-rate signaling. For example, the maximum correlation between the sequences corresponding to users at data rates 4 and 6 bits per sequence period, corresponding to QAM constellations of size 16 and 64, respectively, is bounded above (after energy normalization) by $2.05\sqrt{N}$. Note that this value lies in between the values $\theta_{\max} = 1.8\sqrt{N}$ and $\theta_{\max} = 2.33\sqrt{N}$ of the 16-QAM and 64-QAM families, respectively. A similar variable-data-rate capability is also possessed by interleaved Family $\mathbb{IQ}$.

    Increased Data Rate. It is possible to further increase the data rate of each user, at the cost of some reduction in family size, by assigning multiple sequences to each user in Family $\mathbb{CQ}$. Details may be found in [42].

    Lowered Correlation Values. The value of $\theta_{\max}$ can be lowered by judicious assignment of sequences from Family $\mathbb{A}$ to each user. This, however, comes at the cost of reduced data rate. For example, it is possible to construct a 16-QAM family having $\theta_{\max} = 1.17\sqrt{N}$ with a data rate of 3 bits per sequence period while in comparison, Family $\mathbb{CQ}$ has $\theta_{\max} \lesssim 1.8\sqrt{N}$ but permits 4 bits of data to be transmitted per sequence period.

    The properties of Families $\mathbb{CQ}$ and $\mathbb{IQ}$ over M^2-QAM for various values of constellation size M^2 are summarized in Table 3.1.

Table 3.1 Families of sequences over M^2-QAM

Family	Constellation	Period N	Data rate	Euclidean distance	θ_{max}
CQ	16-QAM	$2^r - 1$	4	$0.4N$	$1.80\sqrt{N}$
CQ	64-QAM	$2^r - 1$	6	$0.10N$	$2.33\sqrt{N}$
CQ	M^2-QAM	$2^r - 1$	$2\log_2 M$	$6N/(M^2 - 1)$	$3.00\sqrt{N}$
IQ	16-QAM	$2(2^r - 1)$	4	N	$1.41\sqrt{N}$
IQ	64-QAM	$2(2^r - 1)$	6	$0.10N$	$1.95\sqrt{N}$
IQ	M^2-QAM	$2(2^r - 1)$	$2\log_2 M$	$15N/(M^2 - 1)$	$2.12\sqrt{N}$

For further details the reader is referred to [42–44].

3.4.5 Example: Generation of a 16-QAM Sequence

Let $f(x) = x^3 + x + 1$ be the characteristic polynomial [38] of an m-sequence $\{a(t)\}$. Then, over $\mathbb{Z}_4$

$$g(x^2) = (-1)^3 f(x) f(-x) = x^6 + 2x^4 + x^2 + 3 \tag{3.8}$$

so that $g(x) = x^3 + 2x^2 + x + 3$. Thus, the sequences in Family $\mathbb{A}$ are generated by the recursion $s(t + 3) + 2s(t + 2) + s(t + 1) + 3s(t) = 0 \pmod 4$. The top shift-register in Fig. 3.2 corresponds to this sequence. Since we need two quaternary sequences to generate a 16-QAM sequence, we use a copy of the same shift-register to produce a second quaternary sequence $\{b(t)\}$. By appropriately weighting and combining the component sequences $\{a(t)\}$ and $\{b(t)\}$, we generate the desired 16-QAM sequence.

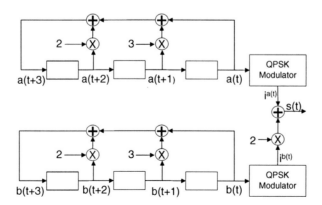

Fig. 3.2 Shift-register-based generation of 16-QAM sequences

3.5 Low-Correlation Zone Sequences

This section deals with sets of sequences, known as low-correlation zone (LCZ) sequence families, that have low values of correlation when approximately synchronized. More precisely, an LCZ sequence family $\{s_i(t)\}_{i=1}^{M}$ is one in which the non-trivial auto- and cross-correlation values are negligibly small (typically 0 or -1) for small values of the timing offset, i.e.,

$$\theta_{i,j}(\tau) \leq \delta \ \text{ for } |\tau| < L, \ i \neq j, \text{ and for } 1 \leq |\tau| < L, \text{ when } i = j.$$

Such families are of interest in CDMA settings where the signals of the various users are approximately (but not perfectly) in synchronism [45]. We provide an overview of some constructions of LCZ families to be found in the literature. While we have attempted to provide a reference to most papers on this subject, the discussion presented here is not claimed to be comprehensive.

Four parameters (N, M, L, δ) characterize an LCZ sequence family, namely the size M of the family, the common period N of each sequence in the family, the width L of the low-correlation zone, and the upper bound δ on correlation magnitudes within the low-correlation zone. An upper bound to the size M of an (N, M, L, δ) LCZ sequence family was derived by Tang et al. [46]:

$$ML - 1 \leq \frac{N-1}{1 - \delta^2/N} \cdot \tag{3.9}$$

Let l, m, n be integers with $l|m|n$, $n > m > l \geq 1$. Let $\mathbb{F}_{p^l} \subseteq \mathbb{F}_{p^m} \subseteq \mathbb{F}_{p^n}$ be a finite field chain and let $\psi(\cdot) : \mathbb{F}_{p^l} \to \mathbb{A} \subseteq \mathbb{C}$ represent the mapping carried out by the modulator, where $\mathbb{A} = \psi(Im(f))$ is the signal constellation. We make the following definitions:

1. A function $f(\cdot) : \mathbb{F}_{p^m} \to \mathbb{F}_{p^l}$ is balanced if, as x varies over $\mathbb{F}_{p^m}$, $f(x)$ runs through all of $\mathbb{F}_{p^l}$ an equal number p^{m-l} of times.
2. A function $g(\cdot) : \mathbb{F}_{p^n} \to \mathbb{F}_{p^m}$ is difference balanced if $g(x) - g(\delta x)$ is balanced for all $\delta \neq 1$, $\delta \in \mathbb{F}_{p^n}$.
3. A function $h(\cdot) : \mathbb{F}_{p^n} \to \mathbb{F}_{p^m}$ is said to have typical two-tuple balance if for every $\lambda \in \mathbb{F}_{p^n} \setminus \mathbb{F}_{p^m}$, the pair $(h(\lambda x), h(x))$ assumes every two-tuple in $\mathbb{F}_{p^m}^2$ an equal number p^{n-2m} of times.

We now present a general construction for LCZ sequences, which draws upon the constructions by Gong et al. [47], Jang et al. [48], and Tang and Fan [49]. As noted in [47, 50], special instances of this construction yield several of the constructions appearing previously in the literature.

Lemma 3.4. *Let* $l, m, n,$ $n > m > l \geq 1$ *be integers with* $l|m|n$. *Let* $h(\cdot)$: $\mathbb{F}_{p^n} \to \mathbb{F}_{p^m}$ *have balance as well as typical two-tuple balance. Let the functions* $f_i(\cdot) : \mathbb{F}_{p^m} \to \mathbb{F}_{p^l}$, $1 \leq i \leq M$ *satisfy (see Fig. 3.3)*

$$\begin{array}{c} \mathbb{F}_{p^n} \\ \Big\downarrow h \\ \mathbb{F}_{p^m} \\ \Big\downarrow f_i \\ \mathbb{F}_{p^l} \xrightarrow{\ \psi\ } \mathbb{A} \end{array}$$

Fig. 3.3 Mappings in the LCZ construction of Lemma 3.4

$$\sum_{x \in \mathbb{F}_{p^m}} \psi(f_i(x)) = 0 \textit{ for all } i \textit{ and} \tag{3.10}$$

$$\sum_{x \in \mathbb{F}_{p^m}} \psi(f_i(x))\psi^*(f_j(x)) = 0 \textit{ for } i \neq j. \tag{3.11}$$

Then,

$$\sum_{x \in \mathbb{F}_{p^n}} \psi(f_i[h(\lambda x)])\psi^*(f_j[h(x)]) = 0 \tag{3.12}$$

whenever $i \neq j$ *and* $\lambda \in \{1\} \cup \mathbb{F}_{p^n} \setminus \mathbb{F}_{p^m}$.

Proof. Let $\lambda \in \mathbb{F}_{p^n} \setminus \mathbb{F}_{p^m}$. Typical two-tuple balance property of $h(\cdot)$ guarantees that

$$\begin{aligned} \sum_{x \in \mathbb{F}_{p^n}} \psi(f_i[h(\lambda x)])\psi^*(f_j[h(x)]) &= p^{n-2m} \sum_{y,\, z \in \mathbb{F}_{p^m}} \psi(f_i(y))\psi^*(f_j(z)) \\ &= p^{n-2m} \left[\sum_{y \in \mathbb{F}_{p^m}} \psi(f_i(y)) \right] \left[\sum_{z \in \mathbb{F}_{p^m}} \psi(f_j(z)) \right]^* \\ &= 0 \end{aligned}$$

from (3.10). Next, when $\lambda = 1$ we have

$$\begin{aligned} \sum_{x \in \mathbb{F}_{p^n}} \psi(f_i[h(x)])\psi^*(f_j[h(x)]) &= p^{n-m} \sum_{y \in \mathbb{F}_{p^m}} \psi(f_i(y))\psi^*(f_j(y)) \\ &= 0 \end{aligned}$$

from (3.11) and the balance property of the function $h(\cdot)$. □

Note that (3.11) would hold, for example, when the modulator function ψ is such that $\psi(x)\psi^*(y) = \psi(x-y)$, when every difference function $(f_i - f_j)(\cdot)$, $i \neq j$ is balanced and when the constellation has zero-sum, i.e., $\sum_{x \in \mathbb{F}_{p^l}} \psi(x) = 0$.

Theorem 3.8. *Given a function $h(\cdot) : \mathbb{F}_{p^n} \to \mathbb{F}_{p^m}$ with balance and typical two-tuple balance and a collection of M functions $f_i(\cdot) : \mathbb{F}_{p^m} \to \mathbb{F}_{p^l}$, $1 \le i \le M$ satisfying (3.10) and (3.11), let $\{s_i(t)\}_{i=1}^{M}$, $s_i(t) \triangleq \psi(f_i(h(\alpha^t)))$ be a family of M sequences where α is a primitive element of $\mathbb{F}_{p^n}$. Then from Lemma 3.4, it follows that this family is an LCZ family with parameters (N, M, L, δ), where $N = p^n - 1$, $L = \frac{p^n-1}{p^m-1}$, and $\delta = \max_i\{|\psi(f_i[h(0)])|^2\}$.*

(Construction taken from [48,51]): a function $g(\cdot) : \mathbb{F}_{p^n} \to \mathbb{F}_{p^m}$, is said to be a d-form if $g(\lambda x) = \lambda^d g(x)$, for all $\lambda \in \mathbb{F}_{p^m}^*$, $x \in \mathbb{F}_{p^n}$. The construction by Jang, No, and Chung [48] provides a set of functions that satisfy the conditions of Theorem 3.8. In the construction, $h(\cdot)$ and $g(\cdot)$ are 1-form functions with difference balance: $h(\cdot)$ maps from $\mathbb{F}_{p^n}$ to $\mathbb{F}_{p^m}$, whereas $g(\cdot)$ maps from $\mathbb{F}_{p^m}$ to $\mathbb{F}_p$. This implies that $h(\cdot)$ has typical two-tuple balance [52–54]. The set of functions $f_i(\cdot) : \mathbb{F}_{p^m} \to \mathbb{F}_{p^2}$ is given by

$$f_i(y) = \begin{cases} \rho g\left([a_i y]^r\right) & a_i \in \mathbb{F}_p^*, \\ g\left([y]^r\right) + \rho g\left([a_i y]^r\right) & a_i \in \mathbb{F}_{p^m} \setminus \mathbb{F}_p, \end{cases} \tag{3.13}$$

where $\gcd(r, p^m - 1) = 1$, $1 \le r \le p^m - 2$, and $\rho \in \mathbb{F}_{p^2} \setminus \mathbb{F}_p$. The modulator mapping is given by $\psi(a + \rho b) = \omega_{p^2}^a \omega_p^b$, where ω_k is a kth root of unity. It can be verified that these functions satisfy (3.10) and (3.11) and hence generate an $(p^n - 1, p^m - 1, \frac{p^n-1}{p^m-1}, 1)$ LCZ sequence set. By specializing the parameters of this construction to $p = 2$, $r = 1$, $h(x) = tr_m^n(x)$, $g(y) = tr_1^m(y)$, $\rho \in \mathbb{F}_4 \setminus \mathbb{F}_2$, $\psi(a + \rho b) = \iota^a(-1)^b$, $a_i = \beta^i$ where β is primitive in $\mathbb{F}_{2^m}$, we get one of the constructions by Kim et al. [51], which yields a $(2^n - 1, 2^m - 1, \frac{2^n-1}{2^m-1}, 1)$ LCZ sequence set. The sequence families constructed in both [48,51] meet the bound (3.9) and are hence optimal.

(Construction taken from [45,55]): a construction by Tang and Fan [55] provides a second set of functions that satisfy Theorem 3.8. Here $h(x) = tr_m^n(x)$, $f_k(y) = tr_1^m(y^r - y^s \beta^{ks})$, $\psi(a) = \omega_p^a$, and $\gcd(r, p^m - 1) = \gcd(s, p^m - 1) = 1$. Shifts k are chosen such that the cross-correlation at shift k of the m-sequences $\{tr_1^m(\beta^{rt})\}$ and $\{tr_1^m(\beta^{st})\}$ equals -1, where β is primitive in $\mathbb{F}_{p^m}$. In particular, by setting $p = 2$, we recover the construction by Long et al. [45].

3.6 Additional Notes

3.6.1 Merit Factor

In 1953 Barker defined an ideal sequence as one with the property that $\rho_s(\tau) = -1$ or 0 for $\tau = 1, 2, \ldots, n-1$. However, such sequences only exist for lengths $n = 3, 7$, and 11, and this is the reason for the relaxation in aperiodic correlation found in Definition 3.6.

In 2007, Yu and Gong [56] constructed sequences of period $4p$ as a direct sum of the sequence (0111) of length 4 and a Legendre sequence of length p, as well as a direct sum of the sequence (0111) of length 4 and a modified Jacobi sequence of length p. They gave numerical results indicating that these constructions lead to an asymptotic merit factor of 6 after a rotation of $r = 1/4$. Similarly using the method of Borwein, Choi, and Jedwab and Kristiansen and Parker they showed that numerically their sequences of length $4p + \lfloor 4tp \rfloor$, with same $r \approx 0.22$ and $t \approx 0.06$ as in previous constructions, lead to an asymptotic merit factor greater than 6.34.

For further details on the merit factor, the excellent surveys by Høholdt [30] and Jedwab [57] are highly recommended.

3.6.2 QAM Sequences

Boztaş [33] was the first to view the QAM constellation in the form given by (3.3). The same observation was independently made by Rößing and Tarokh [34] and Lu and Kumar [35] subsequently.

In the same paper, Boztaş [33] also proposed a 16-QAM CDMA sequence family built from quaternary sequences drawn from Family $\mathbb{A}$. However, the first concerted look at low-correlation sequence families over the QAM constellation is provided by Anand and Kumar [42]. More recently, Garg, Kumar and Madhavan [43, 44] provide improved constructions of low-correlation QAM sequence families through interleaving.

3.6.3 Low-Correlation Zone Sequences

Additional optimal constructions of LCZ sequence families appear in [50, 58–61] and were not included for lack of space. Some other constructions appear in [62–64].

Lack of time and space precluded us from including discussion of the subclass of the LCZ sequences known as sequences with zero correlation zone or ZCZ sequences. In these sequence sets $\delta = 0$, i.e., the non-trivial auto- and cross-correlation values are 0 for small values of the timing offset. The parameters of a ZCZ family are (N, M, Z), where N is the period of the sequences in the family, M is the number of sequences in the family, and Z is the length of the zero correlation zone. The bound in (3.9) specializes to $N \leq M(Z + 1)$ for ZCZ sequence families [46]. Most constructions of optimal ZCZ sequence families involve interleavings of a perfect sequence [65–70]. A perfect sequence [71, 72] is one in which *all* the non-trivial autocorrelation values (not just those in a zone) are 0. These include as a subclass the entire class of one-dimensional bent functions [73].

Additional constructions of ZCZ sequence families may be found in [74–85].

3.7 Conclusions

This chapter provides a glimpse of some recent developments relating to low-correlation sequences. The particular advances chosen here pertain to cyclic Hadamard difference sets, high merit factor binary sequences, low-correlation sequences over the QAM constellations, and sequences having a low-correlation zone. The developments with respect to cyclic Hadamard difference sets are noteworthy for they involve the discovery of new families of Hadamard difference sets after a gap of some 40 years. In similar vein, the section on sequences with merit factor highlights the recent discovery of sequences with merit factor larger than had hitherto been suspected. The third topic on low-correlation QAM sequences is of practical interest as the symbol alphabet corresponds to a commonly employed signal constellation. A description of sequences with low-correlation zone has been included as these offer the promise of improved performance in the presence of approximate synchronization.

References

1. Simon Litsyn, *Peak Power Control in Multicarrier Communications*, New York, Cambridge University Press, 2007.
2. Solomon W. Golomb and Guang Gong, Signal Design for Good Correlation: For Wireless Communication, Cryptography, and Radar, New York, *Cambridge University Press*, 2005.
3. D.V. Sarwate and Pursley, M.B., Crosscorrelation properties of pseudorandom and related sequences, *Proceedings of the IEEE*, May, vol. 68, no. 5, pp. 593-619, 1980.
4. Pingzhi Fan and Mike Darnell, *Sequence Design for Communications Applications*, Michigan Research Studies Press, 1996.
5. B. Gordon, W. H. Mills, and L. R. Welch, "Some new difference sets," *Canad. J. Math*, vol. 14, pp. 614–625, 1962.
6. R. A. Scholtz and L. R. Welch, "GMW sequences," *IEEE Transactions on Information Theory*, vol. 30, no. 3, pp. 548–553, 1984.
7. L. D. Baumert, *Cyclic Difference Sets*, vol. 182 of *Lecture Notes in Mathematics*. Berlin-New York: Springer-Verlag, 1971.
8. S. W. Golomb, *Shift Register Sequences*. Laguna Hills, CA: Aegean Press, 1982.
9. J.-S. No, S. Golomb, G. Gong, H.-K. Lee, and P. Gaal, "Binary pseudorandom sequences of period $2^n - 1$ with ideal autocorrelation," *IEEE Transactions on Information Theory*, vol. 44, pp. 814–817, Mar. 1998.
10. H. Dobbertin, "Kasami power functions, permutation polynomials and cyclic difference sets," *Difference Sets, Sequences and Their Correlation Properties*, Eds. A. Pott et. al., pp. 133–158, 1999, The Netherlands Kluwer Academic Publishers.
11. J. Dillon and H. Dobbertin, "New cyclic difference sets with Singer parameters," *Finite Fields and Their Applications*, vol. 10, no. 3, pp. 342–389, 2004.
12. J. No, H. Chung, and M. Yun, "Binary pseudorandom sequences of period $2^m - 1$ with ideal autocorrelation generated by the polynomial $z^d + (z+1)^d$," *IEEE Transactions on Information Theory*, vol. 44, no. 3, pp. 1278–1282, 1998.
13. J. Dillon, "Multiplicative difference sets via additive characters," *Designs, Codes and Cryptography*, vol. 17, no. 1, pp. 225–235, 1999.
14. A. Maschietti, "Difference sets and hyperovals," *Designs, Codes and Cryptography*, vol. 14, no. 1, pp. 89–98, 1998.

15. R. Turyn and J. Storer, "On binary sequences," *Proc. Amer. Math. Soc*, vol. 12, no. 3, pp. 394–399, 1961.
16. S. Eliahou and M. Kervaire, "Barker sequences and difference sets," *Énseign. Math.*, vol. 38, pp. 345–382, 1992.
17. J. Jedwab and S. Lloyd, "A note on the nonexistence of Barker sequences," *Designs, Codes and Cryptography*, vol. 2, no. 1, pp. 93–97, 1992.
18. M. J. E. Golay, "A class of finite binary sequences with alternate autocorrelation values equal to zero," *IEEE Transactions on Information Theory*, vol. 18, no. 3, pp. 449–450, 1972.
19. D. J. Newman and J. S. Byrnes, "The L^4 norm of a polynomial with coefficients ± 1," *Amer. Math. Monthly*, vol. 97, pp. 42–45, 1990.
20. J. M. Jensen, H. E. Jensen, and T. Høholdt, "The merit factor of binary sequences related to difference sets," *IEEE Transactions on Information Theory*, vol. 37, no. 3, pp. 617–626, 1991.
21. S. Mertens and H. Bauke, "Ground States of the Bernasconi model with open boundary conditions," *available online http://odysseus.nat.uni-magdeburg.de/~mertens/bernasconi/open.dat*, November 2004.
22. J. Knauer, "Merit factor records," *available online http://www.cecm.sfu.ca/~jknauer/labs/records.html*, Nov. 2004.
23. M. J. E. Golay, "Sieves for low autocorrelation binary sequences," *IEEE Transactions on Information Theory*, vol. 23, no. 1, pp. 43–51, 1977.
24. M. J. E. Golay, "The merit factor of long low autocorrelation binary sequences," *IEEE Transactions on Information Theory*, vol. 28, no. 3, pp. 543–549, 1982.
25. T. Høholdt, H. E. Jensen, and J. Justesen, "Aperiodic correlations and the merit factor of a class of binary sequences," *IEEE Transactions on Information Theory*, vol. 31, no. 4, pp. 549–552, 1985.
26. T. Høholdt and H. E. Jensen, "Determination of the merit factor of Legendre sequences," *IEEE Transactions on Information Theory*, vol. 34, no. 1, pp. 161–164, 1988.
27. P. Borwein and K.-K. S. Choi, "Merit factors of polynomials formed by Jacobi symbols," *Canadian Journal of Mathematics*, vol. 53, no. 1, pp. 33–50, 2001.
28. M. J. E. Golay, "The merit factor of Legendre sequences," *IEEE Transactions on Information Theory*, vol. 29, no. 6, pp. 934–936, 1983.
29. M. G. Parker, "Even length binary sequence families with low negaperiodic autocorrelation," *Applied Algebra, Algebraic Algorithms and Error-Correcting Codes, AAECC-14 Proceedings*, vol. 2227, pp. 200–210, 2001.
30. T. Høholdt, "The merit factor of binary sequences," *Difference Sets, Sequences and Their Correlation Properties*, Eds. A. Pott et. al., pp. 227–237, 1999, The Netherlands, Kluwer Academic Publishers.
31. P. Borwein, K.-K. S. Choi, and J. Jedwab, "Binary sequences with merit factor greater than 6.34," *IEEE Transactions on Information Theory*, vol. 50, no. 12, pp. 3234–3249, 2004.
32. R. A. Kristiansen and M. G. Parker, "Binary sequences with merit factor >6.3," *IEEE Transactions on Information Theory*, vol. 50, no. 12, pp. 3385–3389, 2004.
33. S. Boztaş, "CDMA over QAM and other arbitrary energy constellations," *Communication Systems, IEEE International Conference on*, vol. 2, pp. 21.7.1–21.7.5, 1996.
34. C. RoBing and V. Tarokh, "A construction of OFDM 16-QAM sequences having low peak powers," *IEEE Transactions on Information Theory*, vol. 47, no. 5, pp. 2091–2094, 2001.
35. H. Lu and P. V. Kumar, "A unified construction of space-time codes with optimal rate-diversity tradeoff," *IEEE Transactions on Information Theory*, vol. 51, no. 5, pp. 1709–1730, 2005.
36. S. Boztaş, R. Hammons, and P. V. Kumar, "4-Phase sequences with near-optimum correlation properties," *IEEE Transactions on Information Theory*, vol. 38, no. 3, pp. 1101–1113, 1992.
37. A. R. Hammons Jr, P. V. Kumar, A. R. Calderbank, N. J. A. Sloane, and P. Sole, "The Z_4-linearity of Kerdock, Preparata, Goethals, and related codes," *IEEE Transactions on Information Theory*, vol. 40, no. 2, pp. 301–319, 1994.
38. T. Helleseth and P. V. Kumar, "Sequences with low correlation," in *Handbook of Coding Theory*, Eds. V. Pless and C. Huffman, 1998, New York, Elsevier Science Publishers.

39. P. V. Kumar, T. Helleseth, and A. R. Calderbank, "An upper bound for Weil exponential sums over Galois rings and applications," *IEEE Transactions on Information Theory*, vol. 41, no. 2, pp. 456–468, 1995.

40. P. Sole, "A quaternary cyclic code, and a family of quadriphase sequences with low correlation properties," *Proceedings of the Third International Colloquium on Coding Theory and Applications*, pp. 193–201, 1989.

41. K. Yang, T. Helleseth, P. V. Kumar, and A. G. Shanbhag, "On the weight hierarchy of Kerdock codes over Z_4," *IEEE Transactions on Information Theory*, vol. 42, no. 5, pp. 1587–1593, 1996.

42. M. Anand and P. V. Kumar, "Low-correlation sequences over the QAM constellation," *IEEE Transactions on Information Theory*, vol. 54, no. 2, pp. 791–810, 2008.

43. G. Garg, P. V. Kumar, and C. E. V. Madhavan, "Low correlation interleaved QAM sequences," *Information Theory, 2008. Proceedings. IEEE International Symposium on*, 2008.

44. G. Garg, P. V. Kumar, and C. E. V. Madhavan, "Two new families of low correlation interleaved QAM sequences," *Sequences and Their Applications, International Conference on*, 2008.

45. B. Long, P. Zhang, and J. Hu, "A generalized QS-CDMA system and the design of new spreading codes," *IEEE Transactions on Vehicular Technology*, vol. 47, no. 4, pp. 1268–1275, 1998.

46. X. H. Tang, P. Z. Fan, and S. Matsufuji, "Lower bounds on correlation of spreading sequence set with low or zero correlation zone," *Electronics Letters*, vol. 36, no. 6, pp. 551–552, 2000.

47. G. Gong, S. Golomb, and H.-Y. Song, "A note on low correlation zone signal sets," *IEEE Transactions on Information Theory*, vol. 53, no. 7, pp. 2575–2581, 2007.

48. J. Jang, J. No, and H. Chung, "A new construction of optimal p^2-ary low correlation zone sequences using unified sequences," *IEICE Transactions on Fundamentals of Electronics, Communications and Computer Sciences*, vol. 89, no. 10, pp. 2656–2661, 2006.

49. X. H. Tang and P. Z. Fan, "Large families of generalized d-form sequences with low correlations and large linear span based on the interleaved technique," *preprint*, 2004.

50. J. Chung and K. Yang, "New design of quaternary low-correlation zone sequence sets and quaternary hadamard matrices," *IEEE Transactions on Information Theory*, vol. 54, no. 8, pp. 3733–3737, 2008.

51. S. Kim, J. Jang, J. No, and H. Chung, "New constructions of quaternary low correlation zone sequences," *IEEE Transactions on Information Theory*, vol. 51, no. 4, pp. 1469–1477, 2005.

52. G. Gong and H.-Y. Song, "Two-tuple-balance of nonbinary sequences with ideal two-level autocorrelation," *Information Theory, 2003. Proceedings. IEEE International Symposium on*, p. 404, 29 Jun.–4 Jul. 2003.

53. S.-H. Kim, J.-S. No, H. Chung, and T. Helleseth, "New cyclic relative difference sets constructed from d-homogeneous functions with difference-balanced property," *IEEE Transactions on Information Theory*, vol. 51, pp. 1155–1163, March 2005.

54. G. Gong and H.-Y. Song, "Two-tuple balance of non-binary sequences with ideal two-level autocorrelation," *Discrete Applied Mathematics*, vol. 154, no. 18, pp. 2590–2598, 2006.

55. X. Tang and P. Fan, "A class of pseudonoise sequences over GF (P) with low correlation zone," *IEEE Transactions on Information Theory*, vol. 47, no. 4, pp. 1644–1649, 2001.

56. N. Y. Yu and G. Gong, "The perfect binary sequence of period 4 for low periodic and aperiodic autocorrelation," *Lecture Notes in Computer Science (LNCS)*, vol. 4893, pp. 37–49, 2007.

57. J. Jedwab, "A survey of the merit factor problem for binary sequences," *Sequences and their Applications - Proceedings of SETA*, vol. 3486, pp. 30–55, 2004.

58. Y. Kim, J. Jang, J. No, and H. Chung, "New design of low-correlation zone sequence sets," *IEEE Transactions on Information Theory*, vol. 52, no. 10, pp. 4607–4616, 2006.

59. X. Tang and P. Udaya, "New construction of low correlation zone sequences from Hadamard matrices," *preprint*, 2007.

60. J. Jang, J. No, H. Chung, and X. Tang, "New sets of optimal p-ary low-correlation zone sequences," *IEEE Transactions on Information Theory*, vol. 53, no. 2, pp. 815–821, 2007.

61. J. Chung, J. No, Y. Kim, J. Jang, and H. Chung, "Generalized extending method for construction of q-ary low correlation zone sequence sets," *Information Theory, 2008. Proceedings. IEEE International Symposium on*, pp. 1927–1930, 2008.

62. R. De Gaudenzi, C. Elia, and R. Viola, "Bandlimited quasi-synchronous CDMA: A novel satellite access technique for mobile and personal communication systems," *IEEE Journal on Selected Areas in Communications*, vol. 10, no. 2, pp. 328–343, 1992.
63. J. Jang, J. Chung, and J. No, "Quaternary low correlation zone sequence set with flexible parameters," *Information Theory, 2008. Proceedings. IEEE International Symposium on*, pp. 2767–2771, 2008.
64. J. Yang, X. Jin, K. Song, J. No, and D. Shin, "Multicode MIMO systems with quaternary LCZ and ZCZ sequences," *IEEE Transactions on Vehicular Technology*, vol. 57, no. 4, pp. 2334–2341, 2008.
65. H. Torii, M. Nakamura, and N. Suehiro, "A new class of polyphase sequence sets with optimal zero-correlation zones," *IEICE Transactions on Fundamentals of Electronics, Communications and Computer Sciences*, vol. 88, no. 7, pp. 1987–1994, 2005.
66. T. Hayashi and S. Matsufuji, "On optimal construction of two classes of ZCZ codes," *IEICE Transactions on Fundamentals of Electronics, Communications and Computer Sciences*, vol. 89, no. 9, pp. 2345–2350, 2006.
67. T. Hayashi, "Zero-correlation zone sequence set construction using an even-perfect sequence and an odd-perfect sequence," *IEICE Transactions on Fundamentals of Electronics, Communications and Computer Sciences*, vol. 90, no. 9, pp. 1871–1875, 2007.
68. T. Hayashi, "A novel class of zero-correlation zone sequence sets constructed from a perfect sequence," *IEICE Transactions on Fundamentals of Electronics, Communications and Computer Sciences*, vol. 91, no. 4, pp. 1233–1237, 2008.
69. Z. Zhou, X. Tang, and G. Gong, "A new class of sequences with zero or low correlation zone based on interleaving technique," *IEEE Transactions on Information Theory*, vol. 54, no. 9, pp. 4267–4273, 2008.
70. X. Tang and W. H. Mow, "A new systematic construction of zero correlation zone sequences based on interleaved perfect sequences," *preprint*, 2008.
71. F. MacWilliams and N. Sloane, "Pseudo-random sequences and arrays," *Proceedings of the IEEE*, vol. 64, pp. 1715–1729, Dec. 1976.
72. M. Antweiler, L. Bomer, and H.-D. Luke, "Perfect ternary arrays," *IEEE Transactions on Information Theory*, vol. 36, pp. 696–705, May 1990.
73. P. V. Kumar, R. A. Scholtz, and L. R. Welch, "Generalized bent functions and their properties," *Journal of Combinatorial Theory. Series A*, vol. 40, pp. 90–107, 1985.
74. N. Suehiro, "A signal design without co-channel interference for approximately synchronized CDMA systems," *IEEE Journal on Selected Areas in Communications*, vol. 12, no. 5, pp. 837–841, 1994.
75. P. Z. Fan, N. Suehiro, N. Kuroyanagi, and X. M. Deng, "Class of binary sequences with zero correlation zone," *Electronics Letters*, vol. 35, no. 10, pp. 777–779, 1999.
76. H. Torii, M. Nakamura, and N. Suehiro, "A new class of zero-correlation zone sequences," *IEEE Transactions on Information Theory*, vol. 50, pp. 559–565, Mar. 2004.
77. H. Torii and M. Nakamura, "Enhancement of ZCZ sequence set construction procedure," *IEICE Transactions on Fundamentals of Electronics, Communications and Computer Science*, vol. 90, no. 2, pp. 535–538, 2007.
78. D. Peng, P. Fan, and N. Suehiro, "Construction of sequences with large zero correlation zone," *IEICE Transactions on Fundamentals of Electronics, Communications and Computer Sciences*, vol. 88, no. 11, pp. 3256–3259, 2005.
79. X. Tong and Q. Wen, "New constructions of zcz sequence set with large family size," *Signal Design and Its Applications in Communications, 2007. IWSDA 2007. 3rd International Workshop on*, pp. 99–103, Sept. 2007.
80. T. Hayashi, "Binary zero-correlation zone sequence set construction using a primitive linear recursion," *IEICE Transactions on Fundamentals of Electronics, Communications and Computer Sciences*, no. 7, pp. 2034–2038, 2005.
81. T. Hayashi, "Ternary sequence set having periodic and aperiodic zero-correlation zone," *IEICE Transactions on Fundamentals of Electronics, Communications and Computer Sciences*, vol. 89, no. 6, pp. 1825–1831, 2006.

82. T. Hayashi, "Binary zero-correlation zone sequence set construction using a cyclic hadamard sequence," *IEICE Transactions on Fundamentals of Electronics, Communications and Computer Sciences*, vol. 89, no. 10, pp. 2649–2655, 2006.
83. T. Hayashi, "Binary zero-correlation zone sequence set constructed from an M-sequence," *IEICE Transactions on Fundamentals of Electronics, Communications and Computer Sciences*, vol. 89, no. 2, pp. 633–638, 2006.
84. T. Hayashi, "An integrated sequence construction of binary zero-correlation zone sequences," *IEICE Transactions on Fundamentals of Electronics, Communications and Computer Sciences*, vol. 90, no. 10, pp. 2329–2335, 2007.
85. T. Hayashi, "Zero-correlation zone sequence set constructed from a perfect sequence," *IEICE Transactions on Fundamentals of Electronics, Communications and Computer Sciences*, vol. 90, no. 5, pp. 1107–1111, 2007.

Chapter 4
Resource Allocation in Wireless Systems

Jon W. Mark and Lian Zhao

4.1 Introduction

With the increasing demand for wireless communication services, efficient and
effective radio resource management strategies that can cope with the time-varying
nature of the wireless channel are essential. Two of the main resource management
functions in a CDMA system are power control and rate allocation. This chapter
presents a unified approach for power distribution and allocation in a multirate wideband CDMA system. Adaptive schemes to perform power control and rate allocation
to enhance resource utilization are described.

It is well known that CDMA is interference limited [1–4]. Multiple-access interference (MAI) is the dominant parameter that governs the system capacity. High
target received power requires high transmit power, which leads to increased MAI.
Therefore, effective power distribution and allocation can minimize the effect of
MAI, thereby enhancing system utilization.

Power distribution and allocation as a research problem has been receiving much
attention in recent years, see, e.g., [4–6]. In these earlier works, relatively simple approaches have been used to solve the power distribution problem for uplink CDMA
transmission where the total disturbance[1] is assumed to obey a Gaussian distribution. Transmit power control incorporating throughput maximization is investigated
in [7]. In [8], the geometrical and topological properties of the capacity region are

Jon W. Mark
Department of Electrical and Computer Engineering
University of Waterloo, 200 University Ave., W., Waterloo, ON, Canada,
e-mail: jwmark@bbcr.uwaterloo.ca

Lian Zhao
Department of Electrical and Computer Engineering
Ryerson University, 350 Victoria Street, Toronto, Canada,
e-mail: lzhao@ee.ryerson.ca

[1] The term disturbance is used here to represent the intercell interference plus ambient noise.

V. Tarokh (ed.), *New Directions in Wireless Communications Research*,
DOI 10.1007/978-1-4419-0673-1_4,
© Springer Science+Business Media, LLC 2009

investigated in the context of optimal power allocation. In [9], the optimal power allocation problem is investigated using a utility function approach.

In this chapter, we present a unified approach to solve the power distribution problem using matrix analysis. It is shown that finding a solution for the power allocation problem involves the inversion of the traffic matrix. This is done by invoking the well-known Sherman–Morrison inverse formula for rank-1 updated matrices in linear algebra to derive closed-form expressions for the power distribution and the corresponding convergent conditions. Our results give the necessary and sufficient conditions to solve the power control problem. The approach and the results obtained are general in the sense that they are equally applicable to both uplink and downlink transmissions, even though in the downlink, the aggregate disturbance seen by the individual mobile users is normally different. The conditions for convergence of the resultant power distribution are then used to evaluate the capacity region of the system. Our results show that any form of power control or power distribution is indeed a function of the spread spectrum bandwidth, user data rates, and user quality of service (QoS) specifications [10–12]. Our results are quite generic, special cases of which are consistent with works reported in the literature. The feasibility analysis using the traffic demand approach instead of the conventional spectral analysis (eigenvalue problem) is shown to be more tractable, easier, and possesses good physical interpretations.

In a time-varying wireless channel, adaptive transmission that responds to the actual channel condition is effective and efficient. In general, the transmitter may adaptively calculate its data rate and/or transmission power according to the channel state information (CSI). Typical adaptive techniques include power adaptation to maintain the received power of each mobile at a desired level [13–15], coding adaptation [16, 17], modulation adaptation [18], spreading factor (data rate) adaptation [19], or any combination of these approaches [19–21].

The purpose of using closed-loop power control (CLPC) is to regulate the received power/SIR (signal-to-interference ratio) to be around a target level. CLPC enables the transmit power to track the time-varying channel to compensate for channel impairments. The price to pay for the improved tracking is higher average transmit power [15]. On the other hand, rate adaptation attempts to enhance throughput by changing the transmission rate while keeping the transmit power fixed. A higher data rate or a lower spreading factor will lead to throughput improvement. When the carrier-to-interference ratio (CIR) of the desired user is high, a lower spreading factor should be used to enhance throughput. As a result, we expect that joint rate/power adaptation will yield excellent system performance and transmit power consumption trade-off.

From an adaptive processing point of view, rate adaptation is normally done at the frame level, while CLPC adaptation is at the slot level [22], so that the frequency of power adaptation is higher than that of rate adaptation. The complementary roles of power control and rate allocation for system performance enhancement motivate our work on joint power and rate adaptation, recognizing that rate allocation saves transmission power and enhances throughput while power control tunes the received CIR statistics around a target level.

The remainder of the chapter is organized as follows. The power distribution problem and generalized closed-loop control system are described in Section 4.2. It is shown that the key to power allocation involves the inversion of the traffic matrix. An explicit expression of the inverse of the traffic demand matrix needed to solve the power allocation problem is obtained in Section 4.3. The results are applied in Section 4.4 to solve the power allocation problem and its convergent conditions for a system under different operating conditions. In Section 4.5, the optimal spreading factor, the basis for rate allocation in OSF-PC (optimal spreading factor-power control), is derived using a maximum throughput criterion. The greedy rate packing (GRP) algorithm, which sequentially allocates data rate as high as possible by sorting the channel gains of the users, is also described. In Section 4.6, we present the joint rate/power adaptation algorithms: OSF-PC and GRP-PC. Numerical results to illustrate the applications of our approach and algorithm are presented in Section 4.7. Finally, concluding remarks are given in Section 4.8.

4.2 System Model

• Model for power distribution law

In the first part of the chapter, the power distribution law and system feasibility are investigated. We are concerned with a generic cell in a wideband CDMA cellular system, which supports M users. The ensuing results are applicable to both uplink and downlink transmissions. For the uplink case, the transmitters are the mobiles and the receiver is the base station (BS), and for the downlink, the transmitter is the BS and the receivers are the mobiles. In the uplink case, the parameter S_j denotes the received power at the BS from mobile j. In the downlink, it denotes the transmission power dedicated to mobile j by the BS. Besides the power level, the jth mobile is specified by its required data rate, R_j, and a target SIR, denoted as γ_j^{tgt}. QoS satisfaction requires that the received $\gamma_j \geq \gamma_j^{tgt}$, with

$$\gamma_j = \frac{W}{R_j} \cdot \frac{S_j}{I_j}, \tag{4.1}$$

where W is the spread spectrum bandwidth and I_j is the total interference imposed on the desired signal.

For given values of R_j and W, a specification on γ_j is equivalent to a specification on S_j/I_j. Let S_j^* be the target received power at the receiver to achieve γ_j^{tgt}. Then S_j^* can be expressed as

$$S_j^* = \frac{1}{W} R_j \gamma_j^{tgt} I_j. \tag{4.2}$$

For satisfactory operation, the received signal power for the jth user must satisfy the inequality

$$S_j \geq S_j^* = \frac{1}{W} R_j \gamma_j^{tgt} I_j = \Gamma_j I_j, \tag{4.3}$$

where

$$\Gamma_j = \frac{1}{W} R_j \gamma_j^{tgt} \tag{4.4}$$

is the traffic demand of user j. The total interference on the jth user's signal, I_j, can be expressed as

$$I_j = \sum_{l=1, l \neq j}^{M} S_l + n_j = \sum_{l=1}^{M} S_l - S_j + n_j, \tag{4.5}$$

where n_j is the aggregate disturbance consisting of additive white Gaussian noise (AWGN) and intercell MAI. For downlink transmission, different users do not necessarily have the same AWGN level. Furthermore, different users are expected to experience a different level of intercell interference due to their different physical locations. As a result, the disturbance seen at different mobile users is normally different [23, 24]. For uplink transmission, the BS is the common receiver for all M mobile transmitters, with the same disturbance, n_j, for all $j = 1, \ldots, M$. However, for notational convenience for the unified matrix operation, we still use a heterogeneous vector, $\mathbf{n}$, to denote the disturbances. Substituting (4.5) into (4.3) and manipulating, we have the following inequalities:

$$S_1 - \Gamma_1 S_2 - \cdots - \Gamma_1 S_M \geq \Gamma_1 n_1$$
$$-\Gamma_2 S_1 + S_2 - \cdots - \Gamma_2 S_M \geq \Gamma_2 n_2$$
$$\vdots \quad \vdots$$
$$-\Gamma_M S_1 - \Gamma_M S_2 - \cdots + S_M \geq \Gamma_M n_M. \tag{4.6}$$

The right-hand sides of (4.6) represent the disturbance, while the left-hand sides represent signal strengths and intracell interferences. The inequalities mean that, in order for the signals to be discernable, the total strength generated by the transmitted signals must exceed the total disturbance. In matrix form (4.6) becomes

$$\Gamma_S \mathbf{S} \geq \Gamma_D \, \mathbf{n}, \tag{4.7}$$

which implies that

$$\mathbf{S} \geq (\Gamma_S)^{-1} \Gamma_D \, \mathbf{n} \tag{4.8}$$

if the matrix Γ_S is invertible and $(\Gamma_S)^{-1}$ is nonnegative, where

$$\Gamma_S = \begin{bmatrix} 1 & -\Gamma_1 & \cdots & -\Gamma_1 \\ -\Gamma_2 & 1 & \cdots & -\Gamma_2 \\ & & \vdots & \\ -\Gamma_M & -\Gamma_M & \cdots & 1 \end{bmatrix}, \tag{4.9}$$

with $\Gamma_1, \Gamma_2, \ldots, \Gamma_M$ all positive, is the traffic matrix, and $\Gamma_D = \mathrm{diag}[\Gamma_1, \Gamma_2, \ldots, \Gamma_M]$. The power vector, $\mathbf{S} = [S_1, S_2, \ldots, S_M]^T$, needs to be specified. The nonnegative disturbance vector is denoted as $\mathbf{n} = [n_1, n_2, \ldots, n_M]^T$.

The objective of power control is (a) to find a solution for vector $\mathbf{S}$ to specify the power distribution among the M users and (b) to minimize the total power in the cell, i.e., minimize $\sum_{j=1}^{M} S_j$. As indicated in (4.8), finding a solution for $\mathbf{S}$ may involve the inversion of the traffic matrix Γ_S.

Let $\mathbf{I}$ be the $M \times M$ identity matrix and $\Gamma_P = \mathbf{I} - \Gamma_S$ be defined as the complementary traffic matrix. It turns out that solving the power distribution problem can be transformed into one of examining the spectral radius, $r(\Gamma_P)$, of Γ_P. Let $\mathbf{e} = [1, 1, \ldots, 1]^T$ be an M-dimensional column vector of all 1's. The nonnegative complementary traffic matrix can be expressed as

$$\Gamma_P = \Gamma_D(\mathbf{e}\mathbf{e}^T - \mathbf{I}) = \begin{bmatrix} 0 & \Gamma_1 & \cdots & \Gamma_1 \\ \Gamma_2 & 0 & \cdots & \Gamma_2 \\ & & \vdots & \\ \Gamma_M & \Gamma_M & & 0 \end{bmatrix}. \qquad (4.10)$$

Then (4.7) can be written as

$$(\mathbf{I} - \Gamma_P)\mathbf{S} \geq \Gamma_D \mathbf{n}. \qquad (4.11)$$

Since Γ_P is a nonnegative matrix, the classical Perron–Frobenius theorem implies that the spectral radius $r(\Gamma_P)$ of Γ_P is an eigenvalue of Γ_P with a nonnegative eigenvector. Moreover, since Γ_S has negative off-diagonal entries and positive diagonal entries, the theory of M-matrices [25] implies that if $r(\Gamma_P) < 1$, then a nonnegative solution of (4.11) is given by [10]

$$\mathbf{S}^* = (\mathbf{I} - \Gamma_P)^{-1}\Gamma_D \mathbf{n}. \qquad (4.12)$$

In this case any solution $\mathbf{S}$ of (4.11) is bounded from below by $\mathbf{S}^*$, i.e., $\mathbf{S} \geq \mathbf{S}^*$. Hence, $\mathbf{S}^*$ gives the *minimal solution* of inequality (4.11). Using the concept of M-matrices, [11] succeeded in exploring the spectrum for the $M \times M$ matrix Γ_P and obtained an equivalent condition for $r(\Gamma_P) < 1$ in terms of the traffic demand for all the users in the system.

In this chapter, we present a complete solution of the power distribution problem using an approach based on the Sherman–Morrison formula in linear algebra to find a necessary and sufficient condition for the nonnegativity of the inverse of Γ_S. We avoid the intractable eigenvalue problem for a more friendly traffic demand manipulation and provide explicit expressions for the solution of the power control problem (4.7) with or without power constraints. We then apply the convergent conditions to specify the capacity region of the system.

- Model for closed-loop control system

As the second part of this chapter, we will apply the power distribution law to a closed-loop control system in cooperation with rate adaptation/scheduling. The power distribution model is generalized as follows.

At each mobile station (MS), the source at the MS generates a sequence of fixed-length packets, each L_p bits in length. After error control coding, the bits in each of the packets generated by each source are randomized using an interleaver. The rate-controlled symbols are then transmitted at the symbol rate W/N, where W is the chip rate of the spreading function and N is the spreading factor. Assuming a fixed chip rate, the symbol duration is directly a function of the spreading factor N. When power control is applied, the transmit power is updated by a fixed step size and a power control command at the end of each power control interval, T_p. Otherwise, the transmit power is kept constant. For the analysis in the sequel, the buffer is assumed non-empty at all times, so that the users are persistently transmitting.

At the BS, the received signal first goes through a matched filter receiver for despreading. Reproductions of the transmitted information bits are retrieved after deinterleaving and decoding. When power control is applied, the estimated CIR is compared with a target, ξ^*, to generate a power control command.

We consider error-sensitive traffic, which is assumed to be error free by the use of retransmissions. Therefore, after decoding the received packet, the BS may request the transmitter to retransmit the packet if it contains errors. The throughput can be controlled through the selection of the spreading gain and the retransmission probability. The estimate of CIR averaged over a frame interval is used to specify whether the transmission rate needs to be updated in the next frame transmission if rate adaptation is applied.

There are three possible feedback variables from the BS to the MS. The first is the power control command at an interval T_p (equal to one slot). The second is the CIR estimate at a frame length T_f ($= 15\,T_p$), which is used to specify the transmission rate if rate adaptation is applied. The third feedback is an indicator to request retransmission when errors are detected after decoding.

The Suzuki process [26] is a statistical model that has been developed for the land mobile radio channel on the assumption that the local mean of the Rayleigh process follows a log-normal statistic and accounts for the effects of shadowing. Thus, the stationary Suzuki process $z(n)$ is a product of a Rayleigh process $\psi(n)$ and a log-normal process $\zeta(n)$ at time instant n,

$$z(n) = \psi(n) \cdot \zeta(n), \tag{4.13}$$

where $\psi(n)$ and $\zeta(n)$ are called the small-scale-fading component and the large-scale-fading component, respectively. A sample of the composite probability density function of the received signal envelope over time is given by [27]

$$f(z) = \int_0^\infty \frac{z}{P_0} \exp\left(-\frac{z^2}{2P_0}\right) \frac{1}{\sqrt{2\pi}\sigma_\zeta P_0} \exp\left[-\frac{\left(\ln(P_0) - m_\zeta\right)^2}{2\sigma_\zeta^2}\right] dP_0, \tag{4.14}$$

where m_ζ and σ_ζ are, respectively, the values (in dB) of the area mean signal power and standard deviation for log-normal process.

4.3 The Inverse of Γ_S

To calculate the inverse of Γ_S, we use the following Sherman–Morrison formula [25], the proof of which is via direct computation.

Lemma 4.1. *Let A be an invertible $n \times n$ matrix and let x and y be n-dimensional vectors. Then $A + xy^T$ is invertible if and only if $y^T A^{-1} x \neq -1$. Moreover, if $y^T A^{-1} x \neq -1$, then*

$$(A + xy^T)^{-1} = A^{-1} - \frac{1}{1 + y^T A^{-1} x} A^{-1} xy^T A^{-1}. \qquad (4.15)$$

Let

$$c \equiv 1 - \sum_{i=1}^{M} \frac{\Gamma_i}{\Gamma_i + 1} \qquad (4.16)$$

be a parameter that captures the traffic demand. The parameter c governs the invertibility of Γ_S.

Theorem 4.1. *Γ_S is invertible if and only if $c \neq 0$. Moreover, if $c \neq 0$, then*

$$(\Gamma_S)^{-1} = \left[I + \frac{1}{c}(I + \Gamma_D)^{-1} \Gamma_D ee^T \right](I + \Gamma_D)^{-1}. \qquad (4.17)$$

Proof. Since

$$\Gamma_S = I - \Gamma_P = I - \Gamma_D(ee^T - I) = I + \Gamma_D - \Gamma_D ee^T,$$

using Lemma 4.1 with $A = I + \Gamma_D$, $x = -\Gamma_D e$, and $y = e$ so that $\Gamma_S A + xy^T$, we see that Γ_S is invertible if and only if $e^T(I + \Gamma_D)^{-1}(-\Gamma_D e) \neq -1$ or equivalently, $c \neq 0$. Now, if $c \neq 0$, then formula (4.15) gives

$$(\Gamma_S)^{-1} = \left[(I + \Gamma_D) - \Gamma_D ee^T \right]^{-1} = (I + \Gamma_D)^{-1}$$

$$- \frac{(I + \Gamma_D)^{-1}(-\Gamma_D e)e^T(I + \Gamma_D)^{-1}}{1 + e^T(I + \Gamma_D)^{-1}(-\Gamma_D e)}$$

$$= (I + \Gamma_D)^{-1} + \frac{(I + \Gamma_D)^{-1} \Gamma_D ee^T(I + \Gamma_D)^{-1}}{1 - \sum_{i=1}^{M} \frac{\Gamma_i}{\Gamma_i + 1}}$$

$$= \left[I + \frac{1}{c}(I + \Gamma_D)^{-1} \Gamma_D ee^T \right](I + \Gamma_D)^{-1}.$$

Carrying out the right-hand side multiplication of (4.17), we obtain an explicit expression for $(\Gamma_S)^{-1}$ as

$$(\Gamma_S)^{-1} = \frac{1}{c} \begin{bmatrix} (\frac{\Gamma_1}{\Gamma_1+1}+c)\frac{1}{\Gamma_1+1} & \cdots & \frac{\Gamma_1}{(\Gamma_1+1)(\Gamma_M+1)} \\ \frac{\Gamma_2}{(\Gamma_2+1)(\Gamma_1+1)} & \cdots & \frac{\Gamma_2}{(\Gamma_2+1)(\Gamma_M+1)} \\ \vdots & & \\ \frac{\Gamma_M}{(\Gamma_M+1)(\Gamma_1+1)} & \cdots & (\frac{\Gamma_M}{\Gamma_M+1}+c)\frac{1}{\Gamma_M+1} \end{bmatrix}. \tag{4.18}$$

A consequence of Theorem 4.1 is the following:

Corollary 4.1. Γ_S *is invertible and* $(\Gamma_S)^{-1}$ *is a positive matrix if and only if* $c > 0$.

4.4 Convergence of Power Distribution Law

Now we apply the result in the previous section to solve (4.7) for the optimal power vector. In the following we investigate the solvability of (4.7) for different cases, including the cases where the system has (a) zero disturbance ($\mathbf{n} = \mathbf{0}$), (b) nonzero disturbance ($\mathbf{n} \neq \mathbf{0}$), and (c) different power constraints. We evaluate the capacity region based on the convergent conditions.

4.4.1 With Zero Disturbance

In the absence of any disturbance, inequality (4.7) becomes

$$\Gamma_S S \geq \mathbf{0}. \tag{4.19}$$

Solving (4.19) with strict inequality, the existence of a positive power vector is equivalent to the condition $c > 0$ from Theorem 1.5.2 of [28] by using the theory of M-matrices. However, we give a direct proof of the following theorem in which an explicit solution is also obtained.

Theorem 4.2. $\Gamma_S S > \mathbf{0}$ *for a positive power vector* S *if and only if* $c > 0$. *In this case, each solution* S *to the inequality* $\Gamma_S S > \mathbf{0}$ *is positive, and the vector* $S = (\Gamma_S)^{-1}\mathbf{u}$ *is a solution for each positive vector* $\mathbf{u}$. *In particular, a positive solution* S *is given by* $S = (\Gamma_S)^{-1}\mathbf{e}$.

Proof. If $c > 0$, then $(\Gamma_S)^{-1}$ exists and is a positive matrix by Corollary 4.1. Let $S = (\Gamma_S)^{-1}\mathbf{e}$. Then $S > \mathbf{0}$ and $\Gamma_S S = \Gamma_S(\Gamma_S)^{-1}\mathbf{e}\mathbf{e} > \mathbf{0}$.

Now suppose that there is a positive power vector S such that $\Gamma_S S > \mathbf{0}$. Then $S > \Gamma_P S$, which implies that $(\mathbf{I} + \Gamma_D)S > \Gamma_D \mathbf{e}\mathbf{e}^T S = (\mathbf{e}^T S)\Gamma_D \mathbf{e}$. Thus, $(\mathbf{I} + \Gamma_D)^{-1}\Gamma_D \mathbf{e} < (\mathbf{e}^T S)^{-1}S$. It follows that

$$\mathbf{e}^T(\mathbf{I} + \Gamma_D)^{-1}\Gamma_D \mathbf{e} < (\mathbf{e}^T S)^{-1}\mathbf{e}^T S = 1,$$

which means that $c > 0$.

Finally, suppose that $c > 0$. If $\Gamma_S S > 0$ for a vector S, then $S = (\Gamma_S)^{-1}\Gamma_S S > 0$ since $(\Gamma_S)^{-1}$ is positive. Given any vector $u > 0$, then $\Gamma_S(\Gamma_S)^{-1}u = u > 0$. This means that $S = (\Gamma_S)^{-1}u$ is a solution. This completes the proof. $\qquad\square$

The following theorem gives more general results.

Theorem 4.3. *The homogeneous inequality system (4.19) has a non-trivial nonnegative solution S if and only if $c \geq 0$. In the case of $c > 0$, all the vectors of the form $(\Gamma_S)^{-1}u$ with $u \geq 0$ are solutions of (4.19). In particular, every column vector of $(\Gamma_S)^{-1}$ is a solution of (4.19).*

Proof. For the sufficiency part, it is enough to assume that $c = 0$ since the case $c > 0$ has been proved in Theorem 4.2. Then $S = (I + \Gamma_D)^{-1}\Gamma_D e$ gives a positive solution of (4.19) since $\Gamma_S S = (I + \Gamma_D - \Gamma_D e e^T)(I + \Gamma_D)^{-1}\Gamma_D e = \Gamma_D e - \Gamma_D e e^T(I + \Gamma_D)^{-1}\Gamma_D e = \Gamma_D e - \Gamma_D e = 0$.

Now suppose that there is a nonzero and nonnegative power vector S such that $\Gamma_S S \geq 0$. Then $(I + \Gamma_D)S \geq \Gamma_D e e^T S$, which implies that $(I + \Gamma_D)^{-1}\Gamma_D e \leq (e^T S)^{-1}S$ since $e^T S > 0$, so

$$e^T(I + \Gamma_D)^{-1}\Gamma_D e \leq (e^T S)^{-1}e^T S = 1.$$

That is, $c \geq 0$. The last conclusion of the theorem is obvious from the proof of Theorem 4.2. $\qquad\square$

Strictly speaking, there is no minimal solution to the homogeneous inequality (4.19). Theorems 4.2 and 4.3 allow us to formulate the following procedure to perform power allocation.

Power Allocation Procedure

Step 1. From the traffic matrix, calculate $(\Gamma_S)^{-1}$ using (4.18) and (4.16).
Step 2. Let g_{ij} denote the value at the ith row and jth column of $(\Gamma_S)^{-1}$.
Step 3. Sum over the rows,

$$G_j = \sum_{i=1}^{M} g_{ij}, \qquad j = 1, \ldots, M.$$

Step 4. Find the minimal value of $G_j, j = 1, \ldots, M$, say $j = k$ and return the index k.
Step 5. Then the kth column is the optimal power solution: $S_i^* = g_{ik}$.

If s_0 is the minimal required power level for a proper detection of the desired signal, then the power vector can be scaled as

$$S = s_0 \cdot \frac{S^*}{S_{min}^*}, \tag{4.20}$$

where S_{min}^* is the minimum value of the components of the vector S^*.

4.4.2 With Nonzero Disturbance

Let Φ_i be defined as the *normalized traffic demand*:

$$\Phi_i = \frac{\Gamma_i}{\Gamma_i + 1}, \quad i = 1, 2, \ldots, M, \tag{4.21}$$

which is a monotonically increasing function of Γ_i. When Γ_i changes in the range $(0, \infty)$, Φ_i has a corresponding value in the range $(0, 1)$.

The following result gives an equivalent condition for inequality (4.7) to have a positive solution and an explicit expression of its optimal solution.

Theorem 4.4. *Suppose* $\mathbf{n} \neq \mathbf{0}$. *Then (4.7) has a positive solution if and only if* $c > 0$. *If* $c > 0$, *then any solution* $\mathbf{S}$ *to (4.7) satisfies the inequality* $\mathbf{S} \geq \mathbf{S}^*$, *where the minimal positive solution* $\mathbf{S}^*$ *of (4.7) has the expression*

$$\mathbf{S}^* = \left[\mathbf{I} + \frac{1}{c}(\mathbf{I} + \Gamma_D)^{-1}\Gamma_D \mathbf{e}\mathbf{e}^{\mathbf{T}} \right] (\mathbf{I} + \Gamma_D)^{-1}\Gamma_D \mathbf{n} \tag{4.22}$$

and so its ith component is

$$S_i^* = \frac{\Phi_i}{c} \left[(\Phi_i + c)\, n_i + \sum_{j \neq i} \Phi_j n_j \right], \quad i = 1, 2, \ldots, M. \tag{4.23}$$

Proof. If $\mathbf{S}$ is a positive solution of (4.7), then $(\mathbf{I} + \Gamma_D - \Gamma_D \mathbf{e}\mathbf{e}^{\mathbf{T}})\mathbf{S} \geq \Gamma_D \mathbf{n} \geq 0$. Thus,

$$(\mathbf{I} + \Gamma_D)\mathbf{S} \geq \Gamma_D \mathbf{e}\mathbf{e}^T \mathbf{S} = (\mathbf{e}^T \mathbf{S})\Gamma_D \mathbf{e}$$

and the strict inequality holds for at least one component in the above inequality since $\Gamma_D \mathbf{e} \neq \mathbf{0}$, that is,

$$[(\mathbf{I} + \Gamma_D)\mathbf{S}]_i > \left[(\mathbf{e}^T \mathbf{S})\Gamma_D \mathbf{e} \right]_i$$

for at least one i. It follows that $(\mathbf{I}+\Gamma_D)^{-1}\Gamma_D \mathbf{e} \leq (\mathbf{e}^T\mathbf{S})^{-1}\mathbf{S}$ and $\left[(\mathbf{I} + \Gamma_D)^{-1}\Gamma_D \mathbf{e} \right]_i < \left[(\mathbf{e}^T\mathbf{S})^{-1}\mathbf{S} \right]_i$. Therefore

$$c = 1 - \mathbf{e}^{\mathbf{T}}(\mathbf{I} + \Gamma_D)^{-1}\Gamma_D \mathbf{e} > 1 - (\mathbf{e}^T\mathbf{S})^{-1}\mathbf{e}^T\mathbf{S} = 0.$$

Conversely, suppose that $c > 0$. Then $(\Gamma_S)^{-1}$ is a positive matrix. Let $\mathbf{S}^* = (\Gamma_S)^{-1}\Gamma_D \mathbf{n}$. Then $\mathbf{S}^*$ is a positive solution of (4.7) since $\Gamma_S \mathbf{S}^* = \Gamma_S(\Gamma_S)^{-1}\Gamma_D \mathbf{n} = \Gamma_D \mathbf{n}$, and the expressions (4.22) and (4.23) are direct results from Theorem 4.1.

Finally, if $c > 0$ and $\mathbf{S}$ is any solution of (4.7), then

$$\mathbf{S} = (\Gamma_S)^{-1}\Gamma_S\mathbf{S} \geq (\Gamma_S)^{-1}\Gamma_D \mathbf{n} \equiv \mathbf{S}^*,$$

so $\mathbf{S}$ is actually a positive solution and $\mathbf{S}^*$ is the minimal positive solution of (4.7).

Remark 1. For the uplink transmission, the disturbance vector can be written as $\mathbf{n} = \xi \mathbf{e}$ for some positive number ξ, i.e., all the signals are experiencing the same level of disturbance, then the minimal power solution is

$$\mathbf{S}^* = \frac{\xi}{c}(\mathbf{I} + \Gamma_D)^{-1}\Gamma_D \mathbf{e},$$

a result which is consistent with Lemma 3 in [6].

4.4.3 With Power Constraints

In practice we often need to solve (4.7) with some constraints. Here we give the results obtained by solving (4.7) with different power constraints on uplink and downlink transmissions. We will show that with the addition of these power constraints, the necessary and sufficient condition for the existence of a feasible power solution becomes more stringent. However, as long as the system is feasible, the optimal power levels are allocated in the same way as those without these power constraints.

- Uplink Transmission

For uplink transmission, the maximum transmit power of a mobile is normally constrained by

$$S_i \le \bar{S}_i, \quad 1 \le i \le M, \tag{4.24}$$

where $\bar{S}_i$ is the maximum allowable power for the ith user. Under constraint (4.24) optimal power allocation is given by the following theorem.

Theorem 4.5. *Suppose $\mathbf{n} \ne \mathbf{0}$. Then system (4.7) constrained by (4.24) has a positive solution if and only if for $i = 1, 2, \ldots, M$,*

$$0 < \frac{\sum_{j=1}^{M} \Phi_j n_j}{1 - \sum_{j=1}^{M} \Phi_j} \le \frac{\bar{S}_i}{\Phi_i} - n_i. \tag{4.25}$$

Proof. Let $\mathbf{S} > \mathbf{0}$ satisfy both (4.7) and (4.24). Then $c > 0$ by Theorem 4.4. Since $\mathbf{S}^*$ given by (4.22) is the minimal solution of (4.7), $\mathbf{S}^* \le \mathbf{S} \le \bar{\mathbf{S}}$. Thus,

$$\left[\mathbf{I} + \frac{1}{c}(\mathbf{I} + \Gamma_D)^{-1}\Gamma_D \mathbf{e}\mathbf{e}^{\mathsf{T}} \right] (\mathbf{I} + \Gamma_D)^{-1}\Gamma_D \mathbf{n} \le \bar{\mathbf{S}}.$$

Now (4.25) follows immediately from the definition of c given by (4.16). Conversely, if (4.25) is true, then $c > 0$. It is clear that $\mathbf{S}^*$ is a positive solution of (4.7) subject to the constraint of (4.24). $\qquad \square$

Remark 2. When we apply the condition $\mathbf{n} \xi \mathbf{e}$ for uplink transmission, the necessary and sufficient condition for (4.7) and (4.24) to have a positive solution is reduced to

$$\sum_{j=1}^{M} \Phi_j \leq 1 - \max_i \left(\frac{\xi \Phi_i}{\bar{S}_i} \right), \qquad i = 1, 2, \ldots, M. \qquad (4.26)$$

The right-hand side of inequality (4.26) is constrained by the maximal ratio of $\Phi / \bar{S}$, while the left-hand side is the sum of the normalized traffic demand. The higher the normalized traffic demand, the higher the amount of system resources it occupies, leaving a smaller space for the system to support other users.

- Downlink Transmission

  For downlink transmission, the total transmit power of a BS is limited by

$$\sum_{i=1}^{M} S_i \leq S_T, \qquad (4.27)$$

where $S_T > 0$ is the maximum total power of a BS. Then, the necessary and sufficient condition for the existence of a feasible solution is given by the following theorem.

Theorem 4.6. *Suppose* $\mathbf{n} \neq \mathbf{0}$. *Then system (4.7) constrained by (4.27) has a positive solution if and only if*

$$0 < \frac{\sum_{i=1}^{M} \Phi_i n_i}{1 - \sum_{i=1}^{M} \Phi_i} \leq S_T. \qquad (4.28)$$

In particular, if $\mathbf{n} = \xi \mathbf{e}$, *then (4.28) becomes*

$$\sum_{i=1}^{M} \Phi_i \leq \frac{S_T}{S_T + \xi}. \qquad (4.29)$$

Proof. Let $\mathbf{S} > \mathbf{0}$ satisfy both (4.7) and (4.27). Then $c > 0$ by Theorem 4.4. Since $\mathbf{S}^*$ given by (4.22) is the minimal solution of (4.7), $\mathbf{e}^T \mathbf{S}^* \leq \mathbf{e}^T \mathbf{S} \leq \mathbf{S}_T$. Thus,

$$\mathbf{e}^T \left[\mathbf{I} + \frac{1}{c} (\mathbf{I} + \Gamma_D)^{-1} \Gamma_D \mathbf{e} \mathbf{e}^T \right] (\mathbf{I} + \Gamma_D)^{-1} \Gamma_D \mathbf{n} \leq S_T.$$

Since the left-hand side of the above inequality equals

$$\mathbf{e}^T (\mathbf{I} + \Gamma_D)^{-1} \Gamma_D \mathbf{n} \left[1 + \frac{1}{c} (1 - c) \right] = \frac{\mathbf{e}^T (\mathbf{I} + \Gamma_D)^{-1} \Gamma_D \mathbf{n}}{c},$$

(4.28) is true. Conversely, if (4.28) is satisfied, then $c > 0$, and so the minimal solution $\mathbf{S}^*$ of (4.7) is well defined. It is obvious that $\mathbf{S}^*$ satisfies (4.27).   □

4.4.4 Capacity Analysis

It is shown in Section 4.3 that the necessary and sufficient condition for the existence of a positive inverse matrix for Γ_S is that $c > 0$, which implies that $\sum_{i=1}^{M} \Phi_i < 1$. If all the M users in the system have the same traffic demand, i.e., $\Phi_i = \Phi$ for all i, then the above condition implies that $\Phi < 1/M$. Therefore, for this system, each user is allowed to occupy less than $1/M$ of the system resource. For this reason Φ is defined as the normalized traffic demand.

Suppose that there are K classes among the M users and the traffic demand (Φ or Γ) is the same for all the users within each class. Let Φ_k denote the normalized traffic demand for the kth traffic class. The capacity $\mathcal{C}$ is defined as the K-dimensional space spanned by the maximum number of users supported in each class:

$$\mathcal{C} = [N_1, \ldots, N_k, \ldots, N_K]^T,$$

where N_k is the number of users in the kth class. For simplicity, we further suppose that in downlink transmission, all the users are experiencing the same disturbance level ξ, the same situation as in the uplink. With the condition $c > 0$, (4.26) and (4.29) can be used to evaluate the system capacity for different situations.

- When there are no power constraints, the power control convergent condition $c > 0$ is equivalent to the inequality

$$\sum_{k=1}^{K} N_k \Phi_k < 1. \tag{4.30}$$

- When the system is constrained by an upper power level as in (4.24) for uplink transmission, by applying (4.26), we have

$$\sum_{k=1}^{K} N_k \Phi_k < 1 - \max_j \left(\frac{\xi \Phi_j}{\bar{S}_j} \right), \qquad 1 \le j \le K. \tag{4.31}$$

- When the system is constrained by the total transit power as in (4.27) for downlink transmission, using (4.29), we have

$$\sum_{k=1}^{K} N_k \Phi_k \le \frac{S_T}{S_T + \xi}. \tag{4.32}$$

It is clear that the right-hand side of (4.31) and (4.32) is less than that of (4.30). Furthermore, (4.31) and (4.32) reduce to (4.30) as $\bar{S}$ and S_T tend to infinity, respectively. Therefore, power constraints have the effect of reducing the system capacity.

4.5 Optimal Data Rate Allocation

System throughput is a function of the spreading factor, N, or equivalently, the applied data rate. In the following two sections, we will analyze data rate allocation/scheduling in a closed-loop control system for the purpose of maximizing system throughput.

4.5.1 Assumptions

Assuming that user 1 is the desired user and users 2 to M are interferers, the received SIR for the desired user is given by

$$\gamma_1 = N_1 \cdot \frac{h_1 P_1}{\sum_{j=2}^{M} h_j P_j + \sigma^2} = N_1 \cdot \xi_1 = \frac{W}{R_1} \cdot \xi_1, \qquad (4.33)$$

where P_j and h_j are, respectively, the transmit power and the channel gain of user j. $N_1 = W/R_1$ is the spreading factor for the desired user and σ^2 is the ambient noise power. The term ξ_1, which represents the CIR of the desired user, is given by

$$\xi_1 = \frac{h_1 P_1}{\sum_{j=2}^{K} h_j P_j + \sigma^2},$$

which depends only on the received power vector and the background noise level. The difference between SIR and CIR is the processing gain N (i.e., $\Gamma = \xi N$).

Let $g(\gamma)$ denote the relationship between BER (P_b) and SIR (γ), which is a function of the modulation and coding schemes used. We model the BER as an exponentially decaying function [29]:

$$P_b = g(\gamma) = c_1 \exp(-c_2 \gamma), \qquad (4.34)$$

where c_1 and c_2 are parameters which can be adjusted to match a particular modulation/coding scheme. With different modulation/coding schemes, the mapping between SIR and BER may change; but this does not have an effect on the ensuing analytical results. With interleaving/deinterleaving, the bit errors are assumed to be independent. The packet retransmission probability for error-free packet reception is thus

$$P_r = 1 - (1 - P_b)^{L_p r_c}, \qquad (4.35)$$

where r_c is the channel coding code rate and L_p is the packet length.

4.5.2 Optimal Spreading Factor (OSF) Selection

The OSF algorithm presented in this section is based on the assumption that rate adaptation is continuous. The throughput, η, is defined as the average number of information bits successfully transmitted per second and is given by

$$\eta = \frac{r_c W}{N}(1 - P_r) = \frac{r_c W}{N}(1 - c_1 \exp(-c_2 \xi N))^{L_p r_c},\qquad(4.36)$$

where $(1 - P_r)$ is the probability of error-free packet reception. Let $g(\xi N) = c_1 \exp(-c_2 \xi N)$. The optimal spreading factor, N^*, can be obtained by taking the first derivative of (4.36) with respect to N and equating the result to zero:

$$\frac{\partial \eta}{\partial N} = -\frac{r_c W}{N^*}[1 - g(\xi N^*)]^{L_p r_c} \times \left[\frac{1}{N^*}(1 - g(\xi N^*)) + L_p r_c \cdot \xi \cdot g'(\xi N^*)\right] = 0.$$

Since the first term is strictly non-zero, N^* can be solved by equating the second term to zero:

$$1 - g(\xi N^*) + L_p r_c \cdot \xi N^* \cdot g'(\xi N^*) = 0.\qquad(4.37)$$

Substituting $g(\xi N^*) = c_1 \exp(-c_2 \xi N^*)$ and $\gamma^* = \xi N^*$ in (4.37) yields

$$\exp(c_2 \gamma^*) = c_1 + L_p r_c \cdot c_1 c_2 \cdot \gamma^*.\qquad(4.38)$$

Note that the left-hand side is an exponential function, while the right-hand side is a linear function of γ^*. For a given set of system specifications, γ^* can be obtained by solving (4.38), and treating it as a constant. The optimal spreading factor is uniquely specified as

$$N^* = \frac{\gamma^*}{\xi}.\qquad(4.39)$$

For practical applications, the value of the applied spreading factor is normally restricted to a set of integer numbers. In this case, the sub-optimal spreading factor is selected from the available set, which is the closest integer to the optimal solution given by (4.39).

4.5.3 Rate Selection for GRP

The greedy rate packing (GRP) algorithm described in [30] is a sequential heuristic method suitable for a discrete rate situation. A user with a high link gain is assigned as high a rate as possible. The GRP yields high system throughput. Assume that each user can be assigned a discrete rate from the set $\mathbf{R} = \{r^{(1)*}, r^{(2)*}, \ldots, r^{(k)*}\}$, with the condition, $r^{(1)*} < r^{(2)*} < \cdots < r^{(k)*}$, where k is the number of available rates which the users can be assigned. The corresponding set of discrete target CIR is $\Omega = \{\xi^{(1)*}, \xi^{(2)*} \ldots \xi^{(k)*}\}$, with the condition $\xi^{(1)*} < \xi^{(2)*} < \cdots < \xi^{(k)*}$.

The relationship between $\mathbf{R}$ and the target CIR can be obtained from (4.33) and shown as

$$\frac{r^{(1)*}}{\xi^{(1)*}} = \frac{r^{(2)*}}{\xi^{(2)*}} = \cdots = \frac{r^{(k)*}}{\xi^{(k)*}} = \frac{W}{\gamma^*}, \qquad (4.40)$$

where γ^* is the common target SIR, which is a tunable parameter. In this chapter, we use the optimal SIR value given by (4.38). From (4.40), we can see that the target CIR, $\xi^{(i)*}$, is another parameter to match the effective data rate. Thus, the rate assignment and target CIR assignment are interchangeable.

Based on earlier power distribution analysis, for uplink transmission, the necessary and sufficient condition for nonnegative power allocation is

$$\sum_{j=1}^{K} \frac{\xi_j}{1 + \xi_j} \leq 1 - \max_{1 \leq j \leq K} \left[\frac{\frac{\xi_j}{1+\xi_j}}{\frac{h_j P_{\max}}{\sigma^2}} \right], \qquad (4.41)$$

where $P_{\max}$ is the maximum allowed transmit power level.

The GRP algorithm [30] uses the necessary and sufficient condition (4.41) to allocate the target CIR, ξ^*, and the corresponding data rate of the users with the aid of channel state information. The principle is to give a higher priority to the user(s) with better channel condition to enhance system throughput.

4.6 Joint Rate and Power Adaptation

Closed-loop power control (CLPC) can be regarded as an inner-loop control mechanism, while target CIR allocation, which achieves rate adaptation at the frame level, serves as an outer-loop control mechanism to enhance the efficiency of radio resource management.

4.6.1 OSF-PC

Let ρ_1 denote the CIR of the pilot signal of the desired user, which is not power controlled and is specified by the channel fading gains of the users:

$$\rho_1 = \frac{h_1}{\sum_{i=2}^{K} h_i}. \qquad (4.42)$$

For rate adaptation, at the end of the mth frame, the average pilot CIR is estimated by averaging the pilot CIR over the most recent frame (15 slots in length),

$$\bar{\rho}_1 = \frac{1}{15} \sum_{i=1}^{15} \rho_1^i, \qquad (4.43)$$

where the superscript i denotes the index of the slot. The resultant average CIR is used as ξ in (4.39) to calculate the spreading factor for the next transmission frame and feeds it back to the MS. This value, $\xi^*[m+1]$, where ($[m+1]$ denotes the index of the frame, will be set as the target CIR value for the next frame.

At the nth slot, the received data CIR (ξ) is estimated at the receiver and compared to the target ξ^* to generate power control command to exercise closed-loop power control. When the CIR value is above the target, the power control command (PCC) sent to the transmitter is $pcc[n] = -1$; when it is below the target, the PCC sent is $pcc[n] = +1$. The transmit power at the $(n+1)$th slot, $P[n+1]$, can be computed iteratively at the beginning of this slot using the following steps:

$$pcc[n] = \text{sgn}(\xi^*[m] - \text{CIR}[n]) \text{ and } P[n+1] = P[n] + \Delta \cdot pcc[n], \quad (4.44)$$

where Δ is the applied step size in dB, and $\text{sgn}(\cdot)$ is the signum function. Furthermore, the updated transmit power level is limited by

$$P[n+1] = \begin{cases} P_{\max} & \text{if} \quad P[n+1] > P_{\max} \\ P_{\min} & \text{if} \quad P[n+1] < P_{\min} \end{cases}, \quad (4.45)$$

where $P_{\max}$ and $P_{\min}$ are, respectively, the maximum and minimum allowed transmit powers for desirable signal detection. Adaptive processing involves (i) the determination of the spreading factor based on the pilot CIR to achieve data rate adaptation, and (ii) the estimation of the received data CIR at each slot to implement CLPC. The rationale is that by using CLPC, the resultant CIR is forced to be close to the target within each frame. This target has been used to specify the optimal spreading factor for this frame. In this way, we make the rate adaptation works as an outer-loop control and CLPC works as an inner-loop control.

Compared with the conventional way of doing closed-loop power control, there are two differences. The first is that the CIR target varies, leading to a replacement of ξ^* in (4.44) by $\xi^*[m]$ for the mth frame. This is a consequence of using rate adaptation at the frame level. The second difference is that we reset the transmit power to the default power level, assumed to be 0 dBm, at the beginning of each frame when the CIR target is changed. This is also a consequence of changing the target CIR in each frame. In this way, we can control the dynamic range of the transmit power and ensure that the computation of the transmit power will not diverge. Furthermore, this power resetting in conjunction with rate adaptation brings a dramatic average transmit power reduction, which will be illustrated in the numerical results.

4.6.2 GRP-PC

Power adaptation serves a dual role in CDMA systems: *power allocation* and *closed-loop power control*. Power allocation is an important resource management function for multiclass systems and is used to specify the target received power levels to satisfy the QoS requirement and peak power restrictions of all the users in the system, whereas CLPC is to compensate for the signal impairment to make the received signal strength (or SIR level) to maintain at the target level. In OSF-PC, the functionality of power allocation has not been exploited, whereas in GRP-PC, at the beginning of each frame, an initial power vector is specified by solving the power allocation problem while satisfying the target SIR requirement.

GRP judiciously allocates data rate and target CIR value (ξ^*) to each user based on their channel gains, $h_1, h_2, \ldots, h_K$. With a certain target CIR assignment, we can obtain the optimal power vector that supports every user with the required SIR target by solving the linear inequalities

$$\begin{cases} (\mathbf{I} - \mathbf{F})\mathbf{P}^* \geq \mathbf{U} \\ \mathbf{P}^* \geq \mathbf{0} \end{cases}, \tag{4.46}$$

where $\mathbf{P}^* = [P_1^*, P_2^*, \ldots, P_K^*]^T$ is the transmission power vector,

$$\mathbf{U} = \sigma^2 \cdot \left(\frac{\xi_1}{h_1}, \frac{\xi_2}{h_2}, \cdots, \frac{\xi_K}{h_K} \right)^T \tag{4.47}$$

is the normalized noise power vector, $\mathbf{I}$ is a $K \times K$ identity matrix, and $\mathbf{F}$ is the normalized cross-link gain matrix with (i, j)th element given by

$$\mathbf{F}_{ij} = \begin{cases} \frac{\xi_i h_j}{h_i} & i \neq j \\ 0 & i = j \end{cases}. \tag{4.48}$$

By manipulating (4.46) and (4.47), the feasible minimum power solution of the ith user is given by

$$P_i^* = \frac{\sigma^2}{1 - \sum_{j=1}^K \frac{\xi_j^*}{1+\xi_j^*}} \frac{\xi_i^*}{h_i(1 + \xi_i^*)}. \tag{4.49}$$

After the BS assigns the target ξ^*, data rates, and initial optimal powers to the users in the system, closed-loop power control is triggered by comparing the received CIR with the target CIR in the same way as that in the OSF-PC algorithm. The GRP-PC algorithm is implemented by the following procedure.

Assignment Procedure

Step 1. Based on the available spreading factor values, specify the available data rate $r^{(i)*}$, and the corresponding CIR value, $\xi^{(i)*}$.

Step 2. At the end of the mth frame, the base station computes the average channel gains of all users and then sorts them in a decreasing order: $h_{(1)} \geq h_{(2)} \geq \cdots \geq h_{(K)}$, where the subscript (k) denotes the index of the sorted sequence. Initialize the target CIRs of every user, $\xi_{(i)}^*, i \in [1, K]$, to the minimal value of the CIR set, $\xi^{(1)*}$.

Step 3. For $i = 1$ to K, do

$$\xi_{(i)}^* = \max\{\xi^{(1)*}, \xi^{(2)*}, \ldots, \xi^{(k)*}\}$$

while satisfying the constraint in (4.41).

Step 4. At the beginning of the $(m + 1)$th frame, the base station assigns the transmission rates to the users as follows:

$$r_{(i)}^* = \xi_{(i)}^* \cdot \frac{W}{\gamma^*}. \tag{4.50}$$

Step 5. At the mobile station, the user's initial power at the beginning of the frame is assigned using (4.49). Then closed-loop power control starts based on the target CIR (ξ_i^*) for every time slot.

Step 6. Repeat steps 2–5 for the next adaptation cycle.

From this assignment procedure, we can see that the GRP algorithm allocates the maximum feasible rate to each mobile, starting with the mobile experiencing the best channel condition. As a result, we have $\xi_{(1)}^* \geq \xi_{(2)}^* \geq \cdots \geq \xi_{(K)}^*$ and $r_{(1)}^* \geq r_{(2)}^* \geq \cdots \geq r_{(K)}^*$. It is clear that the user with the best channel condition (highest channel gain) will be assigned the highest data rate.

Compared with OSF-PC, where the transmission power is reset to 0 dBm at the beginning of each frame, GRP-PC makes use of the global channel state information to allocate data rate, and assigns a feasible minimum power to users based on their allocated target CIR (ξ^*). It is expected that GRP-PC could achieve performance gain above that of OSF-PC.

4.7 Numerical Results

As a numerical example, consider a system supporting two classes of services: voices and data. The parameters used are spread bandwidth $W = 5\,\text{MHz}$, noise power spectral density $N_0 = 10^{-6}$, so that, neglecting intercell interference, the noise power seen by each user is $\xi = N_0 \cdot W = 5\,\text{W}$. For the voice users, the transmission rate is 8 kbps, the target $\gamma_v^{tgt} = 6$ (corresponding to 7.8 dB), and the maximal power level is 0.5 W. For the data users, the data rate is 24 kbps, the target is $\gamma_d^{tgt} = 10$ (10 dB), and the maximum power level is 1 W.

Figure 4.1(a) shows the capacity curves for the cases with and without the peak power constraint. The result for the unconstrained case, shown by the circled markers, is obtained using the condition $c > 0$, given by (4.30); the lower curve, for the constrained case, is obtained using (4.31). It is observed that the capacity curves are parallel to one another. The capacity loss is five data users for a given number of voice users with the peak power restriction. It is expected that with the increase in the peak power limit, the bottom curve will shift upward and eventually approach the top curve without constraints.

It can also be observed that the capacity for the voice traffic is much higher than that of the data traffic. Consider the normalized traffic demands: $\Phi_v = 0.0095$ and $\Phi_d = 0.0458$. That is, Φ_d is roughly five times greater than Φ_v, leading to roughly five times capacity loss for a data user compared to a voice user.

Figure 4.1(b) shows the capacity curves under the no-constraint case and the total power constraint case using (4.30) and (4.32). From the top to the bottom, the

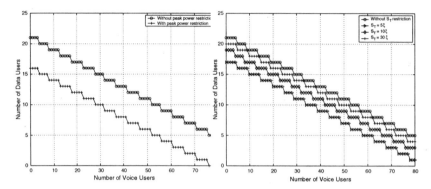

Fig. 4.1 Capacity analysis: (a) with and without peak power constraint and (b) with and without sum power constraint

curves represent the results for no constraint, $S_T = 30\xi$, 10ξ, and 5ξ, respectively, where ξ is the noise level. It can be observed that all the curves are parallel to each other. This is true since the capacity conditions (4.30) and (4.32) have the same slope. Figure 4.1(b) shows that with increasing sum power restrictions, the capacity curves shift upward.

Figure 4.2(a) shows the optimal allocated power levels for voice (lower group) and data (upper group) users as a function of the number of voice users, with the number of data users, $N_d = [2, 6, 10, 14]$, as a parameter. The results are obtained by using Theorem 4.4 with a nonzero background disturbance, where the background noise seen by each active user is $\xi = 5\,W$. It is clear that data users need a much higher power because the data users have a higher traffic demand than that of voice users. It also shows that when the number of voice users increases, the required power for data users increases more quickly than that for voice users. Moreover, it can be observed that the imposition of power restriction reduces the capacity, but does not affect the power allocation. Within the capacity of the system, the power allocation is the same for both restricted and unrestricted cases.

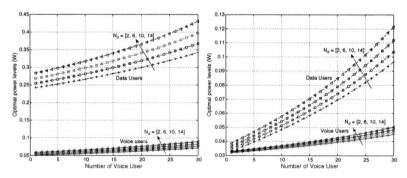

Fig. 4.2 Optimal allocated power levels vs number of voice users: (a) with background disturbance and (b) no background disturbance

Figure 4.2(b) shows the optimal power allocation for the voice and data traffic when the background disturbance is ignored. The results are obtained by using Theorem 4.2 with a minimum required power level for proper detection to be 0.0316 W (corresponding to 15 dBm). It can be observed that the data users require a higher power level than that of voice users. With an increase in the number of active users, the allocated power level for both voice and data users increases and the data users exhibit a faster increasing trend.

In our treatment of the feasibility condition, we have used $c > 0$, which implies $\sum \Phi_i < 1$ as an equivalent condition of $r(\Gamma_P) < 1$, where $r(\Gamma_P)$ is the spectral radius of the matrix Γ_P. Figure 4.3(a–d) compares the values of $\sum \Phi_i$ with those of $r(\Gamma_P)$ when the numbers of users are $M = [10, 20, 30, 40]$. For a given number of users, we randomly generate the normalized traffic demand, Φ_i, to be uniformly distributed between (0, 1). Then we properly scale the vector Φ, making $\sum \Phi_i$ uniformly distributed in (0, 1). The spectral radius of Γ_P can be calculated based on the traffic demand of each user. In order to have a good visual effect, we finally sort the generated $\sum \Phi_i$ and $r(\Gamma_P)$ and sketch them as a function of the simulation index. It shows that $\sum \Phi_i$ is always a little larger than $r(\Gamma_P)$. With an increase of the number of users, the difference becomes smaller. In the simulation (300 runs), the mean difference between $\sum \Phi_i$ and $r(\Gamma_P)$ is [0.0351, 0.0296, 0.0225, 0.0218] when $M = [10, 20, 30, 40]$, respectively. It can also be observed that when the value approaches 1, the difference becomes negligible, which is consistent with our analytical expectation.

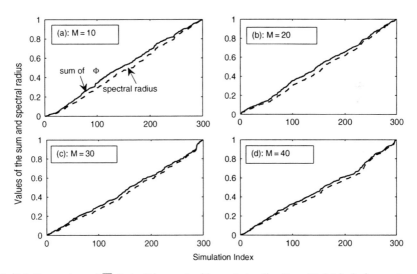

Fig. 4.3 Comparison of $\sum \Phi_i$ (*solid curves*) with spectral radius Γ_P, $r(\Gamma_P)$ (*dashed curves*)

By using the condition $c > 0$ ($\sum \Phi_i < 1$), instead of looking for all the eigenvalues, the power allocation and system feasibility problem is more tractable and has a good physical interpretation of the system demand and the service provisioning.

Simulation results obtained using 400,000 frames are presented to show the throughput, average transmit power consumption of different adaptation schemes in the presence of channel fading. The parameter values used in the calculation are selected as follows: the available discrete spreading factor set [4, 8, 16, 32, 64, 128, 256, 512]. The carrier frequency is assumed to be 2 GHz, the power control cycle is $T_p = 1/1500$ s, which is equivalent to a power control frequency of 1500 Hz,[2] and the background noise power, σ^2, is -10 dBm, which is 10 dB lower than the default transmit power level. The applied step size is 1 dB for all CLPC. The maximum and minimum transmit powers of the mobile stations are 50 and -50 dBm, respectively. The performance presented is for one typical user in the system, on the assumption that all users in the system are equipped to run the same adaptive protocol.

Figure 4.4(a) compares the average throughput for different schemes. From the bottom to the top, the curves are the simulated average throughput for rate adaptation, CLPC, OSF-PC, and GRP-PC, respectively. It shows that as $f_d T_p$ increases, the throughput decreases, except for rate adaptation where the Doppler effect does not have a significant impact due to adaptation saturation. For CLPC, even when the spreading factor does not change, because of the increased retransmission, the throughput still decreases as $f_d T_p$ increases. It can also be observed that GRP-PC achieves much higher throughput compared to the other three algorithms: an average gain of 37.6% over the second best scheme, OSF-PC, which has another 6.9% gain over CLPC. Combining rate allocation and power control judiciously creates great potential for system performance improvement.

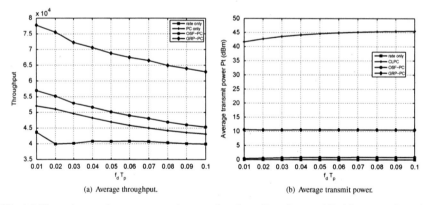

(a) Average throughput.   (b) Average transmit power.

Fig. 4.4 Throughput and average transmit power (number of interferers = 10): (a) average throughput and (b) average transmit power

The improvement of GRP-PC over OSF-PC can be explained as follows: global user channel state information and available spreading factor set are used in rate allocation for GRP, while OSF allocates data rate independently for each user based on its own CIR, and each user truncates its data rate inside the available spreading

---

[2] This is the power control frequency specification in UMTS-WCDMA [22].

factor set, leading to performance degradation compared with GRP. Furthermore, compared to OSF-PC, GRP-PC exploits power allocation at the beginning of each frame for initial power level setting. OSF-PC resets the initial power level as the default value at the beginning of each adaptation frame.

A comparison of the power consumption for these four adaptation approaches is shown in Fig. 4.4(b). By using CLPC, the throughput is significantly improved over rate adaptation. However, the price paid is the dramatic increase in the transmit power in order to track and compensate for the signal fluctuation due to fading. Figure 4.4(b) shows that the power consumption for CLPC is significantly higher than the other three counterparts. It also shows that the average transmit power of OSF-PC is close to that of adaptive rate-only scheme, where the transmit power is kept at the default value, 0 dBm. The average transmission power of GRP-PC is higher than that of OSF-PC. However, if background noise power is lowered, the allocated power level for GRP-PC decreases correspondingly. Nevertheless, the average transmit power is not sensitive to background noise level for the other three schemes. With an increase in $f_d T_p$, the average transmit power for CLPC increases, which implies that it is more demanding to track and compensate for the fading effects at higher Doppler frequencies. However, for GRP-PC and OSF-PC, the average transmit power is not sensitive to the changes in Doppler frequency.

4.8 Conclusions

We have investigated the power distribution problem for multiclass services in a wideband DS-CDMA system. The power vector is expressed as a function of the traffic demand, which depends on the target data rate, SIR requirement, and the spread spectrum bandwidth. Optimal power distribution is solved via a decomposition of the traffic matrix and application of the well-known Sherman–Morrison inverse formula for rank-one updated matrices, subject to some power constraints. We have presented a unified approach to solve the matrix inequality and generic results for the power distribution law; the power distribution law is applicable to both uplink and downlink transmissions. The convergent conditions for the existence of nonnegative power vector are applied to evaluate the capacity region of the system. With the introduction of power restrictions, the capacity regions shrink correspondingly. Within the capacity region, optimal power allocation has the same solution as those without power restrictions. Our analysis shows that the traffic demand fully determines the capacity region. The traffic demand approach makes the system feasibility study easily tractable with a good physical interpretation.

Radio resource management is further investigated in the presence of channel fading. It is shown that the joint GRP (greedy rate packing)-CLPC scheme is the most efficient adaptive approach in terms of system throughput enhancement and transmit power reduction.

References

1. A. J. Viterbi, *CDMA: Principles of Spread Spectrum Communications*, Addison-Wesley Publishing Company, Reading, MA, 1995.
2. W. C. Y. Lee, "Overview of cellular CDMA," *IEEE Trans. Vehicular Tech.*, vol. 40, no. 2, pp. 291–301, May 1991.
3. K. S. Gilhousen, I. M. Jacobs, R. Padovani, and A. J. Viterbi, "On the capacity of a cellular CDMA system," *IEEE Trans. Vehicular Tech.*, vol. 40, no. 2, pp. 303–311, May 1991.
4. L. C. Yun and D. G. Messerschmitt, "Power control for variable QoS on a CDMA channel," *Proc. IEEE Military Communications Conf.*, pp. 178–182, Oct. 1994.
5. A. Sampath, P. S. Kumar, and J. M. Holtzman, "Power control and resource management for a multimedia CDMA wireless system," *in Proc. IEEE Intl. Symposium on Personal, Indoor and Mobile Radio Communications*, vol. 1, pp. 21–25, 1995.
6. S. J. Lee, H. W. Lee, and D. K. Sung, "Capacities of single-code and multicode DS-CDMA systems accommodating multiclass services," *IEEE Trans. Vehicular Tech.*, vol. 48, no. 2, pp. 376–384, Mar. 1999.
7. S. Choi and K. G. Shin, "An uplink CDMA system architecture with diverse QoS guarantees for heterogeneous traffic," *IEEE/ACM Trans. Netw.*, vol. 7, no. 5, pp. 616–628, Oct. 1999.
8. L. A. Imhof and R. Mathar, "Capacity regions and optimal power allocation for CDMA cellular radio," *IEEE Trans. Inform. Theory*, vol. 51, no. 6, pp. 2011–2019, Jun. 2005.
9. P. Liu, P. Zhang, S. Jordan, and M. L. Honig, "Single-cell forward link power allocation using pricing in wireless network," *IEEE Trans. Wireless Commun.*, vol. 3, no. 2, pp. 533–543, Mar. 2004.
10. J. W. Mark and S. Zhu, "Power control and rate allocation in multirate wideband CDMA systems," *in Proc. IEEE Wireless Communications and Networking Conf.*, pp. 168–172, 2000, (invited).
11. L. Zhao, J. W. Mark, J. Ding, and W. Pye, "Power control and call admission in multirate wideband CDMA systems," *in Proc. IEEE Wireless Communications and Networking Conf.*, 2004.
12. L. Zhao, J. W. Mark, and J. Ding, "Power distribution/allocation in multirate wideband CDMA systems," *IEEE Trans. Wireless Commun.*, vol. 5, no. 9, pp. 2458–2467, 2006.
13. S. Ariyavisitakul and L. F. Chang, "Signal and interference statistics of a CDMA system with feedback power control," *IEEE Trans. Commun.*, vol. 41, no. 11, pp. 1626–1634, Nov. 1993.
14. A. Chockalingam, P. Dietrich, L. B. Milstein, and R. R. Rao, "Performance of closed-loop power control in DS-CDMA cellular systems," *IEEE Trans. Vehicular Tech.*, vol. 47, no. 3, pp. 774–789, Aug. 1998.
15. L. Zhao and J. W. Mark, "Multi-step closed-loop power control using linear receivers for DS-CDMA systems," *IEEE Trans. Wireless Commun.*, vol. 3, no. 6, pp. 2141–2155, Nov. 2004.
16. B. Vucetic, "An adaptive coding scheme for time-varying channels," *IEEE Trans. Commun.*, vol. 39, no. 5, pp. 653–663, May 1991.
17. ETSI EN301 709, "Digital cellular telecommunications system (phase 2+); link adaptation," May 2000.
18. W. T. Webb and R. Steel, "Variable rate QAM for mobile radio," *IEEE Trans. Commun.*, vol. 43, no. 7, pp. 2223–2230, Jul. 1995.
19. S. Nanda, K. Balachandran, and S. Kumar, "Adaptation techniques in wireless packet data services," *IEEE Commun. Mag.*, vol. 38, pp. 54–64, Jan. 2000.
20. A. Goldsmith and S. G. Chua, "Variable-rate variable-power MQAM for fading channels," *IEEE Trans. Commun.*, vol. 45, no. 10, pp. 1218–1230, Oct. 1997.
21. K. K. Leung and L. C. Wang, "Integrated link adaptation and power control to improve error and throughput performance in broadband wireless packet networks," *IEEE Trans. Wireless Commun.*, vol. 1, no. 4, pp. 619–629, Oct. 2002.
22. UMTS, *UMTS Overview*, http://www.umtsworld.com/technology/technology.htm.

23. F. Berggren, S. L. Kim, R. Jantti, and J. Zander, "Joint power control and intracell scheduling of DS-CDMA nonreal time data," *IEEE J. Select. Areas Commun.*, vol. 19, pp. 1860–1869, Oct. 2001.

24. S. Kahn, M. K. Gurean, and O. O. Oyefuga, "Downlink throughput optimization for wideband CDMA systems," *IEEE Comm. Lett.*, vol. 7, no. 5, pp. 251–253, May 2003.

25. J. M. Ortega and W. C. Rheinboldt, *Iterative Solutions of Nonlinear Equations in Several Variables*, Academic Press, New York, 1970.

26. M. Patzold, U. Killat, and F. Laue, "A deterministic digital simulation model for Suzuki processes with application to a shadowed Rayleigh land mobile radio channel," *IEEE Trans. Vehicular Tech.*, vol. 45, no. 2, pp. 318–331, May 1996.

27. G. L. Stuber, *Principles of Mobile Communication*, Kluwer Academic Publisher, Boston, MA, Second edition, 2001.

28. R. B. Bapat and T. E. S. Raghavan, *Nonnegative Matrices and Applications*, Cambridge University Press, New York, 1997.

29. J. B. Kim and M. L. Honig, "Resource allocation for multiple classes of DS-CDMA traffic," *IEEE Trans. Vehicular Tech.*, vol. 49, no. 2, pp. 506–519, Mar. 2000.

30. F. Berggren and S. L. Kim, "Energy-efficient control of rate and power in DS-CDMA systems," *IEEE Trans. Wireless Commun.*, vol. 3, no. 3, pp. 725–733, May 2004.

Chapter 5
Iterative Receivers and Their Graphical Models

Ezio Biglieri

5.1 Introduction

In a number of communication systems, optimum receiver design requires joint demodulation and decoding. The complexity problem arising in practical implementation has led to an interest in iterative receivers. This chapter introduces iterative receiver algorithms based on their graphical models. These consist of representing the factorization of a function of several variables into a product of functions of a lower number of variables. Using this representation, efficient algorithms are derived for computing the a posteriori probabilities to be used for optimal symbol-by-symbol detection of the transmitted data. The unified approach presented here allows one to observe how seemingly different transmission schemes share many common features, and hence solutions that were devised for one problem can easily be adapted to a different one. Our presentation is tutorial in nature and relies heavily on graphical descriptions.

5.2 MAP Symbol Detection

Consider the transmission of an n-tuple $\mathbf{x} \triangleq (x_1, \ldots, x_n)$ of discrete-valued symbols through a random channel, modeled by the conditional probability density function $p(\mathbf{y}|\mathbf{x})$ of its output vector $\mathbf{y}$ (we leave the dimension of $\mathbf{y}$ unspecified for the moment). Vector $\mathbf{x}$ is a word chosen from code $\mathcal{C}$. Maximum a posteriori (MAP) detection consists of solving, for $i = 1, \ldots, n$,

$$\hat{x}_i = \arg \max p(x_i \mid \mathbf{y}), \tag{5.1}$$

Ezio Biglieri
Universitat Pompeu Fabra, Barcelona, Spain,
e-mail: e.biglieri@ieee.org.

V. Tarokh (ed.), *New Directions in Wireless Communications Research*,
DOI 10.1007/978-1-4419-0673-1_5,
© Springer Science+Business Media, LLC 2009

where p denotes the "a posteriori" probability (APP) of x_i given the observation of channel output $\mathbf{y}$. This detection strategy minimizes the symbol error probability. Clearly, the solution of (5.1) is exceedingly simple whenever x_i can take on a small number of values: actually, it is the calculation of $p(x_i \mid \mathbf{y})$, based on the observation of $\mathbf{y}$ and the knowledge of the channel and of the code structure, that makes the problem hard to solve. In fact, $p(\mathbf{x} \mid \mathbf{y})$ can be determined in a fairly easy way (as we shall see soon), and we can obtain the required APP through an operation called a "marginalization"

$$p(x_i \mid \mathbf{y}) = \sum_{x_1} \cdots \sum_{x_{i-1}} \sum_{x_{i+1}} \cdots \sum_{x_n} p(\mathbf{x} \mid \mathbf{y}) \qquad (5.2)$$

and this marginalization is responsible for the problem's complexity. MAP symbol detection can be viewed as consisting of two parts, as shown in Fig. 5.1, and from now on we shall focus on the hard part of it, that is, the marginalization.

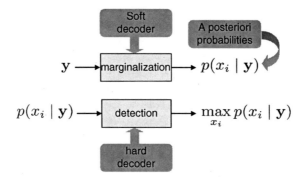

Fig. 5.1 The two steps of MAP symbol detection

In general, marginalization of the function $f(x_1, \ldots, x_n)$ of n variables with respect to one of its variables, say x_i, consists of computing

$$f_i(x_i) \triangleq \sum_{x_1} \cdots \sum_{x_{i-1}} \sum_{x_{i+1}} \cdots \sum_{x_n} f(x_1, \ldots, x_n). \qquad (5.3)$$

For each value taken on by x_i, these "marginals" are obtained by summing the function f over all of its arguments consistent with the value of x_i. It is convenient to introduce the compact notation $\sim x_i$ to denote the set of indices $x_1, \ldots, x_{i-1}, x_{i+1}, \ldots, x_n$ to be summed over, so that we can write

$$f_i(x_i) = \sum_{\sim x_i} f(x_1, \ldots, x_n). \qquad (5.4)$$

If $x_i \in \mathcal{X}$, $i = 1, \ldots, n$, then the complexity of this computation grows as $|\mathcal{X}|^{n-1}$. A simplification can be achieved when f can be factored as a product of functions, each with less than n arguments. We describe this factorization using a graphical form.

5.2.1 Factor Graphs and the Sum–Product Algorithm

Consider an n-tuple of variables $X \triangleq \{x_1, \ldots, x_n\}$, and a function $f(X)$. A factorization of f is a decomposition of f as a product

$$f(X) = \prod_{k=1}^{K} g_k(X_k), \qquad (5.5)$$

where g_k, $1 \leq k \leq K$, are functions of proper subsets of X, denoted as $X_k \subset X$. A *factor graph* is an undirected bipartite graph illustrating (5.5). Its nodes correspond to variables x_i and to functions g_k, while its edges correspond to the pairs (g_k, x_j), $x_j \in X_k$. As a simple example, the factor graph corresponding to the function $f(x_1, x_2, x_3, x_4) = g_1(x_1, x_2)g_2(x_1, x_3)g_3(x_1)$ is shown in Fig. 5.2(a). The nodes here can be viewed as processors, computing function whose arguments label the incoming edges, and the edges as channels by which these processors exchange data.

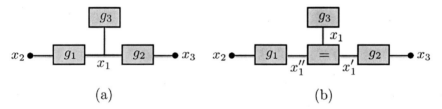

(a)   (b)

Fig. 5.2 Factor graphs of the function $f(x_1, x_2, x_3, x_4) = g_1(x_1, x_2)g_2(x_1, x_3)g_3(x_1)$. Graph (b) on the right is in normal form

Given a factor graph, the sum–product algorithm (SPA) operates on it to generate marginal functions. The SPA is a "belief-propagation" algorithm [29] which, under certain conditions, allows exact and efficient computation of the marginals of the function represented by the graph. It works by propagating, along the edges associated with each argument of the function, "messages" whose values when the algorithm terminates yield the marginals sought. Each message is a function of a variable x_i, and hence, under the assumption of variables taking on a finite number of values, it can be expressed in the form of a vector, each of whose components correspond to one value taken on by x_i. If the graph has cycles, i.e., closed paths, message propagation along the graph must be done iteratively, leading to "turbo" algorithms that may not have a natural termination nor converge to the exact marginals. However, a large set of simulation results show excellent behavior of the algorithm even in the presence of cycles, provided that the girth of the graph is large enough (in general, a bipartite graph has girth ≥ 4). More precisely, belief-propagation algorithms (BPA) converge in a finite number of steps to the correct marginals when the factor graph has no cycles [29]. Convergence and correctness in graphs with a single cycle are studied in [1], and in graphs with a large girth in [30]. A recent result shows that a BPA used to solve the maximum weight matching problem converges to the correct solution as long as the latter is unique [4].

The description of the SPA is especially simple if we restrict ourselves (yet without any loss of generality) to the consideration of *normal* factor graphs (also called "Forney-style" factor graphs). Normality assumes that no variable appears in more than two factors of the function to be marginalized. For example, the graph of Fig. 5.2(a) does not satisfy our definition: in fact, the variable x_1 appears as a factor of g_1, g_2, and g_3, and hence it corresponds to more than one edge. This graph can be normalized as in Fig. 5.2(b) by introducing a "cloning function" $f_=$ taking value 1 when all of its arguments are equal, and value 0 otherwise. This corresponds to the factorization of f written in the form

$$f(x_1, x_2, x_3, x_4) = g_1(x_1'', x_2)g_2(x_1', x_3)g_3(x_1)f_=(x_1, x_1', x_1'').$$

The example of Fig. 5.3 introduces the SPA through the simple example of the marginalization, with respect to x_5, of the function

$$f(x_1, x_2, x_3, x_4, x_5) = g_1(x_1, x_2, x_5)g_2(x_3, x_4, x_5).$$

$$f_5(x_5) = \sum_{x_1}\sum_{x_2}\sum_{x_3}\sum_{x_4} g_1(x_1, x_2, x_5)g_2(x_3, x_4, x_5)$$

$$= \underbrace{\sum_{x_1}\sum_{x_2} g_1(x_1, x_2, x_5)}_{\mu_{g_1 \to x_5}(x_5)} \cdot \underbrace{\sum_{x_3}\sum_{x_4} g_2(x_3, x_4, x_5)}_{\mu_{g_2 \to x_5}(x_5)}$$

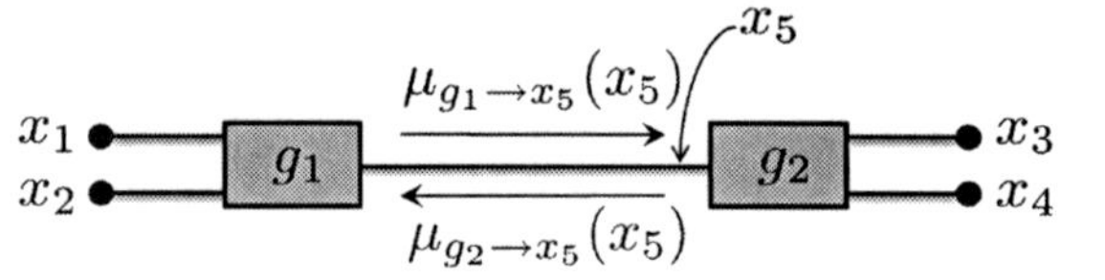

Fig. 5.3 Marginalization of a simple function: a special case of the SP algorithm

It consists of first splitting the multiple sum into two simpler sums, next realizing that the marginal can be computed by taking the product of two simpler marginals, the functions of x_5 called "messages." Observe how the messages can be computed locally and independently at nodes g_1 and g_2.

The general step of the SPA is illustrated in Fig. 5.4 for a normal graph. The figure shows how the message along the edge x_i can be computed, once the messages along all other edges ending in node g are known. This is obtained by marginalizing, with respect to x_i, the product of g and of the messages entering node g.

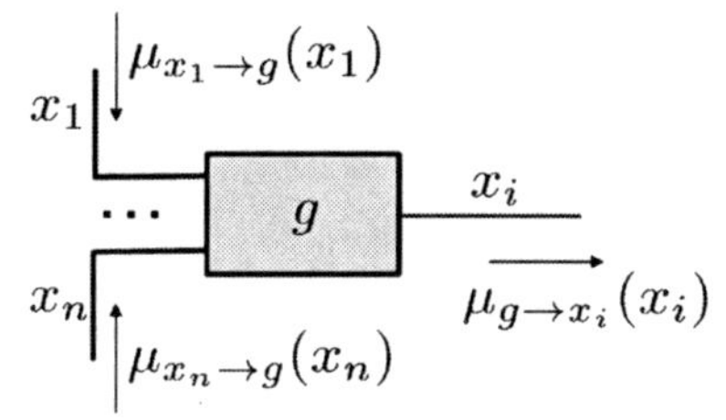

$$\mu_{g \to x_i}(x_i) = \sum_{\sim x_i} g(x_1, \ldots, x_n) \prod_{\ell \neq i} \mu_{x_\ell \to g}(x_\ell)$$

Fig. 5.4 Basic step of the SP algorithm

Two special, yet important, cases of the general SPA are the following:

1. If g is a function of only one argument x_i, then the product Fig. 5.4 is empty, and we simply have

$$\mu_{g \to x_i}(x_i) = g(x_i). \tag{5.6}$$

2. If g is the repetition function $f_=$, then

$$\mu_{f_= \to x_i}(x_i) = \prod_{\ell \neq i} \mu_{x_\ell \to f_=}(x_i). \tag{5.7}$$

5.2.2 The Basic Factorization

Let us now return to our original problem of detecting a vector $\mathbf{x}$ after the observation of the channel output $\mathbf{y}$. We can write

$$p(\mathbf{x} \mid \mathbf{y}) \propto p(\mathbf{x})p(\mathbf{y} \mid \mathbf{x}). \tag{5.8}$$

Here, $p(\mathbf{x})$ denotes the probability that $\mathbf{x} \in C$ be transmitted. Under the assumption that all words in C are transmitted with equal probability $1/|C|$, we have

$$p(\mathbf{x}) = \begin{cases} 1/|C|, & \mathbf{x} \in C \\ 0, & \text{otherwise.} \end{cases}$$

Since the actual value of $1/|C|$ is irrelevant to the decoding process, we can simply write

$$p(\mathbf{x}) \propto [\mathbf{x} \in C], \tag{5.9}$$

where, given a proposition P that may be either true or false, $[P]$ denotes the *Iverson function*

$$[P] \triangleq \begin{cases} 1, & P \text{ is true} \\ 0, & P \text{ is false.} \end{cases} \tag{5.10}$$

Thus, we can rewrite (5.8) in the form

$$p(\mathbf{x} \mid \mathbf{y}) \propto [\mathbf{x} \in \mathcal{C}] \, p(\mathbf{y} \mid \mathbf{x}), \qquad (5.11)$$

which states that $p(\mathbf{x} \mid \mathbf{y})$ is the product of two functions, one, $[\mathbf{x} \in \mathcal{C}]$, describing the code, and the other, $p(\mathbf{y} \mid \mathbf{x})$, describing the channel. Thus, at a high level of abstraction, we can describe the generation of a posteriori probabilities of each x_i as the application of the SPA to a graph of the form shown in Fig. 5.5. The same figure also shows the steps of the SPA necessary to generate, for all symbols x_i, the messages going upward (denoted $i(x_i)$ and called "intrinsic") and those going downward (denoted $e(x_i)$ and called "extrinsic").

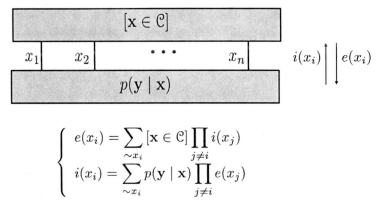

Fig. 5.5 General SP algorithm for detecting a coded signal

5.3 Channel and Codes: A Menagerie of Factor Graphs

We shall now see how, depending on the problem at hand, we can factorize $p(\mathbf{y} \mid \mathbf{x})$ and $[\mathbf{x} \in \mathcal{C}]$, thus obtaining a factorization of the objective function $p(\mathbf{x} \mid \mathbf{y})$ finer than the one in (5.11).

5.3.1 Modeling the Channel

The simplest example of a channel is that of a stationary memoryless channel. By its definition, we have

$$p(\mathbf{y} \mid \mathbf{x}) = \prod_{i=1}^{n} p(y_i \mid x_i) \qquad (5.12)$$

The corresponding factor graph includes n functional blocks, each generating a message $p(y_i \mid x_i)$, $i = 1, \ldots, n$ (Fig. 5.6). The SPA is schematized in Fig, 5.7, where property (5.6) is used.

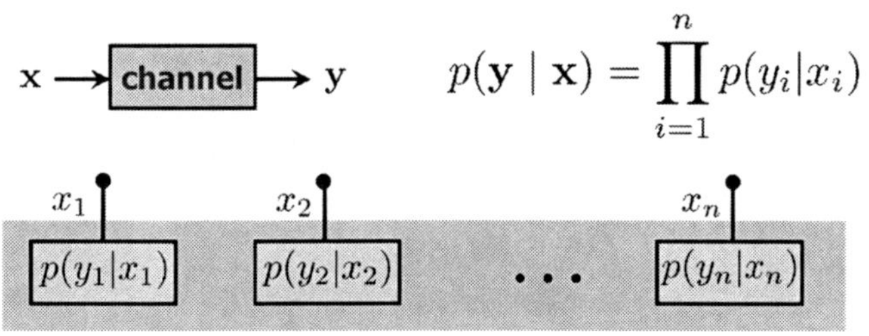

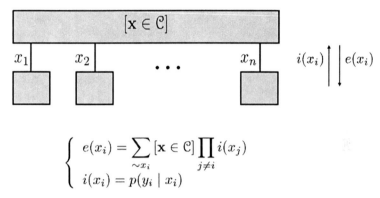

$$x \longrightarrow \boxed{\textbf{channel}} \longrightarrow y \qquad p(\mathbf{y} \mid \mathbf{x}) = \prod_{i=1}^{n} p(y_i \mid x_i)$$

Fig. 5.6 Factor graph of a memoryless channel

$$[\mathbf{x} \in \mathcal{C}]$$

$$\begin{cases} e(x_i) = \sum_{\sim x_i} [\mathbf{x} \in \mathcal{C}] \prod_{j \neq i} i(x_j) \\ i(x_i) = p(y_i \mid x_i) \end{cases}$$

Fig. 5.7 Special case: SP algorithm for detecting a coded signal sent through a memoryless channel

A more complex example of a channel factor graph is shown in Fig. 5.8. Here, owing to the independence of the noise components, and under the assumption that the observed vector **y** has N components, we have the factorization

$$p(\mathbf{y} \mid \mathbf{x}) = \prod_{i=1}^{N} p(y_i \mid \mathbf{x}). \qquad (5.13)$$

This expresses the fact that each observed component y_i depends on the whole vector **x**.

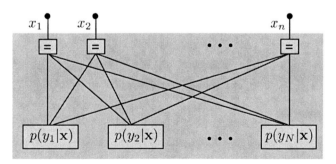

Fig. 5.8 Factor graph of a dispersive channel

5.3.2 Modeling the Code

If uncoded symbols are transmitted, and hence the components of vector **x** are assumed to be independent, we have

$$p(\mathbf{x}) = \prod_{i=1}^{n} p(x_i)$$

and hence the SPA takes the form illustrated in Fig. 5.9.

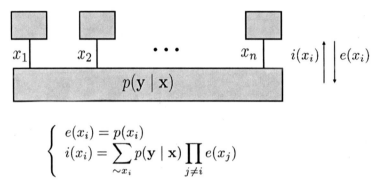

$$\begin{cases} e(x_i) = p(x_i) \\ i(x_i) = \sum_{\sim x_i} p(\mathbf{y} \mid \mathbf{x}) \prod_{j \neq i} e(x_j) \end{cases}$$

Fig. 5.9 Special case: SP algorithm for detecting an uncoded signal

The simplest case of a factor graph modeling an actual code comes under the assumption of a binary linear code described through its parity-check matrix. By definition, every row of this matrix describes a constraint that must be satisfied by the code word components, and hence corresponds to a factor of the function $[\mathbf{x} \in C]$. For example, a code whose parity-check matrix is

$$\mathbf{H} = \begin{bmatrix} 1 & 1 & 0 & 0 & 1 & 1 & 0 \\ 1 & 0 & 0 & 1 & 0 & 1 & 1 \\ 1 & 0 & 1 & 0 & 1 & 0 & 1 \end{bmatrix}$$

has words $\mathbf{x} = (x_1, \ldots, x_7)$ satisfying the constraints

$$\begin{aligned} x_1 + x_2 + x_5 + x_6 &= 0 \\ x_1 + x_4 + x_6 + x_7 &= 0 \\ x_1 + x_3 + x_5 + x_7 &= 0 \end{aligned} \tag{5.14}$$

so that we can write

$$[\mathbf{x} \in C] = [x_1 + x_2 + x_5 + x_6 = 0] \times [x_1 + x_4 + x_6 + x_7 = 0] \times [x_1 + x_3 + x_5 + x_7 = 0].$$

This factorization yields the factor graph shown in Fig. 5.10, where the function blocks describe the constraint that the (modulo-2) sum of the arguments must equal 0. Several other examples of factor graphs describing a given code, and in particular one defined on a trellis, can be found in [6, 30, 33].

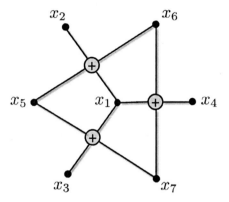

Fig. 5.10 Factor graph of the code whose parity-check matrix is (5.14)

A simple example of the combination of the factor graphs of a code and of a stationary memoryless channel is shown in Fig. 5.11. The code is the binary single-parity-check code with length 3, described by the constraint $[x_1 + x_2 + x_3 = 0]$. Consider decoding of symbol x_1: the message $p(y_1 \mid x_1)$ (to be viewed as a function of x_1, parametrized by the observed value y_1) is called "intrinsic," as it depends only on the observation of the channel output corresponding to the transmitted symbol x_1. The message in the opposite direction carries information on x_1 derived from the observed values y_2 and y_3, and on the structure of the code (this is why it is called "extrinsic").

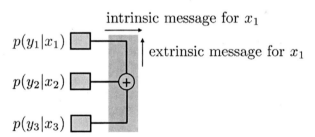

Fig. 5.11 Decoding the $(3, 2)$ single-parity-check code on a memoryless channel: intrinsic and extrinsic messages for symbol x_1

5.4 Equalization

Consider an intersymbol interference channel with channel memory L and impulse response $\mathbf{h}$. Its response to a code word $\mathbf{x}$ with length n is a vector with length $N \geq n$, whose components are linear combinations of those of $\mathbf{x}$:

$$y_i = \mathbf{h}_i' \mathbf{x} + z_i, \quad i = 1, \ldots, N$$

with z_i samples of a white Gaussian noise process. For example, in the simple case of a causal channel with memory 2, impulse response $\mathbf{h} = (h_0, h_1, h_2)$, and whose input is a vector with with block length 2, say $\mathbf{x} = (x_1, x_2)$, the output vector has the form

$$\mathbf{y} = \begin{bmatrix} y_1 \\ y_2 \\ y_3 \\ y_4 \end{bmatrix} = \begin{bmatrix} h_0 & 0 \\ h_1 & h_0 \\ h_2 & h_1 \\ 0 & h_2 \end{bmatrix} \begin{bmatrix} x_1 \\ x_2 \end{bmatrix} + \mathbf{z}. \tag{5.15}$$

In general, we have

$$p(\mathbf{x} \mid y_1, \ldots, y_N) \propto [\mathbf{x} \in \mathcal{C}] \times \prod_{i=1}^{N} p(y_i \mid \mathbf{x}) \tag{5.16}$$

and hence the factor graph like the one shown in Fig. 5.12, which refers to a code word with block length 6, a causal channel with memory 2, and a received vector with length 6. It can be observed that the channel graph has girth 4, which may impair the convergence properties of the iterative decoding algorithm. It has been observed that the length-4 cycles may not affect the convergence of the algorithm [3, p. 1285]. If this is not the case, certain graph transformations may be introduced to increase the girth [14, 24].

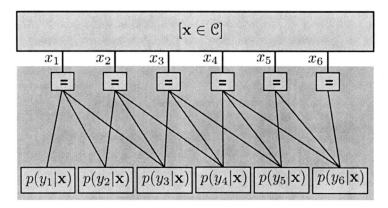

Fig. 5.12 Factor graph for equalization of coded signals

Zero-forcing equalization can be interpreted as removal of all cross-paths in Fig. 5.12. Denoting by $\mathbf{H}^+$ the Moore–Penrose pseudoinverse of $\mathbf{H}$, and defining

$$\tilde{\mathbf{y}} \triangleq \mathbf{H}^+ \mathbf{y}$$

the resulting factor graph becomes the one depicted in Fig. 5.13, in which the cross-paths are removed. More generally, if

$$\tilde{\mathbf{y}} \triangleq \mathbf{A} \mathbf{y}$$

where $\mathbf{A}$ denotes the minimum-mean-square-error filter $\mathbf{A} \triangleq (\mathbf{I} + \delta \mathbf{H}^{\dagger}\mathbf{H})^{-1}\mathbf{H}^{\dagger}$, with δ denoting the inverse of the signal-to-noise ratio, the cross-paths become irrelevant, and can be omitted from the algorithm. The situation is depicted in Fig. 5.14.

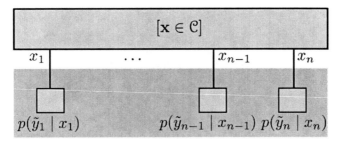

Fig. 5.13 ZF equalization of coded signals

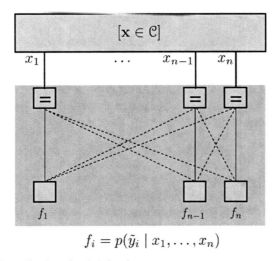

$$f_i = p(\tilde{y}_i \mid x_1, \dots, x_n)$$

Fig. 5.14 MMSE equalization of coded signals

Similarly, decision-feedback equalization is illustrated in Fig. 5.15. Here, only the paths joining channel node f_k to variable nodes x_j, $j \geq k$, are retained. $\hat{x}_i$ denote preliminary decisions made sequentially on x_i, $i = n, n-1, \dots$.

Another approximation, obtained without modifying the factor graph topology, consists of replacing the original messages with vectors having only one nonzero component, which takes on value 1 in correspondence of the most likely value of the variable (*hard decisions*). Yet another way of simplifying the message structure consists of approximating the messages with Gaussian distributions, having the same mean and variance of the original (discrete) distribution. For an in-depth treatment and additional results on iterative equalization, the reader is addressed to [3, 14, 16, 21, 23, 31] and the references therein.

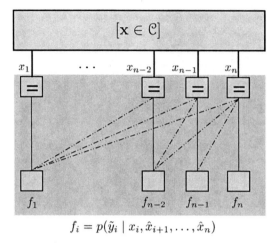

$$f_i = p(\tilde{y}_i \mid x_i, \hat{x}_{i+1}, \dots, \hat{x}_n)$$

Fig. 5.15 Decision-feedback equalization of coded signals

5.5 Multiuser Detection

Application of factor graph techniques to the multiuser-detection problem provides a unified framework for many of the algorithms proposed in the literature for joint decoding and multiuser detection. The balance of this section is based on [11].

We have equations

$$\mathbf{y}_i = \mathbf{SW}_i\mathbf{x}_i + \mathbf{z}_i, \qquad i = 1, \dots, N, \tag{5.17}$$

where N is the common block length of the users' codes, $\mathbf{y}_i \in \mathbb{C}^L$, L the spreading factor, $\mathbf{S} \in \mathbb{C}^{L \times K}$ is the correlation matrix of users' signatures, K the number of users, and $\mathbf{W}_i \in \mathbb{C}^{K \times K}$ the diagonal matrix of user amplitudes. Denoting by $\mathcal{C}_k$ the code of user k, and by $\mathbf{X}$ the $K \times N$ matrix whose rows $\mathbf{x}^k$ are the users' code words and whose columns are denoted $\mathbf{x}_i$, $i = 1, \dots, N$, we can write

$$p(\mathbf{X} \mid \mathbf{y}_1, \dots, \mathbf{y}_N) \propto \prod_{k=1}^{K} \left[\mathbf{x}^k \in \mathcal{C}_k \right] \prod_{i=1}^{N} p(\mathbf{y}_i \mid \mathbf{x}_i). \tag{5.18}$$

For example, over the AWGN channel with noise power spectral density N_0 we have

$$p(\mathbf{y}_i \mid \mathbf{x}_i) \propto \exp{-|\mathbf{y}_i - \mathbf{SW}_i\mathbf{x}_i|^2/N_0}.$$

Figure 5.16 shows the corresponding factor graph. A special case, corresponding to $K = 2$ and $N = 3$, is illustrated in Fig. 5.17.

A number of low-complexity implementations of the optimum detector can be derived in a unified way. Within this framework, one can observe that multiaccess interference is formed by the messages shown in Fig. 5.18. Reference [11] suggests two simplifications based on the scheme of the figure. The first one

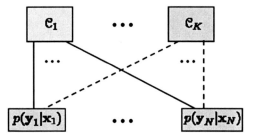

Fig. 5.16 Factor graph for multiuser detection

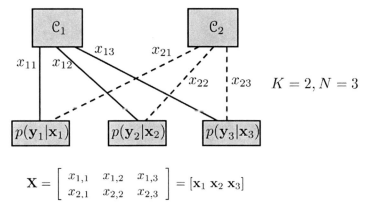

Fig. 5.17 An example of an actual factor graph for multiuser detection

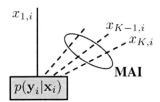

Fig. 5.18 Graphical interpretation of multiaccess interference

consists of replacing the messages of multiaccess interference with hard decisions on the symbols to which they refer ("hard-interference-cancellation receiver"). Another one replaces the multiaccess-interference messages with Gaussian random variables ("soft-interference-cancellation receiver").

5.6 MIMO Detection

This section follows [7, Chapter 5], to which the reader is referred for additional details. Figure 5.19 shows the MIMO model, as defined by the equation $\mathbf{y} = \mathbf{Hx} + \mathbf{z}$ with $\mathbf{H}$ the $r \times t$ "channel matrix" whose entries are the fading gains of the paths

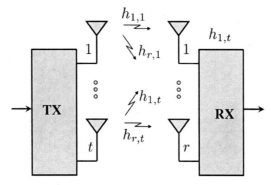

Fig. 5.19 Model of MIMO channel

leading from the t transmit to the r receive antennas. The corresponding factor graph is shown in Fig. 5.20. Notice that, as observed for intersymbol interference channels, the graph has girth 4, which may impair the convergence properties of the iterative detection algorithm. A number of suboptimum solutions are available, similar in spirit to those described for equalization in Section 5.4. As an example of these, consider the VBLAST receiver architecture [10, 18]. This is based on the following three operations:

1. *Nulling of spatial interference.* This is achieved by simplifying the structure of the factor graph through linear preprocessing, along the lines described in Section 5.4. The components y_i of the received vector are replaced by $\tilde{y}_i$.
2. *Cancellation of spatial interference.* The function nodes correspond to the modified densities

$$p(\tilde{y}_i \mid x_i, x_{i+1}, \ldots, x_t) \approx p(\tilde{y}_i \mid x_i, \hat{x}_{i+1}, \ldots, \hat{x}_t)$$

sequentially for $i = t - 1, t - 2, \ldots, 1$, where $\hat{x}$ denotes a hard decision made on x.
3. *Ordering.* This consists of selecting the order of the antennas on which the two steps above are performed.

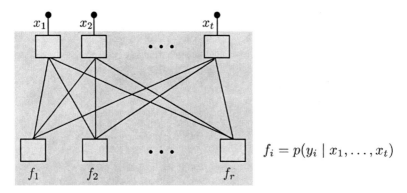

$$f_i = p(y_i \mid x_1, \ldots, x_t)$$

Fig. 5.20 Factor graph of MIMO channel with t transmit and r receive antennas

An implementation of VBLAST is described in Fig. 5.21, where the dashed lines correspond to paths that are disregarded in message passing.

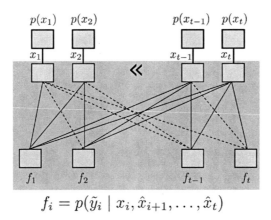

$$f_i = p(\tilde{y}_i \mid x_i, \hat{x}_{i+1}, \ldots, \hat{x}_t)$$

Fig. 5.21 Factor graph corresponding to a MIMO receiver with VBLAST architecture

5.7 Multilevel Coded Modulation

In multilevel coded modulation [5, Chapter 12], we start from an "elementary" signal constellation S with 2^L signals, and its L-level partition. Specifically, at level 0, S is partitioned into two constellations S_1 of 2^{L-1} signals each. At level 1, each S_1 is partitioned into two constellations with 2^{L-2} signals each, and so forth, until at level $L - 1$ we obtain "constellations" S_{L-1} with one signal each. Labeling the constellations at each partition level, we obtain a one-to-one correspondence between binary L-tuples and signals in S. Figure 5.22 shows partitioning and labeling of 4PSK. Consider next L binary encoders, generating words with equal length n denoted as $c_0, \ldots, c_{L-1}$. The first, second, ..., L-tuples of binary symbols are mapped to signals $x_0, \ldots, x_n$ using the correspondence generated by the partition. Figure 5.23 shows the multilevel encoder, while Fig. 5.24 shows the corresponding factor graph in a special case. Comparing this figure with Fig. 5.17, one can immediately realize the similarity between the two factor graphs, and hence between the two problems. (Observe, however, that the order of magnitude of the parameters may be different in the two cases.)

5.8 Convergence of the Iterative Algorithm

A rigorous analysis of the convergence of the iterative receivers described in this chapter is based on the density evolution (DE) approach. This was introduced to study iterative decoders for low-density parity-check codes [30] and can be

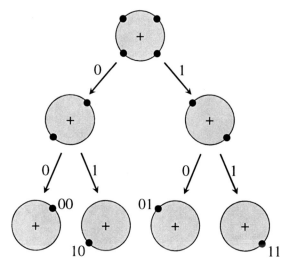

Fig. 5.22 Partitioning and labeling 4PSK

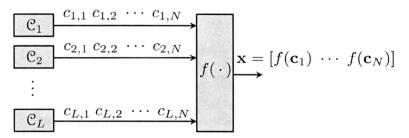

Fig. 5.23 Multilevel coded modulation

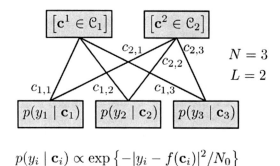

$$p(y_i \mid \mathbf{c}_i) \propto \exp\left\{-|y_i - f(\mathbf{c}_i)|^2/N_0\right\}$$

Fig. 5.24 Graphical interpretation of multilevel demodulation

extended, mutatis mutandis, to more general iterative receivers (for example, see [11] for the application of DE to iterative multiuser detection). This analysis assumes a very large value for the block length n. As $n \to \infty$, a *concentration theorem* shows that the probability density of the messages generated by the iterative decoder after any fixed number of iterations approaches the expected density resulting from a random *cycle-free* factor graph. As error probabilities can be derived from these densities, one can follow the evolution of the former with the number of iterations. A computationally efficient simplification of the DE procedure can be used under the assumption that the message densities are Gaussian. If this holds, since it can be shown that mean and variance of this density are proportional to each other, the evolution of the density is reduced to the evolution of a single parameter, the most convenient among them being the average mutual information carried by the messages about the actual values of the transmitted symbols [32]. This analysis is carried on graphically, using plots referred to as EXIT charts ("EXIT" stands for extrinsic information transfer). Although some inaccuracies must be expected from EXIT-chart analyses, their practical usefulness is out of the question (see, for example, the EXIT-chart analysis of the MIMO detectors carried out in [7, Chapter 5; 8]).

5.9 Conclusions

This chapter was intended as a primer on analysis and design of iterative receivers based on their graphical models. For several further details that could not be included here, the reader may refer, for example, to the recent book [35].

Among the possible extensions of the theory of iterative receivers outlined in the previous sections, one well-worth mentioning consists of accounting for an unknown random channel state **u**. In this case, the channel is described by the density $p(\mathbf{y} \mid \mathbf{u}, \mathbf{x})$, and the objective function to factor becomes $p(\mathbf{y} \mid \mathbf{u}, \mathbf{x})p(\mathbf{u})[\mathbf{x} \in \mathcal{C}]$. In this case, since the channel state may be continuous valued, the messages propagating from the channel-state nodes become probability density functions, and as such complex to manipulate. One solution to this complexity problem, advocated in [34], consists of using parametrized canonical densities for the messages associated with continuous variables. A simple application of this concept makes use of the density consisting of an impulse centered at the estimated value of a parameter of the channel state, $\delta(u_i - \hat{u}_i)$. This corresponds to using an estimate of the unknown channel state as if it were the correct value. Another solution [34] uses a canonical density consisting of a weighted sum of impulses. For applications of the factor-graph approach to receivers with unknown channel states, the reader is referred to [3, 12, 17, 19, 22].

Other applications of the factor-graph approach, not considered in this chapter (Kalman filtering, image segmentation, phase unwrapping, and network tomography) can be found in [13, 15, 16, 20, 25–28]. For aspects concerning the implementation of message-passing in the logarithmic domain, see [14] and references therein. Other applications of message-passing algorithms are described in [13, 15, 20, 25].

The computation of the messages coming from the code block may in certain cases require additional iterations. If this occurs, a doubly iterative algorithm should be set up (see [9] for an example).

The formulation of the sum–product algorithm depends on the fact that two operations are available (sum, product) and that one is distributive with respect to the other:

$$(a \cdot b) + (a \cdot c) = a \cdot (b + c).$$

Generalizations of the sum–product algorithm are possible for all problems in which two operations are defined for which distributivity holds: for example, *max–product* and *min–sum* algorithms can be defined [2].

Acknowledgments This work was supported by the Spanish Ministry of Education and Science under Project TEC2006-01428/TCM

References

1. S. M. Aji, G. B. Horn, and R. J. McEliece, "On the convergence of iterative decoding on graphs with a single cycle," in *Proc. IEEE Int. Symp. Inform. Theory (ISIT 1998)*, Cambridge, MA, p. 276, August 1998.
2. S. M. Aji and R. J. McEliece, "The generalized distributive law," *IEEE Trans. Inform. Theory*, vol. 46, no. 2, pp. 325–343, March 2000.
3. A. Anastasopoulos, K. M. Chugg, G. Colavolpe, G. Ferrari, and R. Raheli, "Iterative detection for channels with memory," *Proc. IEEE*, vol. 95, no. 6, pp. 1272–1294, June 2007.
4. M. Bayati, D. Shah, and M. Sharma, "Max–product for maximum weight matching: Convergence, correctness, and LP duality," *IEEE Trans. Inform. Theory*, vol. 54, no. 3, pp. 1241–1251, March 2008.
5. S. Benedetto and E. Biglieri, *Principles of Digital Transmission*. New York, NY: Kluwer/ Plenum, 1999.
6. E. Biglieri, *Coding for Wireless Channels*. New York: Springer, 2005.
7. E. Biglieri, R. Calderbank, A. Constantinides, A. Goldsmith, A. Paulraj, and H. V. Poor, *MIMO Wireless Communications*. Cambridge, UK: Cambridge University Press, 2007.
8. E. Biglieri, A. Nordio, and G. Taricco, "Iterative receivers for coded MIMO signaling," *Wireless Commun. Mob. Comput.*, vol. 4, no. 7, pp. 697–710, November 2004.
9. E. Biglieri, A. Nordio, and G. Taricco, "MIMO Doubly-iterative receivers: Pre- vs. post-cancellation filtering ," *IEEE Commun. Lett.*, vol. 9, no. 2, pp. 106–108, February 2005.
10. E. Biglieri, G. Taricco, and A. Tulino, "Decoding space–time codes with BLAST architectures," *IEEE Trans. Signal Process.*, vol. 50, no. 10, pp. 2547–2552, October 2002.
11. J. J. Boutros and G. Caire, "Iterative multiuser joint decoding: Unified framework and asymptotic analysis," *IEEE Trans. Inform. Theory*, vol. 48, no. 7, pp. 1772–1793, July 2002.
12. G. Caire, A. Tulino, and E. Biglieri, "Iterative multiuser joint detection and parameter estimation: A factor-graph approach," *IEEE Information Theory Workshop (ITW2001)*, Cairns, Australia, September 2–7, 2001.
13. M. Coates and R. Nowak, "Network for networks: Internet analysis using graphical statistical models," *Proc. IEEE Workshop Neural Netw. Signal Process.*, vol. 2, pp. 755–764, December 11–13, 2000.
14. G. Colavolpe and G. Germi, "On the application of factor graphs and the sum-product algorithm to ISI channels," *IEEE Trans. Commun.*, vol. 53, no. 5, pp. 818–825, May 2005.

15. R. Drost and A. C. Singer, "Factor graph methods for three-dimensional shape reconstruction as applied to LIDAR imaging," *J. Opt. Soc. Amer. A, Opt. Image Sc.*, vol. 21, no. 10, pp. 1855–1868, October 2004.
16. R. Drost and A. C. Singer, "Factor graph algorithms for equalization," *IEEE Trans. Signal Process.*, vol. 55, no. 5, pp. 2052–2065, May 2007.
17. A. W. Eckford and S. Pasupathy, "Iterative multiuser detection with graphical modeling," *IEEE Int. Conf. Personal Wireless Communications*, Hyderabad, India, pp. 454–458, December 17–20, 2000.
18. G. J. Foschini, "Layered space–time architecture for wireless communication in a fading environment when using multi-element antennas," *Bell Labs. Tech. J.*, vol. 1, no. 2, pp. 41–59, Autumn 1996.
19. B. Frey, *Graphical Models for Machine Learning and Digital Communication*. Cambridge, MA: MIT Press, 1998.
20. B. J. Frey, R. Koetter, and A. C. Singer, " 'Codes' on images and iterative phase unwrapping," in *Proc. IEEE Inform. Theory Workshop*, Cairns, Australia, pp. 9–11, September 2–7, 2001.
21. Q. Guo, L. Ping, and H.-A. Loeliger, "Turbo equalization based on factor graphs," in *Proc. IEEE Int. Symp. Information Theory*, Adelaide, Australia, September 4–9, 2005, pp. 2021–2025.
22. C. Herzet, N. Noels, V. Lottici, H. Wymeersch, M. Luise, M. Moeneclaey, and L. Vandendorpe, "Code-aided turbo synchronization," *Proc. IEEE*, vol. 95, no. 6, pp. 1255–1271, June 2007.
23. R. Koetter, A. C. Singer, and M. Tüchler, "Turbo equalization," *IEEE Signal Process. Mag.*, Vol. 21, no. 1, pp. 67–80, January 2004.
24. F. R. Kschischang, B. J. Frey, and H.-A. Loeliger, "Factor graphs and the sum–product algorithm," *IEEE Trans. Inform. Theory*, vol. 47, no. 2, pp. 498–519, February 2001.
25. H.-A. Loeliger, "Least squares and Kalman filtering," in *Codes, Graphs, and Systems*, R. Blahut and R. Koetter, Eds. Boston, MA: Kluwer, 2002, pp. 113–135.
26. H.-A. Loeliger, "An introduction to factor graphs," *IEEE Signal Proc. Mag.*, vol. 21, no. 1, pp. 28–41, January 2004.
27. H.-A. Loeliger, J. Dauwels, J. Hu, S. Korl, L. Ping, and F. R. Kschischang, "The factor graph approach to model-based signal processing," *Proc. IEEE*, vol. 95, no. 6, pp. 1295–1322, June 2007.
28. H.-A. Loeliger, J. Dauwels, V. M. Koch, and S. Korl, "Signal processing with factor graphs: Examples," in *Proc. First Int. Symp. on Control, Communications and Signal Processing*, Hammamet, Tunisia, March 21–24, 2004, pp. 571–574.
29. J. Pearl, *Probabilistic Reasoning in Intelligent Systems: Networks of Plausible Inference*. San Francisco, CA: Morgan Kaufmann, 1988.
30. T. Richardson and R. Urbanke, *Modern Coding Theory*. New York, NY: Cambridge University Press, 2008.
31. M. Tüchler, R. Koetter, and A. C. Singer, "Turbo equalization: Principles and new results," *IEEE Trans. Commun.*, vol. 50, no. 5, pp. 754–766, May 2002.
32. M. Tüchler, S. ten Brink, and J. Hagenauer, "Masures for tracing convergence of iterative decoding algorithms," in *Proc. 4th IEEE/ITG Conf. Source and Channel Coding*, Berlin, Germany, pp. 53–60, January 2002.
33. N. Wiberg, *Codes and Decoding on General Graphs*, Ph.D. Dissertation 440, Univ. Linköping, Linköping, Sweden, Linköping Studies in Science and Technology, 1996.
34. A. P. Worthen and W. E. Stark, "Unified design of iterative receivers using factor graphs," *IEEE Trans. Inform. Theory*, vol. 47, no. 2, pp. 843–849, February 2001.
35. H. Wymeersch, *Iterative Receiver Design*. Cambridge, UK. Cambridge University Press, 2007.

Chapter 6
Fundamentals of Multi-user MIMO Communications

Luca Sanguinetti and H. Vincent Poor

6.1 Introduction

In recent years, the remarkable promise of multiple-antenna techniques has motivated an intense research activity devoted to characterizing the theoretical and practical issues associated with multiple-input multiple-output wireless channels. This activity was first focused primarily on single-user communications but more recently there has been extensive work on multi-user settings. The aim of this chapter is to provide an overview of the fundamental information-theoretic results and practical implementation issues of multi-user multiple-antenna networks operating under various conditions of channel state information.

The contents of this chapter are as follows. In Section 6.1 we introduce basic notation and describe the system model of interest. The latter includes both uplink and downlink models for a general mobile communication system operating with multiple antennas at both the base station and mobile terminals. In Section 6.2 we concentrate on channel capacity as a means of characterizing such systems, and review basic results under various operating conditions, including patterns of knowledge of information about the state of the channel between transmitter(s) and receiver(s). In Section 6.3 we address the problem of acquisition of channel state information at the transmitter and receiver, and describe the distinctive features of open-loop and closed-loop systems. In Section 6.4 we provide an overview of system design issues and discuss techniques that require channel state information at the transmitter, while in Section 6.5 we briefly review some recent work on techniques that allow

Luca Sanguinetti
Dipartimento di Ingegneria dell'Informazione, Universita di Pisa,
Via Caruso 16, Pisa 56122, Italy,
e-mail: luca.sanguinetti@iet.unipi.it

H. Vincent Poor
Department of Electrical Engineering, Princeton University,
Olden Street, Princeton, NJ 08544, USA,
e-mail: poor@princeton.edu

V. Tarokh (ed.), *New Directions in Wireless Communications Research*,
DOI 10.1007/978-1-4419-0673-1_6,
© Springer Science+Business Media, LLC 2009

achievement of the potential benefits of multiple antennas using limited feedback links. Finally, we draw some conclusions in Section 6.6.

Due to the considerable amount of work in this field, this exposition is necessarily incomplete, and rather reflects the subjective taste and inclination of the authors. To compensate for this partial view, a list of references is included as an entree into the extensive literature available on the subject.

6.2 System Model

We consider both the uplink and downlink of a flat-fading multi-user multiple-input multiple-output (MIMO) network[1] in which the base station (BS) is endowed with M antennas, and K mobile terminals (MTs) are simultaneously active. Without loss of generality, we assume that all the MTs are equipped with the same number N of antennas.[2]

We denote by $\mathbf{H}_{ul,k} \in \mathbb{C}^{M \times N}$ the *uplink* channel matrix whose entries represent the channel gains from the transmit antennas at the kth MT to the receive antennas of the BS and are modeled as independent complex circularly symmetric Gaussian random variable with zero-mean and unit variance.[3] Moreover, we assume that the users are randomly located within the cell so as to experience independent fading channels. Such a model is known in the literature as independent and identically distributed (i.i.d) Rayleigh fading (see, for example, the tutorial paper of Biglieri et al. in [1] for a detailed discussion of other channel models). At the BS, the discrete-time received signal $\mathbf{y} \in \mathbb{C}^{M \times 1}$ can be written as

$$\mathbf{y} = \sum_{k=1}^{K} \mathbf{H}_{ul,k} \mathbf{x}_k + \mathbf{n} \qquad (6.1)$$

where $\mathbf{x}_k \in \mathbb{C}^{N \times 1}$ is the signal transmitted by the kth user while $\mathbf{n} \in \mathbb{C}^{M \times 1}$ is the receiver noise modeled as a complex Gaussian vector with zero mean and covariance

---

[1] Although specific for a flat-fading channel, the model adopted throughout the chapter can easily be extended to frequency-selective environments using orthogonal frequency-division multiplexing (OFDM) as a transmission technique.

[2] We adopt the following notation: boldface uppercase and lowercase letters denote matrices and vectors. We use $\mathbf{A} = \text{diag}\{a(n); n = 1, 2, \ldots, N\}$ to indicate an $N \times N$ diagonal matrix with entries $a(n)$ and $\mathbf{B} = \text{diag}\{\mathbf{B}(1), \mathbf{B}(2), \ldots, \mathbf{B}(Q)\}$ to represent a block-diagonal matrix. We use, respectively, $\mathbf{A}^{-1}$, $\text{tr}\{\mathbf{A}\}$, and $|\mathbf{A}|$ to denote the inverse, trace, and determinant of a matrix $\mathbf{A}$. We denote by $\mathbf{I}_K$ and $\mathbf{0}_K$ the identity and null matrices of order K, respectively, while we use $\text{E}\{\cdot\}$ for expectation, $\|\cdot\|$ for the Euclidean norm of the enclosed vector and the superscripts $*$, T and H for complex conjugation, transposition, and Hermitian transposition, respectively. The notation $\mathbf{A} \geq 0$ indicates that $\mathbf{A}$ is positive semidefinite while $[\cdot]_{k,\ell}$ denote the (k, ℓ)th entry of the enclosed matrix. Finally, we use $\text{conv}\{\cdot\}$ for the convex hull operator.

[3] This model can be reasonably adopted whenever the transmit and receive antennas operate in a scattering environment and the antenna spacing is larger than the spatial coherence distance.

matrix $\mathbf{I}_M$. We denote by $\mathbf{X}_k = \mathrm{E}\{\mathbf{x}_k \mathbf{x}_k^H\}$ the transmit covariance matrix of the kth MT and assume that it is constrained to satisfy the following inequality:

$$\mathrm{tr}\,(\mathbf{X}_k) \le \rho_k, \tag{6.2}$$

where ρ_k is the power available for transmission at the kth MT. Letting $\mathbf{H}_{ul} = [\mathbf{H}_{ul,1}\mathbf{H}_{ul,2}\cdots\mathbf{H}_{ul,K}]$, we may rewrite (6.1) in matrix notation as

$$\mathbf{y} = \mathbf{H}_{ul}\mathbf{x} + \mathbf{n}, \tag{6.3}$$

where we have defined $\mathbf{x} = [\mathbf{x}_1^T, \mathbf{x}_2^T, \ldots, \mathbf{x}_K^T]^T$.

Similarly, we denote by $\mathbf{H}_{dl,k} \in \mathbb{C}^{N \times M}$ the *downlink* channel matrix whose (j, i)th entry now represents the channel gain from the ith transmit antenna at the BS to the jth receive antenna of the kth MT and again is modeled as a complex circularly symmetric Gaussian random variable with zero mean and unit variance. The discrete-time signal at the kth MT can be written as

$$\mathbf{r}_k = \mathbf{H}_{dl,k}\mathbf{s} + \mathbf{z}_k, \tag{6.4}$$

where $\mathbf{s} \in \mathbb{C}^{M \times 1}$ is the signal transmitted by the BS while $\mathbf{z}_k \in \mathbb{C}^{N \times 1}$ accounts for the receiver noise modeled as a complex Gaussian vector with zero mean and covariance matrix $\mathbf{I}_N$. We denote by $\mathbf{S} = \mathrm{E}\{\mathbf{s}\mathbf{s}^H\}$ the transmit covariance matrix and assume that the BS is subject to the following power constraint:

$$\mathrm{tr}\,(\mathbf{S}) \le p. \tag{6.5}$$

Collecting the signals received at all MTs into a single vector $\mathbf{r} \in \mathbb{C}^{KN \times 1}$ yields

$$\mathbf{r} = \mathbf{H}_{dl}\mathbf{s} + \mathbf{z}, \tag{6.6}$$

where $\mathbf{H}_{dl} = [\mathbf{H}_{dl,1}^T\mathbf{H}_{dl,2}^T\cdots\mathbf{H}_{dl,K}^T]^T$ while $\mathbf{z} = [\mathbf{z}_1^T, \mathbf{z}_2^T, \ldots, \mathbf{z}_K^T]^T$ is Gaussian with zero mean and covariance matrix $\mathbf{I}_{KN}$.

The systems described by (6.3) and (6.6) are, respectively, known in the technical literature as the Gaussian MIMO multiple-access channel (MAC) and the Gaussian MIMO broadcast channel (BC), respectively. They essentially represent an extension of the single antenna uplink and downlink systems first formally introduced by Ahslwede [2] and Cover [3] in the early 1970s. Since then they have attracted a great deal of attention in the research community as they model the communication links of a large variety of practical systems such as cellular networks, wireless local area networks (WLANs) and digital subscriber line (DSL) links (to name only a few).

6.3 Capacity

In the sequel, we consider the capacity of the two multi-user MIMO networks described above. Although not the only way of characterizing these channels, the capacity is without doubt the most important information-theoretic measure that drives the design of communication systems. In a single-user system, it is operationally defined as the maximum data rate that the channel can support with an arbitrarily low error probability while it is mathematically computed maximizing the mutual information $I(\mathbf{x}, \mathbf{y})$ between the input $\mathbf{x}$ and the output $\mathbf{y}$ of the channel over all the possible choices of $P_\mathbf{x}$

$$C = \max_{P_\mathbf{x}} I(\mathbf{x}, \mathbf{y}). \tag{6.7}$$

On the other hand, a multi-user channel with K users is characterized by a K-dimensional capacity region $\mathcal{C} \in \mathbb{R}_+^K$ (we denote by $\mathbb{R}_+^K$ the set of K-tuples with non-negative real-valued entries). Each point in this region is identified by a K-tuple $\mathcal{R} = (R_1, R_2, \dots, R_K)$ and represents a combination of rates at which users can send information with an arbitrarily low error probability.

6.3.1 Capacity Region of the Gaussian MIMO MAC

We begin by reviewing the basic results on the channel capacity of the Gaussian MIMO MAC (MIMO GMAC) described by (6.2) and (6.3). In particular, we consider two different situations. The first is referred to as the *deterministic channel* and relies on the assumption that $\mathbf{H}_{ul}$ is constant and known at the receiver and transmitters. The second, known as *fading channel*, is the case in which $\mathbf{H}_{ul}$ is time varying and *ergodic* [1]. We also assume that in this latter case the transmitters are subject to short-term power constraints meaning that each MT must meet the constraint in (6.2) for every channel realization. This is equivalent to saying that the available power cannot be adaptively allocated over time. Moreover, we will consider the situation in which the channel is either known or not at the MTs. For notational simplicity, in this section we drop the subscript ul.

6.3.1.1 Deterministic Channel

To ease understanding, we begin by considering a two-user scenario (i.e., $K = 2$). Moreover, we assume that each user is equipped with a single-transmit antenna (i.e., $N = 1$). In this case, the capacity region is a pentagon in the positive quadrant of the (R_1, R_2) plane, which can be written as follows [4]:

$$\mathcal{C}_{MAC}(\mathbf{H}, \boldsymbol{\rho}) = \left\{ \mathcal{R} \in \mathbb{R}_+^2 : \begin{array}{c} 0 \le R_1 \le \log \left| \mathbf{I}_M + \rho_1 \mathbf{h}_1 \mathbf{h}_1^H \right| \\[2mm] 0 \le R_2 \le \log \left| \mathbf{I}_M + \rho_2 \mathbf{h}_2 \mathbf{h}_2^H \right| \\[2mm] R_1 + R_2 \le \log \left| \mathbf{I}_M + \rho_1 \mathbf{h}_1 \mathbf{h}_1^H + \rho_2 \mathbf{h}_2 \mathbf{h}_2^H \right| \end{array} \right\}, \tag{6.8}$$

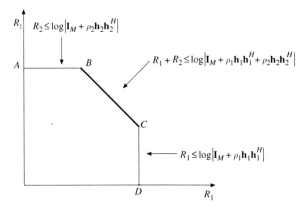

Fig. 6.1 Example of the capacity region in a two-user scenario with single-transmit antennas

where $\rho = [\rho_1, \rho_2]$ while $\mathbf{h}_k \in \mathbb{C}^{M \times 1}$ ($k = 1, 2$) represents the kth column of the uplink channel matrix, i.e., $\mathbf{H} = [\mathbf{h}_1, \mathbf{h}_2]$. The above equation leads to an interesting physical interpretation. The achievable rate of each user cannot exceed that of a single-user system in which the other user is turned off. This is made evident by the first two terms on the right-hand side (RHS) of (6.8). On the other hand, the last term on the RHS of (6.8) indicates that the sum rate cannot be larger than that of a system in which the two active users act as a single user equipped with two transmit antennas sending independent signals and subject to different power constraints [4].

An example of the capacity region for a two-user scenario is depicted in Fig. 6.1. The point A corresponds to the maximum rate at which user 1 can reliably send information over the channel when user 2 is not transmitting. The corner point D is the complementary rate for user 2. The points B and C are of particular interest because they represent the maximum achievable sum rate given by

$$R_{MAC}^{max} = \max_{(R_1, R_2) \in \mathcal{C}_{MAC}(\mathbf{H}, \rho)} R_1 + R_2 \tag{6.9}$$

and they can be achieved by resorting to a two-stage *interference cancellation* (known also as *successive decoding*) scheme, which operates as follows. In the first stage, it detects the data stream of user 1 treating the signal from user 2 as interference. In the second stage, the contribution of user 1 is reconstructed and cancelled out from the received signal before detection of user 2. Based on the above procedure, it can be easily found that the maximum rate at which user 1 can transmit is given by

$$R_1 = \log\left(1 + \rho_1 \mathbf{h}_1^H \left(\mathbf{I}_M + \rho_2 \mathbf{h}_2 \mathbf{h}_2^H\right)^{-1} \mathbf{h}_1\right), \tag{6.10}$$

where $\rho_1 \mathbf{h}_1^H \left(\mathbf{I}_M + \rho_2 \mathbf{h}_2 \mathbf{h}_2^H\right)^{-1} \mathbf{h}_1$ represents the signal-to-noise-plus-interference ratio (SINR) for user 1 when user 2 is treated as colored Gaussian noise [5]. On the other hand, the maximum rate at which user 2 can transmit is given by

$$R_2 = \log\left(1 + \rho_2 \mathbf{h}_2^H \mathbf{h}_2\right). \tag{6.11}$$

The tuple (R_1, R_2) of (6.10) and (6.11) is precisely the point B in Fig. 6.1. The corresponding sum rate can be easily computed as follows. We begin by using the identity $(1 + \mathbf{BA}) = |\mathbf{I}_M + \mathbf{AB}|$ with $\mathbf{A} = \rho_1\left(\mathbf{I}_M + \rho_2 \mathbf{h}_2 \mathbf{h}_2^H\right)^{-1} \mathbf{h}_1$ and $\mathbf{B} = \mathbf{h}_1^H$ so as to rewrite (6.10) as follows:

$$R_1 = \log\left|\mathbf{I}_M + \rho_1\left(\mathbf{I}_M + \rho_2 \mathbf{h}_2 \mathbf{h}_2^H\right)^{-1} \mathbf{h}_1 \mathbf{h}_1^H\right| \tag{6.12}$$

or, equivalently,

$$R_1 = \log\left|\mathbf{I}_M + \rho_1 \mathbf{h}_1 \mathbf{h}_1^H + \rho_2 \mathbf{h}_2 \mathbf{h}_2^H\right| - \log\left|\mathbf{I}_M + \rho_2 \mathbf{h}_2 \mathbf{h}_2^H\right|. \tag{6.13}$$

Similarly, (6.11) becomes

$$R_2 = \log\left|\mathbf{I}_M + \rho_2 \mathbf{h}_2 \mathbf{h}_2^H\right|. \tag{6.14}$$

Collecting all the above results togheter, it follows that

$$R_1 + R_2 = \log\left|\mathbf{I}_M + \rho_1 \mathbf{h}_1 \mathbf{h}_1^H + \rho_2 \mathbf{h}_2 \mathbf{h}_2^H\right|, \tag{6.15}$$

which is exactly the maximum sum rate as depicted in Fig. 6.1. Clearly, inverting the cancellation order allows one to achieve the other corner point C. Any other point on the segment BC guarantees the same maximum achievable sum rate and can be attained by means of *time sharing* between the two different strategies at points B and C or by using an alternative technique known as *rate splitting* proposed by Rimoldi and Urbanke [6].

When $N > 1$, each MT has more degrees of freedom that can be used to improve the system performance. To see how this comes about, we decompose the covariance matrices as $\mathbf{X}_k = \mathbf{U}_k \mathbf{D}_k \mathbf{U}_k^H$ ($k = 1, 2$) where $\mathbf{U}_k$ is unitary (i.e., $\mathbf{U}_k \mathbf{U}_k^H = \mathbf{I}$) and $\mathbf{D}_k$ is a diagonal matrix whose entries represent the power allocated to the different data streams and have to satisfy the following inequality $\text{tr}(\mathbf{D}_k) \le \rho_k$ (obtained from (6.2) after substituting $\mathbf{X}_k$ with $\mathbf{U}_k \mathbf{D}_k \mathbf{U}_k^H$). Thus, it follows that when $N > 1$ each MT can arbitrarily choose between different power allocations and rotations before sending the data streams out of the transmit antennas. This is in contrast to the case of $N = 1$ where each MT has no other choice than allocating all the available power to the single-transmit data stream. In general, different strategies result into different pairs $(\mathbf{X}_1, \mathbf{X}_2)$ so that the capacity region is a convex set given by the union of an infinite number of rate regions, each corresponding to a different pair $(\mathbf{X}_1, \mathbf{X}_2)$ and representing a pentagon in the (R_1, R_2) plane. Mathematically, it can be described as follows:

$$\mathcal{C}_{MAC}(\mathbf{H}, \rho) =$$

$$\bigcup_{\{\mathbf{X}_k \geq 0, \, \mathrm{tr}(\mathbf{X}_k) \leq \rho_k\}} \left\{ \mathcal{R} \in \mathbb{R}_+^2 : \begin{array}{c} 0 \leq R_1 \leq \log \left| \mathbf{I}_M + \mathbf{H}_1 \mathbf{X}_1 \mathbf{H}_1^H \right| \\ 0 \leq R_2 \leq \log \left| \mathbf{I}_M + \mathbf{H}_2 \mathbf{X}_2 \mathbf{H}_2^H \right| \\ R_1 + R_2 \leq \log \left| \mathbf{I}_M + \mathbf{H}_1 \mathbf{X}_1 \mathbf{H}_1^H + \mathbf{H}_2 \mathbf{X}_2 \mathbf{H}_2^H \right| \end{array} \right\},$$

$$(6.16)$$

where $\mathbf{X}_k$ for $k = 1, 2$ have to be positive semidefinite and to satisfy the power constraints given in (6.2). An example of capacity region for the two-user scenario with multiple-transmit antennas is sketched in Fig. 6.2. For the sake of explanation, a few rate regions corresponding to different sets of covariance matrices are depicted with dashed lines. The bold line represents the boundary of the capacity region and is obtained as the union of all the possible rate regions. As discussed before, the corner points A and D indicate the maximum rate at which users 1 and 2 can transmit, respectively, while the points within the segment connecting B and C represent the maximum achievable sum rate and can be achieved using a specific set of covariance matrices and applying the same techniques discussed before.

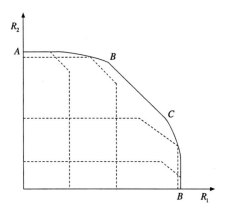

Fig. 6.2 A two-user capacity region as a union of capacity regions, each one corresponding to a feasible set of covariance matrices

At this stage, we are left with the problem of characterizing the capacity region for the general case of a K-user MIMO GMAC. Following the above line of reasoning, such a region can be mathematically described as follows:

$$\mathcal{C}_{MAC}(\mathbf{H}, \rho) = \bigcup_{\{\mathbf{X}_k \geq 0, \, \mathrm{tr}(\mathbf{X}_k) \leq \rho_k\}} \left\{ \mathcal{R} \in \mathbb{R}_+^K : \begin{array}{c} \forall \mathcal{S} \subseteq \{1, 2, \ldots, K\} \\ \sum_{k \in \mathcal{S}} R_k \leq \log \left| \mathbf{I}_M + \sum_{k \in \mathcal{S}} \mathbf{H}_k \mathbf{X}_k \mathbf{H}_k^H \right| \end{array} \right\},$$

$$(6.17)$$

where ρ is now given by $\rho = [\rho_1, \rho_2, \ldots, \rho_K]$ while $\mathcal{S}$ refers to an arbitrary subset of $\{1, 2, \ldots, K\}$. Each set of covariance matrices results in a K-dimensional *polymatroid* of achievable rates as shown by Tse and Hanly in [7], i.e.,

$$
\left\{
\mathcal{R} \in \mathbb{R}_+^K :
\begin{array}{c}
\forall \mathcal{S} \subseteq \{1, 2, \ldots, K\} \\
\sum_{k \in \mathcal{S}} R_k \leq \log \left| \mathbf{I}_M + \sum_{k \in \mathcal{S}} \mathbf{H}_k \mathbf{X}_k \mathbf{H}_k^H \right|
\end{array}
\right\},
\tag{6.18}
$$

while the capacity region corresponds to the union of such polymatroids over all sets of covariance matrices. An interesting problem is how to design the set $\{\mathbf{X}_k\}$ that achieves any point on the boundary of the capacity region. This problem can be addressed as follows. We begin by observing that (6.17) is a convex set and, thus, its boundary points can be achieved by maximizing a linear combination of the user rates [8]:

$$
\mu_1 R_1 + \mu_2 R_2 + \cdots + \mu_K R_K,
\tag{6.19}
$$

where μ_k are non-negative real-valued parameters satisfying $\mu_1 + \mu_2 + \cdots + \mu_K = 1$. The μ_k's are known as *user priorities* due to the fact that for a given set of covariance matrices and user priorities, the corner points of the corresponding polymatroid are attained using successive decoding in order of *increasing priority*, i.e., the user with the highest priority is decoded last. Assume, for example, that $\mu_1 \leq \mu_2 \leq \cdots \leq \mu_K$. In this case the signal of user k is decoded before any other user with priority $\mu_i > \mu_k$ so that its achievable rate is given by

$$
R_k = \log \left| \mathbf{I}_M + \left(\mathbf{I}_M + \sum_{i=k+1}^{K} \mathbf{H}_i \mathbf{X}_i \mathbf{H}_i^H \right)^{-1} \mathbf{H}_k \mathbf{X}_k \mathbf{H}_k^H \right|,
\tag{6.20}
$$

where $(\mathbf{I}_M + \sum_{i=k+1}^{K} \mathbf{H}_i \mathbf{X}_i \mathbf{H}_i^H)^{-1}$ denotes the covariance matrix of user k when users $\mu_i > \mu_k$ are treated as colored Gaussian noise. Using simple mathematical derivations, the above equation can be rewritten as follows:

$$
R_k = \log \left| \mathbf{I}_M + \sum_{i=k}^{K} \mathbf{H}_k \mathbf{X}_k \mathbf{H}_k^H \right| - \log \left| \mathbf{I}_M + \sum_{i=k+1}^{K} \mathbf{H}_k \mathbf{X}_k \mathbf{H}_k^H \right|.
\tag{6.21}
$$

Substituting (6.21) into (6.19), it follows that the optimal set $\{\mathbf{X}_k\}$ maximizing (6.19) can be found by solving the following problem [9]:

$$
\begin{aligned}
&\max_{\mathbf{X}} \quad f(\{\mu_k\}) \\
&\text{s.t.} \quad \mathbf{X}_k \geq 0 \text{ and } \operatorname{tr}\{\mathbf{X}_k\} \leq \rho_k \text{ for } k = 1, 2, \ldots, K,
\end{aligned}
\tag{6.22}
$$

where $\mathbf{X} = \operatorname{diag}\{\mathbf{X}_1, \mathbf{X}_2, \ldots, \mathbf{X}_K\}$ and $f(\{\mu_k\})$ is given by

$$f(\{\mu_k\}) = \mu_1 \log \left| \mathbf{I}_M + \sum_{k=1}^{K} \mathbf{H}_k \mathbf{X}_k \mathbf{H}_k^H \right| + \sum_{k=2}^{K} (\mu_k - \mu_{k-1}) \log \left| \mathbf{I}_M + \sum_{i=k}^{K} \mathbf{H}_i \mathbf{X}_i \mathbf{H}_i^H \right|.$$
(6.23)

Finding the exact solution of (6.22) is in general a computationally expensive task. However, we observe that we are maximizing a linear combination of concave functions subject to constraints that are convex in the set of positive-semidefinite matrices [10]. Hence, the problem is convex and the solution can be efficiently found using numerical optimization tools [8].

When the objective is to maximize the sum rate of the system (i.e., $\mu_1 = \mu_2 = \cdots = \mu_K$), we may rewrite (6.22) as

$$R_{MAC}^{\max} = \max_{\{\mathbf{X}_k \geq 0, \ \mathrm{tr}(\mathbf{X}_k) \leq \rho_k \ \forall k\}} \log \left| \mathbf{I}_M + \sum_{k=1}^{K} \mathbf{H}_k \mathbf{X}_k \mathbf{H}_k^H \right|$$
(6.24)

and the solution can be found by using the following generalization of the single-user water-filling algorithm as shown by Yu et al. in [11]:

Multi-user iterative water-filling algorithm

(1)   Initialization

    (a)   Set $\mathbf{X}_k = \mathbf{0}_N$ for $k = 1, 2, \ldots, K$.

(2)   Until the sum rate converges

    (a)   Until $k < K$

        (i)   Compute $\mathbf{X}' = \mathbf{I}_M + \sum_{j \neq k} \mathbf{H}_j \mathbf{X}_j \mathbf{H}_j^H$.

        (ii)   Set $\mathbf{X}_k = \arg \max_{\bar{\mathbf{X}}} \ \log \left(\mathbf{H}_k \bar{\mathbf{X}} \mathbf{H}_k^H + \mathbf{X}' \right)$.

As seen, the optimal $\{\mathbf{X}_k\}$ is computed iteratively and at each iteration the covariance matrix of one user is given by the single-user water-filling solution obtained by treating the others as Gaussian noise. In [11], the authors show that the above procedure converges to the maximum sum rate solution irrespective to the starting point. All this can be formally characterized by the following theorem.

Theorem 6.1. [65] *In a K-user MIMO GMAC, $\mathbf{X}_k$ is an optimal solution to the sum rate maximization problem if, and only if, $\mathbf{X}_k$ is the single-user waterfilling solution covariance matrix of the channel with matrix $\mathbf{H}_k$ and with covariance matrix $\mathbf{I}_M + \sum_{i=1, i \neq k}^{K} \mathbf{H}_i \mathbf{X}_i \mathbf{H}_i^H$ of the Gaussian noise, for all $k = 1, 2, \ldots, K$.*

6.3.1.2 Fading Channel

We begin by considering the case of perfect channel knowledge at both the transmitter and receiver and then focus on the situation in which only the receiver has this information. In the former case, Yu et al. [12] show that the system under

investigation can be thought of as a set of *parallel non-interfering* MIMO GMACs.
In particular, each element of the set corresponds to a different channel realization.
The ergodic capacity[4] region is thus computed as an average of the capacity regions
characterizing the parallel MIMO GMACs. Mathematically, we have that

$$\mathcal{C}_{MAC}(\rho) = \mathrm{E_H}\left\{\mathcal{C}_{MAC}(\mathbf{H}, \rho)\right\}, \tag{6.25}$$

where $\mathcal{C}_{MAC}(\mathbf{H}, \rho)$ is given in (6.17) while the statistical expectation must be computed with respect to the channel distribution. Similar to the deterministic channel,
we concentrate on the ergodic sum rate and aim at computing the optimal set of
covariance matrices. This is tantamount to solving

$$R_{MAC}^{\max} = \mathrm{E_H}\left\{\max_{\{\mathbf{X}_k \geq 0,\ \mathrm{tr}(\mathbf{X}_k) \leq \rho_k\ \forall k\}} \log\left|\mathbf{I}_M + \sum_{k=1}^{K} \mathbf{H}_k \mathbf{X}_k \mathbf{H}_k^H\right|\right\}. \tag{6.26}$$

As discussed by Rhee and Cioffi in [12], the above problem is convex and its solution can be computed by resorting to an extension of the iterative water-filling
algorithm (described previously), which operates independently for each channel
realization. On the other hand, when channel state information (CSI) is available only at the receiver, the MTs cannot adapt their covariance matrices to the
specific channel realization so that they are forced to use the same transmission
strategy over all fading states. In this case, the ergodic capacity region is still
computed as an average of the capacity regions corresponding to the different
channel realizations but now with a given transmission strategy. Mathematically, we
have that

$$\mathcal{C}_{MAC}(\rho) =$$

$$\bigcup_{\{\mathbf{X}_k \geq 0,\ \mathrm{tr}(\mathbf{X}_k) \leq \rho_k\ \forall k\}} \left\{\mathcal{R} \in \mathbb{R}_+^K : \begin{array}{c} \forall \mathcal{S} \subseteq \{1, 2, \ldots, K\} \\ \sum_{k \in \mathcal{S}} R_k \leq \mathrm{E_H}\left\{\log\left|\mathbf{I}_M + \sum_{k \in \mathcal{S}} \mathbf{H}_k \mathbf{X}_k \mathbf{H}_k^H\right|\right\} \end{array}\right\}, \tag{6.27}$$

which is equivalent to (6.17) except for the statistical expectation. Then, its boundary and optimal set $\{\mathbf{X}_k\}$ can be computed via the maximization problem given in
(6.22) even though the computation of the statistical expectation makes the problem
a bit more challenging. However, in the particular case of independent and identically distributed (i.i.d.) Rayleigh channels, this optimization is no longer required as
the boundary of the capacity region can be achieved by means of Gaussian codes and
the same uniform power allocation strategy over all fading states. This is formalized
in the following theorem (see, for example, [13, 14]).

---

[4] The ergodic capacity is the direct extension of capacity to fading channels. It corresponds to the
maximum mutual information averaged over all channel realizations. However, there are a number
of different ways to define the capacity of fading channels [1].

Theorem 6.2. *For an i.i.d. Rayleigh channel, the ergodic capacity region of a K-user MIMO GMAC is a polymatroid given by*

$$
\mathcal{C}_{MAC}(\rho) = \left\{ \mathcal{R} \in \mathbb{R}_+^K : \begin{array}{c} \forall S \subseteq \{1, 2, \dots, K\} \\ \sum_{k \in S} R_k \leq \mathrm{E}_\mathbf{H} \left\{ \log \left| \mathbf{I}_M + \sum_{k \in S} \frac{\rho_k}{N} \mathbf{H}_k \mathbf{H}_k^H \right| \right\} \end{array} \right\} \tag{6.28}
$$

and any point on its boundary is achieved when each user uses a Gaussian code and a uniform power allocation strategy.

Based on the above result, it follows that in these circumstances the maximum ergodic sum rate takes the form

$$
R_{MAC}^{max} = \mathrm{E}_\mathbf{H} \left\{ \log \left| \mathbf{I}_M + \sum_{k=1}^{K} \frac{\rho_k}{N} \mathbf{H}_k \mathbf{H}_k^H \right| \right\}. \tag{6.29}
$$

6.3.1.3 Asymptotic Analysis

In the following, we are interested in evaluating how the sum rate of a MIMO GMAC scales with respect to some system parameters such as the number of transmit and receive antennas, the number of users and the signal-to-noise ratios (SNRs).[5] To this end, we consider the fading channel and focus on the case where CSI is available only at the receiver. Assuming that the same amount of power is allocated to the different users, i.e., $\rho_k = \rho/K$, from (6.29) we have

$$
R_{MAC}^{max} = \mathrm{E}_\mathbf{H} \left\{ \log \left| \mathbf{I}_M + \frac{\rho}{KN} \mathbf{H}\mathbf{H}^H \right| \right\}. \tag{6.30}
$$

Interestingly, the RHS of the above equation is equivalent to the ergodic capacity of a single-user MIMO system equipped with KN transmit and M receive antennas, when only the receiver has channel knowledge [13]. This means that in these circumstances the lack of cooperation between the MTs does not reduce the capacity, which is exactly the same of a fully cooperative system. Hence, we may conclude that in an i.i.d. Rayleigh fading model in which all users have the same power constraint, CSI at the receiver is enough to achieve the potential benefits of a fully cooperative multiple antenna system. Similar to a single-user MIMO system, channel knowledge at the transmitters leads to an improvement of the performance in the low SNR regime but becomes irrelevant when the SNR increases as illustrated by Viswanath et al. [15]. The authors show also that such an information provides vanishing advantages even in a system in which the SNR is fixed but the number of transmit antennas and the number of users is taken to infinity. These results allow

---

[5] As the noise power is normalized to 1, the SNRs coincide with the transmit powers.

us to characterize the asymptotic performance of a MIMO GMAC on the basis of (6.30)[6]:

1. If M is fixed and $KN \to \infty$, the law of large numbers yields $\frac{1}{KN}\mathbf{HH}^H \to \mathbf{I}_M$ (almost surely) so that (6.30) tends to an upper bound given by

$$R_{MAC}^{\max} = \log|\mathbf{I}_M + \rho\mathbf{I}_M| = \log(1+\rho)^M = M\log(1+\rho) \qquad (6.31)$$

   from which it follows that asymptotically (in KN) the sum rate grows linearly with M. This is equivalent to saying that a 3 dB increase in the SNR gives an M bit/s/Hz increase in the sum rate. As increasing KN is essentially equivalent to increasing the number of independent paths, this is a simple example of the fact that fading is a resource to be *exploited* rather than a detrimental effect to be mitigated [17].

2. According to [14], in the high-SNR regime and for any finite values of M, K, and N the ergodic sum rate in (6.30) is well approximated by

$$R_{MAC}^{\max} = \min(M, KN)\log(1+\rho) + o(1), \qquad (6.32)$$

   where $o(1)$ denotes a term not increasing with ρ. From the above equation it follows that

$$A_{MAC}^{\infty} = \lim_{\rho\to\infty} \frac{R_{MAC}^{\max}}{\log\rho} = \min(M, KN), \qquad (6.33)$$

   which is commonly called the *multiplexing gain* and essentially measures the *degrees of freedom* available for reliable communication [18].

A close inspection of (6.33) leads to the following interesting practical observation. Increasing K while keeping N fixed makes the sum rate grow linearly with M as long as $KN \geq M$. This is in sharp contrast to a single-user MIMO system in which the multiplexing gain would be constant and bounded by $\min(M, N)$. From a practical point of view, this is very interesting since increasing the number of antennas at the BS is not an issue of concern as it is at the MTs. Moreover, we observe that this situation is even more realistic than $KN \leq M$ as many networks already operate with a number of users that is much larger than the number of transmit antennas.

6.3.2 Gaussian MIMO Broadcast Channel

The capacity region of a general BC is still unknown. Up to now, only an achievable rate region computed by Marton [19] is available, but it is not clear whether it coincides or not with the capacity region. On the other hand, for those BCs (known as

---

[6] We stress again the fact that most of the results reported in this chapter depend heavily on i.i.d Rayleigh fading. Different models such as spatially correlated fading, Ricean fading and so forth may lead to different conclusions. A good survey of the results obtained on these subjects can be found in [16].

degraded) in which the active users can be essentially ordered from the strongest to the weakest the capacity region is a well-known result and can be achieved using a technique known as *superposition coding* discussed by Bergman [20] and outlined as follows. At the transmitter, the signal is obtained as a linear combination of the different data streams and is decoded at the receivers using successive cancellation, where each user decodes and removes from the received signal the contributions of the weaker users before decoding its own signals.

The single antenna Gaussian BC (GBC) (i.e., $M = N = 1$) belongs to the class of degraded BCs as the users can be naturally ordered according to their respective SINRs. In the two-user case (assuming without loss of generality $|h_1| > |h_2|$), the capacity region can be described as follows [4]:

$$\mathcal{C}_{BC}(\mathbf{H}, p) =$$

$$\bigcup_{\{p_1,p_2\}:p_1+p_2=p} \left\{ \mathcal{R} \in R_+^2 : \begin{array}{c} R_1 \leq \log\left(1 + |h_1|^2 \, p_1\right) \\ R_2 \leq \log\left(1 + |h_2|^2 \, p_2 \left(1 + |h_2|^2 \, p_1\right)^{-1}\right) \end{array} \right\}, \quad (6.34)$$

where $\mathbf{H} = [h_1, h_2]^T$ while p_1 and p_2 denote the powers allocated to users 1 and 2, respectively, and must satisfy the equality $p_1 + p_2 = p$.

Although the MIMO GBC does not fall within the class of degraded BC channels,[7] its capacity region has recently been computed by Weingarten et al. [21]. Such a result is probably one of the major achievements in information theory of the recent years as it establishes that there exists a non-degraded BC whose capacity region can be fully characterized. In the following, we first consider the deterministic channel model and overview the fundamental steps that have led to the computation of the capacity region in this scenario. Then, we concentrate on the fading model and summarize some of the more important asymptotic results. To simplify notation, in the sequel we drop the subscript *dl*.

6.3.2.1 Deterministic Channel

In the next, we begin by introducing two basic concepts that have driven the research activity in this area over the last years, namely dirty paper coding (DPC) and uplink–downlink duality. Then, we illustrate how they have been employed to compute the achievable sum rate of a MIMO-GBC. Finally, we give a short description of the main result stated in [21].

Dirty Paper Coding

DPC is a channel coding technique closely related to superposition coding in which the user data streams are encoded sequentially in such a way that at the receive side

---

[7] Roughly speaking, the main reason is that the different propagation channels in a MIMO BC are described by matrices and no natural ordering exists for matrices.

each user sees no interference from the others that have been previously encoded. To highlight the effectiveness of a DPC-based strategy in the downlink of a multi-user scenario, we again consider the two-user single-antenna GBC. As for superposition coding, we assume that the transmit signal is computed as a linear combination of the user data streams and denote by p_k ($k = 1, 2$) the power allocated to the kth user. The data stream for user 2 is encoded using Gaussian coding while DPC is used to encode that for user 1 treating user 2 as non-casually known interference (this is equivalent to saying that the received signal at user 1 is not affected by the contribution of user 2).

At this stage, in order to compute the achievable rates of DPC we make use of a surprising result known as *writing on dirty paper* which was presented by Costa [22].

Theorem 6.3. [16] *Consider any channel whose output signal y takes this form* $y = x + i + w$ *where i and w are independent Gaussian random variables. If i is known non-casually at the transmitter and not at the receiver then the capacity of the channel is the same as if i were not present. Moreover, the capacity-achieving x is statistically independent of i.*

Based on the above theorem, the achievable rate of user 1 is given by

$$R_1 = \log\left(1 + |h_1|^2\, p_1\right) \tag{6.35}$$

and it is equivalent to the capacity of a system in which user 2 is not present. On the other hand, user 2 decodes its data stream treating user 1 as Gaussian interference so that its achievable rate takes the form

$$R_2 = \log\left(1 + |h_2|^2\, p_2 \left(1 + |h_2|^2\, p_1\right)^{-1}\right). \tag{6.36}$$

The tuple (R_1, R_2) is precisely a point on the boundary of the capacity region given in (6.34) which is known to be achievable using superposition coding and successive cancellation. Thus, it follows that the above DPC-based transmission technique represents an alternative capacity-achieving solution for the single-antenna GBC.

We now turn to the general case described by (6.6) and extend the above technique according to a result of Yu et al. [23], which essentially generalizes Theorem 6.3 to the vector case. Specifically, the transmit signal is given by $\mathbf{x} = \sum_{k=1}^{K} \mathbf{x}_k$ where the kth stream $\mathbf{x}_k$ is Gaussian with covariance matrix $\mathbf{S}_k = \mathrm{E}\{\mathbf{x}_k \mathbf{x}_k^H\}$ and it is encoded according to DPC so as to remove the interference of the streams with indices $i < k$. The transmit covariance matrix takes the form $\mathbf{S} = \sum_{k=1}^{K} \mathbf{S}_k$ (the user signals are uncorrelated by construction [22]) and the power constraint in (6.5) becomes

$$\sum_{k=1}^{K} \mathrm{tr}\,(\mathbf{S}_k) \leq p. \tag{6.37}$$

According to [23] for a given user ordering $\pi = [\pi(1), \pi(2), \ldots, \pi(K)]^T$ and a given set $\{\mathbf{S}_{\pi(k)}\}$ satisfying (6.37), the following rates are achievable:

$$R_{\pi(k)} = \log \frac{\left| \mathbf{H}_{\pi(k)} \sum_{i=k}^{K} \mathbf{S}_{\pi(i)} \mathbf{H}_{\pi(k)}^{H} + \mathbf{I}_N \right|}{\left| \mathbf{H}_{\pi(k)} \sum_{i=k+1}^{K} \mathbf{S}_{\pi(i)} \mathbf{H}_{\pi(k)}^{H} + \mathbf{I}_N \right|}, \quad k = 1, 2, \dots, K, \quad (6.38)$$

where $\pi(i)$ denotes the ith user to be encoded. The DPC rate region is given by the convex hull of the union of the above tuples over all possible sets of permutations and covariance matrices satisfying (6.37). Thus, we have

$$\mathcal{R}_{DPC}(\mathbf{H}, p) = \text{conv} \left\{ \bigcup_{\{\pi\}} \bigcup_{\{\mathbf{S}_k\}: \sum_{k=1}^{K} \text{tr}(\mathbf{S}_k) \leq p} \left\{ (R_{\pi(1)}, R_{\pi(2)}, \dots, R_{\pi(K)}) \in \mathbb{R}_+^K \right\} \right\}. \quad (6.39)$$

It is worth observing that the rates given in (6.38) are neither a concave nor a convex function of the correlation matrices. This makes the computation of the DPC region given in (6.39) a very computationally demanding task as the optimal set of covariance matrices can be found only by means of an exhaustive search.

Uplink–Downlink Duality

The *uplink–downlink duality* has been observed in seemingly different contexts and forms in the literature. It was first pointed out by Telatar in [13], where it is shown that for a single-user MIMO system interchanging the transmitter and receiver does not change the capacity of the system. On the other hand, Jindal et al. [24] demonstrate that the capacity region of the single-antenna GBC is equal to the capacity region of the *dual* (i.e., $h_{ul,k} = h_{dl,k}^* \ \forall k$) Gaussian MAC subject to the same *sum* power constraint instead of the common individual constraints. The general MIMO GBC has been later investigated by Viswanath and Tse [25] and Vishwanath et al. [26] and their major results can be formalized as follows.

Theorem 6.4. [56,57] *The DPC rate region of the MIMO GBC given in (6.6) subject to a power constraint p is equivalent to*

$$\mathcal{R}_{DPC}(\mathbf{H}, p) = \mathcal{C}_{D-MAC}(\mathbf{H}, p), \quad (6.40)$$

where $\mathcal{C}_{D-MAC}(\mathbf{H}, p)$ is the capacity region of the dual MIMO GMAC:

$$\mathcal{C}_{D-MAC}(\mathbf{H}, p) =$$

$$\bigcup_{\{\mathbf{X}_k \geq 0\}: \sum_{k=1}^{K} \text{tr}(\mathbf{X}_k) \leq p} \left\{ \mathcal{R} \in \mathbb{R}_+^K : \begin{array}{c} \forall \mathcal{S} \subseteq \{1, 2, \dots, K\} \\ \sum_{k \in \mathcal{S}} R_k \leq \log \left| \mathbf{I}_M + \sum_{k \in \mathcal{S}} \mathbf{H}_k^H \mathbf{X}_k \mathbf{H}_k \right| \end{array} \right\}. \quad (6.41)$$

A close observation of (6.41) indicates that $\mathcal{C}_{D-MAC}$ is equal to (6.17) after replacing $\mathbf{H}_k$ with $\mathbf{H}_k^H$ and imposing a sum power constraint p. This is not only extremely interesting from an information-theoretic point of view but it is also very appealing for practical purposes. In fact, it provides a powerful tool to numerically evaluate the DPC rate region. This is due to the fact that in contrast to (6.39) the boundary of the dual MAC capacity region (as discussed in Section 6.3.1.1) can be easily computed by solving a convex problem. Each corner point can then be attained by using the corresponding optimal set of covariance matrices $\{\mathbf{X}_k\}$ and re-sorting to successive decoding with a specific order. A possible example of the DPC rate region is shown in [16, Fig. 9].

The question, though, is now how to achieve the same data rates in the downlink. The answer is provided by Vishwanath et al. [26] where the authors propose a computationally efficient transformation that takes as inputs the optimal uplink set $\{\mathbf{X}_k\}$ and the corresponding decoding order and returns as outputs the set of downlink matrices $\{\mathbf{S}_k\}$ satisfying

$$\sum_{k=1}^{K} \text{tr}(\mathbf{X}_k) = \sum_{k=1}^{K} \text{tr}(\mathbf{S}_k) \tag{6.42}$$

and achieving the same data rates by means of DPC in the reverse order.

Achievable Sum Rate

The pioneering work in the computation of the capacity region of a MIMO GBC is due to Caire and Shamai [27]. Therein, the authors consider the simple case of a system with two users ($K = 2$) each equipped with a single-receive antenna ($N = 1$) and compute through direct calculation the maximum sum rate of DPC. Then, its optimality is proved making use of the Sato upper bound [28] which refers to the capacity of a system where the users are allowed to cooperate.

The above result has been later extended to the general case ($K > 2$ and/or $N > 1$) by several authors simultaneously. In particular, Yu et al. [29] compute the achievable sum rate of the system as the saddle-point of a Gaussian mutual information game, where a player chooses a transmit covariance matrix to maximize the mutual information $I(\mathbf{x}, \mathbf{y})$ and a *malicious nature* chooses a fictitious noise correlation matrix to minimize $I(\mathbf{x}, \mathbf{y})$. Once computed, this upper bound is used to prove that the structure of the optimal coding technique takes the form of DPC. Independent and different proofs of such a result are also given by Viswanath and Tse [25] and Vishwanath et al. [26]. Both essentially rely on the *uplink–downlink duality* and, instead of proving directly the optimality of DPC aim at demonstrating that the maximum sum rate of the dual MAC is equivalent to the Sato upper bound with the same constraint on the transmit power.

All these results can be collected into the following theorem which represents a fundamental step toward the computation of the entire capacity region.

Theorem 6.5. ([8, 56, 57, 65]) *The maximum sum rate of a MIMO GBC is achieved by means of DPC, i.e.,*

$$R_{BC}^{\max} = \max_{(R_1, R_2, \dots, R_K) \in \mathcal{R}_{DPC}(\mathbf{H}, p)} R_1 + R_2 + \dots + R_K. \tag{6.43}$$

From the above theorem and using (6.38), it follows that the problem of computing the maximum sum rate can be formulated as

$$\max_{\pi} \max_{\mathbf{S}} \sum_{k=1}^{K} \log \frac{\left| \mathbf{H}_{\pi(k)} \sum_{i=k}^{K} \mathbf{S}_{\pi(i)} \mathbf{H}_{\pi(k)}^{H} + \mathbf{I}_N \right|}{\left| \mathbf{H}_{\pi(k)} \sum_{i=k+1}^{K} \mathbf{S}_{\pi(i)} \mathbf{H}_{\pi(k)}^{H} + \mathbf{I}_N \right|} \tag{6.44}$$

s.t. $\mathbf{S}_k \geq 0 \ \forall k$ and $\sum_{k=1}^{K} \mathrm{tr}\{\mathbf{S}_k\} \leq p,$

and its solution can be found by resorting to two efficient iterative algorithms presented by Jindal et al. [30]. Both rely on the uplink–downlink duality to reformulate the problem into the following convex form:

$$R_{BC}^{\max} = \max_{\{\mathbf{X}_k \geq 0\}: \sum_{k=1}^{K} \mathrm{tr}(\mathbf{X}_k) \leq p} \log \left| \mathbf{I}_M + \sum_{k=1}^{K} \mathbf{H}_k^{H} \mathbf{X}_k \mathbf{H}_k \right|, \tag{6.45}$$

which is a subset of (6.41) and differs from (6.24) only in the sum power constraint. Hence, iterative algorithms inspired by the water-filling policy discussed in [11] can be used to converge to the optimal set $\{\mathbf{X}_k\}$ which is then mapped to its corresponding dual $\{\mathbf{S}_k\}$ using the transformation discussed previously [26].

The Capacity Region

Up to now, we have seen that DPC is the optimal strategy to achieve the maximum sum rate of a MIMO GBC and also that its rate region coincides with the dual MIMO GMAC capacity region. Moreover, we have recalled that the capacity region of the single-antenna GBC is equal to the capacity region of its dual GMAC [24]. Collecting all these facts together, the most obvious idea that comes to mind is either to find a better achievable rate region that may not be attained through DPC or to prove that the DPC region is indeed the capacity region of the MIMO GBC. There have been many attempts in these directions but only Weingarten et al. [21] have finally provided an answer to these claims. In particular, it is found that the DPC region is *exactly* the capacity region of the MIMO GBC as stated in the following theorem.

Theorem 6.6. ([60]) *Let $\mathcal{C}_{BC}(\mathbf{H}, p)$ denote the capacity region of the MIMO GBC given in (6.6) and subject to the power constraint p. Then*

$$\mathcal{C}_{BC}(\mathbf{H}, p) = \mathcal{R}_{DPC}(\mathbf{H}, p). \tag{6.46}$$

This important result is achieved by making use of all the key ideas described thus far plus several new concepts such as *degraded* and *enhanced* MIMO BCs and an extension of Bergmans' proof. Due to space limitations, we cannot provide further details about these results or their proofs; we invite the interested reader to refer to [21].

6.3.2.2 Fading Channel

When perfect CSI is available at both transmitter and receiver, the computation of the ergodic capacity region can be addressed following the same arguments as in Section 6.3.1.2. In fact, in this case, similar to the uplink, the system reduces to a set of parallel non-interfering MIMO GBCs, each one corresponding to a different channel realization (see, for example, Yu and Rhee [31] and Yu [32]). Then, the ergodic capacity region is given by

$$C_{BC}(p) = C_{D-MAC}(p), \tag{6.47}$$

where $C_{D-MAC}(p)$ is now computed from (6.25) using (6.17) after replacing $\mathbf{H}_k$ with $\mathbf{H}_k^H$ and imposing the sum power constraint p. Following the same arguments as before, the computation of the ergodic sum rate can be formulated as follows:

$$R_{BC}^{\max} = \mathrm{E_H} \left\{ \max_{\{\mathbf{X}_k \geq 0\}: \sum_{k=1}^K \mathrm{tr}(\mathbf{X}_k) \leq p} \log \left| \mathbf{I}_M + \sum_{k=1}^K \mathbf{H}_k^H \mathbf{X}_k \mathbf{H}_k \right| \right\} \tag{6.48}$$

and the solution can be computed by resorting to the algorithms discussed in [26,30].

On the other hand, when CSI is not available at the BS, the ergodic capacity region is still unknown as it is the optimal coding strategy, since DPC relies on perfect channel knowledge. However, there exists one special setting in which this does not hold true: the case in which the user channels are *statistically* equivalent and the same number of antennas is employed at all MTs (as the system investigated here) [33]. In fact, in these circumstances, if we assume that any one of the K users can reliably decode its received signal, then the same can be done by any other user. Hence, the sum rate of the system is bounded by the capacity of the channel between the BS and any single MT. For the i.i.d. Rayleigh fading model, this can be mathematically formulated as

$$C_{BC}(p) = \left\{ \mathcal{R} \in \mathbb{R}_+^K : \sum_{k=1}^K R_k \leq \mathrm{E_H} \left\{ \log \left| \mathbf{I}_M + \frac{p}{M} \mathbf{H}_1 \mathbf{H}_1^H \right| \right\} \right\} \tag{6.49}$$

from which it follows that the optimal coding strategy, in this case, is time-division multiple access (TDMA), i.e., transmit to only one user at a time. Hence, the ergodic sum rate is given by

$$R_{BC}^{\max} = \mathrm{E_H} \left\{ \log \left| \mathbf{I}_M + \frac{P}{M} \mathbf{H}_1 \mathbf{H}_1^H \right| \right\}. \tag{6.50}$$

The RHS of the above equation is equivalent to the ergodic capacity of a single-user system with M transmit and N receive antennas. This implies that, as already pointed out by Caire and Shamai [27] for the single-receive antenna case, the complete lack of channel knowledge at the transmitter reduces the multiplexing gain of the system to $\min(M, N)$. This observation has been extended to a general *isotropic*[8] channel model by Jafar and Goldsmith [34]. It is worth noting that such a result is in sharp contrast to the uplink where the multiplexing gain is always the same as that of a fully cooperative system, i.e., $\min(M, KN)$, independently from the CSI at the transmitters. We elaborate further on this in the next section.

6.3.2.3 Asymptotic Results

As for the MIMO GMAC, we now analyze how the sum rate of the MIMO GBC scales with the system parameters. In particular, we first focus on the case in which the transmitter and receiver both have perfect channel knowledge, and then consider the situation in which only the receivers have this information. When CSI is available at both BS and MTs, the following asymptotic results are in order.

1. If KN and p are fixed and $M \to \infty$, the ergodic sum rate grows as [35]

$$R_{BC}^{\max} = KN \log \left(1 + \frac{M}{KN} p \right). \tag{6.51}$$

This result can be easily proven as follows. From (6.48), the maximum achievable sum rate is obviously lower bounded by the sum rate achieved using a uniform power allocation for each user and each fading state:

$$R_{BC}^{\max} \geq E_{\mathbf{H}} \left\{ \log \left| \mathbf{I}_{KN} + \frac{p}{KN} \mathbf{H}^H \mathbf{H} \right| \right\}. \tag{6.52}$$

As M goes to infinity applying the law of large number yields $\mathbf{H}^H \mathbf{H} \to M \mathbf{I}_{KN}$ (almost surely) so that (6.51) easily follows from the RHS of the above equation using the same arguments leading to (6.31). Moreover, such a lower bound can be shown to be tight when M becomes large [36].

2. If M and p are fixed and $K \to \infty$, then for any N the ergodic sum rate grows as [37]

$$R_{BC}^{\max} = M \log \log KN. \tag{6.53}$$

The intuition behind this result is that, when K grows and CSI is available at the BS, it is likely to find M channels to transmit over. On the other hand, the doubly logarithmic increase with KN is essentially due to the inherent *multi-user diversity*. The latter can be thought of as a form of *selection diversity* and it was originally introduced by Knopp and Humblet [38]. It is essentially based on the following idea. As the users' signals undergo *independent* time-varying fading

---

[8] A set of channels $\mathcal{H}$ is said to be isotropic, if for each channel realization $\mathbf{H}_k \in \mathcal{H}$, we have that $\mathbf{H}_k \mathbf{U}_k \in \mathcal{H}$ where $\mathbf{U}_k$ is unitary [34].

channels, it is likely that, at a given time, there are users whose channels are not
faded. Intuitively, the maximization of the sum rate can be achieved by allocating,
at any time, the available resources to the "best" set of users. As we select these
users as the maximum among KN possible choices in an i.i.d. Rayleigh fading
channel, the effective SNR benefits by a factor $\log KN$ [37].

3. If M, K, and N are kept fixed, then the multiplexing gain is given by [35]

$$A_{BC}^{\infty} = \lim_{\rho \to \infty} \frac{R_{BC}^{\max}}{\log \rho} = \min(M, KN), \qquad (6.54)$$

which is the same as that of a system in which the users are allowed to cooperate.

On the other hand, when CSI is available only at the MTs the following results
apply:

1. If M, N, and p are fixed, the ergodic sum rate does not increase as $K \to \infty$,
thereby meaning that

$$\lim_{K \to \infty} \frac{R_{BC}^{\max}}{\log \log K} = 0. \qquad (6.55)$$

This is due to the fact that the optimal strategy in these circumstances (as discussed in Section 6.3.2.2) is to transmit to a randomly selected user during each
transmission time slot.

2. If M, K, and N are kept fixed, then the multiplexing gain is given by [35]

$$A_{BC}^{\infty} = \min(M, N), \qquad (6.56)$$

which is the same as that of a single-user system.

From the above results, it follows that CSI at the transmitter plays a key role for
downlink transmissions. This is in contrast to the uplink where CSI is not mandatory
to guarantee maximum capacity scaling.

6.4 Open- and Closed-Loop Systems

As seen from the above discussion, in order to exploit the potential benefits of multiple antennas, explicit knowledge of the channel parameters is required at the receiver
in the uplink and at both receiver and transmitter in the downlink.

Channel estimation at the receiver has received significant attention in the literature and several different solutions are already available for both uplink and
downlink.[9] Common approaches are the *training*-aided schemes where a known

---

[9] It is worth observing that channel estimation in uplink transmissions is a more demanding task.
The main problem is that users transmit from different locations and the signals arrive at the BS
after passing through different multipath channels. This means that the BS must estimate a large
number of parameters in the presence of multiple access interference, which is inevitably expected
to degrade the accuracy of the channel estimates. On the other hand, in downlink transmissions, all
the signals arriving from the transmitter to a given terminal propagate through the same channel.
This facilitates the channel estimation task, which can be accomplished with the same methods
employed for single-user applications.

sequence of symbols called *pilots* is periodically inserted within the transmitted signal (see [39] and the references therein). Although easy to implement, the use of pilot-based schemes inevitably reduces the spectral efficiency of the system. This has inspired considerable research on either *blind* or *semi-blind* channel estimation techniques, which are largely categorized into subspace-based or decision-directed methods. Both solutions lead to a considerable saving in training overhead but increase the required computational burden. An excellent overview of the results obtained in this area can be found in [40, Chapters 14 and 17].

Channel acquisition at the transmitter is in general a more demanding task. In the sequel, we discuss two different operating techniques that have to be used on the basis of the characteristics of the system under investigation.

6.4.1 Open-Loop Systems

This technique relies on the channel reciprocity between alternate uplink and downlink transmissions and, thus, it is suitable for time-division-duplex systems (such as Hyperlan/2 and IEEE 802.11a), in which the same frequency band is alternatively used for transmission and reception. In particular, if the channel variations are sufficiently slow (as occurs in indoor applications) or the time lag between two consecutive time slots is much smaller than the channel coherence time, the uplink and downlink propagation channels can reasonably be considered to be reciprocal. Under this assumption, the transmitter can estimate the channel during the reception phase and employ this estimate in the subsequent transmission time slot. We refer to this strategy as *open loop* since CSI is acquired without requiring any feedback information.

Despite its appealing features, the design of an open-loop system poses several technical challenges that, if not properly addressed, may preclude the validity of the reciprocity principle. Apart from the channel estimation errors that may occur during the reception phase, the main impairment is due to the fact that, in practical applications, channel estimation is usually performed at the baseband level after the radio-frequency (RF) chain. This calls for efficient calibration schemes able to compensate for any possible amplitude and phase error between the transmit and receive RF chains. If not properly mitigated, these errors might result in severe degradation of the system performance [41]. From a practical point of view, however, this calibration is not easy to achieve and, thus, open-loop systems are nowadays considered unsuitable for commercial multi-user systems.

An alternative solution to overcome these impairments is to use a form of *long-term reciprocity* known as *statistical reciprocity*. The latter is intrinsically immune to the aforementioned drawbacks as it relies on the fading distribution of the two-way communication channel rather than on its instantaneous realization. In those applications characterized by a strong correlation either in space or time, the fading distribution between uplink and downlink transmissions is usually the same and it can be employed to perform some form of statistical adaptation at the transmitter. Unfortunately, although easier to implement, this approach comes at the price of non-negligible performance loss with respect to a system that uses instantaneous channel knowledge.

6.4.2 Closed-Loop Systems

An effective way to overcome most of the drawbacks of open-loop systems consists of performing the estimation process at the receiver and feeding channel measurements back to the transmitter through a reliable reverse link. This is a reasonable alternative since control channels are already used in systems such as 3G cellular networks to perform power control. We refer to this strategy as *closed loop* and observe that it is generally suitable for those applications in which the reciprocity condition cannot be guaranteed or does not hold a priori. Among others, this is the case of frequency-division-duplex systems (such as cellular networks), in which transmission and reception take place over different frequency bands whose separation is typically larger than the coherence bandwidth of the channel. Albeit intrinsically robust to non-reciprocal effects, a closed-loop system presents the following major technical challenges. First, it suffers from an inherent feedback delay, which might result in outdated CSI at the time of transmission. Second, the feedback channel must be characterized by a high level of reliability, thereby making the link design task much more challenging than in open-loop systems. Third, the amount of information exchanged increases linearly with the total number of channel propagation paths, i.e., $M \times KN$, and counts against the overall spectral efficiency of the system.

6.5 System Design

In this section, we provide an overview of system design issues in multi-user MIMO systems. We focus on the downlink for which transmitter design is the key issue while we touch on the uplink (in which receiver design is paramount) only briefly. An extensive treatment of this later case can be found in [42] and elsewhere.

6.5.1 Receiver Design for Uplink Transmissions

We begin by observing that, in the uplink, the received signal in (6.1) results in a non-orthogonal superposition of different user data streams, thereby meaning that the data detection process can be accomplished by means of the well-known multi-user detection techniques [43].[10] In particular, the optimal receiver architectures are derived on the basis of the maximum-likelihood (ML) principle or the maximum a posteriori probability (MAP) criterion. They offer performance close to that of an interference-free system, but at the expense of substantial complexity, which increases exponentially with the number of users and data streams. Iterative multi-user detection techniques based either on the expectation-maximization (EM) algorithm or the turbo principle can be employed to reduce the computational burden [44]. Another well-studied class of suboptimal solutions is represented by the linear receivers in the form of decorrelating or minimum mean square error (MMSE) detectors which achieve a reasonable trade-off between performance and complexity. Alternatively, non-linear techniques based on interference cancellation (IC) (for example, the Bell Labs space-time architecture) can be adopted to obtain better performance [45].

6.5.2 Transmitter Design for Downlink Transmissions

As discussed in Section 6.3.1, in case of channel state information at the BS, the optimal capacity-achieving coding strategy for MIMO GBCs is DPC. Despite its significance from the information-theoretic point of view, this result has almost no practical relevance since the implementation of DPC requires a tremendous computational complexity at both transmitter and receiver and the problem of finding feasible codes close to the capacity of DPC is still much open. Attempts in this direction are discussed by Erez and Brink [46] and Yu et al. [47]. Both rely on a generalization of Tomlinson-Harashima precoding (THP) to a multidimensional vector quantization scheme, which is, however, achieved by means of complicated concatenated coding structures. On the other hand, a practical, although suboptimal, transmission technique that has received considerable attention due to its simplicity is represented by *linear pre-precoding*, also known as *linear transmit beamforming*. The latter refers to any transmission strategy that relies only on linear processing to mitigate the interference among the active users, and it has been proven to be asymptotically optimal by Sharif and Hassibi in [37] at least for the case of $K \gg M$.

Motivated by the above discussion, in the sequel we concentrate on two well-known beamforming techniques and revise their basic ideas and performance. In particular, we first focus on the case that has received more attention in the technical literature: MTs equipped with a single-receive antenna. In such a scenario, interference mitigation can be accomplished only at the BS. The simplest approach in this case is referred to as *channel inversion* or zero-forcing (ZF) linear beamforming (LB) and essentially relies on the idea of *pre-inverting* the channel matrix at the transmitter so as to completely remove the interference at all remote units. Obviously, this approach can be easily applied when the number of users is smaller

---

[10] Due to space limitations, we provide only a brief description of the possible receiver architectures and refer the interested reader to [42] (and references therein) for a comprehensive overview of the literature available on this subject (see also Chapters 5 and 6 of [48]).

than the number of transmit antennas ($K \leq M$) while it can be extended to the case $K > M$ provided that appropriate user selection algorithms are employed. Albeit simple and easy to implement, such schemes achieve good performance especially when the number of users is large [36]. However, in the case of $N > 2$ their application would be equivalent to considering the multiple-receive antennas of each remote device as individual users not cooperating among each other. While this has the advantage of leading to simple receiver architectures, it does not allow exploitation of the potential benefits of multiple-receive antennas in the data detection processing. One approach to overcoming this problem is represented by the block-diagonalization ZF (BD-ZF) scheme proposed independently by Haardt et al. [49] and Choi and Murch [50]. It is essentially based on the idea of completely removing the multi-user interference at the transmitter while leaving to each receiver the task of mitigating the interference among its own data streams.

To proceed further, we assume that a given set of users $\mathcal{A} \subseteq \{1, 2, \ldots, K\}$ with cardinality $|\mathcal{A}|$ is active and that the signal $\mathbf{s}$ in (6.6) takes the following general structure:

$$\mathbf{s} = \sum_{k=1}^{|\mathcal{A}|} \mathbf{W}_k \mathbf{u}_k, \tag{6.57}$$

where $\mathbf{W}_k \in \mathbb{C}^{M \times N}$ is the beamforming matrix associated with the kth data stream $\mathbf{u}_k \in \mathbb{C}^{N \times 1}$.

6.5.2.1 Single-Antenna receivers

We denote by $\mathbf{h}_k \in \mathbb{C}^{M \times 1}$ the vector collecting the channel coefficients between the BS array and the single-receive antenna at the kth MT so that $\mathbf{H}$ can be written as $\mathbf{H} = [\mathbf{h}_1, \mathbf{h}_2, \ldots, \mathbf{h}_{|\mathcal{A}|}]^T$. Similarly, we denote by $\mathbf{w}_k \in \mathbb{C}^{M \times 1}$ the kth beamforming vector associated with the kth data symbol u_k. Substituting (6.57) into (6.6) produces

$$\mathbf{r} = \mathbf{H} \sum_{k=1}^{|\mathcal{A}|} \mathbf{w}_k u_k + \mathbf{z} \tag{6.58}$$

or, equivalently, letting $\mathbf{W} = [\mathbf{w}_1, \mathbf{w}_2, \ldots, \mathbf{w}_{|\mathcal{A}|}]$ and $\mathbf{u} = [u_1, u_2, \ldots, u_{|\mathcal{A}|}]^T$,

$$\mathbf{r} = \mathbf{H}\mathbf{W}\mathbf{u} + \mathbf{z}. \tag{6.59}$$

As mentioned before, the ZF-LB technique aims at the complete elimination of the interference at all MTs. Assuming $|\mathcal{A}| \leq M$, from (6.59) we see that this can be easily achieved by setting $\mathbf{W}$ equal to the Moore–Penrose pseudo-inverse of the channel matrix $\mathbf{H}$, i.e.,

$$\mathbf{W} = \mathbf{H}^H \left(\mathbf{H}\mathbf{H}^H \right)^{-1}. \tag{6.60}$$

In doing so, the received signal at the kth MT reduces to

$$r_k = u_k + z_k, \qquad k = 1, 2, \ldots, |\mathcal{A}|, \tag{6.61}$$

from which it follows that the multi-user system has been decoupled into a set of $|\mathcal{A}|$ Gaussian single-user links. The sum rate of the system is thus given by

$$R_{ZF-LB} = \sum_{k=1}^{|\mathcal{A}|} \log\left(1 + \lambda_k\right), \tag{6.62}$$

where $\lambda_k = \mathrm{E}\{|u_k|^2\}$ is the power allocated to the kth user. When the objective is the maximization of R_{ZF-LB} the optimal power allocation strategy is achieved by means of the water-filling algorithm. This yields [5]

$$\lambda_k = \left[\mu\gamma_k - 1\right]_+, \tag{6.63}$$

where γ_k represents the effective channel gain at the kth MT:

$$\gamma_k = \frac{1}{\left[\left(\mathbf{H}^H\mathbf{H}\right)^{-1}\right]_{k,k}}, \tag{6.64}$$

while μ is computed as the solution of the following equation:

$$\sum_{k=1}^{|\mathcal{A}|} \left[\mu - \frac{1}{\gamma_k}\right]_+ = p. \tag{6.65}$$

Intuitively, different selections of sets $\mathcal{A}$ leads to different values of R_{ZF-LB} in (6.62) so that the maximum achievable sum rate of the system is obtained by considering all the possible sets $|\mathcal{A}| \subseteq \{1, 2, \ldots, K\}$ of cardinality $|\mathcal{A}| \leq M$, i.e.,

$$R_{ZF-LB}^{\max} = \max_{\mathcal{A} \subset \{1,2,\ldots,K\},|\mathcal{A}| \leq M} R_{ZF-LB}. \tag{6.66}$$

Analytical and numerical results show that the lower complexity of ZF-LB comes at the price of a non-negligible loss in terms of sum rate with respect to the optimal DPC technique especially when $K \leq M$ [27]. The main reason for this penalty is essentially due to the power boosting effect, which occurs in the pseudo-inverse computation of ill-conditioned channel matrices. This translates into high power consumption, which inevitably reduces the SNRs at the MTs, with ensuing degradation of the system performance. A possible way to alleviate this problem is represented by the scheme independently proposed by Peel et al. [51] and Joham et al. [52] which is known as transmit Wiener filtering and is based on minimization of the sum of the mean square errors at all MTs.

An alternative way to improve the performance of ZF-LB is to take advantage of the multi-user diversity. This is possible in those applications in which $K \gg M$. However, in these circumstances the computation of the solution in (6.66) becomes prohibitive even in the presence of a small number of users. To overcome this problem, a pseudo-random selection algorithm is proposed in [27] but simulation and analytical results show that it does not provide any gain in an i.i.d. Rayleigh fading channel. An other suboptimal approach is presented by Viswanath et al. [53] where they propose to first select a set of $\bar{K} \leq K$ users exhibiting the highest SINRs and then perform an exhaustive search over this smaller set to select the best M users. Numerical results show that the sum rate of this scheme comes close to the capacity of DPC as $\bar{K}$ approaches K. Alternative schemes can be derived using greedy user selection strategies where the optimal set is computed incrementally one user at a time according to some optimization criterion. Among them, a feasible scheme is the algorithm discussed by Dimić and Sidiropoulos [54], which operates as follows.

A greedy ZF-LB (GZF-LB) algorithm

(1) Initialization:

    (a) Set $n = 1$ and find a user k_1 such that

$$k_1 = \arg \max_{i \in \{1,2,...,K\}} \mathbf{h}_i^H \mathbf{h}_i.$$

    (b) Set $\mathcal{A}_1 = \{k_1\}$ and denote the achieved rate by $R_{ZF-LB}^{\max}(\mathcal{A}_1)$.

(2) While $k < K$:

    (a) Find a user k_n such that

$$k_n = \arg \max_{i \in \{1,2,...,K\} \setminus \mathcal{A}_{n-1}} R_{ZF-LB}^{\max}(\mathcal{A}_{n-1} \cup \{i\}).$$

    (b) Set $\mathcal{A}_n = \mathcal{A}_n \cup \{k_n\}$ and denote the achieved rate by $R_{ZF-LB}^{\max}(\mathcal{A}_n)$.

    (c) If $R_{ZF-LB}^{\max}(\mathcal{A}_n) < R_{ZF-LB}^{\max}(\mathcal{A}_{n-1})$ stop and decrease n by 1.

(3) Compute the beamforming matrix $\mathbf{W}$ corresponding to the selected set.

(4) Allocate the transmission power according to (6.63)–(6.65).

Figure 6.3 illustrates the achievable sum rate of GZF-LB versus the SNR when $K = 16$ and $M = 4$. Comparisons are made with respect to DPC and ZF-LB with pseudo-random selection. As seen, GZF-LB achieves excellent results even for a relatively small number of users. Moreover, its complexity is in the order of $O(KM^3)$, thereby leading to a substantial computational saving with respect to the brute-force implementation.

Another attempt to improve the effectiveness of ZF-LB when $K \gg M$ is represented by the scheme discussed by Yoo and Goldsmith [55], where ZF-LB is coupled with a user selection algorithm operating on the basis of successive projections and selecting the set of active users according to a semi-orthogonality principle. Briefly, among all possible users it chooses as active only those which are nearly orthogonal to each other. The same technique is used by Viswanathan

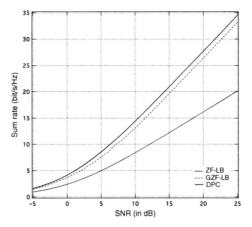

Fig. 6.3 Sum rate of GZF-LB versus SNR when $K = 16$ and $M = 4$

and Kumaran [56]. Analytical results show that the proposed approach, although suboptimal, achieves the *same asymptotic sum rate* of DPC (given in (6.56)) as the number of users goes to infinity. Thus, even a simple strategy like ZF-LB with a heuristic selection algorithm becomes asymptotically optimal in terms of sum rate, when K is relatively large. The reason is due to the fact that in these circumstances there is a high probability of finding a set of orthogonal users. Then, computing the pseudo-inverse of the channel matrix becomes merely a rotation operation and no increase of the transmit power occurs. Numerical results illustrate that satisfactory performance can be achieved for $K < 100$.

6.5.2.2 Multiple-Antenna Receivers

We begin by substituting (6.57) into (6.4) to obtain

$$\mathbf{r}_k = \mathbf{H}_k \mathbf{W}_k \mathbf{u}_k + \sum_{i=1, i \neq k}^{|\mathcal{A}|} \mathbf{H}_k \mathbf{W}_i \mathbf{u}_i + \mathbf{z}_k \qquad (6.67)$$

from which it follows that $\mathbf{H}_k \mathbf{W}_i \mathbf{u}_i$ is the interference induced at the kth receiver by the signal transmitted from the BS to user i. Therefore, its complete elimination at all the other MTs implies that

$$\mathbf{H}_k \mathbf{W}_i = \mathbf{0}_N \quad \text{for } k \neq i, \qquad (6.68)$$

which is equivalent to saying that $\mathbf{W}_i$ must be constrained to lie in the null space of the following $N(|\mathcal{A}| - 1) \times M$ matrix:

$$\tilde{\mathbf{H}}_i = [\mathbf{H}_1^T, \ldots, \mathbf{H}_{i-1}^T, \mathbf{H}_{i+1}^T, \ldots, \mathbf{H}_{|\mathcal{A}|}^T]^T. \qquad (6.69)$$

As discussed in [49, 50], this can be achieved by setting $\mathbf{W}_i = \tilde{\mathbf{V}}'_i$ where $\tilde{\mathbf{V}}'_i$ is obtained from the singular value decomposition (SVD) of $\tilde{\mathbf{H}}_i$. It is worth noting that the same goal can be obtained with lower complexity using a QR-based decomposition approach as shown by Chen et al. [57]. In the sequel, however, we resort to the SVD which produces

$$\tilde{\mathbf{H}}_i = \begin{bmatrix} \tilde{\mathbf{U}}_i & \tilde{\mathbf{U}}'_i \end{bmatrix} \begin{bmatrix} \tilde{\mathbf{D}}_i & \mathbf{0} \\ \mathbf{0} & \mathbf{0} \end{bmatrix} \begin{bmatrix} \tilde{\mathbf{V}}_i & \tilde{\mathbf{V}}'_i \end{bmatrix}^H , \tag{6.70}$$

where $\tilde{\mathbf{V}}'_i$ is composed by the right singular vectors associated with the $M - N$ $(|\mathcal{A}| - 1)$ null singular values of $\tilde{\mathbf{H}}_i$. Replacing $\mathbf{W}_i$ with $\tilde{\mathbf{V}}'_i$ in (6.67) produces

$$\mathbf{r}_k = \tilde{\mathbf{H}}'_k \mathbf{u}_k + \mathbf{z}_k \qquad \text{for } k = 1, 2, ..., |\mathcal{A}|, \tag{6.71}$$

where we have defined $\tilde{\mathbf{H}}'_k = \mathbf{H}_k \tilde{\mathbf{V}}'_k$. From the above equation, it is seen that the multi-user interference has been completely removed at each MT. It is worth observing that a sufficient condition to guarantee the existence of such a solution is that the number $M - N(|\mathcal{A}| - 1)$ of non-zero singular values is larger than N. This implies $M \geq N|\mathcal{A}|$.

Letting $\tilde{\mathbf{H}}' = \text{diag}\{\tilde{\mathbf{H}}'_1, \tilde{\mathbf{H}}'_2, ..., \tilde{\mathbf{H}}'_{|\mathcal{A}|}\}$, the maximum sum rate under the BD constraint can be computed as

$$R_{BD-ZF} = \max_{\{\mathbf{S} \geq 0, \, \text{tr}(\mathbf{S}) \leq p\}} \log \left| \mathbf{I}_N + \tilde{\mathbf{H}}' \mathbf{S} \tilde{\mathbf{H}}'^H \right| \tag{6.72}$$

from which it easily follows that the optimal $\mathbf{S}$ can be computed by means of the water-filling algorithm over the effective block-diagonal channel matrix $\tilde{\mathbf{H}}'$ [13]. Similar to ZF-LB, $R_{BD-ZF}^{\max}$ is obtained maximizing over all possible sets of $\mathcal{A}$ satisfying $N|\mathcal{A}| \leq M$, i.e.,

$$R_{BD-ZF}^{\max} = \max_{\mathcal{A} \subset \{1,2,...,K\}, N|\mathcal{A}| \leq M} R_{BD-ZF}. \tag{6.73}$$

As observed for ZF-LB, a brute force search of the optimal user set is prohibitive as it has combinatorial complexity. To reduce this complexity, two suboptimal greedy algorithms have been recently proposed by Shen et al. [58]. The first selects the user with the highest sum rate and then chooses from the remaining ones that provide the largest sum rate increase with the others already selected. The algorithm terminates when $|\mathcal{A}| = M/N$ users are taken or when the sum rate drops if more users are chosen. Although reasonable, such a solution is still too computationally intensive as it requires the computation of an SVD at each iteration in order to evaluate the sum rate of each possible user. The second approach has lower complexity as it is based on the maximization of the Frobenius norm of the effective channel. The rationale behind this scheme is that the sum rate is closely related to the eigenvalues of the effective channel whose sum is exactly the Frobenius norm. Numerical results show that both algorithms achieve about 95% of the sum rate of the solution obtained through complete search even though their complexities increase only linearly with

K instead of exponentially as the optimal one does. Both algorithms, however, incur a non-negligible loss with respect to DPC.

6.6 Limited Feedback Systems

The main drawbacks of the aforementioned transmission schemes are that they require perfect CSI at the transmitter. In Section 6.4, we have discussed how this information can be achieved in open- and closed-loop systems and we have also pointed out the large number of technical challenges arising in such systems. All this has recently stimulated a substantial interest toward the development of different solutions that may attain the potential gains of MIMO GBCs using a relatively small amount of information. They are commonly referred to as *limited* or *finite-rate feedback* systems [59]. Albeit this topic is still much open, some interesting solutions have recently been proposed (see, for example, the tutorial paper of Love et al. [60] and the corresponding special issue [61] for a fairly complete list of references).

6.6.1 Channel Quantization

The simplest approach to limit the feedback information relies on the efficient quantization of the channel estimates before transmission. In systems with a limited number of users (in the order of M) a well-studied approach is based on feeding back only the directions of the estimated channel vectors, while no information about the channel magnitudes is conveyed to the transmitter. The rationale behind this approach is that channel magnitudes play a key role when K is much larger than M. In fact, only in this case they can potentially be used by a user selection algorithm to exploit the multi-user diversity of the system. To obtain further insights into this strategy, in the following we review its basic ideas in the single-receive antenna scenario. The kth generic user ($k = 1, 2, \ldots, K$) computes an estimate $\tilde{\mathbf{h}}_k$ of $\mathbf{h}_k$ and quantizes the corresponding direction $\tilde{\mathbf{h}}_k / \left\| \tilde{\mathbf{h}}_k \right\|$ to the channel quantization vector $\hat{\mathbf{h}}_k$. The latter is chosen from a codebook $\mathcal{J}_k$ different from one user to the other and composed of $J = 2^B$ unit norm vectors, i.e., $\mathcal{J}_k = \{\mathbf{f}_{k,1}, \mathbf{f}_{k,2}, \ldots, \mathbf{f}_{k,J}\}$ with $\left\| \mathbf{f}_{k,i} \right\| = 1$. As in the single-user case, the kth MT computes $\hat{\mathbf{h}}_k$ as follows:

$$\hat{\mathbf{h}}_k = \mathbf{f}_{k,n}, \tag{6.74}$$

where the index n is chosen according to the minimum distance criterion

$$n = \arg \max_{i=1,2,\ldots,J} \left| \hat{\mathbf{h}}_k^H \mathbf{f}_{k,i} \right|. \tag{6.75}$$

Since the codebook can be designed off-line and known to the transmitter and receiver, each user needs to send back to the transmitter only the selected index n, meaning that the total number of feedback bits is given by B. In single-user applications with multiple-transmit and multiple-receive antennas, it has been shown that selecting B equal to the number of transmit antennas is enough to achieve almost the same performance as a system operating with perfect channel knowledge at the transmitter. A different result has been found by Jindal et al. [62] for a multi-user setting in which K is equal to M and the J quantization vectors are chosen from the isotropic distribution on the M-dimensional unit sphere. In particular, the author shows that, in such a case, the sum rate of a ZF-LB scales optimally only when the number of feedback bits per user increases *linearly* with both the number of transmit antennas and the SNR (in dB), i.e.,

$$B \approx \rho \, (M - 1) \log \text{SNR}, \qquad (6.76)$$

where $\rho \geq 1$ is a design parameter. The above result is motivated by the fact that in a multi-user environment (unlike the single-user case) any mismatch in the CSI at the transmitter leads not only to a reduction of the received SNR at the MTs but also to an increase of the multiple access interference since the ZF condition is no longer valid. In the high SNR regime, this effect is much more harmful than a reduction of the desired signal power (as it happens in the single-user case) and, thus, a larger number of bits is required.

A possible way to alleviate this problem is discussed in [63] where each user sends back to the transmitter not only the directions of the estimated channel vectors but also an unquantized channel quality indicator based on the SINR. This is then used at the BS to perform user selection so as to exploit the multi-user diversity, thereby leading to a substantial improvement of the system performance. In particular, it turns out that the number of feedback bits required to achieve the same asymptotic performance of a system with perfect CSI can be computed as follows:

$$B \approx \rho \, (M - 1) \log \text{SNR} + \log K \qquad (6.77)$$

from which it follows that, for a given value of SNR, it largely decreases as the number of user K increases.

6.6.2 Random Beamforming

Alternative solutions are based on the *random beamforming* (RBF) technique, which was originally proposed in [17] for the single-user scenario and later extended to the multi-user case in [64]. The basic idea behind this solution is to process the user data streams with M orthonormal beamforming vectors, which are randomly generated at each transmission time according to an isotropic distribution. At the remote ends, each receive antenna acts as an *effective* user (no cooperation is assumed), which measures the received SINRs corresponding to all the M transmit beamforming vectors (this can be easily achieved by means of a properly designed training phase). Once all these estimates are obtained, each effective user selects and conveys back to the transmitter its maximum SINR along with the corresponding beamforming index. This information is then used at the BS to decide, according to the maximum sum rate criterion, which set of effective users must be served or not. Surprisingly, when the number of users is relatively large, these bandwidth-efficient scheme (only one real-valued parameter and an integer number per effective user) is asymptotically optimal as the sum rate of the system scales like $M \log \log K$. This is made possible by multi-user diversity, which enables even a random beamforming vector to be closely matched to certain users. In [65], a further reduction of feedback information is achieved using random beamforming in conjunction with a receive antenna selection algorithm and an efficient SINR quantization technique. Numerical results show that both the above solutions work properly when K is on the order of hundreds. In practical applications, however, K is likely to be much smaller and a large degradation of the system performance may occur. Possible solutions to alleviate this problem are discussed in [66, 67]. Both schemes rely on the idea of using the feedback SINRs not only to select the active users but also to dynamically allocate the transmit power among the random beamforming directions. In this way, less or even no power can be assigned to those beams that are not aligned to any user channel.

6.6.3 Transceiver Optimization

In addition to the aforementioned techniques, considerable recent interest has been devoted to alternative solutions that are essentially based on the joint optimization of transmitter and receivers [68] through the use of limited feedback and feedforward links (see, for example, [69, 70]). Although powerful, these approaches often either lead to a complicated iterative procedure or require an excessive overhead of exchanged information.

6.7 Conclusions

In this chapter, we have reviewed a number of key concepts for understanding the uplink and downlink issues of a general multi-user mobile communication system operating with multiple antennas at both the BS and MTs. The first part of the chapter has focused on the computation of the capacity as a means for characterizing such systems. To this end, we have considered various operating conditions that substantially account for different degrees of CSI available at the transmitter(s) and/or receiver(s). In particular, it has been shown that in the uplink the capacity is, in most of the cases under investigation, a well-established result whose computation can be formulated as a convex optimization problem. On the other hand, it has been pointed out that in the downlink the problem is a more challenging task whose solution has recently been computed using the duality principle and the DPC technique, but only for the case of perfect channel knowledge at the transmitter. Unfortunately, much is still unknown in the other cases of interest. On the basis of the information-theoretic results provided, we have also made evident the fact that, in the downlink, CSI at the transmitter plays a key role in exploiting the promising gains of multiple antennas. For this purpose, we have briefly described how this information can be acquired by resorting to different operating techniques, which are suited for open- or closed-loop systems.

The second part of the chapter has been devoted to analyzing practical issues in the design of these systems. Most of this analysis has focused on the downlink, as the uplink has been shown to be reminiscent of other well-investigated communication systems. Among the large number of existing solutions, we have focused our attention on two simple linear transmit beamforming techniques based on the zero-forcing criterion. Both require perfect CSI at the transmitter and exhibit good trade-offs between performance and complexity, especially when the number of users is large.

Finally, in the last part of the chapter, we have examined briefly the issue of limited feedback, which has been the topic of considerable recent research activity. In particular, we have described some techniques that may provide benefits nearly identical to systems with perfect CSI at the transmitter, and we have made some comments on promising directions for future research in this area.

References

1. Biglieri, E., Proakis, J., and Shamai (Shitz), S. (1998) Fading channels: Information-theoretic and communications aspects. *IEEE Transactions on Information Theory*, vol. 44 (6) pp. 2619–2692.
2. Ahslwede, R. (1971) Multi-way communication channels. In *Proceedings of the IEEE International Symposium on Information Theory (Tsahkadsor, Armenian S.S.R.)*, Budapest, Hungary: Akademiai Kiado, June, pp. 23–52.
3. Cover, T., (1972) Broadcast channels. *IEEE Transactions on Information Theory*, vol. 18 (1) pp. 2–14.
4. Cover, T. M., and Thomas, J. A. (1991) *Elements of Information Theory*. New York: Wiley-Interscience.
5. Tse, D., and Viswanath, P. (2005) *Fundamentals of Wireless Communications*. Cambridge, UK: Cambridge University Press.
6. Rimoldi, R., and Urbanke, R. (1996) A rate-splitting approach to the Gaussian multiple-access channel. *IEEE Transactions on Information Theory*, vol. 46 (2) pp. 364–375.
7. Tse, D., and Hanly, S. V. (1998) Multiaccess fading channels – Part I : Polymatroid structure, optimal resource allocation and throughput capacities. *IEEE Transactions on Information Theory*, vol. 44 (7) pp. 2796–2815.
8. Boyd, S., and Vanderberghe, L. (2003) *Convex Optimization*. Cambridge, UK: Cambridge University Press.
9. Cheng, R., and Verdú, S. (1993) Gaussian multiaccess channels with ISI: Capacity region and multiuser water-filling. *IEEE Transactions on Information Theory*, vol. 21 (5) pp. 684–702.

10. Vandenberghe, L., Boyd, S., and Wu, S. P. (1998) Determinant maximization with linear matrix inequality constraints. *SIAM Journal Matrix Analytical Applications,* vol. 19 (2) pp. 499–533.
11. Yu, W., Rhee, W., Boyd, S., and Cioffi, J. M. (2004) Iterative water-filling for Gaussian vector multiple-access channels. *IEEE Transactions on Information Theory,* vol. 50 (1) pp. 145–152.
12. Yu, W., Rhee, W., and Cioffi, J. M. (2001) Optimal power control in multiple-access fading channels with multiple antennas. In *Proceedings of IEEE International Conference on Communications,* Helsinki, Finland, June 11–14, pp. 575–579.
13. Telatar, E. (1999) Capacity of multi-antenna Gaussian channels. *European Transactions on Telecommunications,* vol. 10 (6) pp. 585–595.
14. Rhee, W., and Cioffi, J. M. (2003) On the capacity of multi-user wireless channels with multiple antennas. *IEEE Transactions on Information Theory,* vol. 49 (10) pp. 2580–2595.
15. Viswanath, P., Tse, D., and Anantharam, V. (2001) Asymptotically optimal water-filling in vector multiple-access channels. *IEEE Transactions on Information Theory,* vol. 47 (1) pp. 241–267.
16. Goldsmith, A., Jafar, S. A., Jindal, N., and Vishwanath, S. (2003) Capacity limits of MIMO channels. *IEEE Journal on Selected Areas in Communications,* vol. 21 (5) pp. 684–702.
17. Viswanath, P., Tse, D., and Laroia, R. (2002) Opportunistic beamforming using dumb antennas. *IEEE Transactions on Information Theory,* vol. 48 (6) pp. 1277–1294.
18. Foschini, G. J. and Gans, M. J. (1998) On limits of wireless communications in a fading environment when using multiple antennas. *Wireless Personal Communications,* vol. 6 (2) pp. 331–320.
19. Marton, K. (1979) A coding theorem for the discrete memoryless broadcast channel. *IEEE Transactions on Information Theory,* vol. 23 (3) pp. 306–311.
20. Bergman, P. (1973) Random coding theorem for broadcast channels with degraded components. *IEEE Transactions on Information Theory,* vol. 19 (3), pp. 197–207.
21. Weingarten, H., Steinberg, Y., and Shamai (Shitz), S. (2006) The capacity region of the Gaussian multiple-input multiple-output broadcast channel. *IEEE Transactions on Information Theory,* vol. 50 (9) pp. 3936–3964.
22. Costa, M. (1983) Writing on dirty paper. *IEEE Transactions on Information Theory,* vol. 29 (3) pp. 439–441.
23. Yu, W., Sutivong, A., Julian, D., Cover, T. M., and Chiang, M. (2001) Writing on colored paper. In *Proceedings of IEEE International Symposium on Information Theory,* Washington, DC, USA, June 24–29, pp. 302–311.
24. Jindal, N., Vishwanath, S., and Goldsmith, A. (2004) On the duality of Gaussian multiple-access and broadcast channels. *IEEE Transactions on Information Theory,* vol. 50 (5) pp. 768–783.
25. Viswanath, P., and Tse, D. (2003) Sum capacity of the vector Gaussian channel and uplink-downlink duality. *IEEE Transactions on Information Theory,* vol. 49 (8) pp. 1912–1921.
26. Vishwanath, S., Jindal, N., and Goldsmith, A. (2003) Duality, achievable rates and sum rate capacity of Gaussian MIMO broadcast channels. *IEEE Transactions on Information Theory,* vol. 49 (10) pp. 2658–2668.
27. Caire, G., and Shamai (Shitz), S. (2003) On the achievable throughput of a multi-antenna Gaussian broadcast channel. *IEEE Transactions on Information Theory,* vol. 49 (7) pp. 1691–1706.
28. Sato, H. (1978) An outer bound on the capacity region of broadcast channels. *IEEE Transactions on Information Theory,* vol. 24 (3) pp. 374–377.
29. Yu, W., and Cioffi, J. M. (2004) Sum capacity of the vector Gaussian channels. *IEEE Transactions on Information Theory,* vol. 50 (9) pp. 1875–1892.
30. Jindal, N., Rhee, W., Vishwanath, S., Jafar, S. A., and Goldsmith, A. (2005) Sum power iterative water-filling for multi-antenna Gaussian broadcast channels. *IEEE Transactions on Information Theory,* vol. 51 (4) pp. 1570–1580.
31. Yu, W., and Rhee, W. (2006) Degrees of freedom in wireless multiuser spatial multiplex systems with multiple antennas. *IEEE Transactions on Communications,* vol. 54 (10) pp. 1747–1753.

32. Yu, W. (2006) Sum-capacity computation for the Gaussian vector broadcast channel via dual decomposition. *IEEE Transactions on Information Theory,* vol. 52 (2) pp. 754–759.
33. Amraoui, A., Kramer, G. and Shamai (Shitz), S. (2003) Coding for the MIMO broadcast channel. In *Proceedings of the IEEE International Symposium on Information Theory,* Pacifico Yokohama, Kanagawa, Japan, June 29–July 4, p. 296.
34. Jafar, S., and Goldsmith, A. (2004) Isotropic fading vector broadcast channels: The scalar upper bound and loss in degrees of freedom. *IEEE Transactions on Information Theory,* vol. 51 (3) pp. 848–857.
35. Jindal, N., and Goldsmith, A. (2005) Dirty-paper coding versus TDMA for MIMO broadcast channels. *IEEE Transactions on Information Theory,* vol. 51 (5) pp. 1783–1794.
36. Lee, J., and Jindal, N. (2007) High SNR Analysis for MIMO broadcast channels: Dirty paper coding versus linear precoding. *IEEE Transactions on Information Theory,* vol. 53 (12) pp. 4787–4792.
37. Sharif, M., and Hassibi, B. (2007) A comparison of time-sharing, DPC, and beamforming for MIMO broadcast channels with many users. *IEEE Transactions on Communications,* vol. 55 (1) pp. 11–15.
38. Knopp, R., and Humblet, P. A. (1995) Information capacity and power control in single-cell multi-user communications. In *Proceedings of the IEEE International Conference on Communications,* Seattle, WA, June, vol. 1, pp. 331–335.
39. Tong, L., Sadler, B. M. and Dong, M. (2004) Pilot-assisted wireless transmissions. *IEEE Signal Processing Magazine,* vol. 21 (6) pp. 12–25.
40. Bolcskei, H., Gesbert, D., Papadias, C. B., and Van Der Veen, A.-J. (2006) *Space-Time Wireless Systems: From Array Processing to MIMO Communications,* Cambridge, UK: Cambridge University Press.
41. Dias, A. R., Bateman, D., and Gosse, K. (2004) Impact of RF front-end impairments and mobility on channel reciprocity for closed-loop multiple antenna techniques. In *Proceedings of the IEEE International Symposium on Personal, Indoor and Mobile Radio Communications,* Barcelona, Spain, September 05–08, vol. 2, pp. 1434–1438.
42. Wang, X., and Poor, H. V. (2004) *Wireless Communication Systems: Advanced Techniques for Signal Reception.* Upper Saddle River, NJ: Prentice-Hall.
43. Verdú, S. (1998) *Multiuser Detection.* Cambridge, UK: Cambridge University Press.
44. Poor, H. V. (2004) Iterative multi-user detection. *IEEE Signal Processing Magazine,* vol. 21 (1) pp. 81–88.
45. Foschini, G. J. (1996) Layered space-time architecture for wireless communications in fading environments when using multi-element antennas. *Bell Labs Technical Journal,* vol. 1 (2) pp. 41–59.
46. Erez, U., and ten Brink, S. (2005) A close to capacity dirty paper coding scheme. *IEEE Transactions on Information Theory,* vol. 51 (10) pp. 3417–3432.
47. Yu, W., Varodayan, D. P., and Cioffi, J. M. (2005) Trellis and convolutional precoding for transmitter-based interference pre-cancellation. *IEEE Transactions on Information Theory,* vol. 53 (7) pp. 1220–1230.
48. Biglieri, E., Calderbank, R., Constantinides, A., Goldsmith, A., Paulraj, A., and Poor, H. V. (2007) *MIMO Wireless Communications,* Cambridge, UK: Cambridge University Press.
49. Spencer, Q. H., Swindlehurst, A. L. and Haardt, M. (2004) Zero-forcing methods for the downlink spatial multiplexing in multi-user MIMO channels. *IEEE Transactions on Signal Processing,* vol.52 (2) pp. 461–471.
50. Choi, L., and Murch, R. D. (2004) A transmit preprocessing technique for multiuser MIMO systems using a decomposition approach. *IEEE Transactions on Wireless Communications,* vol. 3 (1) pp. 20–24.
51. Peel, C. B., Hochwald, B. M., and Swindlehurst, A. L. (2005) A vector perturbation technique for near capacity multiantenna multiuser communication – Part I: Channel inversion and regularization. *IEEE Transactions on Communications,* vol. 53 (1) pp. 195–202.
52. Joham, M., Kusume, K., Gzara, M. H., and Utschick, W. (2004) Transmit Wiener filter for the downlink of TDD DS-CDMA systems. In *Proceedings of the IEEE Symposium on Spread-Spectrum Technologies and Applications,* Lisbon, Portugal, September 15–18, pp. 9–13.

53. Viswanathan, H., Venkatesan, S., and Huang, H. (2003) Downlink capacity evaluation of cellular networks with known-interference cancellation. *IEEE Journal on Selected Areas of Communications*, vol. 21 (6) pp. 802–811.
54. Dimić, G., and Sidiropoulos, N. D. (2005) On downlink beamforming with greedy user selection: Performance analysis and a simple new algorithm. *IEEE Transactions on Signal Processing*, vol. 53 (10) pp. 3857–3868.
55. Yoo, T., and Goldsmith, A. (2006) On the optimality of multiantenna broadcast scheduling using zero-forcing beamforming. *IEEE Journal on Selected Areas in Communications*, vol. 25 (7) pp. 1478–1491.
56. Viswanathan, H., and Kumaran, K. (2001) Rate scheduling in multiple antenna downlink wireless systems. In *Proceedings of the Allerton Conference on Communications, Control and Computing*, Monticello, IL, USA, September 29 – October 1, vol. 39, pp. 747–756.
57. Chen, R., Heath, R. W., Jr., and Andrews, J. G. (2007) Transmit selection diversity for unitary precoded multiuser spatial multiplexing systems with linear receivers. *IEEE Transactions on Information Theory*, vol. 55 (3) pp. 1159–1171.
58. Shen, Z., Chen, R., Andrews, J. G. and Heath, R. W., Jr. (2006) Low complexity user selection algorithm for multiuser MIMO systems with block diagonalization. *IEEE Transactions on Signal Processing*, vol. 54 (9) pp. 3658–3663.
59. Love, D. J., Heath, R. W., Jr., Santipach, W. and Honig, M. L. (2003) What is the value of limited feedback for MIMO channels? *IEEE Communications Magazine*, vol. 42 (10) pp. 54–59.
60. Love, D. J., Heath, R. W., Jr., Lau, V. K. N., Gesbert, D., Rao, B. D., and Andrews, M. (2008) An overview of limited feedback in wireless communication systems. *IEEE Journal on Selected Areas in Communications*, vol. 26 (8) pp. 1341–1365
61. *IEEE Journal on Selected Areas of Communications. Special issue on Limited Feedback*, vol. 26 (8), October 2008.
62. Jindal, N. (2006) MIMO broadcast channels with finite rate feedback. *IEEE Transactions on Information Theory*, vol. 52 (11) pp. 5045–5059.
63. Yoo, T., Jindal, N., and Goldsmith, A. (2007) Multi-antenna downlink channels with limited feedback and user selection. *IEEE Journal on Selected Areas in Communications*, vol. 25 (7) pp. 1478–1491.
64. Sharif, M., and Hassibi, B. H. (2005) On the capacity of MIMO broadcast with partial side information. *IEEE Transactions on Information Theory*, vol. 51 (2) pp. 506–522.
65. Zhang, W., and Letaief, K. B. (2007) MIMO broadcast scheduling with limited feedback. *IEEE Journal on Selected Areas in Communications. Special issue on MIMO Transceivers for Realistic Communication Networks: Challenges and Opportunities*, vol. 25 (7) pp. 1457–1467.
66. Kountouris, M. and Gesbert, D. (2005) Robust multi-user opportunistic beamforming for sparse networks. In *Proceedings of the IEEE Workshop on Signal Processing Advances in Wireless Communications*, New York, NY, June 5–8, pp. 975–979.
67. Wagner, J., Liang, Y.-C., and Zhang R. (2007) On the balance of multiuser diversity and spatial multiplexing gain in random beamforming. *IEEE Transactions on Wireless Communications*, vol. 7 (7) pp. 2512–2525.
68. Pan, Z., Wong, K.-K., and Ng, T.-S. (2004) Generalized multiuser orthogonal space-division multiplexing. *IEEE Transactions on Wireless Communications*, vol. 3 (6) pp. 1969–1973.
69. Chae, C.-B., Mazzarese, D., and Heath, R. W., Jr. (2006) Coordinated beamforming for multiuser MIMO systems with limited feedforward. In *Proceedings of the Asilomar Conference on Signals, Systems and Computers*, Pacific Grove, CA, USA, October 30 – November 1, pp. 1511–1515.
70. Chae, C.-B., Mazzarese, D., Jindal, N., and Heath, R. W., Jr. (2008) Coordinated beamforming with limited feedback in the MIMO broadcast channel. *IEEE Journal on Selected Areas in Communications*, vol. 26 (8) pp. 1505–1515.

Chapter 7
Collaborative Beamforming

Hideki Ochiai and Hideki Imai

7.1 Introduction

One of the most challenging issues in wireless communications is how to cope with
a reduction of the received signal-to-noise power ratio (SNR) at the remote receiver.
The fluctuation of the SNR can be caused by several phenomena such as path loss,
shadowing, and multipath fading [19]. *Beamforming* is one of the techniques that
can cope with this problem. Suppose that the transmitter is equipped with multiple
antennas. If the same signal is transmitted from geometrically separated multiple
antennas, even without multipath the instantaneous power of the received signal
can vary depending on the geometrical locations of the transmit antennas and the
receive antenna. This is due to the same mechanism as that of multipath fading. In
other words, by sending the same signal from the multiple antennas, the transmitter
is artificially creating multipath fading. On the other hand, if the transmitter knows
the direction of the intended receiver a priori, it is possible for the transmitter to
send signals such that the received SNR of the intended receiver is maximized in
that direction. This can be achieved by manipulating the transmitted signals from
each antenna such that their signals are added coherently at the receiver (i.e., the
resulting channel coefficient of the artificial fading is maximized).

In many wireless ad hoc networks, battery-driven mobile terminals are likely to
be equipped with a single antenna and this assumption precludes the use of beamforming. Nevertheless, if the nearby users are transmitting their data to the distant receivers, it would be much more efficient if they cooperate each other such
that they share their transmitting data a priori and then synchronously transmit the

Hideki Ochiai
Yokohama National University, 79-5 Tokiwadai, Hodogaya, Yokohama, Japan,
e-mail: hideki@ynu.ac.jp

Hideki Imai
Chuo University, 1-13-27 Kasuga, Bunkyo, Tokyo, Japan,
e-mail: h-imai@elect.chuo-uac.jp

V. Tarokh (ed.), *New Directions in Wireless Communications Research*,
DOI 10.1007/978-1-4419-0673-1_7,
© Springer Science+Business Media, LLC 2009

compound data to each destination receiver as in the recent cooperative diversity literature (e.g., [8, 16]). By appropriately setting the initial phase of the transmitting signal of each user, they can also cooperatively perform beamforming assuming the other users as virtual antennas.

Since the communicating users in wireless ad hoc networks are randomly distributed by nature, beamforming must be performed in a distributed manner. Such beamforming is often referred to as a *distributed beamforming*. The two main challenges for distributed beamforming are how to share each message efficiently and how to perform synchronization among the cooperative users.

The idea of distributed beamforming can be naturally incorporated into the framework of wireless ad hoc sensor networks. Let us consider the sensor network where the nearby multiple sensor nodes, each collecting their own data, form a cluster and transmit to the same destination receiver as sketched in Fig. 7.1. Since the sensor nodes in this setting must be tiny and inexpensive, it is very likely that the node is equipped with only a single antenna. Nevertheless, if the nearby sensor nodes share their information a priori and collaboratively transmit in a synchronous manner, it is possible to form a beam in the intended direction. This particular approach is referred to as a *collaborative beamforming* [13] (rather than cooperative beamforming) as all the nodes in the cluster collaboratively send their shared messages to the same destination.

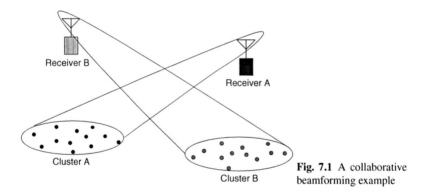

Fig. 7.1 A collaborative beamforming example

There are largely two main issues when implementing collaborative beamforming. The feasibility of precise synchronization and accurate channel estimation is one critical issue. In principle, beamforming is achieved by manipulating the initial phase of synchronized transmitting signals with identical message. This is possible only if the virtual antennas (i.e., sensor nodes) are synchronous and channel impulse response is known at the transmitter. This issue is studied in [11].

The other issue is the beam pattern formed by the distributed sensors. If the nice narrow beam with low sidelobes can be formed by these nodes, it provides another dimension of communication channel: space division multiple access (SDMA), which is a potential technique that significantly boosts multiple access capacity. In fact, it will be shown that if the sensors are randomly distributed and fully synchronized, the resulting beam pattern formed by these sensors has, with high probability, a nice sharp mainlobe with low sidelobe.

In this chapter, with the scenario of ad hoc sensor networks in mind, we review the beam patterns of various distributed antennas as well as the main results explored in [13]. The following section describes the beamforming system model considered throughout this chapter and reviews several known results on the beam patterns of antenna arrays. Section 7.3 derives several statistical properties of the beam pattern in random arrays. The average and distribution of the beam pattern of randomly generated antenna arrays are derived, and the distribution of the maximum sidelobe peak is also studied. Finally, conclusions are given in Section 7.4.

7.2 System Model and Beam Patterns of Fixed Nodes

In this section, we begin with the description of a collaborative beamforming system model and also review beam patterns of several antenna arrays well studied in antenna theory [3, 6] as a reference.

Figure 7.2 illustrates the geometrical configuration of the distributed nodes and destination. All the sensor nodes (or, equivalently, antenna elements) are assumed to be located on the $x-y$ plane and thus form a planar array. For analytical convenience, we designate each node location in polar coordinates. The location of the kth node is thus denoted by (r_k, ψ_k). The location of the destination is given in spherical coordinates by (A, ϕ_0, θ_0). Following the standard notation in antenna theory (e.g., [3]), the angle $\theta \in [0, \pi]$ denotes the elevation direction and the angle $\phi \in [0, 2\pi)$ represents the azimuthal direction. In order to make the subsequent analyses tractable, we make the following assumptions:

1. Each sensor node is equipped with an ideal isotropic antenna element.
2. All nodes transmit with identical energies, and the path losses of all nodes are also identical.
3. There is no reflection or scattering of the signal.
4. The sensor nodes are sufficiently separated such that mutual coupling effects are negligible.
5. All the nodes are perfectly synchronized (or connected) such that no phase offset or jitter occurs.

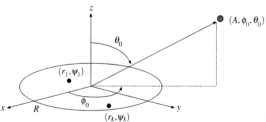

Fig. 7.2 Definitions of node location

In short, the above assumptions guarantee that the resulting array formed by the sensor nodes is an ideal planar phased array.

7.2.1 Array Factor and Beam Pattern

Suppose we have N sensor nodes within the disk of radius R and let $d_k(\phi, \theta)$ denote the Euclidean distance between the kth node and the reference location (A, ϕ, θ), where $k \in \{1, 2, \ldots, N\}$. It then follows that

$$d_k(\phi, \theta) = \sqrt{A^2 + r_k^2 - 2r_k A \sin \theta \cos(\phi - \psi_k)} . \tag{7.1}$$

Our primary concern of beamforming performance is the radiation pattern in the far-field region. Therefore, assuming $A \gg r_k$, the distance $d_k(\phi, \theta)$ in (7.1) is approximated by

$$d_k(\phi, \theta) = A \left(1 - 2\frac{r_k}{A} \sin \theta \cos(\phi - \psi_k) + \left(\frac{r_k}{A}\right)^2\right)^{\frac{1}{2}} \approx A - r_k \sin \theta \cos(\phi - \psi_k). \tag{7.2}$$

Let Θ_k denote the initial phase of the transmit signal carrier of the node k. Given the realization of node locations $\mathbf{r} = [r_1, r_2, \ldots, r_N] \in [0, R]^N$ and $\boldsymbol{\psi} = [\psi_1, \psi_2, \ldots, \psi_N] \in [0, 2\pi)^N$, the *array factor* can be defined as

$$F(\phi, \theta | \mathbf{r}, \boldsymbol{\psi}) \triangleq \frac{1}{N} \sum_{k=1}^{N} e^{j\Theta_k} e^{j\frac{2\pi}{\lambda} d_k(\phi, \theta)} = \frac{1}{N} \sum_{k=1}^{N} e^{j\left\{\frac{2\pi}{\lambda} d_k(\phi, \theta) + \Theta_k\right\}}, \tag{7.3}$$

where λ is the wavelength of the radio frequency (RF) carrier.

Given the direction of the target receiver by (A, ϕ_0, θ_0), the node k sets its initial phase as

$$\Theta_k = -\frac{2\pi}{\lambda} d_k(\phi_0, \theta_0). \tag{7.4}$$

The corresponding array factor can be written as

$$F(\phi, \theta | \mathbf{r}, \boldsymbol{\psi}) = \frac{1}{N} \sum_{k=1}^{N} e^{j\Theta_k} e^{j\frac{2\pi}{\lambda} d_k(\phi, \theta)} = \frac{1}{N} \sum_{k=1}^{N} e^{j\frac{2\pi}{\lambda}[d_k(\phi, \theta) - d_k(\phi_0, \theta_0)]}.$$

With the far-field distance approximation (7.2), the far-field radiation pattern can be expressed as

$$F(\phi, \theta | \mathbf{r}, \boldsymbol{\psi}) = \frac{1}{N} \sum_{k=1}^{N} e^{j\frac{2\pi}{\lambda} r_k[\sin \theta_0 \cos(\phi_0 - \psi_k) - \sin \theta \cos(\phi - \psi_k)]}. \tag{7.5}$$

Apparently, the magnitude of the above array factor is maximized (with maximum value 1) when the observation angle is equal to the target angle

($\phi = \phi_0$ and $\theta = \theta_0$). To achieve this, the initial phase Θ_k of (7.4) should be calculated at each node, which implies that each node must have a precise knowledge of its relative location to the destination. By broadcasting the beacon from the destination, it is possible for each node to adjust its initial phase (similar to self-phasing arrays) as sketched in Fig. 7.3(a). This scenario is referred to as a *closed-loop* case [13]. Since all the nodes and the remote receiver have to operate in a synchronous manner in this case, they must share an identical clock. Therefore, the closed-loop system should be implemented with the help of outer clock such as that of the global positioning system (GPS) as illustrated in Fig. 7.3(a).

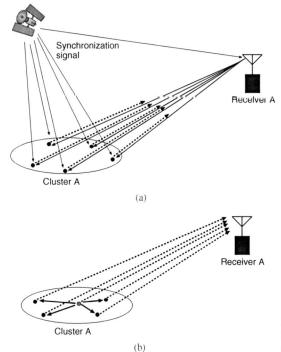

Fig. 7.3 Two collaborative beamforming scenarios: (a) closed loop and (b) open loop

Alternatively, instead of using (7.4) if we choose the initial phase of the node k as

$$\Theta_k = \frac{2\pi}{\lambda} r_k \sin\theta_0 \cos(\phi_0 - \psi_k),\tag{7.6}$$

the array factor of (7.3) with the far-field distance approximation (7.2) is given by

$$F(\phi, \theta | \mathbf{r}, \boldsymbol{\psi}) = \frac{1}{N} \sum_{k=1}^{N} e^{j\frac{2\pi}{\lambda}\{A - r_k \sin\theta \cos(\phi - \psi_k) + r_k \sin\theta_0 \cos(\phi_0 - \psi_k)\}}$$

$$= \frac{1}{N} e^{j\frac{2\pi}{\lambda}A} \sum_{k=1}^{N} e^{j\frac{2\pi}{\lambda}r_k\{\sin\theta_0 \cos(\phi_0 - \psi_k) - \sin\theta \cos(\phi - \psi_k)\}},\tag{7.7}$$

which has the same magnitude as that of the array factor (7.5). Therefore, each node can use (7.6) as an initial phase for collaborative beamforming. The calculation of (7.6) requires the precise location (r_k, ϕ_k) of each node, which can be achieved, for example, by coordination of the master node (or cluster head) as sketched in Fig. 7.3(b). This scenario, which does not require any synchronization signal from the destination, is referred to as an *open-loop* case [13] and this is useful in the case that the network has receivers in multiple directions. In either scenario, practical implementation of collaborative beamforming should require efficient joint design of medium access control (MAC) layer and physical (PHY) layer. This cross-layer design issue is discussed in [4].

Finally, we define the far-field beam pattern as a square of the array factor magnitude

$$P(\phi, \theta | \mathbf{r}, \boldsymbol{\psi}) \triangleq |F(\phi, \theta | \mathbf{r}, \boldsymbol{\psi})|^2 . \tag{7.8}$$

7.2.2 Beam Patterns of Linear Arrays

Let us first consider the beam patterns of linear arrays. In a conventional linear array, the antenna elements are located in a line with equal spacing. Let the nodes be located on the x-axis in our planar phased array model. A linear array with N antenna elements, where N is assumed to be an even number for simplicity, can be expressed, for instance, as

$$\begin{cases} r_k &= \left(\lfloor \frac{k-1}{2} \rfloor + \frac{1}{2} \right) d, \\ \psi_k &= ((k-1) \bmod 2)\, \pi, \end{cases} \tag{7.9}$$

where d is a spacing between neighboring nodes, $k \in \{1, 2, \ldots, N\}$, and $\lfloor x \rfloor$ denotes the greatest integer no greater than x. This sensor node allocation is sketched in Fig. 7.4.

The corresponding far-field array factor defined in (7.5) is given by

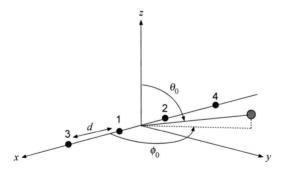

Fig. 7.4 Linear array model with $N = 4$

$$F(\phi, \theta | \mathbf{r}, \boldsymbol{\psi}) = \frac{1}{N} \sum_{k=1}^{N/2} \{ e^{j \frac{2\pi}{\lambda} r_{2k} [\sin \theta_0 \cos \phi_0 - \sin \theta \cos \phi]}$$

$$+ e^{-j \frac{2\pi}{\lambda} r_{2k} [\sin \theta_0 \cos \phi_0 - \sin \theta \cos \phi]} \} . \tag{7.10}$$

By noticing that $r_{2k} = \left(k - \frac{1}{2} \right) d$ and defining

$$u \triangleq \pi \frac{d}{\lambda} [\sin \theta_0 \cos \phi_0 - \sin \theta \cos \phi] , \tag{7.11}$$

we obtain

$$F(\phi, \theta | \mathbf{r}, \boldsymbol{\psi}) = \frac{1}{N} \left\{ \sum_{k=1}^{N/2} e^{j(2k-1)u} + \sum_{k=1}^{N/2} e^{-j(2k-1)u} \right\} = \frac{2}{N} \sum_{k=1}^{N/2} \cos \left((2k-1) u \right)$$

$$= \frac{2}{N} \Re \left\{ \sum_{k=1}^{N/2} e^{j(2k-1)u} \right\} = \frac{2}{N} \Re \left\{ e^{ju} \frac{1 - e^{jNu}}{1 - e^{j2u}} \right\}$$

$$= \frac{2}{N} \Re \left\{ e^{j \frac{Nu}{2}} \frac{\sin \frac{Nu}{2}}{\sin u} \right\} = \frac{2 \cos \frac{Nu}{2} \sin \frac{Nu}{2}}{N \sin u} = \frac{\sin (Nu)}{N \sin u}. \tag{7.12}$$

The far-field beam pattern is then given by

$$P(\phi, \theta) = \frac{1}{N^2} \frac{\sin^2 (Nu)}{\sin^2 u}. \tag{7.13}$$

If the target is located on the positive side of y-axis in Fig. 7.4, we have $\phi_0 = \theta_0 = 90°$ and thus $u = -\pi \frac{d}{\lambda} \sin \theta \cos \phi$. This type of linear array is often called a *broad-side array*. The beam pattern on the x–y plane is given by

$$P(\phi, \theta = 90°) = \frac{1}{N^2} \frac{\sin^2 \left(N\pi \frac{d}{\lambda} \cos \phi \right)}{\sin^2 \left(\pi \frac{d}{\lambda} \cos \phi \right)}. \tag{7.14}$$

The beam patterns of (7.14) are plotted[1] for several cases of d/λ in Fig. 7.5 with $N = 20$. We observe that increasing the antenna spacing d/λ makes the mainlobe narrower. However, when d is as large as its wavelength λ, there appear sidelobes that have power level as large as its mainlobe. This undesirable large sidelobe is often referred to as a *grating lobe*.

On the other hand, if the target is located on the x-axis (i.e., in the same direction of the array line), the resulting array is called an *end-fire array*. In this case, u of (7.11) is given by $u = \pi \frac{d}{\lambda} (1 - \sin \theta \cos \phi)$. Thus, the beam pattern of this linear array with $\phi_0 = 0°$ is given by

---

[1] We plot beam patterns with (radius, angle) = ($P(\phi)$ in dB, ϕ in degree) in polar coordinates throughout this chapter.

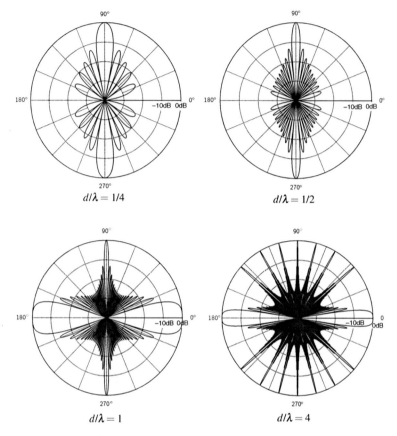

Fig. 7.5 Beam patterns of linear arrays with $N = 20$ and $\phi_0 = 90°$ (broadside)

$$P(\phi, \theta = 90°) = \frac{1}{N^2} \frac{\sin^2\left(N\pi\frac{d}{\lambda}(1 - \cos\phi)\right)}{\sin^2\left(\pi\frac{d}{\lambda}(1 - \cos\phi)\right)}. \tag{7.15}$$

Plotted in Fig. 7.6 are the beam patterns of end-fire arrays (7.15) for several cases of d/λ with $N = 20$. As observed, grating lobes appear as d/λ increases. In fact, by noticing that

$$\sin(N\pi\gamma(1 - \cos\phi)) = \sin(N\pi\gamma)\cos(N\pi\gamma\cos\phi) - \cos(N\pi\gamma)\sin(N\pi\gamma\cos\phi)$$
$$= (-1)^{N\gamma+1}\sin(N\pi\gamma\cos\phi) \quad \text{for } \gamma = \text{integer}, \quad (7.16)$$

the beam patterns of broadside and end-fire arrays become identical if $\gamma = d/\lambda$ is integer, as apparent from Figs. 7.5 and 7.6. In summary, the linear array model is not flexible enough for our collaborative beamforming purpose where the location of the target can be in the arbitrary direction; allocation of sensor nodes in line should thus be avoided.

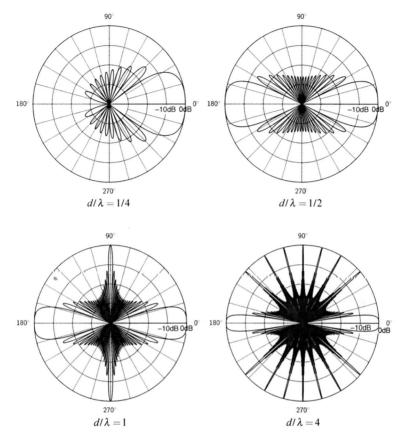

Fig. 7.6 Beam patterns of linear arrays with $N = 20$ and $\phi_0 = 0°$ (end fire)

7.2.3 Beam Patterns of Circular Arrays

Let us next consider the case where the nodes are located in a circular manner. Specifically, consider N_R concentric rings where each ring has N/N_R nodes located with an equal spacing in the azimuthal direction as sketched in Fig. 7.7.

We begin with the single ring case ($N_R = 1$) and derive its beam pattern expression following [10]. We first rewrite the far-field radiation pattern of (7.5) as

$$F(\phi, \theta | \mathbf{r}, \boldsymbol{\psi}) = \frac{1}{N} \sum_{k=1}^{N} e^{j 2\pi \frac{r_k}{\lambda} \rho \cos(\psi_k - \xi)}, \qquad (7.17)$$

where

$$\rho = \sqrt{(\sin\theta \cos\phi - \sin\theta_0 \cos\phi_0)^2 + (\sin\theta \sin\phi - \sin\theta_0 \sin\phi_0)^2}, \qquad (7.18)$$

$$\xi = \arctan \frac{\sin\theta \sin\phi - \sin\theta_0 \sin\phi_0}{\sin\theta \cos\phi - \sin\theta_0 \cos\phi_0}. \qquad (7.19)$$

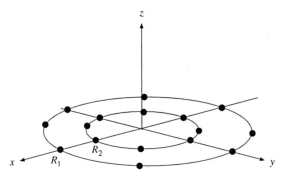

Fig. 7.7 Circular array model with $N = 16$ and $N_R = 2$

Note that ρ can be seen as a measure that represents the difference between the observed angle from the target angle. Thus, $\rho = 0$ if the target and observation directions coincide, and $\rho > 0$ otherwise. By assumption, we have

$$\begin{cases} r_k = R, \\ \psi_k = 2\pi \frac{k}{N}, \end{cases} \tag{7.20}$$

where R is the radius of the ring. Substituting this into (7.17) and using the Bessel function series expansion

$$e^{jx\cos\theta} = \sum_{n=-\infty}^{\infty} j^n J_n(x) e^{jn\theta}, \tag{7.21}$$

we obtain

$$\begin{aligned} F(\phi, \theta | \mathbf{r}, \boldsymbol{\psi}) &= \frac{1}{N} \sum_{n=-\infty}^{\infty} j^n J_n\left(2\pi\rho\frac{R}{\lambda}\right) e^{-jn\xi} \left\{ \sum_{k=1}^{N} e^{j2\pi\frac{n}{N}k} \right\} \\ &= \frac{1}{N} \sum_{n=-\infty}^{\infty} j^n J_n\left(2\pi\rho\frac{R}{\lambda}\right) e^{-jn\xi} N\delta_{n \bmod N}, \end{aligned} \tag{7.22}$$

where δ_i is the Kronecker delta function. Furthermore, noticing that $J_{-n}(x) = (-1)^n J_n(x)$, the above series can also be expressed as

$$\begin{aligned} F(\phi, \theta | \mathbf{r}, \boldsymbol{\psi}) &= \sum_{l=-\infty}^{\infty} e^{j(\frac{\pi}{2}-\xi)lN} J_{lN}\left(2\pi\rho\frac{R}{\lambda}\right) \\ &= J_0\left(2\pi\rho\frac{R}{\lambda}\right) + \sum_{l=1}^{\infty} J_{lN}\left(2\pi\rho\frac{R}{\lambda}\right) \left\{ e^{j(\frac{\pi}{2}-\xi)lN} + e^{j(\frac{\pi}{2}+\xi)lN} \right\}. \end{aligned} \tag{7.23}$$

The above series rapidly converges with the first few terms (especially when $\frac{R}{\lambda}$ is small) and thus can be easily calculated numerically.

In the case of multiple rings, let $R_1, R_2, \ldots, R_{N_R}$ denote the radii of the rings with $R_1 > R_2 > \cdots > R_{N_R}$. For simplicity, we assume that the number of nodes in each ring is identical and given by $N' = N/N_R$. The resulting array factor can be expressed as

$$F(\phi, \theta | \mathbf{r}, \boldsymbol{\psi}) = \frac{1}{N_R} \sum_{m=1}^{N_R} \left\{ J_0\left(2\pi\rho\frac{R_m}{\lambda}\right) + \sum_{l=1}^{\infty} J_{lN'} \right.$$
$$\left. \times \left(2\pi\rho\frac{R_m}{\lambda}\right) \left\{ e^{j(\frac{\pi}{2}-\xi)lN'} + e^{j(\frac{\pi}{2}+\xi)lN'} \right\} \right\}. \qquad (7.24)$$

Finally, the far-field beam pattern can be obtained by substituting (7.24) into (7.8).

Without loss of generality, let us consider the case with the target direction $(\phi_0, \theta_0) = (0°, 90°)$ and consider the beam patterns on the x–y plane ($\theta = 90°$). In this case, ρ and ξ are given by

$$\begin{cases} \rho = 2\left|\sin\frac{\phi}{2}\right|, \\ \xi = -\arctan\cot\frac{\phi}{2} = \frac{\phi-\pi}{2}, \quad 0 < \phi < 2\pi, \end{cases} \qquad (7.25)$$

respectively. In Fig. 7.8, several beam patterns of circular arrays with $N = 20$ and $N_R = 1, 2$ are plotted. In the case of $N_R = 2$, the inner radius is chosen as $R_2 = R_1/2$. It is seen that the mainlobe becomes sharp as the R_1/λ increases similar to those of linear arrays, while the peaks of the grating lobes are smaller in all the cases examined. However, we still observe sidelobes as large as -4 dB in the case of $N_R = 1$ and $R_1/\lambda = 4$. Therefore, node allocation similar to circular arrays may not necessarily yield good beam patterns that have low sidelobes.

7.3 Collaborative Beamforming by Randomly Distributed Nodes

In the previous section, we have seen that when the antennas are aligned with equal spacing (i.e., periodic array), the resulting beam patterns tend to have large grating lobes. This is a well-known characteristic in the antenna theory literature [3], and from this viewpoint unequally spaced arrays [7] have been preferred.

When we deal with a distributed beamforming in wireless ad hoc sensor networks, it may be reasonable to assume that the distributed antenna nodes are located randomly by nature. Therefore, the beam patterns of these random arrays are determined by particular realizations of randomly chosen node locations. Consequently, it may be natural to treat the beam pattern with probabilistic arguments. The question is whether one can form a nice beam pattern with a narrow mainlobe and low sidelobes with certain probability.

The probabilistic analysis of random arrays can be found in the history of antenna theory literature. In the framework of linear arrays with a large number of antenna elements, Lo [9] has developed a comprehensive theory of random arrays. The main

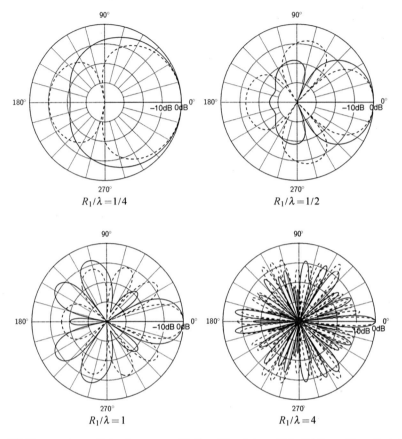

Fig. 7.8 Beam patterns of circular arrays with $N = 20$ and $\phi_0 = 0°$ (solid lines: $N_R = 2$, dashed lines: $N_R = 1$)

result is that a random array can form a good beam pattern with high probability and if the number of antenna elements is N, the directivity of resulting random arrays approaches N asymptotically, thus yielding an antenna gain of order N. In the framework of ad hoc sensor networks, statistical properties of beam patterns with random arrays have been studied in [13]. This section serves as an overview of the main results in [13].

7.3.1 Definition

Let us go back to the node location model of Fig. 7.2 and assume that the node locations (r_k, ψ_k) are a pair of random variables (RVs) that follow some probability distributions. If we randomly drop the sensor nodes over a given disk of radius R, it is reasonable to assume that the node location RVs follow uniform distribution.

Specifically, the joint probability density function (pdf) of (x_k, y_k) in the x–y (Cartesian) coordinates is given by

$$f_{x_k, y_k}(x, y) = \begin{cases} \frac{1}{\pi R^2}, & x^2 + y^2 < \pi R^2, \\ 0, & \text{otherwise.} \end{cases} \tag{7.26}$$

By changing the variables from Cartesian to polar coordinates (r_k, ψ_k), the joint pdf is given by

$$f_{r_k, \psi_k}(r, \psi) = r\, f_{x_k, y_k}(x, y)\big|_{x = r\cos\psi, y = r\sin\psi} = \frac{r}{\pi R^2} \tag{7.27}$$

from which we obtain

$$\begin{cases} f_{r_k}(r) = \int_0^{2\pi} f_{r_k, \psi_k}(r, \psi)\, d\psi = \dfrac{2r}{R^2}, & 0 \le r < R, \\ f_{\psi_k}(\psi) = \int_0^R f_{r_k, \psi_k}(r, \psi)\, dr = \dfrac{1}{2\pi}, & 0 \le \psi < 2\pi. \end{cases} \tag{7.28}$$

Recall that the far-field beam pattern for a given constellation (r_k, ψ_k) in polar coordinates can be expressed from (7.17) as

$$F(\phi, \theta | \mathbf{r}, \boldsymbol{\psi}) = \frac{1}{N} \sum_{k=1}^N e^{j 2\pi \frac{R}{\lambda} \rho \frac{r_k}{R} \cos(\psi_k - \xi)}, \tag{7.29}$$

where ρ and ξ are defined in (7.18) and (7.19), respectively. The above equation can also be expressed as

$$F(\phi, \theta | \mathbf{z}) \triangleq \frac{1}{N} \sum_{k=1}^N e^{j \alpha_{\phi, \theta} z_k}, \tag{7.30}$$

where $\alpha_{\phi, \theta}$ is defined as

$$\begin{aligned} \alpha_{\phi, \theta} &\triangleq 2\pi \frac{R}{\lambda} \rho \\ &= 2\pi \frac{R}{\lambda} \sqrt{(\sin\theta \cos\phi - \sin\theta_0 \cos\phi_0)^2 + (\sin\theta \sin\phi - \sin\theta_0 \sin\phi_0)^2}, \end{aligned} \tag{7.31}$$

and z_k is a new RV expressed as

$$z_k = \frac{r_k}{R} \cos(\psi_k - \xi), \quad -1 \le z_k \le 1. \tag{7.32}$$

Since ξ is a constant offset and the location angle ψ_k is a uniform RV, $\tilde{\psi}_k \triangleq \psi_k - \xi$ also follows a uniform distribution *for a fixed observation angle* (ϕ, θ).

Using (7.28), the pdf of the RV z_k can be obtained as

$$f_{z_k}(z) = \frac{2}{\pi}\sqrt{1-z^2}, \quad |z| \leq 1. \tag{7.33}$$

Given the realization of $\mathbf{z} = [z_1, z_2, \ldots, z_N]$, the far-field beam pattern can be expressed from (7.30) as

$$P(\phi, \theta | \mathbf{z}) = |F(\phi, \theta | \mathbf{z})|^2 = \frac{1}{N^2} \sum_{k=1}^{N} \sum_{l=1}^{N} e^{j\alpha_{\phi,\theta}(z_k - z_l)}$$

$$= \frac{1}{N} + \frac{1}{N^2} \sum_{k=1}^{N} e^{j\alpha_{\phi,\theta}z_k} \sum_{\substack{l=1 \\ l\neq k}}^{N} e^{-j\alpha_{\phi,\theta}z_l}. \tag{7.34}$$

7.3.2 Average Beam Patterns

We first consider the statistical average of beam patterns of random arrays defined in the previous section. The average beam patterns of random arrays can be defined as

$$P_{\mathrm{av}}(\phi, \theta) \triangleq E_{\mathbf{r},\boldsymbol{\psi}}\{P(\phi, \theta | \mathbf{r}, \boldsymbol{\psi})\} = E_{\mathbf{z}}\{P(\phi, \theta | \mathbf{z})\}, \tag{7.35}$$

where $E_{\mathbf{x}}\{\cdot\}$ denotes expectation with respect to the random variables $\mathbf{x}$. Assuming that the r_k's and ψ_k's (and thus also z_k's) are independent and identically distributed (i.i.d.), it can be easily shown from (7.33) and (7.35) [or directly from (7.28) and (7.29)] that

$$P_{\mathrm{av}}(\phi, \theta) = \frac{1}{N} + \left(1 - \frac{1}{N}\right)\left|2\frac{J_1\left(\alpha_{\phi,\theta}\right)}{\alpha_{\phi,\theta}}\right|^2. \tag{7.36}$$

The width of the mainlobe in the average beam patterns can be characterized by the minimum nonzero value of ρ that minimizes (7.36). The term $|J_1(x)/x|$ that appears in (7.36) is oscillatory and the minimum positive value of x that has $J_1(x)/x = 0$ is numerically found as $x_0 = 3.8317$. Let ρ_0 represent the value of ρ at $\alpha_{\phi,\theta} = x_0$. Then, since

$$\rho_0 = \frac{x_0}{2\pi R/\lambda}, \tag{7.37}$$

ρ_0 approaches 0 as R/λ increases. This indicates that the width of the mainlobe of the average beam patterns becomes sharper as R/λ increases.

Next, let us consider the sidelobe level of the average beam patterns. Using the approximation of the Bessel function

$$J_1(x) \sim \sqrt{\frac{2}{\pi x}} \cos\left(x - \frac{3}{4}\pi\right),$$ (7.38)

which becomes accurate for $x \gg 1$, the average beam pattern can be approximated as

$$P_{\mathrm{av}}(\phi, \theta) \sim \frac{1}{N} + \left(1 - \frac{1}{N}\right) \frac{8}{\pi \alpha_{\phi,\theta}^3} \cos^2\left(\alpha_{\phi,\theta} - \frac{3}{4}\pi\right).$$ (7.39)

From (7.31), $\alpha_{\phi,\theta}$ linearly increases with R/λ as long as $\rho > 0$. Therefore, (7.39) indicates that the average beam pattern of the sidelobe region (i.e., $\rho > 0$) approaches $1/N$ as R/λ increases.

In Fig. 7.9, the average beam pattern of random arrays on the x–y plane ($\theta = 90°$) is plotted with $N = 20$ and target angle $(\phi_0, \theta_0) = (0°, 90°)$. Note that in this case, ρ is equal to that given in (7.25) and the value of ϕ that corresponds to the first null from the mainlobe is related to R/λ by

$$\rho_0 = 2 \sin\frac{\phi}{2} = \frac{x_0}{2\pi} \frac{1}{R/\lambda}.$$ (7.40)

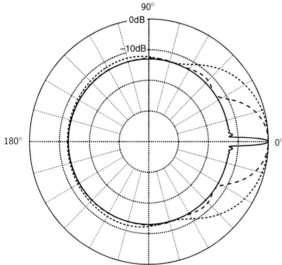

Fig. 7.9 Average beam patterns of random arrays with $N = 20$ (*dotted lines*: $R/\lambda = 0.5$, *dashed lines*: $R/\lambda = 1$, *solid lines*: $R/\lambda = 10$)

Arranging the above equation with respect to ϕ and using the approximation $\arcsin(x) \sim x$ which is valid for $x \ll 1$, we obtain

$$\phi \sim \frac{x_0}{2\pi} \frac{1}{R/\lambda} = \frac{0.61}{R/\lambda} = \frac{35°}{R/\lambda}.$$ (7.41)

Therefore, the width of the mainlobe is (approximately) inversely proportional to R/λ and does not depend on the number of nodes. This tendency agrees with the results obtained in Fig. 7.9.

As for the sidelobe, it is observed that the power level in the sidelobe region is as low as $1/N(= -13\,\mathrm{dB})$ in all the cases, and the mainlobe becomes sharper as R/λ increases. No grating lobe is observed.

Note that if the carrier frequency $f_c = c/\lambda$ is 100 MHz (where c is the speed of light), then the wavelength $\lambda = 3\,\mathrm{m}$ and thus the case with $R/\lambda = 10$ corresponds to $R = 30\,\mathrm{m}$, which may be a realistic value for some sensor network applications such as outdoor field monitoring.

In summary, as R/λ increases, the average beam patterns of the random arrays have the sidelobe as low as $1/N$ and its mainlobe becomes narrower. These are the two most desirable characteristics in the antenna array design. Further aspects of the average beam patterns of random arrays (on the x–y plane) such as an achievable average directive gain can be found in [13].

7.3.3 Distribution of Beam Patterns

In the previous section, we have seen that the average beam patterns of random arrays have ideal characteristics such as sharp mainlobe and low sidelobes. Nevertheless, the average patterns do not represent any particular beam patterns that are realizable. They simply represent the levels averaged over all best- and worst-case beam patterns. As an example, Fig. 7.10 compares the average beam patterns $P_{av}(\phi, \theta)$ and their particular realizations $P(\phi, \theta|\mathbf{r}, \boldsymbol{\psi})$ on the x–y plane with $N = 20$ and 100 randomly generated node locations $\mathbf{r}$ and $\boldsymbol{\psi}$. As can be seen, the realizations of random arrays have relatively high sidelobes. Therefore, it is also important to consider the *statistical distributions* of beam pattern levels in random arrays.

Statistical distributions of linear random arrays are first studied by Lo [9]. We extend the work of Lo to our random array model. The statistical distributions of random arrays can be evaluated exactly as done in [13], but in what follows, we only focus on a Gaussian approximation approach which generally yields simpler mathematical expressions.

From the far-field beam pattern of (7.30), we can write

$$F(\phi, \theta|\mathbf{z}) = \frac{1}{N} \sum_{k=1}^{N} e^{j\alpha_{\phi,\theta} z_k} = \frac{1}{\sqrt{N}} (X + jY), \qquad (7.42)$$

where

$$X \triangleq \frac{1}{\sqrt{N}} \sum_{k=1}^{N} \cos\left(\alpha_{\phi,\theta} z_k\right), \quad Y \triangleq \frac{1}{\sqrt{N}} \sum_{k=1}^{N} \sin\left(\alpha_{\phi,\theta} z_k\right). \qquad (7.43)$$

Since the z_k's are i.i.d. RVs, as N increases, the distributions of X and Y in the direction where $\alpha_{\phi,\theta} \neq 0$ may approach those of Gaussian random variables by the

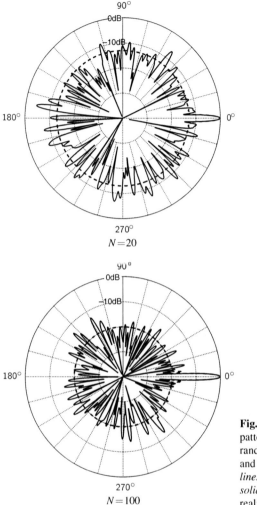

Fig. 7.10 Average beam
patterns and realizations of
random arrays with $\phi_0 = 0°$
and $R/\lambda = 10$ (*dashed
lines*: average beam patterns,
solid lines: their particular
realizations)

central limit theorem. Their means and variances are given by

$$E\{X\} = 2\frac{J_1\left(\alpha_{\phi,\theta}\right)}{\alpha_{\phi,\theta}}\sqrt{N} \triangleq m_x, \tag{7.44}$$

$$\text{Var}\left(X\right) = \frac{1}{2}\left(1 + \frac{J_1\left(2\alpha_{\phi,\theta}\right)}{\alpha_{\phi,\theta}}\right) - \left[2\frac{J_1\left(\alpha_{\phi,\theta}\right)}{\alpha_{\phi,\theta}}\right]^2 \triangleq \sigma_x^2, \tag{7.45}$$

$$E\{Y\} = 0, \tag{7.46}$$

$$\text{Var}\left(Y\right) = \frac{1}{2}\left(1 - \frac{J_1\left(2\alpha_{\phi,\theta}\right)}{\alpha_{\phi,\theta}}\right) \triangleq \sigma_y^2. \tag{7.47}$$

Note that $E\{XY\} = 0$, i.e., X and Y are orthogonal and thus statistically uncorrelated. The joint pdf of X and Y is then given by

$$f_{X,Y}(x, y) = \frac{1}{2\pi\sigma_x\sigma_y} \exp\left(-\frac{|x - m_x|^2}{2\sigma_x^2} - \frac{y^2}{2\sigma_y^2}\right). \tag{7.48}$$

The complementary cumulative distribution function (CCDF) of the beam pattern, i.e., the probability that the instantaneous power of a given realization in the direction (ϕ, θ) exceeds a threshold power, P_0, is given by

$$\Pr[P(\phi, \theta|\mathbf{z}) > P_0] = \Pr\left[\frac{X^2 + Y^2}{N} > P_0\right] = \Pr\left[\sqrt{X^2 + Y^2} > \sqrt{NP_0}\right]$$

$$= \int_{\sqrt{NP_0}}^{\infty} \int_{-\pi}^{\pi} \frac{r}{2\pi\sigma_x\sigma_y} \exp\left(-\frac{|r\cos\varphi - m_x|^2}{2\sigma_x^2} - \frac{r^2\sin^2\varphi}{2\sigma_y^2}\right) d\varphi\, dr$$

$$= \int_{-\pi}^{\pi} \frac{1}{4\pi\sigma_x\sigma_y U_\varphi^2} e^{V_\varphi^2 - \frac{m_x^2}{2\sigma_x^2}} \left[\sqrt{\pi} V_\varphi \operatorname{erfc}\left(W_\varphi - V_\varphi\right) + e^{-(W_\varphi - V_\varphi)^2}\right] d\varphi, \tag{7.49}$$

where

$$U_\varphi \triangleq \sqrt{\frac{\cos^2\varphi}{2\sigma_x^2} + \frac{\sin^2\varphi}{2\sigma_y^2}}, \quad V_\varphi \triangleq \frac{m_x\cos\varphi}{2\sigma_x^2 U_\varphi}, \quad W_\varphi \triangleq \sqrt{NP_0} U_\varphi. \tag{7.50}$$

For the sidelobe regions ($\rho > 0$) and when R/λ is large, we have $\alpha_{\phi,\theta} \gg 1$. In this case, the terms $J_1(2\alpha_{\phi,\theta})/\alpha_{\phi,\theta}$ and $|J_1(\alpha_{\phi,\theta})/\alpha_{\phi,\theta}|^2$ in the variance expressions (7.45) and (7.47) rapidly decrease and both variances are approximately equal as $\sigma_x^2 \approx \sigma_y^2 \approx 1/2$. If this approximation holds, the envelope of a complex Gaussian random variable $W = \sqrt{X^2 + Y^2}$ follows a Nakagami–Rice distribution, which is defined as

$$f_W(r) = 2re^{-(r^2 + m_x^2)} I_0(2m_x r),$$

where I_n is the nth-order modified Bessel function of the first kind.

Consequently, the integral of (7.49) can be expressed as

$$\Pr[P(\phi, \theta|\mathbf{z}) > P_0] = Q\left(\frac{m_x}{\sigma_x}, \frac{\sqrt{NP_0}}{\sigma_x}\right) = Q\left(\sqrt{2}m_x, \sqrt{2NP_0}\right), \tag{7.51}$$

where

$$Q(x, y) = \int_y^{\infty} te^{-\frac{t^2 + x^2}{2}} I_0(xt)dt$$

is the first-order Marcum-Q function.

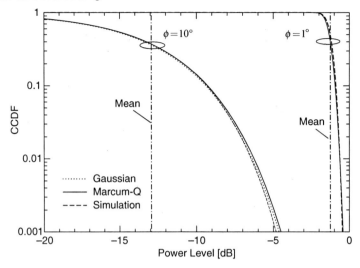

Fig. 7.11 Distribution of beam pattern of random arrays with $N = 20$, $R/\lambda = 10$, and $(\phi_0, \theta_0) = (0°, 90°)$. Observation angle: $(\phi, \theta) = (1°, 90°)$ and $(10°, 90°)$

Furthermore, if the reference angle is well away from the mainlobe such that the mean m_x can also be assumed to be zero, the envelope follows a Rayleigh distribution and (7.51) can be further simplified to

$$\Pr[P(\phi, \theta | \mathbf{z}) > P_0] = e^{-N P_0}, \qquad (7.52)$$

which no longer depends on the observation angles θ and ϕ.

The distribution of beam patterns of random arrays on the x–y plane ($\theta = 90°$) is obtained by computer simulations based on 10^6 random realizations of node locations with $(\phi_0, \theta_0) = (0°, 90°)$. In Fig. 7.11, the results are plotted, along with the theoretical distributions based on Marcum-Q function approximation (7.51) and zero-mean Gaussian approximation (7.52). The vertical lines also indicate the mean values of corresponding beam patterns.

In this example, we have chosen $R/\lambda = 10$ and thus from (7.41) the mainlobe lies approximately within the region $\phi < 3.5°$. Among the two observation angles $\phi = 1°$ and $\phi = 10°$ plotted in this figure, the former corresponds to the mainlobe region and the latter corresponds to the sidelobe region. As is observed, while Marcum-Q function expression can serve as a good reference in the case of the distribution in the mainlobe region, simple zero-mean Gaussian approximation has reasonable accuracy in the case of sidelobe region. Also observed is the fact that while the beam pattern in the mainlobe region shows little fluctuation, the sidelobe region has some noticeable increase of power level. For example, we observe that the sidelobe level exceeds -5 dB (8 dB above the average) with 0.1% of probability. It is apparent from (7.52) that the only way to reduce this sidelobe level is to increase the number of nodes N.

7.3.4 *Distribution of Maxima in Sidelobe*

The analysis of the distribution in the previous section may serve as a reference when the observation angle is fixed. In practice, rather than a particular direction, we wish to reduce the sidelobe level in *any* direction except for the mainlobe region. In this case, the previous analysis of the beam pattern distribution in a given direction may not be sufficient, and we have to analyze the distribution of the maxima in the sidelobe region. In the antenna theory literature, the statistical distribution of the maximum peaks in the sidelobe is analyzed in [1, 5, 17, 18] in the framework of linear arrays.

In this section, we analyze the distribution of maxima in sidelobe for our random array model. Specifically, we are interested in the statistical behavior of the following random variable:

$$\Gamma = \max_{(\phi,\theta)\in \text{ sidelobe region}} P(\phi, \theta|\mathbf{r}, \boldsymbol{\psi}).$$

In what follows, the CCDF of the maximum peak sidelobe, which is the probability that the maximum peak sidelobe exceeds a given power level, P_0, will also be referred to as *outage probability* and denoted by $P_{\text{out}} \triangleq \Pr(\Gamma > P_0)$.

The results in the previous section show that the distribution of the beam pattern within the sidelobe region away from mainlobe can be characterized by a zero-mean Gaussian random variable. We further assume that the beam pattern in the sidelobe region is a zero-mean Gaussian random *process*. The far-field array factor in the sidelobe region is thus characterized by a zero-mean complex Gaussian random process.

Let $\nu(a)$ denote the random variable representing the number of times that the envelope of the array factor crosses upward a given threshold level a per unit interval in the sidelobe region. Then $\nu(a)$ can be calculated using the level crossing theory developed by Rice [14, 15]. For simplicity of analysis, we focus on the beam pattern on the x–y plane ($\theta = \theta_0 = 90°$). The mean of this random variable can then be approximated as [13]

$$E\{\nu(a)\} \approx 4\sqrt{\pi}(R/\lambda)ae^{-a^2} \tag{7.53}$$

Note that a in the above expression corresponds to the envelope level normalized by the square root of the average power (i.e., $P_{\text{av}} = 1/N$ in the sidelobe region of our mathematical model). Also, since we assume in the following argument that (7.53) should be monotonically decreasing with a, this approximation is meaningful only in the range $a \geq 1/\sqrt{2}$.

Using the argument of [12], the cumulative distribution function (CDF) of the maximum sidelobe power, which is a probability that a realization of maximum sidelobe peak is below some normalized threshold, p_0, is approximated as

$$\Pr[\Gamma/P_{av} < p_0] = \Pr\left[\text{all peaks above some threshold } a_0 \text{ is less than } \sqrt{p_0}\right]$$

$$\approx \Pr\left[\text{peak above } a_0 \text{ is less than } \sqrt{p_0}\right]^{E\{v(a_0)\}}$$

$$\approx \left(1 - \frac{E\left\{v(\sqrt{p_0})\right\}}{E\left\{v(a_0)\right\}}\right)^{E\{v(a_0)\}} \rightarrow e^{-E\{v(\sqrt{p_0})\}} \quad (\text{as } E\{v(a_0)\} \to \infty).$$

Consequently, the outage probability is given by

$$P_{out} = \Pr[\Gamma/P_{av} > p_0] \approx 1 - e^{-E\{v(\sqrt{p_0})\}} = 1 - e^{-4\sqrt{\pi}(R/\lambda)\sqrt{p_0}e^{-p_0}}, \quad (7.54)$$

where $p_0 > 1/2$. Substituting $p_0 = P_0/P_{av} = N P_0$ results in the desired outage expression:

$$P_{out} = \Pr[\Gamma > P_0] \approx 1 - e^{-4\sqrt{\pi}(R/\lambda)\sqrt{N P_0}e^{-N P_0}}, \quad \text{for } P_0 > \frac{1}{2N}. \quad (7.55)$$

Figure 7.12 shows a comparison between simulation results and approximation (7.55) for $N = 20$ and 100 with different values of R/λ. Simulation results are based on 10^4 realizations of random node locations generated by computer simulations. Due to a number of approximations involved, the theoretical curves do not match precisely with those of simulations, but they can at least serve for predicting the behavior of the maximum sidelobe peak. From this figure it is observed that the maximum sidelobe peak level is inversely proportional to the number of antenna elements N. In other words, the maximum sidelobe level normalized by its average power does not depend on the value of N as is apparent from (7.54). Therefore, at least in a statistical sense, increasing N always contributes to the reduction of the

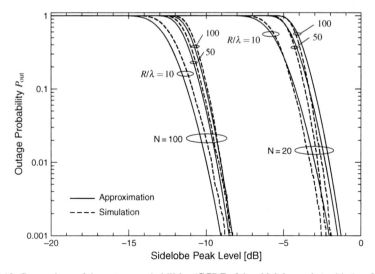

Fig. 7.12 Comparison of the outage probabilities (CCDF of the sidelobe peaks) with $\theta = 90°$

maximum sidelobe level. It is also recognized from this result that increasing the disk radius R/λ for a given number of elements N may increase the maximum sidelobe peak level. This is due to the fact that as R/λ increases, the mainlobe becomes narrow, but the sidelobe peaks also become narrow on average. Consequently, the number of sidelobe peaks per sidelobe interval increases, increasing the likelihood of higher sidelobe peaks.

To further investigate this behavior, Fig. 7.13 shows the amount of peak sidelobe level increase p_0 for a given outage probability and R/λ, derived from (7.54). As can be observed, the maximum sidelobe may grow as R/λ increases, but the amount is below 12 dB for many cases of interest.

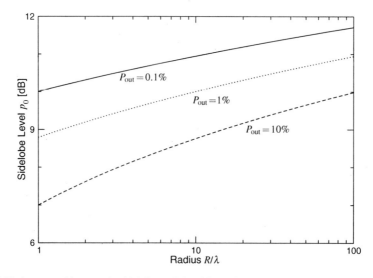

Fig. 7.13 Amount of increase in sidelobe peak level for a given outage probability P_{out}

7.4 Conclusions

Collaborative beamforming has a potential to become one realization of future distributed cooperative/collaborative communications paradigm. Probabilistic studies have shown that if the collaborating nodes are located sparsely with inherent randomness and yet able to synchronize perfectly, they can form a nice beam with sharp mainlobe and low sidelobes with high probability. The SDMA based on collaborative beamforming may significantly increase the multiple-access capacity as they can avoid multiple-access interference by collaboratively forming a sharp narrow beam to the target receiver.

Reference [13] further investigates the effects of the synchronization errors on the average beam patterns for both closed- and open-loop scenarios.

We have considered a uniform distribution as its statistical model throughout this work. Nevertheless, it should be noted that the results in this chapter may not be sensitive to the probabilistic models. For example, the beam patterns with node locations following Gaussian distribution are analyzed and similar results are obtained in [2].

Acknowledgements This work was in part supported by the Strategic Information and Communications R&D Promotion Programme (SCOPE), the Ministry of Internal Affairs and Communications, Japan.

References

1. V. D. Agrawal and Y. T. Lo, "Mutual coupling in the phased arrays of randomly spaced antennas," *IEEE Trans. Antennas Propagat.*, vol. 20, pp. 288–295, May 1972.
2. M. F. A. Ahmed, S. A. Vorobyov, "Collaborative beamforming for wireless sensor networks with Gaussian distributed sensor nodes," *IEEE Trans. Wireless Commun.*, vol. 8, no. 2, pp. 638–643, Feb. 2009.
3. C. A. Balanis, *Antenna Theory: Analysis and Design*. New York: Wiley, 1997.
4. L. Dong, A. P. Petropulu, and H. V. Poor, "A cross-layer approach to collaborative beamforming for wireless ad hoc networks," *IEEE Trans. Signal Process.*, vol. 56, no. 7, pp. 2981–2993, July 2008.
5. M. B. Donvito and S. A. Kassam, "Characterization of the random array peak sidelobe," *IEEE Trans. Antennas Propagat.*, vol. 27, pp. 379–385, May 1979.
6. R. C. Hansen, *Phased Array Antennas*. New York: Wiley, 1998.
7. A. Ishimaru, "Theory of unequally-spaced arrays," *IRE Trans. Antennas Propagat.*, vol. 10, pp. 691–702, 1962.
8. J. N. Laneman, D. Tse, and G. W. Wornell, "Cooperative diversity in wireless networks: Efficient protocols and outage behavior," *IEEE Trans. Inform. Theory*, vol. 50, pp. 3062–3080, Dec. 2004.
9. Y. T. Lo, "A mathematical theory of antenna arrays with randomly spaced elements," *IRE Trans. Antennas Propagat.*, vol. 12, pp. 257–268, May 1964.
10. M. T. Ma, *Theory and Application of Antenna Arrays*. New York: Wiley, 1974.
11. R. Mudumbai, G. Barriac, and U. Madhow," On the feasibility of distributed beamforming in wireless networks," *IEEE Trans. Wireless Commun.*, vol. 6, no. 5, pp. 1754–1763, May 2007.
12. H. Ochiai and H. Imai, "On the distribution of the peak-to-average power ratio in OFDM signals," *IEEE Trans. Commun.*, vol. 49, pp. 282–289, Feb. 2001.
13. H. Ochiai, P. Mitran, H. V. Poor, and V. Tarokh, "Collaborative beamforming for distributed wireless *ad hoc* sensor networks," *IEEE Trans. Signal Process.*, vol. 53, no. 11, pp. 4110–4124, Nov. 2005.
14. S. O. Rice, "Mathematical analysis of random noise – Part I," *Bell Syst. Tech. J.*, vol. 23, pp. 282–332, Jul. 1944.
15. S. O. Rice, "Mathematical analysis of random noise – Part II," *Bell Syst. Tech. J.*, vol. 24, pp. 46–156, Jan. 1945.
16. A. Sendonaris, E. Erkip, and B. Aazhang, "User cooperation diversity – Part I: System description," *IEEE Trans. Commun.*, vol. 51, pp. 1927–1938, Nov. 2003.
17. B. D. Steinberg, "The peak sidelobe of the phased array having randomly located elements," *IEEE Trans. Antennas Propagat.*, vol. 20, pp. 129–136, Mar. 1972.
18. B. D. Steinberg, *Principles of Aperture & Array System Design*. New York: Wiley, 1976.
19. D. Tse and P. Viswanath, *Fundamentals of Wireless Communication*. New York: Cambridge University Press, 2005.

Chapter 8
Cooperative Wireless Networks

Behnaam Aazhang, Chris B. Steger, Gareth B. Middleton, and Brett Kaufman

8.1 Introduction

In this section we will provide an overview of the research area and will review a few issues on physical layer cooperation.

8.1.1 Overview

Our thesis is that two edges exist in modern wireless communication networks. The first is physical, at the fringe of network rollouts in developing nations and underserved areas. The other is in performance, as new users and devices demand more resources from existing infrastructure. Both of these scenarios present significant challenges to communication engineers, requiring new ways of thinking at all layers of the network architecture. In this chapter, we will present a new paradigm to address the issues on both of these edges.

Existing network architectures based on hot spots and cellular structures fail to take advantage of the cooperative possibilities offered by adjacent nodes. Opportunism is similarly limited by a lack of coordination and flexibility [1, 2]. We consider a more organic network model in which the nodes are free to exploit spatio-temporal structures for maximum performance and network utilization [3,4]. Prior work has demonstrated the feasibility and effectiveness of cooperative communications in small networks [5–9]. However, the incorporation of three- and four-node cooperative units into a large network presents significant technical challenges due to more complicated routing and interference patterns. These issues are

Behnaam Aazhang, Chris B. Steger, Gareth B. Middleton, and Brett Kaufman
Center for Multimedia Communication,
Department of Electrical and Computer Engineering, Rice University, Houston, TX 77005, USA
e-mail: \{aaz,gbmidd,bkaufman\}@rice.edu,chrissteger@gmail.com

V. Tarokh (ed.), *New Directions in Wireless Communications Research*,
DOI 10.1007/978-1-4419-0673-1_8,
© Springer Science+Business Media, LLC 2009

magnified when we consider competing flows and heterogeneous nodes, which may have different processing capabilities and queue states. These critical issues must be addressed if we are to fully leverage cooperation and opportunism at all network layers.

To that end, we present a three-pronged solution for developing algorithmic protocols grounded in theoretical research:

- **Node Information Management:** Creating a high-performance network – whether at the geographical or technological edge – demands efficient use of network resources while maintaining fairness. We must therefore gather meaningful information about the network while balancing overhead against potential performance gains. To that end, we consider *network state information*, which reduces potentially highly dynamic network information to much more manageable segments, allowing the network to make informed decisions about its topology without excessive overhead.
- **Novel Network Representations:** We consider a coherent framework for the analysis of cooperative networks, which can capture the timing, interference, and duplexing issues inherent in cooperative architectures. This *trellis* framework will be developed to be flexible and accommodate a wide variety of cooperative schemes. It will also be capable of considering many flows simultaneously and will enable the realization of a cooperative, opportunistic network addressing interference, routing, and fairness in competing flows.
- **Distributed Cooperative Discovery:** An organic network is largely decentralized [10–12], meaning that network state information must be gathered in a distributed manner. Historically, discovery has been primarily concerned with finding topological information without accounting for cooperative or opportunistic communication. We will discuss discovery protocols to fill this gap, not only discovering the network state information but also identifying cooperative architectures in an efficient and decentralized manner.

8.1.2 Physical Layer Cooperation

The notion of a source-relay-destination link, introduced by van der Meulen in 1968 [5], has been studied in isolation, and from a limited physical layer only approach for many years [13]. These results all show significant gains of several dB when we allow one relay to cooperate in a transmission. This naturally led to research into topologies in which more than one node cooperates – degraded Gaussian channels [14, 15], parallel Gaussian channels [8], and models with many transmitters in cooperation [16] have all been analyzed.

At its core, a cooperative strategy relies on one of three different physical layer techniques defining the behavior of the relay: amplify-and-forward, decode-and-forward, or estimate-and-forward. In all three cases, the destination implements a combining scheme at the signal level to maximize the diversity offered by receiving multiple copies of the original information [17]. The key differences among these strategies are as follows:

- **Amplify-and-Forward:** The relay receives the message (coded or uncoded), amplifies, and forwards it to the destination without decoding or detecting.
- **Decode-and-Forward:** The relay receives the packet, decodes the message, maps it to the nearest valid codeword, and sends the re-encoded information to the destination.
- **Estimate-and-Forward:** The relay receives the packet and assuming that it cannot decode the message, sends side information (e.g., quantized parity information) about the codeword it has detected to the destination.

As these cooperative strategies have been developed for three-node relay systems depicted in Fig. 8.2(a), it is easy to envision that we will be able to develop similar practical strategies for parallel (or diamond) relay channels depicted in Fig. 8.2(b). In general, we can divide these protocols into two classes: those which resemble multiple-transmitting antennas (transmitter cooperation) and those which resemble multiple-receiving antennas (receiver cooperation) [7, 18 20].

Transmitter cooperation performs well when the source-to-relay channel is stronger than the source-to-destination channel. This could happen when the relay is physically near the source or as a result of fading. Transmitter cooperation is enabled by the decode-and-forward protocol, where the relaying node(s) first reliably decode the source transmission before using it as the basis for cooperation during multiple access.

Receiver cooperation, on the other hand, is enabled by the amplify-and-forward and estimate-and-forward protocols. The common strategy for this pair of protocols is to forward soft information to the destination, where it is optimally combined prior to decoding [21]. Among the two protocols used for receiver cooperation, amplify-and-forward is more attractive to implement due to its simplicity [22]. Estimate-and-forward calls for Wyner–Ziv coding on the relay-to-destination link, which in theory would involve vector quantization and Slepian–Wolf coding, and promises around 30% gain in achievable rate over direct transmission.

This large body of work has focused entirely on the physical layer issues surrounding cooperative architectures. A true demonstration of cooperative gains is only possible when cooperative techniques are validated in a network setting while considering higher layer interactions. Some recent efforts have considered physical layer cooperation within a network in a limited scope. For example, [23–25] considered cooperation with strict resource constraints, [26] derived the max-flow min-cut bound for cooperative networks, and [27–31] considered MAC protocols in a cooperative setting. We consider cooperative networks capable of addressing cooperative routing, distributed coding across cooperative nodes, and spatial interference with competing flows.

This chapter is organized in the following two sections. The general network model from the physical layer to network flows is introduced in Section 8.2. Section 8.3 includes our methodology to develop protocols to manage network state information (NSI), discover NSI, and Sections 8.4 and 8.5 include concepts that enable cooperative routes through the network.

8.2 System Model

In this section we will detail the model under consideration and emphasize the issues and questions which must be addressed when realizing a truly cooperative network. We will divide our attention into three main areas. First, we will describe the characteristics of the wide area network and the users within it. Next, we will consider the presence of existing traffic flows and their effect on cooperation. Finally, we will focus on a novel network representation based on fundamental cooperative building blocks.

8.2.1 Wide Area Network

In general, we consider a heterogeneous network composed of many different kinds of nodes. As in existing networks, the nodes' physical attributes can be highly diverse in terms of processing power, queue state, transmission power, and other traits. In order to enable cooperation among users, we assume that some nodes in the network are able to execute the cooperative protocols detailed in Section 8.1.2. Our system is agnostic to the radio access technology, but for purposes of exposition we can consider technologies similar to the IEEE 802.16 and IEEE 802.11 standards. Similarly, though our work readily generalizes to multi-band scenarios, we will limit our discussion to nodes operating in a single frequency band. Finally, we make the practical assumption that all radios are half-duplex: nodes cannot transmit and receive simultaneously on the same frequency band. We emphasize that, though this constraint is often viewed as a hindrance in previous works, it is a realistic assumption that allows nodes from competing flows to exploit opportunistic access.

We consider the problem of communicating data between two nodes at the edge of a network. One node might be a gateway node to a trunking line inside a cell, while the other rests beyond the edge of the cellular coverage yet still requires service. We will find the "optimal" multi-hop cooperative link between these two units. We will limit the scope of our network by considering only nodes within a somewhat small area, as might be found on the edge of an active cell. We will therefore consider our network to have two fundamental qualities. First, we assume that the network is *not sparse,* thereby assuming that no nodes in the network are poorly connected to nearby nodes. Second, the network is *medium sized* and will contain tens of nodes, meaning that any path between a source and destination will require few (that is two to four) hops. We can prune a larger network into a smaller one using the concept of an *active topology,* which denotes those nodes willing and able to participate in cooperative communication protocols. It has been shown that considering a network of this size limits the complexity of both analysis and implementation [11, 32, 33].

Because no topology is perpetually fixed, we will allow nodes to join the network at any time, first by listening to their surroundings and then by engaging in a discovery process. We will also permit nodes to check out of the network as required. We refer to the length of time the network is in a given state as its *coherence time*. By listening to existing traffic flows, joining nodes will be able to self-synchronize with the network.

8.2.2 Multiple Flows and Flow Priority

Much of the prior work in physical layer networking analysis has considered networks with traffic flows existing in isolation [34–38]. This assumption is unrealistic in today's networks, which often support tens of flows simultaneously [1, 2]. As such, we cannot realistically consider the problem of determining optimal cooperative routes for links existing in isolation. We must take into account both existing and emerging flows when computing our cooperative routes.

To aid in the discovery and establishment of a new flow in the presence of an existing flow, we consider a novel frame structure. This frame structure, while based on current standards, will have notable differences that permit a new flow to establish itself in a minimally intrusive fashion. A *frame*, shown in Fig. 8.1, will consist of two parts: a control window and a data window, each of which will be of a standard known length by all users. After performing a network discovery phase, the details of which will be discussed in Section III-C, a source node will have all the information necessary to determine feasible routes to the destination. The source node will use the control window to reserve both the medium and the intermediate nodes that will help forward information. This reservation will be accomplished by an RTS/CTS process similar to 802.11. In this system, we envision an RTS/CTS

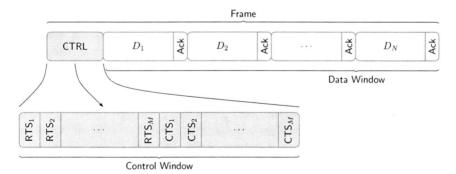

Fig. 8.1 An example frame structure is shown: the CTRL window contains control information, followed by data slots, each of which has an associated acknowledgment message

process in which an RTS, addressed to the destination as well as all intermediate nodes along the intended route, is sent all the way to the destination and likewise a CTS is sent back from the destination to the source. The combination of RTS/CTS informs all nodes along the route of their specific roles, for instance, whether they will be used as a cooperative link or a direct link, as well as the time slots in which nodes will either be transmitting or receiving. Depending on the coherence time of the network, as well as the fairness goals of the system, the source can feasibly send multiple packets along this route during the period of the reservation.

In dealing with competing flows, we adopt the idea of *priority* service [39–44]. In these priority-based models, flows have differing priorities when being scheduled by the network; these priorities can be derived from the pricing structures of the network or the requirements of the service class. As such, an existing high priority flow should not be interrupted by the emergence of a new lower priority flow and an emerging high priority flow must have a mechanism by which to gain access to network services. The particulars of how to assign priorities to different flows are not within the scope of this chapter; however, we emphasize that our proposed trellis and cooperative routing framework are robust to handle flows with different priorities.

8.2.3 Cooperative Building Blocks

Our end goal is to find an optimal cooperative link connecting a source and destination. Traditional approaches to this problem consider a network as a collection of point-to-point links and use a routing protocol to assemble the links and establish a route in order to reach a target end-to-end rate. Node cooperation, where a source employs the assistance of neighboring nodes to help send its message over two separate paths, allows the destination to have two different realizations of the same message. The extra information helps the destination decode the source's message more reliably and increases the set of reliably achievable rates.

A single cooperative relay, combined with the source and destination, forms the well-known triangle relay channel [5]. Two cooperative relays, with the source and destination out of range of each other, form the diamond structure of the parallel relay channel [8]. These two cases, shown in Fig. 8.2, have become the canonical cooperative topologies.

The source and the relay combine their resources to communicate a single symbol to the destination with overlapping transmissions. This fact allows us to view both paths together as a single link from the source to the destination. We will use the canonical topologies in Fig. 8.2 as fundamental building blocks with which to decompose the network. The triangle and diamond structures can be combined with direct links to establish an optimal cooperative route from the source to the destination. However, the interference pattern resulting from a cooperative route will be substantially different from traditional multi-hop routes.

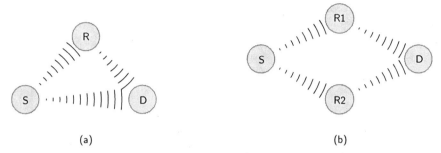

Fig. 8.2 The two fundamental topologies of cooperation: (a) triangular relay channel and (b) parallel, or diamond, relay channel

For the triangle relay topology, there will be two sources of interference, since both the source and the relay will be interfering with surrounding nodes. Likewise, in the diamond relay topology, both relays will cause interference as they transmit simultaneously. To address this issue we define a *metanode* to model both nodes in a cooperative link and the combined effects of their interference. Analyzing interference patterns is particularly important when managing competing flows in our cooperative wireless network.

8.3 Learning About the Environment: Network State Information

To achieve our goal of cooperative routing at all layers, our coding and scheduling algorithms must have access to information about every layer of the network. Channel state information (CSI) is frequently used for path selection in wireless networks [45–47]. However, this approach does not consider other key characteristics of the nodes and channels that influence the quality of the end-to-end path seen by the application layer. For example, a node in the selected path may have a large queue causing unacceptable end-to-end delay. Existing routing schemes, which optimize over only CSI at the physical layer or queue state at the MAC layer, do not account for route coupling or processing power of the nodes. We propose to integrate these and other metrics into a comprehensive informational unit: *network state information* (NSI). The results of this research will enable the source to determine an optimal route while seamlessly incorporating multi-hop, relay, and MIMO schemes.

In order for NSI to be a comprehensive representation of link quality, we consider including several diverse measures from all network and hardware layers. Some examples of NSI components are channel state information, queue state, processor capability, and battery state. Physical layer attribute is commonly used to select data rate and is the input to existing routing solutions, as previously stated. Queue state signifies the additional data packet traffic a node is handling, while processor capability is the speed at which a node can process packets. This characteristic influences the coding scheme used; awareness of NSI can dramatically improve the results of routing optimization in a heterogeneous network. For example, though LDPC is a highly effective coding scheme, insufficient processing power will make it inappropriate for some nodes [48]. A classical routing algorithm would choose a path based on channel quality or queue state and would ignore the processing bottleneck; routing with NSI will account for this bottleneck and will allow us to discover a route with better overall performance.

8.3.1 NSI Overhead Management

While a globally optimal solution would necessarily require full information about the entire network, the overhead could mitigate the potential gains. In order to limit the overhead, we will place practical limits on the amount of information required by the routing algorithm. We can consider three avenues to limit overhead: NSI component selection, component quantization, and topology size limitation.

NSI will capture information from the entire network stack. However, the relative importance of the different NSI components will vary depending on application and topology. For example, time-critical applications such as VoIP require short queues whereas high data rate streaming video applications demand high-quality channels. This motivates us to regulate overhead by limiting the NSI to the parameters that are relevant to the application.

In choosing which NSI components to consider, we must also address the granularity of each component. Depending on the nature of the information, each component will be quantized differently (e.g., battery state and channel state). Components will be quantized based on their relevance and variability, and the precision of the information will be directly proportional to the overhead. The definition of NSI for a given network should contain the minimum number of components with the maximum amount of quantization to achieve performance goals.

Ideally, NSI would be gathered for all nodes in the topology, but this could cause catastrophic amounts of overhead in large networks. We consider limiting the size of the network to a manageable number of nodes, referred to as the *active topology*, for each flow. By restricting discovery to a subset of nodes, overhead can be limited to an acceptable level. During NSI discovery, mechanisms such as hop counts and time limits constrain the active topology to the subset of well-connected nodes in the neighborhood between a source and destination.

By defining the active topology using heuristics that favor nodes near the source and destination, we manage overhead without sacrificing significant performance.

8.3.2 NSI Metric

In this study, we consider finding an optimal dynamic characterization of NSI that is geared toward cooperative networks. We can then derive mappings from vectors of NSI components to scalar quality metrics, such that the mappings provide meaningful indications of application layer link quality. Since NSI components have different relative importance and variability, the various components will be weighted accordingly. In the next section, we will detail the role of quality metrics derived from NSI in the construction of the trellis representation of the active topology.

8.4 Finding the Optimal Cooperative Path

NSI equips our system with a unique and powerful tool with which to schedule the flows in our heterogeneous network. In order to fully leverage the NSI, we must account for the special structure that a cooperative link introduces to the network graph, and find an efficient way to manage the information about the network. In this section, we will present a novel and robust framework specifically suited to that task.

8.4.1 Routing Cooperative Paths

Our goal is to determine an optimal cooperative route through a wireless network, employing a wide variety of network state information. This problem is significantly more difficult than the traditional route-selection problem for wired networks. Many robust algorithms have been developed to determine optimal routes between sources and destinations; Dijkstra's algorithm, the Bellman–Ford algorithm, and their derivatives [49, 50] iteratively search the network graph for a least-cost or maximum-throughput route and are guaranteed to return it in a finite number of iterations.

These algorithms perform route optimization, but do so only for routes composed of single links. This is because the algorithms are solving separable optimization problems, which reduce to local optimizations at each step of the route. If we let $\mathcal{R} = \{e_1, e_2, \ldots, e_N\}$ be a set of edges in a network graph forming a route between a source and destination, then these algorithms can solve the maximum-throughput route problem

$$C_{\max} = \min_{\mathcal{R}} \max_{\{e_l\}} C_l, \tag{8.1}$$

where C_l is the capacity of link e_l. For wireless channels, this capacity is classically a function of bandwidth W and the channel SNR. In the case of a cooperative path, the physical layer capacity is given by a *function* of the SNRs of the constituent channels, depending on the cooperative scheme in use. Mapping this to the optimization problem, we reach

$$C_{\max} = \min_{\mathcal{R}} \max_{\{e_l\}} f(C_l^{SD}, C_l^{SR}, C_l^{RD}). \tag{8.2}$$

Because this function may be nonlinear in link capacities C_l^{SD}, C_l^{SR}, and C_l^{RD}, the optimization problem does not separate across routing steps, leaving it largely intractable. Further, the confluence of multiple flows and relay half-duplex timing constraints require us to define an opportunistic algorithm to establish the routing. Previous maximal-set and load-balancing algorithms schedule flows such that they have equal transmission times over the long run, but do not explicitly define the sequence of packet transmissions in the network. Our optimization problem must therefore account for fine-grained timing control and opportunistic scheduling of users, while managing the interference a wireless relay link can introduce.

We will use the NSI metric, which is influenced by classical capacity, in our optimization to find a route solution which is optimal at all network layers. To determine that solution, we must analyze the network at each time step, charting the path of packets in each flow and the quality of each link as measured with the NSI.

8.4.2 Trellis Representation

We represent the interaction of nodes in our network using a trellis structure. This will enable us to use the tools of graph theory to chart the flow of data through our network, capturing the temporal and interference issues mentioned earlier. Each node in the trellis (hereafter referred to as a "state") represents a source and sink for data, and each edge (referred to as a branch) represents the path data can follow when being sent to the next state and will be weighted by that link's NSI metric. Thus each successive set of states represents a time step in the data flow. Since the states are data sinks, we can immediately include all nodes in the network as states in the trellis.

We must introduce a way to show that a collection of nodes can be cooperating. With the term *metanode* we denote a state in the trellis which represents exactly that behavior: data emanating from a metanode are actually being transmitted by several different nodes cooperating with each other. The branches in our trellis represent how data being transmitted from a particular state access the wireless medium and spread through the network to the next state. We can construct the branches using the network graph, which shows immediately how different nodes are connected with one another and therefore, how packets can flow between states. By overlaying the potential data flows from *each* state on the same trellis, we are able to chart how interference occurs in the network.

We now present an example of how to construct such a trellis from the simple network shown on the left in Fig. 8.3. Here we have two flows and a relay node, denoted R. To create the trellis, we incorporate all network terminals as states in the structure. With full network information, we immediately identify a cooperative link between S2 and D2 via R and add the metanode (S2,R) to our list of trellis states. This indicates that both S2 and R transmit simultaneously, since the cooperative physical layer schemes allow for combining of these messages at D2, such that the interference at D2 is permissible. We can now chart how data potentially spread through the network, as shown on the right in Fig. 8.3. Here the gray solid arrows indicate a wireless transmission, while the green dashed arrows indicate "storing" a packet until the next time step.

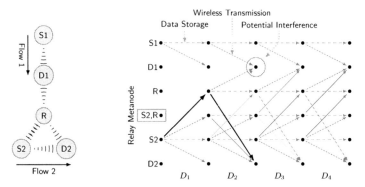

Fig. 8.3 An example network with two flows is shown on the *left*. On the *right* is the trellis derived from it. We have indicated the relay metanode S2,R, along with the potential data propagation paths. Interference from wireless transmissions can occur where *solid arrows* meet, while data storage is possible as indicated by the *broken line*. The *thick line* indicates a potential multi-hop data path through the trellis from S2 to R and then D2, without using cooperative schemes

8.4.3 Timing, Interference, and Duplexing Management

Using this representation, we will be able to design algorithms to find optimal paths through the network, managing interference, timing, and duplexing conflicts. This will enable us to opportunistically schedule multiple flows together, since the half-duplex nature of the radios naturally creates "space" in the medium for other flows to utilize. We capture the half-duplex nature of the nodes by requiring in any given state transition an edge that does not depart from and arrive at the same state. The challenge now becomes how to choose an optimal set of state transitions, known as *traversals*, which route flows through the trellis. Generally, we will be choosing the traversal from a set of feasible traversals; this is illustrated well in the example network. There are two scenarios we can consider: In the first case, the link between S1 and D1 is active in every time step, which precludes opportunistic use of the relay node in any way; using R would cause interference at D1. This restricts the S2–D2 flow to single-hop communications only. This scenario is shown on the left in Fig. 8.4. We have indicated with a red 'X' those states for which the use of the medium explicitly prevents other states from using it due to interference. In this case using S,R and R are not permitted while S1 is transmitting.

If we create an opportunistic schedule, we can allow the relay to be used as shown on the right in Fig. 8.4. Here, S1 and R transmit orthogonally in time, intertwining

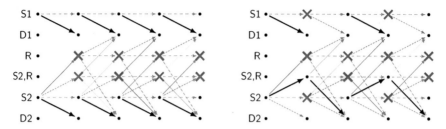

Fig. 8.4 Trellis traversal for which the relay is unusable (*left*) and a traversal which allows for the relay to assist between S2 and D2. Depending on the NSI values for the edges in this trellis and the object of our optimization, one of these two traversals will be preferable

their transmissions such that Flow 2 uses the relay while Flow 1 is idle. Which of these two configurations is optimal depends on the NSI edge weight parameters; if using the S,R metanode in Flow 2 creates a significant increase in throughput and the direct link in Flow 1 is relatively slow, then we would favor the opportunistic schedule. Conversely, if the queue at R is long, we would prefer to use two simultaneous single flows. These situations are fully captured by NSI, and will be incorporated into the trellis as edge weights.

8.4.4 Traversal Algorithms

The trellis captures all of the issues surrounding cooperative route determination in the presence of existing flows: interference, timing, and route alternatives weighted by NSI. As much as the trellis is similar to those used for decoding convolutional codes [51], the algorithms used in that field do not apply in this case. We consider using the trellis and NSI tools we have developed here, in conjunction with our knowledge of physical layer cooperation, to determine ideal traversal algorithms for this trellis. The result of executing these algorithms will be an optimal schedule, making use of cooperative links where possible.

The power of this approach is the way in which it combines traditionally physical layer concepts, such as cooperative communication, with networking concepts, such as routing and scheduling. By merging these in a tractable and usable way, we have created a framework for analyzing the true convergence of wireless technologies.

8.5 Network Discovery

The previous discussions have introduced network state information metric, along with a powerful framework leveraging this metric and capturing the exact interactions between nodes. In a distributed network, we require a way to gather this information in the most efficient way possible, since overhead can potentially mitigate gains our schemes may offer. In this section, we will discuss discovery for cooperative networks.

8.5.1 Filling the Trellis: Gathering States, Edges, and NSI

The trellis, our key data structure, requires three key pieces of information–the states, the edges, and the NSI edge weights. The states include both single nodes and metanodes, the edges detail the node interactions, and the NSI captures the application-dependent quality of using a given edge. Classical discovery algorithms can help us fill in some, but not all, of this information. For the problem of discovering single nodes and the topological connections between them, we can leverage existing discovery results and protocols to fill in these pieces of the trellis.

In a wired network, discovery typically simplifies to a routing problem, where effectiveness is limited by either a large hop count or delays imposed by long queue lengths. Discovery in a wireless network faces the same challenges, along with interference, fading, noise, and others imposed by the wireless medium. Network discovery can be partitioned into local and global discovery problems, and there are bodies of work focused on both aspects. In [52–60], the focus is on global discovery given local connectivity information. Protocols were developed in [61] that enabled users to discover local information. In particular, [62–64] require new users to a network to broadcast their presence as well as a periodic broadcast by existing users to help maintain a fresh view of the network. In an effort to manage the complexities of discovery, protocols in [64,65] take a centralized control approach. While ad hoc, their networks assume a central node connected to all users, similar to cellular networks, which disseminates information about location to aid in discovery. Also, some of these protocols use a wired backbone to connect central nodes to coordinate discovery across several ad hoc networks.

Since overhead is a major concern in many networks, a distributed discovery protocol will be established to enable an efficient discovery process. The main focus area in distributed methods is source-driven on-demand discovery [10,66,67]. Without complete knowledge of the other nodes in the network, discovery messages are often blindly broadcast and flooded through the network. Results in [60,68–71] show that flooding techniques are beneficial in that there is diversity in the discovery message by traversing more than one link. Flooding, however, can have adverse affects on a network, such as increased interference and contention issues as multiple users try to broadcast simultaneously. Controlled flooding techniques in [72–75] attempt to mitigate some of these problems. Two widely used techniques, DSR [10] and AODV [66], establish multi-hop routes along the path of discovery messages. These protocols are later expanded and refined in [34,76,77].

These algorithms efficiently and comprehensively solve the problem of discovering single nodes, their connections, and their NSI. However, they fail to discover the *cooperative building blocks* we seek to leverage.

8.5.2 Filling the Trellis: Metanodes

Discovery methods outlined in previous works provide the information necessary to construct a partial trellis that consists of all the states and branches, except the metanode states and the branches leaving these states. In order to find metanodes and the cooperative links associated with them, we must be able to find sets of nodes that are connected in a way that allows for cooperative communication, i.e., a set of three nodes where the nodes behave as source, relay, and destination. The source and relay node in the grouping will be considered to be a metanode, indicating their role in a cooperative link; these metanodes along with all individual nodes are used together in the trellis formation. As such, the trellis representation will emphasize all cooperative links in addition to single-hop links.

In this section we identify potential cooperative topologies in a distributed and efficient manner. A centralized approach would have access to the network graph, which is easily filtered using a tree search or an analysis of the connectivity matrix. In the distributed approach, the determination of cooperative links is made in the network itself, as illustrated in the following discovery example, which refers to Fig. 8.3.

When S2 initiates discovery, it transmits a discovery packet to both R and D2. In a simple "forward once" discovery protocol, R forwards the discovery packet after having appended its own address. Upon receiving this second discovery packet, D2 combines it with the packet it received directly from S2 to deduce that a cooperative triangle topology exists between itself and S2, with R acting as a relay. In the return phase of the discovery protocol, D2 transmits this knowledge along with its NSI back to S2, so that the appropriately weighted S2,R metanode can be added to the trellis being formed at S2.

This is a simple example of a distributed cooperative discovery protocol, though one which becomes inefficient in larger and more well-connected topologies. As open research area, one can develop discovery protocols which jointly optimize the cooperative discovery problem and the issue of efficiently recovering NSI from the topology and making that information available to a source. After successfully implementing such a protocol, a node in our network will be fully equipped to establish an optimal cooperative route.

8.6 Conclusions

We have now introduced the major thrusts of this exploratory chapter: NSI, the trellis data structure, and distributed cooperative discovery. These three areas, while independent in the tools they require and the optimization problems they solve, are intimately related when viewed from a wide area network perspective. Each depends on the other, and each is required to successfully implement the technologies leading to an optimal cooperative network. Until now, these ideas have existed purely on paper, though we plan to take them to the field in a unique wireless testbed.

References

1. L. Tassiulas and A. Ephremides, "Stability properties of constrained queueing systems and scheduling policies for maximum throughput in multihop radio networks," *Decision and Control, 1990., Proceedings of the 29th IEEE Conference on*, vol. 4, pp. 2130–2132, 1990.
2. X. Liu, E. Chong, and N. Shroff, "Opportunistic transmission scheduling with resource-sharing constraints in wireless networks," *Selected Areas Commun., IEEE J.*, vol. 19, no. 10, pp. 2053–2064, 2001.
3. P. Gupta and P. R. Kumar, "The capacity of wireless networks," *IEEE Trans. Inform. Theory*, vol. 46, pp. 388–404, Mar. 2000.
4. ——, "Towards an information theory of large networks: An achievable rate region," *IEEE Trans. Inform. Theory*, vol. 49, no. 8, pp. 1877–1894, Aug. 2003.
5. E. C. van der Meulen, "Transmission of information in a T-terminal discrete memoryless channel," Ph.D. dissertation, Dept. of Statistics, University of California, Berkeley, 1968.
6. T. M. Cover and A. A. El Gamal, "Capacity theorems for the relay channel," *IEEE Trans. Inform. Theory*, vol. 25, pp. 572–584, Sep. 1979.
7. A. Sendonaris, E. Erkip, and B. Aazhang, "User cooperation diversity. Part I. System description," *IEEE Trans. Commun.*, vol. 51, pp. 1927–1938, Nov. 2003.
8. B. Schein, "Distributed coordination in network information theory," Ph.D. dissertation, Massachusetts Institute of Technology, Sep. 2001.
9. J. N. Laneman, D. N. C. Tse, and G. W. Wornell, "Cooperative diversity in wireless networks: Efficient protocols and outage behavior," *IEEE Trans. Inform. Theory*, vol. 50, pp. 3062–3080, Dec. 2004.
10. D. B. Johnson and D. A. Maltz, "Dynamic source routing in as hoc wireless networks," *IEEE Trans. Mobile Comput.* vol. 353, pp. 153–181, 1996.
11. M. Corson and A. Ephremides, "A distributed routing algorithm for mobile wireless networks," *Wireless Netw.*, vol. 1, no. 1, pp. 61–81, 1995.
12. I. Akyildiz, J. McNair, J. Ho, H. Uzunalioglu, and W. Wang, "Mobility management in next-generation wireless systems," *Proc. IEEE*, vol. 87, no. 8, pp. 1347–1384, 1999.
13. E. C. van der Meulen, "Three-terminal communication channels," *Adv. Appl. Probability*, vol. 3, pp. 120–154, 1971.
14. A. Reznik, S. R. Kulkarni, and S. Verdú, "Degraded Gaussian multiple relay channel: capacity and optimal power allocation," *IEEE Trans. Inform. Theory*, vol. 50, no. 12, pp. 3037–3046, Dec. 2004.
15. M. Gastpar and M. Vetterli, "On the capacity of large Gaussian relay networks," *IEEE Trans. Inform. Theory*, vol. 51, no. 3, pp. 765–779, Mar. 2005.
16. P. Bergmans and T. M. Cover, "Cooperative broadcasting," *IEEE Trans. Inform. Theory*, vol. 20, pp. 317–324, May 1974.
17. M. A. Khojastepour, "Distributed cooperative communications in wireless networks," Ph.D. dissertation, Dept. of Electrical and Computer Engg., Rice University, 2004.
18. A. Sendonaris, E. Erkip, and B. Aazhang, "User cooperation diversity. Part II. Implementation aspects and performance analysis," *IEEE Trans. Commun.*, vol. 51, pp. 1939–1948, Nov. 2003.
19. P. Mitran, H. Ochiai, and V. Tarokh, "Space-time diversity enhancements using collaborative communications," *IEEE Trans. Inform. Theory*, vol. 51, no. 6, pp. 2041–2057, Jun. 2005.
20. I. E. Telatar, "Capacity of multiple-antenna Gaussian channels," *Eur. Trans. Tel.*, vol. 10, no. 6, pp. 585–595, 1999.
21. A. Chakrabarti, A. Sabharwal, and B. Aazhang, "Half-duplex estimate-and-forward relaying: achievable rates and quantizer Design," accepted for publication in *IEEE Trans. Commun.*, 2009.
22. P. Murphy, A. Sabharwal, and B. Aazhang, "On building a cooperative communication system: Testbed implementation and first results," to appear in *EURASIP* 2009.
23. A. Sabharwal and Urbashi Mitra, "Bounds and protocals for a rate-constrained relay channel," *IEEE Trans. Inform. Theory*, 53(7), pp. 2616–2624, July 2007.

24. A. Høst-Madsen, "Cooperation in the low power regime," in *Proceedings of the Allerton Conference*, 2004.

25. F. Li, K. Wu, and A. Lippman, "Energy-efficient cooperative routing in multi-hop wireless ad hoc networks," *Performance, Computing, and Communications Conference, 2006. IPCCC 2006. 25th IEEE International*, 10–12 Apr. 2006.

26. A. El Gamal and S. Zahedi, "Capacity of a class of relay channels with orthogonal components," *Inform. Theory, IEEE Trans.*, vol. 51, no. 5, pp. 1815–1817, May 2005.

27. A. Avestimehr, S. Diggavi, and D. Tse, "Wireless network information flow," in *Proceedings of the 45th Annual Allerton Conference on Signals and Systems*.

28. ——, "A deterministic model for wireless relay networks an its capacity," *Information Theory for Wireless Networks, 2007 IEEE Information Theory Workshop on*, pp. 1–6, 1–6 Jul. 2007.

29. L. L. Xie and P. R. Kumar, "An achievable rate for the multiple-level relay channel," *IEEE Trans. Inform. Theory*, vol. 51, no. 4, pp. 1348–1358, Apr. 2005.

30. F. Xue, L.-L. Xie, and P. R. Kumar, "The transport capacity of wireless networks over fading channels," *IEEE Trans. Inform. Theory*, vol. 51, no. 3, pp. 834–847, Mar. 2005.

31. L.-L. Xie and P. R. Kumar, "A network information theory for wireless communication: Scaling laws and optimal operation," *IEEE Trans. Inform. Theory*, vol. 50, no. 5, pp. 748–767, May 2004.

32. O. Dousse, P. Thiran, and M. Hasler, "Connectivity in ad-hoc and hybrid networks," in *Proceedings of the IEEE Infocom Conference*, Jun. 2002, pp. 1079–1088.

33. L. Xiaojun and N. Shroff, "The impact of imperfect scheduling on cross-layer rate control in wireless networks," *INFOCOM 2005. 24th Annual Joint Conference of the IEEE Computer and Communications Societies. Proceedings IEEE*, vol. 3, pp. 1804–1814, 2005.

34. I. Gerasimov and R. Simon, "A bandwidth-reservation mechanism for on-demand ad hoc path finding," *Simulation Symposium, 2002. Proceedings 35th Annual*, pp. 27–34, 14–18 Apr. 2002.

35. C. Zhu, M. Lee, and T. Saadawi, "On the route discovery latency of wireless mesh networks," *Consumer Communications and Networking Conference, 2005. CCNC. 2005 Second IEEE*, pp. 19–23, 3–6 Jan. 2005.

36. P. Fu, J. Li, and D. Zhang, "Heuristic and distributed qos route discovery for mobile ad hoc networks," *Computer and Information Technology, 2005. CIT 2005. The Fifth International Conference on*, pp. 512–516, 21–23 Sept. 2005.

37. B.-C. Seet, B.-S. Lee, and C.-T. Lau, "Optimisation of route discovery for dynamic source routing in mobile ad hoc networks," *Electron. Lett.*, vol. 39, no. 22, pp. 1606–1607, 30 Oct. 2003.

38. C. Zhu, M. Lee, and T. Saadawi, "Rtt-based optimal waiting time for best route selection in ad hoc routing protocols," *Military Communications Conference, 2003. MILCOM 2003. IEEE*, vol. 2, pp. 1054–1059, 13–16 Oct. 2003.

39. Y. Li and H. Man, "Three load metrics for routing in ad hoc networks," in *Proceedings of the IEEE VTC*, Sept. 2004, pp. 26–29.

40. A. Shaikh, J. Rexford, and K. G. Shin, "Load-sensitive routing of long-lived ip flows," in *Proceedings of ACM Sigcomm*, Sept. 1999, pp. 215–226.

41. X. Fang, X. Hu, J. Zhang, F. Jiang, and P. Zhang, "A priority mac protocol for ad hoc networks with multiple channels," *Personal, Indoor and Mobile Radio Communications, 2007. PIMRC 2007. IEEE 18th International Symposium on*, pp. 1–5, 3–7 Sept. 2007.

42. C.-S. Li, Y. Ofek, and M. Yung, "time-driven priority flow control for real-time heterogeneous internetworking," *INFOCOM '96. Fifteenth Annual Joint Conference of the IEEE Computer Societies. Networking the Next Generation. Proc. IEEE*, vol. 1, pp. 189–197, Mar. 1996.

43. S. J. P. Sheu, C. H. Liu and Y. Tseng, "A priority mac protocol to support real-time traffic in ad hoc networks," in *Wireless Networks*, vol. 10, Jan. 2004, pp. 61–69.

44. Y. Xue, and N. Vaidya, "Priority scheduling in wireless ad hoc networks," in *Wireless Netw.*, vol. 12, Jun. 2006, pp. 273–286.

45. A. K. P. Bhagwat, P. Bhattacharya, and S. Tripathi, "Enhancing throughput over wireless lans using channel state dependent packet scheduling," in *Proceedings of the IEEE Infocom 97*, Apr. 1997.

46. L. G. V. Tsibonis and L. Tassiulas, "Exploiting wireless channel state information for throughput maximization," *IEEE Trans. on Inform. Theory*, vol. 50, pp. 2566–2582, Nov. 2004.
47. S. Adireddy and L. Tong, "Exploiting decentralized channel state information for random access," *IEEE Trans. Inform. Theory*, vol. 51, no. 2, p. 537, Feb. 2005.
48. M. Karkooti, P. Radosavljevic, and J. R. Cavallaro, "Configurable high throughput irregular LDPC decoder architecture: tradeoff analysis and implementation," *IEEE 17th International Conference on Application-specific Systems, Architectures and Processors (ASAP)*, pp. 360–367, Sept. 2006.
49. E. Dijkstra, "A note on two problems in connexion with graphs," *Numerische Mathematik*, vol. 1, pp. 269–271, 1959.
50. R. Bellman, "On a routing problem," *Quartely Applied Mathematics*, vol. 16, no. 1, pp. 87–90, 1959.
51. A. Viterbi, "Error bounds for convolutional codes and an asymptotically optimum decoding algorithm," *Inform. Theory, IEEE Trans.*, vol. 13, no. 2, pp. 260–269, Apr. 1967.
52. H. Ammari and H. El-Rewini, "A location information-based route discovery protocol for mobile ad hoc networks," *Performance, Computing, and Communications, 2004 IEEE International Conference on*, pp. 625–630, 2004.
53. H. Tian and H. Shen, "Mobile agents based topology discovery algorithms and modeling," in *Proceedings of the International Symposium on Parallel Architectures, Algorithms, and Networks*, May 2004, pp. 502–507.
54. W. Peng and X.-C. Lu, "On the reduction of broadcast redundancy in mobile ad hoc networks," in *Proceedings of MobiHoc*, 2000, pp. 129–30.
55. B. Nassu, T. Nanya, and E. Duarte, "Topology discovery in dynamic and decentralized networks with mobile agents and swarm intelligence," in *Proceedings on ISDA*, Oct. 2007, pp. 685–690.
56. P. Papadimitratos and Z. Haas, "Secure route discovery for qos-aware routing in ad hoc networks," *Advances in Wired and Wireless Communication, 2005 IEEE/Sarnoff Symposium on*, pp. 176–179, 18–19 April 2005.
57. J. Gomez, J. M. Cervantes, V. Rangel, R. Atahualpa, and M. Lopez-Guerrero, "Nard: Neighbor-assisted route discovery in wireless ad hoc networks," *Mobile Adhoc and Sensor Systems, 2007. MASS 2007. IEEE International Conference on*, pp. 1–9, 8–11 Oct. 2007.
58. B.-N. Cheng, M. Yuksel, and S. Kalyanaraman, "Orthogonal rendezvous routing protocol for wireless mesh networks," *Network Protocols, 2006. ICNP '06. Proceedings of the 2006 14th IEEE International Conference on*, pp. 106–115, Nov. 2006.
59. K. Zeng, K. Ren, and W. Lou, "Geographic on-demand disjoint multipath routing in wireless ad hoc networks," *Military Communications Conference, 2005. MILCOM 2005. IEEE*, pp. 1–7, 17–20 Oct. 2005.
60. Y. ki Hwang, H. Lee, and P. Varshney, "An adaptive routing protocol for ad-hoc networks using multiple disjoint paths," *Vehicular Technology Conference, 2001. VTC 2001 Spring. IEEE VTS 53rd*, vol. 3, pp. 2249–2253, 2001.
61. S. Penz, "Slp-based service management for dynamic ad hoc networks," *Proc. Internat. Work. MPAC*, vol. 115, Nov. 2005, pp. 1–8.
62. D. Chakraborty, A. Joshi, T. Finin, and Y. Yesha, "Gsd: A novel group-based service discovery protocol for manets," in *Proceedings of IEEE MWCN*, Nov. 2002, pp. 140–144.
63. H.-Y. An, L. Zhong, X.-C. Lu, and W. Peng, "A cluster-based multipath dynamic source routing in manet," *Wireless and Mobile Computing, Networking And Communications, 2005. (WiMob'2005), IEEE International Conference on*, vol. 3, pp. 369–376, 22–24 Aug. 2005.
64. J. Li and P. Mohapatra, "Laker: location aided knowledge extraction routing for mobile ad hoc networks," *Wireless Communications and Networking, 2003. WCNC 2003. 2003 IEEE*, vol. 2, pp. 1180–1184, 16–20 Mar. 2003.
65. F. Sailhan and V. Issarny, "Scalable service discovery for manet," in *Proceedings of IEEE PerCom*, Mar. 2005, pp. 235–244.
66. C. E. Perkins and E. M. Royer, "Ad hoc on-demand distance vector routing," in *Proceedings of IEEE Workshop Mobile Comput. Syst. Appl.*, Nov. 1999, pp. 90–100.

67. C.-S. Oh, Y.-B. Ko, and J.-H. Kim, "A hybrid discovery for improving robustness in mobile ad hoc networks," in *Proceedings of International DSN*, Jul. 2004.
68. A. Tsirigos and Z. J. Haas, "Multipath routing in the presence of frequent topological changes," *IEEE Commun.*, vol. 48, pp. 132–138, Nov. 2001.
69. S.-J. Lee and M. Gerla, "Aodv-br: backup routing in ad hoc networks," *Proc. IEEE WCNC*, vol. 3, pp. 1311–1316, Sept. 2000.
70. F. Xie, L. Du, Y. Bai, and L. Chen, "A novel multiple routes discovery scheme for mobile ad hoc networks," *Communications, 2006. APCC '06. Asia-Pacific Conference on*, pp. 1–5, Aug. 2006.
71. P. Pham and S. Perreau, "Performance analysis of reactive shortest path and multipath routing mechanism with load balance," *INFOCOM 2003. Twenty-Second Annual Joint Conference of the IEEE Computer and Communications Societies. IEEE*, vol. 1, pp. 251–259, 30 Mar.–3 Apr. 2003.
72. M. Abolhasan and J. Lipman, "Efficient and highly scalable route discovery for on-demand routing protocols in ad hoc networks," *Local Computer Networks, 2005. 30th Anniversary. The IEEE Conference on*, pp. 358–366, 15–17 Nov. 2005.
73. J. Sucec and I. Marsic, "An application of parameter estimation to route discovery by on-demand routing protocols," *Distributed Computing Systems, 2001. 21st International Conference on.*, pp. 207–216, Apr. 2001.
74. J. Abdulai, M. Ould-Khaoua, and L. Mackenzie, "Improving probabilistic route discovery in mobile ad hoc networks," *Local Computer Networks, 2007. LCN 2007. 32nd IEEE Conference on*, pp. 739–746, 15–18 Oct. 2007.
75. Q. Zhang and D. Agrawal, "Impact of selfish nodes on route discovery in mobile ad hoc networks," *Global Telecommunications Conference, 2004. GLOBECOM '04. IEEE*, vol. 5, pp. 2914–2918, 29 Nov.–3 Dec. 2004.
76. H.-H. Choi and D.-H. Cho, "Fast and reliable route discovery protocol considering mobility in multihop cellular networks," *Vehicular Technology Conference, 2006. VTC 2006-Spring. IEEE 63rd*, vol. 2, pp. 886–890, 7–10 May 2006.
77. E. Perevalov, R. Blum, X. Chen, and A. Nigara, "On route discovery success in ad hoc networks," *Information Sciences and Systems, 2006 40th Annual Conference on*, pp. 717–722, 22–24 Mar. 2006.

Chapter 9
Interference Rejection and Management

Arun Batra, James R. Zeidler, John G. Proakis, and Laurence B. Milstein

9.1 Introduction

In this chapter, we consider interference suppression in several different wireless communication systems. The first topic treated is self-interference encountered among cooperating systems, for example, the self-interference that is encountered in cognitive radio systems and ultra-wideband (UWB) communication systems. Both single-carrier, direct sequence signals and multicarrier signals are considered, and the effects of the interference on the performance of a direct sequence UWB system that employs channel state estimation in the presence of the interference is evaluated.

The second topic that is treated is the mitigation of narrowband interference in block-modulated multicarrier systems. Two schemes, multicarrier code-division multiple access (MC-CDMA) and orthogonal frequency division multiplexing (OFDM), are investigated. The inherent frequency diversity of MC-CDMA through

Arun Batra
Department of Electrical and Computer Engineering, University of California,
San Diego, La Jolla, CA 92093-0407, USA.
e-mail: abatra@ucsd.edu

James R. Zeidler
Department of Electrical and Computer Engineering, University of California,
San Diego, La Jolla, CA 92093-0407, USA.
e-mail: zeidler@ece.ucsd.edu

John G. Proakis
Department of Electrical and Computer Engineering, University of California,
San Diego, La Jolla, CA 92093-0407, USA.
e-mail: jproakis@ucsd.edu

Laurence B. Milstein
Department of Electrical and Computer Engineering, University of California,
San Diego, La Jolla, CA 92093-0407, USA.
e-mail: milstein@ece.ucsd.edu

V. Tarokh (ed.), *New Directions in Wireless Communications Research*,
DOI 10.1007/978-1-4419-0673-1_9,
© Springer Science+Business Media, LLC 2009

the use of spreading codes allows for robustness against interference. Conversely, OFDM must employ a signal processing technique to suppress the interference due to its lack of frequency diversity. The performance of OFDM is improved with the addition of forward error correction (FEC) coding in the frequency domain, thereby providing the system with frequency diversity.

The third topic is the suppression of interference in multiple-input, multiple-output (MIMO) wireless communication systems that employ multiple-transmit and multiple-receive antennas to increase the data rate and achieve signal diversity in fading multipath channels. Interference in MIMO wireless communication systems usually consists of intersymbol interference (ISI) due to channel multipath dispersion and cross-talk, or interchannel interference, due to the simultaneous transmissions from the multiple-transmit antennas. The focus of the section is on point-to-multipoint (broadcast) MIMO systems in which the channel characteristics are assumed to be known at the transmitter, so that interference mitigation can be performed at the transmitter.

9.2 Self-Interference Among Cooperating Systems

This section is concerned with the operation of systems that have been specifically designed to operate in the same frequency band as do other, preexisting, systems. Such overlay-type operations are often referred to as dynamic spectrum allocation. Because the spectral bands are the same for both systems, some type of interference suppression and/or interference avoidance is typically necessary. Further, the presence of the interference makes virtually all standard system functions, such as synchronization and diversity combining, less accurate. In Section 9.2.1, a brief overview of waveform design appropriate for overlay systems is presented, and in Section 9.2.2, the effects of overlaid interference, both with and without interference suppression at the receiver, on the accuracy of channel state information is presented.

9.2.1 Interference Suppression to Enable Spectrum Sharing

A fairly recent goal is to allow multiple classes of users to share common spectrum in order to increase the spectral efficiency in whatever band is under consideration. This has led to technologies such as cognitive radio and ultra wideband communications, which have the characteristics of enabling secondary users to share spectrum with primary users, providing the former set of users do not impose excessive interference on the latter set. Such scenarios typically, require the use of interference suppression and management techniques.

This involves the design and analysis of spectrum sensing techniques in order to determine where and when primary users are present. It also involves choosing appropriate waveform designs, such as single-carrier DS or multicarrier techniques (e.g., multicarrier DS), and combining them with appropriate signal processing schemes to enable them to both withstand the affects of overlaid interference and intentional jamming and minimize the interference they impose on the overlaid waveforms. For the latter case, the signal processing will take the form of placing appropriate notches in the spectrum of the transmitted DS signals at frequency locations where the spectrum sensing operation indicated that preexisting narrowband users were active. In the former case, the signal processing can again take the form of notch filtering, but now at the DS receivers. In other words, notch filtering at the DS transmitters is designed to protect the overlaid users and notch filtering at the DS receivers is designed to protect the DS signals themselves. Also, the term "notching" is used in a general sense, meaning that, for example, in the case of multicarrier DS, the notching might imply simply not transmitting appropriate subcarriers. Note that the interaction of channel estimation errors (such as those incurred while performing tasks such as spectral sensing and channel state estimation) and interference/jamming is an important consideration and is discussed below.

9.2.1.1 Single-Carrier Direct Sequence

Adaptive MMSE receivers: As is well known, a Wiener filter can greatly improve the performance of DS receivers in the presence of narrowband interference. Based on the a priori knowledge of the channel and narrowband interference (NBI) statistics, a Wiener filter is able to suppress NBI and obtain a good estimate of the channel realization. For instance, the frequency response of a Wiener filter is presented in Section 9.2.2 for different NBI levels, and it will be seen that the filter suppresses the frequencies where the NBI is located. In practice, the channel and narrowband interference statistics are not known and might be changing with time. Therefore, the Wiener filter must be made adaptive. As shown in [49, 53], the LMS filter is very effective in rejecting NBI when the ratio of the NBI bandwidth to the signal bandwidth is small, assuming the LMS filter has had sufficient time to converge.

Tunable notch filters: There are many examples in the literature that illustrate the usefulness of notch filters when trying to suppress narrowband interference. For example, it was shown in [33] that the presence of an interference rejection filter can significantly improve the acquisition system performance of a DS CDMA receiver. And it was shown in [34] that in order to successfully have a CDMA system overlay narrowband users, i.e., to deploy it in a manner such that neither set of users caused excessive interference to the other set, it was desirable to use tunable notch filters at both the CDMA transmitters and the CDMA receivers.

9.2.1.2 Multicarrier Direct Sequence

In a multicarrier DS system, multiple narrowband DS waveforms, each at a distinct carrier frequency, are combined to yield a composite wideband DS signal. Among the advantages of such an approach is the ability to achieve the same type of system performance that a single-carrier DS signal would provide, such as diversity enhancement over a multipath channel; however, this is achieved without the need for a contiguous spectral band. It was shown in [9] that for an exponential multipath intensity profile and imperfect channel estimation, multicarrier DS systems outperform single-carrier DS systems.

In the multicarrier scenario, the receiver first operates in a spectrum sensing mode, detecting the presence and frequency occupancy of possible narrowband users, as well as possible jammers. Then, after appropriate subcarriers are excised, both channel estimation and data detection are performed.

9.2.2 Effects of Interference on Channel State Estimation

One of the key needs of wireless receivers is accurate channel state information (CSI) so that basic functions such as synchronization and diversity combining can be accomplished. In this section, we consider the problem of trying to achieve accurate CSI in the presence of external interference; we also look at the impact on system performance when either the estimation module or the data detection module at the receiver does not take sufficient measures to account for the interference. As a specific system with which to illustrate the concepts, we will use UWB.

Consider a direct sequence UWB system as described in [43, 44]. A block diagram of the low-pass equivalent receiver is shown in Fig. 9.1. The first block is a chip-matched filter which is perfectly synchronized to the desired waveform, and where binary phase-shift keying (BPSK) is the modulation format. The output of the matched filter is sampled at the chip rate, and the samples are then processed by a rake receiver. The received signal consists of the UWB waveform, additive white Gaussian noise (AWGN), and narrowband interference which is modeled as a wide-sense stationary Gaussian random process.

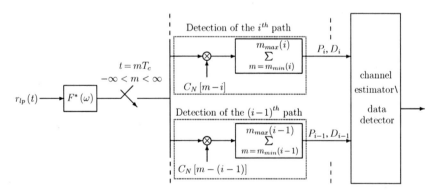

Fig. 9.1 Low-pass equivalent of the DS/CDMA receiver. For the reception of the vth data bit, $m_{\min}(i) = (v-1)N + i$ and $m_{\max}(i) = vN + i - 1$. For the vth channel estimation period, $m_{\min}(i) = (v-1)NN_p + i$ and $m_{\max}(i) = vNN_p + i - 1$. Note that N is the processing gain and N_p is the number of pilot symbols.

There are two types of receiver configurations that are studied. The first one is based upon simplicity of implementation: It uses a sample mean (SM) to estimate the gain of each resolvable path used in the rake receiver of Fig. 9.1 and combines those paths that are resolved using maximal-ratio combining (MRC). Note that neither the estimator nor the detector requires knowledge of the statistics of the channel, and neither of them make any explicit attempt to suppress the interference. The second receiver is more complex. It employs a minimum mean-squared error (MMSE) estimator and a maximum-likelihood (ML) detector. Thus, the second receiver requires knowledge of the second-order statistics of the interference, and those statistics typically are unknown at the receiver and thus requires an estimate. In what follows we assume that this latter information is available noise free.

A detailed analysis of this problem is presented in [43,44]. Here, we just describe some of the results in the context of the effect that the interference has in achieving accurate CSI. Consider first Fig. 9.2. The leftmost curve, labeled (I), displays the performance of the system when perfect CSI is available (i.e., we have a genie-aided receiver) and ML detection is used. This curve corresponds to interference-free conditions, and thus represents a lower bound on the performance of any of the other systems to be discussed. The three curves labeled (II) correspond to imperfect CSI and to signal-to-interference ratios (SIR) of -10, -15, and -20 dB. The receiver is an MMSE/ML, and we observe that, because this receiver has statistical knowledge of the interference, it can configure itself to suppress that interference. This will be

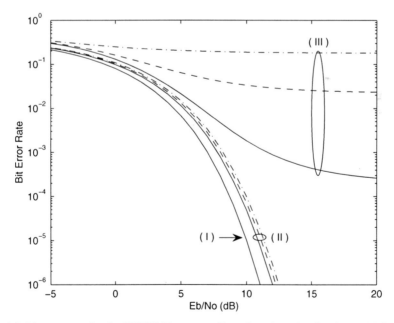

Fig. 9.2 Bit error rate for the DS/CDMA system: (I) perfect channel estimation and ML data detection in the absence of NBI, (II) MMSE/ML receiver, and (III) SM/MRC receiver. SIR = -10 dB (*solid lines*), -15 dB (*dashed*), -20 dB (*dash-dot*). Number of training bits: 50

illustrated below and is the reason the latter three curves of Fig. 9.2 are so close to the interference-free curve.

If we replace the MMSE/ML receiver with an SM/MRC receiver, we get the performance curves shown in Fig. 9.2 that are labeled (III). Note that all three curves now experience an error floor, and that the performance of the curve with an SIR of −20 dB is virtually worthless, notwithstanding the fact that the system is operating with a processing gain of 127 (i.e., roughly 21 dB).

All the estimates used in the curves shown in Fig. 9.2 were made with a training sequence of 50 bits. To see the sensitivity of the system to the number of training bits, consider the curves shown in Fig. 9.3. As in Fig. 9.2, the leftmost curve corresponds to perfect CSI, the curves labeled (II) correspond to the MMSE/ML receiver, and the curves labeled (III) correspond to the SM/MRC receiver. For each of the latter two sets of curves, results are shown for observation intervals corresponding to 10, 30, and 50 bits. Once again, it is seen that the performance of the MMSE/ML receiver in the presence of channel estimation error is very close to the performance achievable with perfect CSI, whereas the performance of the SM/MRC receiver suffers tremendous degradation.

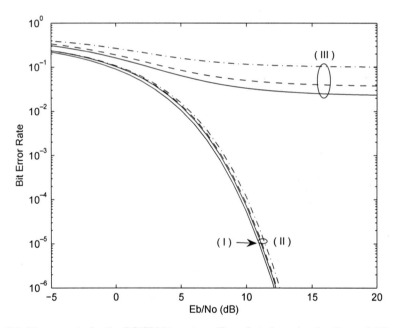

Fig. 9.3 Bit error rate for the DS/CDMA system: (I) perfect channel estimation and ML data detection, (II) MMSE/ML receiver, and (III) SM/MRC receiver. Number of training bits: 10 (*dash-dot lines*), 30 (*dashed*), 50 (*solid*). SIR $= -15$ dB

To obtain an intuitive feeling for what is taking place, consider the three curves shown in Fig. 9.4. All three correspond to the use of the MMSE/ML receiver, and all three show the transfer function of that receiver. A frequency range of −250 to +250 MHz is shown, and there is a narrowband Gaussian interferer located between

−20 and +20 MHz. The curves correspond to SIRs of −20 dB, −10 dB, and infinity (i.e., no interference). Note that the three curves virtually lie on top of one another over the entire frequency range shown, except for the region −20 to +20 MHz. In that region, it can be seen that the receiver is implementing a notch to suppress the interference (when it is present), and the depth of that notch is proportional to the strength of the interference.

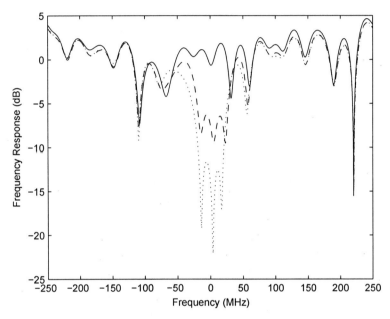

Fig. 9.4 Frequency response of the ML detector in the absence of NBI (*solid line*) and in the presence of NBI with SIR = −10 dB (*dashed line*) and −20 dB (*dotted line*). The NBI has 40 MHz of bandwidth and is centered at 0 Hz

9.3 Interference Mitigation in Block-Modulated Multicarrier Systems

In this section, two multicarrier schemes [7, 15], specifically multicarrier code-division multiple access (MC-CDMA) [10, 16, 23, 59] and orthogonal frequency division multiplexing (OFDM) [11, 54], are evaluated in the presence of narrow-band interference (NBI). This interference can arise when other systems are occupying the same spectrum, as occurs naturally in the unlicensed bands [27] or from intentional jamming. This has led to the emergence of cognitive radio systems [35]. As discussed in the previous section, the transmitter and receiver of these systems must coordinate which frequencies are unaffected by the interference. For effective operation without cooperation, systems are limited to dealing with interference at the receiver, which is what is examined in this section.

MC-CDMA and OFDM are designed as block modulation schemes that convert an input into multiple low-rate streams that are transmitted on separate narrowband subcarriers, allowing the subcarriers to overlap orthogonally in frequency, thereby providing an increase in spectral efficiency. This design also allows the bandwidth of each subcarrier to approximate the coherence bandwidth of the channel, ensuring that each tone experiences flat fading. This also simplifies the equalization process at the receiver. These systems fit naturally in the context of cognitive radios because they have the ability to turn off subcarriers that are affected by interference [3]. In MC-CDMA, each subcarrier of a given user is multiplied by a single chip of a spreading sequence [39]. At the receiver, after demodulation of the subcarriers, the chips are then correlated with a locally generated chip sequence to provide the multiple-access capability. Conversely, OFDM does not spread the data symbols across the subcarriers. OFDM is thus a special case of MC-CDMA, in which the spreading codes are simply the columns of the identity matrix. The result of this is that uncoded OFDM does not provide frequency diversity.

These techniques differ from other wideband systems, such as direct sequence CDMA (DS-CDMA) and multicarrier direct sequence CDMA (MC-DS-CDMA) [13, 28], which were discussed in the previous section. For these modulations the bandwidth of the signal is greater than the coherence bandwidth of the channel, inducing frequency-selective fading. A rake [39] receiver can be used in addition to obtain path diversity, thereby reducing the deleterious effects of the frequency-selective multipath fading channel. A comparison of MC-DS-CDMA and MC-CDMA can be found in [15, 32].

The inherent frequency diversity of MC-CDMA allows for improved performance in interference-limited, frequency-selective channels [57]. Since OFDM lacks frequency diversity, it must employ signal processing and coding techniques to mitigate the interference. There has been a great deal of research concerning mitigation of NBI in an uncoded OFDM system. Proposed techniques range from using orthogonal codes [8, 21, 57], frequency-domain cancellation [14, 37], receiver windowing [41], and excision filtering [12]. More recently there has been some work relating to NBI suppression in coded OFDM systems [12, 45, 57].

In this work, the prediction-error filter (PEF) [30, 60] is selected for interference mitigation in OFDM. This is an example of excision filtering first described in [12]. First, this system is compared with MC-CDMA in an uncoded scenario. Then the addition of forward error correction coding for each system will be examined. In this case, OFDM will now also possess frequency diversity due to the coding across the subcarriers.

9.3.1 Interference Mitigation in an Uncoded Multicarrier System

A block diagram of the system model considered in this section is shown in Fig. 9.5. The code (spreading) matrix, $\mathbf{B}$, is an $N \times N$ unitary matrix whose columns contain the N spreading sequences. The kth column is responsible for spreading the kth data symbol, D_k. Note that when $\mathbf{B}$ is the identity matrix (i.e., $\mathbf{B} = \mathbf{I}$), the system becomes identical to OFDM.

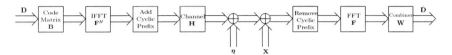

Fig. 9.5 Block diagram of uncoded MC-CDMA

The received signal is given by

$$\mathbf{Y}' = \mathbf{F}\left[\mathbf{H}\mathbf{F}^H\mathbf{B}\mathbf{D} + \mathbf{x} + \boldsymbol{\eta}\right], \tag{9.1}$$

where $\mathbf{F}$ is the $N \times N$ DFT matrix, $\mathbf{B}$ is the $N \times N$ code matrix, $\mathbf{D}$ is the $N \times 1$ vector of data symbols that is drawn from an arbitrary QAM constellation, $\mathbf{x}$ is an $N \times 1$ vector of interference samples, $\boldsymbol{\eta}$ is an $N \times 1$ vector of Gaussian noise samples, and $(\cdot)^H$ is the Hermitian (conjugate transpose) operator. Finally, $\mathbf{H}$ is the $N \times N$ channel matrix whose components are given by the taps of a multipath (frequency-selective) channel, as defined by the impulse response

$$h(t) = \sum_{l=0}^{L-1} h_l \delta(t - lT_s), \tag{9.2}$$

where L is the number of multipath components and T_S is the original symbol period. The channel coefficients, h_l, are i.i.d. zero-mean, circularly Gaussian random variables with variance

$$E\left[|h_l|^2\right] = \frac{1}{L} \quad \forall l, \tag{9.3}$$

where $E[\cdot]$ represents the expectation operator.

A cyclic prefix, composed of the last G samples of $\mathbf{F}^H\mathbf{D}$, is employed to mitigate the effects of intersymbol interference and intercarrier interference (ICI). When the length of the cyclic prefix is greater than the delay spread of the channel (L), linear convolution is equivalent to circular convolution and $\mathbf{H}$ is circulant. A well-known property of circulant matrices is that their eigenvectors are simply the columns of the DFT matrix [22]. Thus, (9.1) can be rewritten as follows:

$$\mathbf{Y}' = \tilde{\mathbf{H}}\mathbf{B}\mathbf{D} + \mathbf{F}\mathbf{x} + \mathbf{F}\boldsymbol{\eta}$$
$$= \tilde{\mathbf{H}}\mathbf{B}\mathbf{D} + \mathbf{X} + \mathbf{N}, \tag{9.4}$$

where $\tilde{\mathbf{H}}$ is an $N \times N$ diagonal matrix of the eigenvalues of $\mathbf{H}$. Coherent demodulation requires removing the phase of the channel, thus let $\tilde{\mathbf{H}}_\theta$ be a diagonal matrix containing the phase information of the channel, that is

$$\tilde{\mathbf{H}}_\theta = e^{j\angle\tilde{\mathbf{H}}}, \tag{9.5}$$

and let $\tilde{\mathbf{H}}_\rho$ be the magnitude of the eigenvalues, given as

$$\tilde{\mathbf{H}}_\rho = |\tilde{\mathbf{H}}| = \begin{bmatrix} \rho_0 & 0 & \cdots & 0 \\ 0 & \rho_1 & \cdots & 0 \\ \vdots & \vdots & \ddots & \vdots \\ 0 & 0 & \cdots & \rho_{N-1} \end{bmatrix}, \tag{9.6}$$

where ρ_k is a Rayleigh random variable. Note that $\tilde{\mathbf{H}} = \tilde{\mathbf{H}}_\theta \tilde{\mathbf{H}}_\rho$.

After coherent detection, the input to the combiner is given as

$$\mathbf{Y} = \tilde{\mathbf{H}}_\theta^H \mathbf{Y}' = \tilde{\mathbf{H}}_\rho \mathbf{BD} + \tilde{\mathbf{H}}_\theta^H (\mathbf{X} + \mathbf{N}). \tag{9.7}$$

The interference samples in the time domain seen in (9.1) are given by

$$\begin{aligned} x_n &= \sqrt{E_x} e^{j(2\pi f_x n T_s + \theta)} \\ &= \sqrt{E_x} e^{j\left(\frac{2\pi}{N}(m+\alpha)n + \theta\right)}, \end{aligned} \tag{9.8}$$

where E_x is the interferer power and θ is a random phase distributed uniformly between $-\pi$ and π. The frequency of the interference, f_x, is defined as a function of the original sampling frequency, f_s, namely $f_x = (m+\alpha)\frac{f_s}{N}$. In this formulation of the interference, m represents the tone closest to the interference, while $\alpha \in [-0.5, 0.5]$ is the offset from tone m. The second line of (9.8) is obtained by noting that $f_s = \frac{1}{T_s}$.

The time-domain interference samples in (9.8) are converted into the frequency domain

$$X_k = [\mathbf{Fx}]_k = \sqrt{\frac{E_x}{N}} \frac{1 - e^{j2\pi\alpha}}{1 - e^{j\frac{2\pi}{N}(m+\alpha-k)}} e^{j\theta}, \tag{9.9}$$

where $[\cdot]_k$ is the kth component of the given vector. The result of (9.9) indicates that the interference per tone is dependent on the relative distance of tone k from the interferer frequency. This is referred to as spectral leakage, as the interference power is leaked into all of the tones. From (9.9), it can be seen that as the interference power increases, so does the number of tones that are significantly interfered with.

In the case of $\alpha = 0$, the interference is orthogonal to all the subcarriers other than the mth subcarrier. The interference after the DFT reduces to

$$X_k = \begin{cases} \sqrt{NE_x} e^{j\theta}, & k = m \\ 0, & k \neq m \end{cases}. \tag{9.10}$$

Thus the interference is limited to tone m in this case.

Finally, the last term of (9.4), $\mathbf{N}$, is an $N \times 1$ vector of noise samples that have the same statistics as in the time domain.

9.3.1.1 MC-CDMA

The code matrix (**B**) employed in this work is defined component wise [36,57] by

$$[B]_{kl} = \frac{1}{\sqrt{N}} e^{j\frac{2\pi}{N}kl}, \quad k,l = 0,\ldots,N-1. \tag{9.11}$$

Note that the code matrix is unitary, that is $\mathbf{BB}^H = \mathbf{B}^H\mathbf{B} = \mathbf{I}_N$, where $\mathbf{I}_N$ is the $N \times N$ identity matrix. Another option that can be used for the orthogonal spreading codes, are the Walsh–Hadamard codes. However, note that these codes are only defined for lengths of $N = 2^n$, where n is an integer greater than zero [15].

The combiner weights required to extract the symbol vector, **D**, are derived under the minimum mean-square error (MMSE) criterion [57,58] and obtained by solving the Wiener–Hopf equation

$$\mathbf{W} = \mathbf{R}_{YY}^{-1}\mathbf{R}_{YD}, \tag{9.12}$$

where $\mathbf{R}_{YY}$ is the autocorrelation matrix of the received signal and $\mathbf{R}_{YD}$ is the cross-correlation matrix between the received signal and the symbol vector, **D**. Assuming that all the received components are independent and that the channel amplitudes (ρ_k) are known and deterministic, the autocorrelation matrix is given by

$$\mathbf{R}_{YY} = E_b\tilde{\mathbf{H}}_\rho\tilde{\mathbf{H}}_\rho^H + N_0\mathbf{I}_N + \mathbf{R}_X, \tag{9.13}$$

where $\mathbf{R}_X$ is the autocorrelation matrix of the interference samples in the frequency domain defined as

$$\mathbf{R}_X = \tilde{\mathbf{H}}_\theta^H\mathbf{F}\mathbf{R}_x\mathbf{F}^H\tilde{\mathbf{H}}_\theta, \tag{9.14}$$

where $\mathbf{R}_x$ is the time-domain correlation matrix of the interference, its entries given by

$$r_x(l) = E_x e^{j\frac{2\pi}{N}(m+\alpha)l}. \tag{9.15}$$

Similarly, the cross-correlation vector is given by

$$\mathbf{R}_{YD} = E_b\tilde{\mathbf{H}}_\rho\mathbf{B}. \tag{9.16}$$

For the case of no interference ($\mathbf{R}_X = \mathbf{0}$), the weights are given as

$$\mathbf{W} = \text{diag}\left(\left[\frac{\rho_0 E_b}{\rho_0^2 E_b + N_0}, \frac{\rho_1 E_b}{\rho_1^2 E_b + N_0}, \ldots, \frac{\rho_{N-1} E_b}{\rho_{N-1}^2 E_b + N_0}\right]\right)\mathbf{B}, \tag{9.17}$$

where $\text{diag}(\cdot)$ returns a diagonal matrix from the inputted vector.

When the interference is orthogonal to the subcarriers (except for the mth subcarrier), $\alpha = 0$, recall that **X** will have one non-zero component, at index m. Thus, $\mathbf{R}_X$ will also only have one non-zero value, equal to NE_x at position (m,m). Let $\mathbf{w} = [w(0),\ldots,w(N-1)]$, where $w(l)$ is given by

$$w(l) = \begin{cases} \frac{\rho_l E_b}{\rho_l^2 E_b + N_0}, & l \neq m \\ \frac{\rho_l E_b}{\rho_l^2 E_b + N_0 + N E_x}, & l = m \end{cases}. \tag{9.18}$$

The MMSE weights are then given as

$$\mathbf{W} = \text{diag}\,(\mathbf{w})\,\mathbf{B}. \tag{9.19}$$

It can be seen from (9.19) that when interference is not present on the tone, the
weight is the same as for the no-interference case. Note that when $l = m$ and the
narrowband interference is strong, then $W_k(l)$ is essentially zero, thus ignoring the
subcarrier that is unreliable.

Finally, for the case of a nonorthogonal interferer, the components of the autocorrelation matrix for the interference are given by

$$R_X(k, l) = \frac{E_x}{N} \frac{2 - 2\cos(2\pi\alpha)}{1 - e^{j\frac{2\pi}{N}(m+\alpha-k)} - e^{-j\frac{2\pi}{N}(m+\alpha-l)} + e^{j\frac{2\pi}{N}(l-k)}}, \quad k, l = 0, \ldots, N-1. \tag{9.20}$$

To find the weights one could use (9.20) and invert (9.12). However, noticing that
$\mathbf{R}_X$ is rank-1, having one dominant eigenvalue, then $\mathbf{R}_X$ can be approximated using
the singular value decomposition (SVD) [25] as follows:

$$\mathbf{R}_X = \mathbf{USV}^H$$
$$\approx \lambda \mathbf{uv}^H, \tag{9.21}$$

where $\lambda = N E_x$ is the non-zero eigenvalue of $\mathbf{R}_X$, $\mathbf{u}$ is the first column of $\mathbf{U}$, and $\mathbf{v}$
is the first column of $\mathbf{V}$. Then using the Sherman–Morrison formula [25], the MMSE
combiner weights are derived to be

$$\mathbf{W} = \left(\mathbf{A} - \frac{\lambda \mathbf{Auv}^H \mathbf{A}}{1 + \lambda \mathbf{v}^H \mathbf{Au}}\right) E_b \tilde{\mathbf{H}}_\rho \mathbf{B}, \tag{9.22}$$

where $\mathbf{A} = \text{diag}\left(\left[\frac{1}{\rho_0^2 E_b + N_0}, \frac{1}{\rho_1^2 E_b + N_0}, \ldots, \frac{1}{\rho_{N-1}^2 E_b + N_0}\right]\right)$.

Finally, the estimates of the transmitted data symbols are given by

$$\hat{\mathbf{D}} = \mathbf{W}^H \mathbf{Y}. \tag{9.23}$$

Note that a consequence of MMSE combining in frequency-selective channels is
that there is a loss of orthogonality among the data symbols,

$$\hat{\mathbf{D}} = \mathbf{D} + \left(\mathbf{W}^H \tilde{\mathbf{H}}_\rho \mathbf{B} - \mathbf{I}_N\right)\mathbf{D} + \mathbf{W}^H \tilde{\mathbf{H}}_\theta^H (\mathbf{X} + \mathbf{N}). \tag{9.24}$$

On the other hand, when the channel is AWGN or flat fading, equal gain combining
(i.e., $\mathbf{W} = \mathbf{B}$) can be utilized to preserve orthogonality, that is

$$\hat{\mathbf{D}} = \mathbf{D} + \mathbf{W}^H \tilde{\mathbf{H}}_\theta^H (\mathbf{X} + \mathbf{N}). \tag{9.25}$$

9.3.1.2 OFDM

Unlike MC-CDMA, OFDM does not provide frequency diversity since the code matrix in Fig. 9.5 is set to the identity matrix. The prediction-error filter (PEF) is considered for this system as a means for removing the interference in the time domain, thereby avoiding the spectral leakage that occurs after demodulation (see block diagram given in Fig. 9.6). The PEF is a well-studied structure that uses the correlation between past samples to form an estimate of the current sample [40]. The PEF has the property that it removes the correlation between samples, thereby whitening the spectrum.

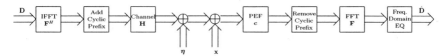

Fig. 9.6 Block diagram of uncoded OFDM with the PEF

This technique has been used to remove narrowband interference in many applications, such as spread-spectrum systems when the processing gain does not provide enough immunity to the interference [49,53]. A block diagram of the PEF is shown in Fig. 9.7. It is simply a transversal filter with M taps. The decorrelation delay (Δ) ensures that the current sample is decorrelated from the samples in the filter when calculating the error term.

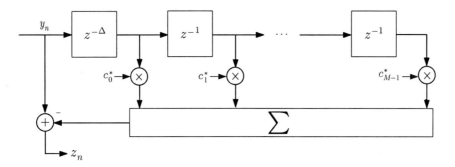

Fig. 9.7 Block diagram of the prediction-error filter (PEF)

The output of the PEF, z_n, is defined as the subtraction of the estimate of the interference from the current input sample,

$$z_n = y_n - \sum_{l=0}^{M-1} c_l^* y_{n-\Delta-l}, \qquad (9.26)$$

where c_l are the weights of the predictor and $(\cdot)^*$ represents complex conjugation. The optimal weights under the minimum mean-squared error (MMSE) criterion are found using the orthogonality principle and the method of undetermined coefficients [61]. The weights are given by

$$c_l = K e^{-j\frac{2\pi}{N}(m+\alpha)(l+\Delta)}, \quad l = 0, \dots, M-1, \tag{9.27}$$

where K is given by

$$K = \frac{E_x}{E_b + N_0 + M E_x}. \tag{9.28}$$

For the scenario of interest in this chapter, the interference power is much larger than both the signal power and the noise power. Therefore, the SIR and the noise to interference (NIR) can be assumed to be very small (i.e., SIR $\ll$ 0 dB, NIR $\ll$ 0 dB [31]) allowing the coefficient, K, to be approximated as $K \approx \frac{1}{M}$.

This filter is easily implemented adaptively using the least mean square (LMS) algorithm. The convergence properties are well described [5, 42]. Note also that z_n is the error term for the adaptive structure, indicating that no training symbols are required when calculating the error term.

The PEF is implemented before the removal of the cyclic prefix. Let the one-step predictor weights be defined as $\mathbf{c} = \left[1, -c_0^*, \dots, -c_{M-1}^*\right]$ and the convolution of the filter and channel be defined as $\mathbf{a} = \mathbf{c} * \mathbf{h}$. It is assumed that the overall length of $\mathbf{a}$ is less than the length of the cyclic prefix (i.e., $L + M - 1 \le G$) to ensure that there is no ISI or ICI. Therefore, the effective channel matrix, $\mathbf{A}$, is circulant. Then, the filtered signal in the frequency domain can be written as

$$\begin{aligned} \mathbf{Z}' &= \mathbf{FAF}^H \mathbf{D} + \mathbf{FC}_{\text{noise}} \left(\mathbf{x}_{\text{cp}} + \boldsymbol{\eta}_{\text{cp}}\right) \\ &= \tilde{\mathbf{A}}\mathbf{D} + \mathbf{FC}_{\text{noise}} \left(\mathbf{x}_{\text{cp}} + \boldsymbol{\eta}_{\text{cp}}\right), \end{aligned} \tag{9.29}$$

where $\tilde{\mathbf{A}}$ is the diagonal matrix of the eigenvalues of $\mathbf{A}$, $\mathbf{x}_{\text{cp}}$ and $\boldsymbol{\eta}_{\text{cp}}$ are vectors of interference and noise samples (length $N + G$), respectively that are not cyclically extended, and $\mathbf{C}_{\text{noise}}$ is the $N \times (N+G)$ filtering matrix for the noise and interference that is defined as

$$\mathbf{C}_{\text{noise}} = \left[\mathbf{0}_{N,G-M} \text{ Toeplitz}\left(\left[c_{M-1}^*, \mathbf{0}_{1,N-1}\right]^T \left[\mathbf{c}^*, \mathbf{0}_{1,N-1}\right]\right)\right], \tag{9.30}$$

where $\mathbf{0}_{i,j}$ is the $i \times j$ zero matrix. The Toeplitz operator, Toeplitz(*column, row*), generates a Toeplitz matrix from a column vector and a row vector. Note that $\tilde{\mathbf{A}}$ in (9.32) can also be defined as

$$\tilde{\mathbf{A}} = \tilde{\mathbf{H}}\tilde{\mathbf{C}}, \tag{9.31}$$

where $\tilde{\mathbf{C}} = \sqrt{N}\mathbf{F}\left[\mathbf{c}^* \ \mathbf{0}_{1,N-(M+1)}\right]^T$. This vector is the sampled frequency response of the notch filter, with the notch located on the tone closest to the interference.

For coherent demodulation, the phase of the effective channel matrix $\tilde{\mathbf{A}}$ must be removed; therefore, similar to (9.5) and (9.6), let $\tilde{\mathbf{A}} = \tilde{\mathbf{A}}_\theta \tilde{\mathbf{A}}_\rho$, where $\tilde{\mathbf{A}}_\rho = |\tilde{\mathbf{A}}|$ and $\tilde{\mathbf{A}}_\theta = e^{\angle \tilde{\mathbf{A}}}$. Thus for coherent demodulation,

$$\mathbf{Z} = \tilde{\mathbf{A}}_\theta^H \mathbf{Z}' = \tilde{\mathbf{A}}_\rho \mathbf{D} + \tilde{\mathbf{A}}_\theta^H \mathbf{F} \mathbf{C}_{\text{noise}} \left(\mathbf{x}_{\text{cp}} + \boldsymbol{\eta}_{\text{cp}} \right) . \tag{9.32}$$

Note that the diagonal elements of $\tilde{\mathbf{A}}_\rho$ are not necessarily distributed as Rayleigh random variables due to the PEF.

Let the uncanceled interference and noise be grouped into one general noise term

$$\tilde{\mathbf{N}} = \tilde{\mathbf{A}}_\theta^H \mathbf{F} \mathbf{C}_{\text{noise}} \left(\mathbf{x}_{\text{cp}} + \boldsymbol{\eta}_{\text{cp}} \right) . \tag{9.33}$$

It is clear from (9.33) that the noise samples, $\tilde{N}_k$, are correlated due to the PEF matrix, $\mathbf{C}_{\text{noise}}$. It is also noted that the noise samples $\tilde{N}_k$ are assumed to be Gaussian random variables. As stated in [31], this system is difficult to analyze when the noise samples are not strictly independent, however, due to the fact that the noise power in the main tap (equal to unity) is much larger than in the remaining taps (approximately $\frac{1}{M}$ for the scenario of interest in this chapter), it is reasonable to assume that the noise samples are independent. Therefore, let $\sigma_{\tilde{N}}^{2(k)}$ be the variance of the noise on tone k, given by

$$\begin{aligned} \sigma_{\tilde{N}}^2(k) &= E\left[\tilde{N}_k \tilde{N}_k^H \right] \\ &= \left[\tilde{\mathbf{A}}_\theta^H \mathbf{F} \mathbf{C}_{\text{noise}} \left(\mathbf{R}_x + \frac{N_0}{2} \mathbf{I}_N \right) \mathbf{C}_{\text{noise}}^H \mathbf{F}^H \tilde{\mathbf{A}}_\theta \right]_{kk} . \end{aligned} \tag{9.34}$$

Note that the variances given in (9.34) are scaled according to the notch filter that is used to suppress the interference.

One-tap equalization is performed instead of MMSE combining as is done for MC-CDMA. Thus the estimates of the transmitted data symbols are given by

$$\hat{\mathbf{D}} = \mathbf{Z}/\tilde{\mathbf{A}}_\rho . \tag{9.35}$$

An analytic expression for the BER of this OFDM system given knowledge of the channel amplitudes is simply the average of the per-tone BER, given by

$$P_e = \frac{1}{N} \sum_{k=0}^{N-1} Q\left(\sqrt{\frac{|\tilde{A}_\rho(k)|^2 E_b}{\sigma_{\tilde{N}}^2(k)}} \right) . \tag{9.36}$$

The analytical results of (9.36) are seen in Fig. 9.8 along with the simulation results for SIR $= -20\,\text{dB}$ and $N = 64$, $G = 16$, $L = 5$, and $M = 12$. The simulation and theoretical results agree quite well, validating the assumption of independent noise samples. Also note that an error floor arises, because the OFDM system is limited by its worst set of tones. In this case, these tones are the ones located around the notch generated by the PEF. This suggests the need for forward error correction coding. The addition of a convolutional code can reduce the BER from 10^{-2} to 10^{-5}. Note that without the use of the PEF, the BER is so poor that forward error correction coding cannot be effectively employed.

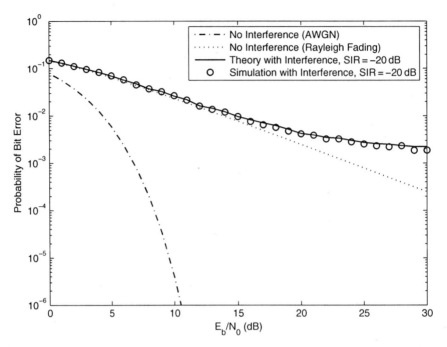

Fig. 9.8 Uncoded BER results for theoretical and simulated OFDM with the prediction-error filter (PEF). Also plotted are the theoretical BER curves for AWGN and Rayleigh fading. Parameters: $N = 64$, $G = 16$, $L = 5$, $M = 12$, SIR $= -20$ dB

9.3.1.3 Results

A comparison of the two techniques described previously is compared in an uncoded scenario in Fig. 9.9. For reference, the theoretical BER curves for both AWGN and Rayleigh fading are provided. Each system is simulated for the cases (i) no interference and (ii) one nonorthogonal interferer. The plot for MC-CDMA in no interference indicates that frequency diversity provides a benefit in a frequency-selective channel. The performance of MC-CDMA with a single nonorthogonal interferer is very close to the case of no interference. In this case, the combiner utilizes the statistics of the channel and interference to successfully recombine the transmitted signal. The deviation from the case of no interference is very small.

The PEF used with OFDM assumes knowledge of the MMSE weights given in (9.27). In the presence of no interference the performance of the system reduces to that of Rayleigh fading, as is expected. When interference is present, the case of a nonorthogonal interference approximates the case of Rayleigh fading at low E_b/N_0 values; however at high E_b/N_0 values, an error floor arises. This is due to the notch filter that removes a few tones when mitigating the interference.

For both MC-CDMA and OFDM with PEF, the interference must be estimated in order to be implemented in a real-world communication system. As mentioned earlier, the PEF is easily implemented using the low-complexity LMS algorithm without training symbols. This allows the interference to be adaptively estimated

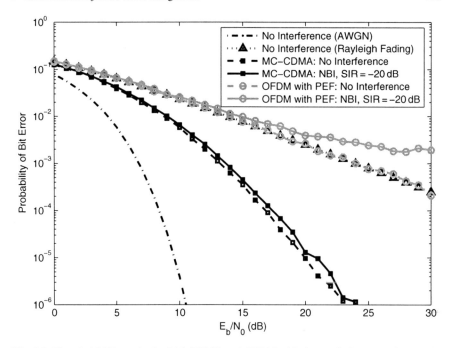

Fig. 9.9 Uncoded BER results for MC-CDMA and OFDM with the prediction-error filter (PEF) for the cases of (i) no interference and (ii) one nonorthogonal interferer. Also plotted are the theoretical BER curves for AWGN and Rayleigh fading. Parameters: $N = 64$, $G = 16$, $L = 5$, $M = 12$, SIR $= -20\,$dB

and removed. In the case of MC-CDMA, the correlation matrix of the interference is required when determining the MMSE combiner weights. This can be accomplished using training symbols to provide an estimate for the received signal correlation matrix.

9.3.2 Interference Mitigation in Coded Multicarrier Systems

In this section, the impact of applying forward error correction to the uncoded systems analyzed above is evaluated. Block diagrams for coded MC-CDMA and coded OFDM with PEF are depicted in Figs. 9.10 and 9.11, respectively. A convolutional code of rate $1/2$, constraint length equal to 7 with generating polynomials $(133, 171)_8$ is used. One codeword spans 50 MC-CDMA/OFDM symbols. At the receiver the Viterbi [39] decoder is used to decode the codeword utilizing soft decisions. Note that a block interleaver is used with OFDM to provide frequency diversity. An interleaver can also be used with MC-CDMA when the data are spread over a subset of the subcarriers.

The soft decisions inputted into the Viterbi decoder encourage the use of erasures when reliable log-likelihood ratios (LLRs) are unavailable. In OFDM, erasures are

Fig. 9.10 Block diagram for coded MC-CDMA

Fig. 9.11 Block diagram for coded OFDM with the PEF

only effective when the interference is weak [4]. However, when the interference is strong, the spectral leakage of interference power causes a large number of tones to be corrupted. Insertion of erasures ends up compromising the code's error correction capability. The use of the PEF can be considered as a form of erasure insertion, whereby the erasures are localized around the interference by the notch that is placed in the frequency spectrum of the received signal. The code is then tasked with correcting any erasures caused by the channel or interference. The full erasures are inserted after the PEF using the Bayesian erasure formula described in [6].

The LLR values [15] that are inputted into the Viterbi decoder for MC-CDMA with BPSK modulation are given as

$$\Lambda_{\text{MC-CDMA}}(D_k|Y_k) = \frac{2\Re\{Y_k\}}{\sigma^2}, \tag{9.37}$$

where $\Re\{\cdot\}$ is the real operator and σ^2 is the variance of the ICI, noise, and interference as defined in (9.24). Assuming these vectors are composed of independent Gaussian random variables, σ^2 is given as follows:

$$\sigma^2 = \text{diag}\left[E_b \left(\mathbf{W}^H \tilde{\mathbf{H}}_\rho \mathbf{B} - \mathbf{I} \right) \left(\mathbf{W}^H \tilde{\mathbf{H}}_\rho \mathbf{B} - \mathbf{I} \right)^H + \frac{N_0}{2} \mathbf{W}\mathbf{W}^H \right.$$
$$\left. + \mathbf{W}^H \tilde{\mathbf{H}}_\theta^H \mathbf{F}\mathbf{R}_x \mathbf{F}^H \tilde{\mathbf{H}}_\theta \mathbf{W} \right]. \tag{9.38}$$

Similarly, for OFDM

$$\Lambda_{\text{OFDM}}(D_k|Z_k) = \frac{2\Re\{Z_k\}}{\sigma^2}, \tag{9.39}$$

where σ^2 is given as

$$\sigma^2 = \frac{\text{diag}\left[\tilde{\mathbf{A}}_\theta^H \mathbf{F}\mathbf{C}_{\text{noise}} \left(\mathbf{R}_x + \frac{N_0}{2}\mathbf{I}_N \right) \mathbf{C}_{\text{noise}}^H \mathbf{F}^H \tilde{\mathbf{A}}_\theta \right]}{\text{diag}\left[\tilde{\mathbf{A}}_\rho \tilde{\mathbf{A}}_\rho^H \right]}. \tag{9.40}$$

9.3.2.1 Results

Figure 9.12 demonstrates the results for the coded simulation of MC-CDMA and OFDM with the PEF for the cases of (i) no interference and (ii) one nonorthogonal interferer, for $N = 64$, $G = 16$, $L = 5$, $M = 12$, SIR $= -20$ dB. Also plotted are the theoretical BER curves for uncoded Rayleigh fading and a BER bound for AWGN. It can be clearly seen that the performance of both systems with interference is very close to the case of no interference. Also note that the performance of OFDM with frequency diversity provided by coding and the PEF as an erasure insertion technique is equivalent to that of coded MC-CDMA for which frequency diversity is obtained through spreading. The coding gain is also apparent when comparing the coded results with Fig. 9.9.

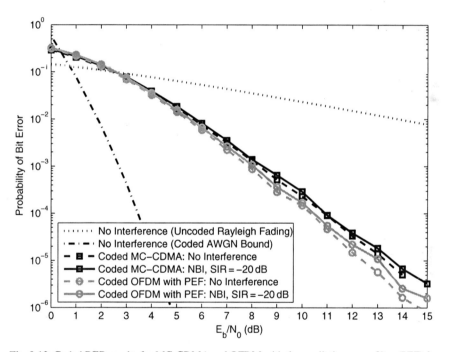

Fig. 9.12 Coded BER results for MC-CDMA and OFDM with the prediction-error filter (PEF) for the cases of (i) no interference and (ii) one nonorthogonal interferer. Also plotted are the theoretical BER curves, uncoded Rayleigh Fading, and a BER bound for AWGN. Parameters: $N = 64$, $G = 16$, $L = 5$, $M = 12$, SIR $= -20$ dB

9.3.3 Doppler Sensitivity of OFDM in Mobile Applications

Mobile systems that employ OFDM may suffer degradations from ISI and ICI in addition to the other forms of interference discussed above. In [50], the Doppler sensitivity of OFDM was investigated for PSK and QAM modulations in the presence of ICI and the loss in performance due to Doppler spreading was evaluated. Also, in [52], comparisons were made between the Doppler sensitivity of OFDM and filtered multitone (FMT) modulation in frequency-selective time-varying channels.

FMT is an alternate form of multicarrier modulation that has been proposed for very high-rate subscriber lines and for wireless communications. FMT avoids spectral overlap between subcarriers by implementing a noncritical sampling technique that satisfies a perfect reconstruction condition, so that the sampling signals at the receiver are free from ISI as well as ICI. The ISI and ICI in an FMT system is suppressed at the expense of choosing a larger upsampling-factor-to-number-of-subcarriers ratio. Due to the phase and amplitude distortion introduced by the fading channel, not only is the orthogonality among different subcarriers destroyed but also the perfect Nyquist sampling condition of the baseband matched filters is no longer valid. Consequently, ICI as well as ISI will cause distortions to the transmitted signal. The interference caused by the channel frequency selectivity and time variance was quantified by analyzing the demodulated signals at the receiver under different fading channel conditions. It was shown [52] analytically that quasi-static frequency flat fading channels do not introduce ICI or ISI to a multicarrier system, while frequency-selective channels introduce both ISI and ICI in an FMT system. The resulting carrier-to-interference (C/I) ratio obtained under different channel conditions was shown to provide a trade-off between spectral efficiency and system performance degradation. Also it was shown [52] that OFDM outperforms FMT in a static or slowly fading and low frequency-selective channel but is inferior to FMT modulation in highly frequency-selective and time-varying fading environments.

This work was extended in [51] to consider a per-channel-equalized FMT architecture. It is shown that with a sufficient number of equalizer coefficients it is possible to mitigate the ISI, but the ICI remains. A closed-form analytical expression for the C/I ratio and its upper bound is derived and the trade-offs in spectral efficiency and system performance are further quantified.

9.4 Interference Suppression in Broadcast MIMO Systems

In this section, we consider multiple-input, multiple-output (MIMO) wireless communication systems that employ multiple-transmit and multiple-receive antennas to increase the data rate and achieve signal diversity in fading multipath channels. The performance of MIMO systems is degraded by two types of interference. One is intersymbol interference due to channel multipath dispersion. The other is crosstalk or interchannel interference due to the simultaneous transmissions from the multiple-transmit antennas. The focus of this section is on equalization and detection algorithms for mitigating these two types of interference. In particular, we consider a point-to-multipoint (broadcast) MIMO transmission system in which the channel characteristics are assumed to be known at the transmitter, so that interference mitigation is performed at the transmitter.

Let us consider a point-to-multipoint MIMO system which transmits data simultaneously to multiple users that are geographically distributed. The transmitter is assumed to employ N_T antennas to transmit data to K receivers, where $N_T \geq K$. Each user is assumed to have a receiver with one or more receiving antennas. This scenario applies, for example, to the downlink of a wireless local area network. The distinguishing feature of the MIMO broadcast system is that the receivers, which are geographically separated, do not employ any coordination in processing the received signals.

In a MIMO broadcast system, there are two possible approaches for dealing with the multiple-access interference (MAI) resulting from the simultaneous transmission to multiple users. One approach is to have each receiver employ interference mitigation in the recovery of its desired signal. In most cases, this approach is impractical because the users lack the processing capability and are constrained by the limited energy resources inherent in the use of battery power. The alternative approach is to employ interference mitigation at the transmitter, which possess significantly more processing capability and energy resources.

To mitigate the MAI at the transmitter, the transmitter must know the channel characteristics, typically the channel impulse response. This channel state information (CSI) may be obtained from channel measurements performed at each of the receivers by means of received pilot signals sent by the transmitter. Then, the CSI must be sent to the transmitter. In such a scenario, the channel time variations must be relatively slow so that a reliable estimate of the channel characteristics is available at the transmitter. In some systems, the uplink and downlink channels are identical, e.g., the same frequency band is employed for both the uplink and the downlink, but separate time slots are used for transmission. This transmission mode is called time-division duplex (TDD). In TDD systems, the pilot signals for channel measurement may be sent by each of the users in the uplink. In our treatment below, we assume that the CSI at the transmitter is perfect.

The suppression of MAI by means of transmitter processing is usually called signal precoding or signal preprocessing. Signal precoding at the transmitter may take one of several forms, depending on the criterion or the method used to perform the precoding. The simplest precoding methods are linear and are based on either the zero-forcing (ZF) criterion or the MSE criterion. On the other hand, there are nonlinear signal precoding methods that result in better system performance. First, we treat linear precoding and, then, we describe nonlinear precoding methods and compare their performance.

9.4.1 Linear Precoding of the Transmitted Signals

In this section, for mathematical convenience, we assume that each user has a single antenna and the number of receivers (users) is $K \leq N_T$. It is also convenient to assume that the channel is nondispersive. The communication system configuration is shown in Fig. 9.13, where the precoding matrix is denoted as $\mathbf{A}_T$. Hence, the received signal vector is

$$\mathbf{y} = \mathbf{HA}_T\mathbf{s} + \boldsymbol{\eta}, \tag{9.41}$$

where $\mathbf{H}$ is an $K \times N_T$ matrix, $\mathbf{A}_T$ is an $N_T \times K$ precoding matrix, $\mathbf{s}$ is a $K \times 1$ data vector, and $\boldsymbol{\eta}$ is a $K \times 1$ Gaussian noise vector. The matrix that eliminates the MAI at each receiver is generally given by the Moore–Penrose pseudo-inverse

$$\mathbf{H}^+ = \mathbf{H}^H(\mathbf{HH}^H)^{-1}. \tag{9.42}$$

Hence, the precoding matrix is

$$\mathbf{A}_T = \gamma\mathbf{H}^+, \tag{9.43}$$

where γ is a scale factor that is selected to satisfy the total transmitted power allocation, i.e., $\| \mathbf{A}_T\mathbf{s} \|^2 = P$. Thus, the precoding matrix in (9.43) allows the individual users to recover their desired symbols without any interference from the signals transmitted to the other users. We also observe that in the special case where $K = N_T$, $\mathbf{A}_T = \gamma\mathbf{H}^{-1}$, so that the precoding matrix is proportional to the inverse channel matrix. This constitutes a zero-forcing equalizer implemented at the transmitter.

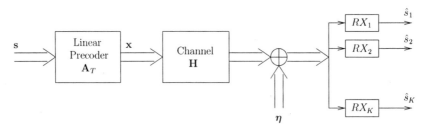

Fig. 9.13 Model of MIMO broadcast system employing linear precoding

Figure 9.14 illustrates the error rate performance of the zero-forcing precoder obtained via Monte Carlo simulation for $K = N_T = 4$, 6, and 10 and QPSK modulation. The channel matrix elements are realizations of complex-valued i.i.d. zero-mean Gaussian random variables with unit variance. We observe that the error rate increases with an increase in the number of users. We attribute this deterioration in performance to the ill-conditioning of the channel matrix $\mathbf{H}$. This is the major drawback with the zero-forcing precoder.

If we relax the condition that the interference be zero at all the receivers, the performance degradation can be reduced. This can be accomplished by using the linear MSE criterion in the design of the precoding matrix $\mathbf{A}_T$. Thus, we select $\mathbf{A}_T$ to minimize the cost function

$$J(\mathbf{A}_T, \gamma) = \underset{\gamma, \mathbf{A}_T}{\arg\min}\ E \left\| \frac{1}{\gamma}(\mathbf{HA}_T\mathbf{s} + \boldsymbol{\eta}) - \mathbf{s} \right\|^2, \tag{9.44}$$

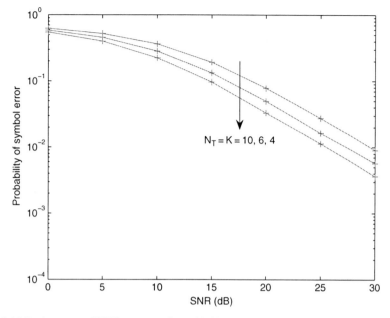

Fig. 9.14 Performance of ZF linear precoding with $N_T = K = 4, 6, 10$

subject to the transmitted power allocation $\|\mathbf{A}_T \mathbf{s}\|^2 = P$ and where the expectation in (9.44) is taken over the noise and signal statistics. The solution to the MSE criterion is the precoding matrix

$$\mathbf{A}_T = \gamma \mathbf{H}^H (\mathbf{H}\mathbf{H}^H + \beta \mathbf{I})^{-1}, \tag{9.45}$$

where γ is the scale factor that is selected to satisfy the power allocation and β is defined as a loading factor, which when selected as $\beta = K/P$ maximizes the signal-to-interference-plus-noise ratio (SINR) at the receiver [38].

The error rate performance of the MMSE linear precoder obtained by Monte Carlo simulation in a frequency-nonselective Rayleigh fading channel is illustrated in Fig. 9.15 for $K = N_T = 4, 6$, and 10. We observe that the error rate performance improves slightly as the number of users K increases and that it exceeds the performance of the zero-forcing precoder.

9.4.2 Nonlinear Precoding of the Transmitted Signals: The QR Decomposition

When the transmitter knows the interference caused on other users by the transmission of a signal to any particular user, the transmitter can design signals for each of the other users to cancel the interference. The major problem with such an

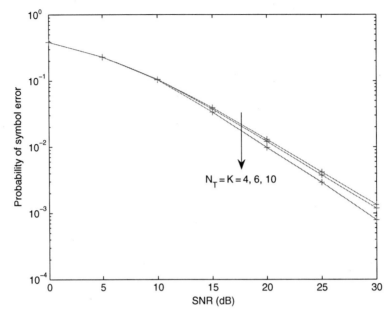

Fig. 9.15 Performance of MMSE linear precoding with $N_T = K = 4, 6, 10$

approach is to perform the interference cancellation without increasing the transmitter power. This same problem is encountered in decision-feedback channel equalization of a single user, where the feedback filter of the DFE is implemented at
the transmitter. In that case, when the difference between the desired transmitted symbol and the ISI exceeds the range of the desired transmitted symbol, the
difference is reduced by subtracting an integer multiple of $2M$ for M-ary PAM,
where $[-M, M)$ is the range of the desired transmitted symbol. This same nonlinear precoding method called Tomlinson-Harashima precoding [24, 47] can be
applied to the cancellation of the MAI in a MIMO broadcast communication system
[17, 18, 56].

Figure 9.16 illustrates the precoding operations for the MIMO multiuser system.
The channel impulse response between the ith transmit antenna and the receive
antenna of the kth user is given by

$$h_{ki}(t) = \sum_{l=0}^{L-1} h_{ki}^{(l)} \delta(t - lT),\qquad(9.46)$$

where L is the number of multipath components in the channel response, T is the
symbol duration, and $h_{ki}^{(l)}$ is the complex-valued channel coefficient for the lth path.
The channel coefficients $\{h_{ki}^{(l)}\}$ are known at the transmitter and are realizations
of i.i.d. zero-mean, circularly symmetric complex Gaussian random variables with
variance

$$E\left[|h_{ki}^{(l)}|^2\right] = \frac{1}{L} \ \forall k, i \text{ and } l. \tag{9.47}$$

It is convenient to arrange these channel coefficients for the lth path in a $K \times N_T$ matrix $\mathbf{H}^{(l)}$, where $[\mathbf{H}^{(l)}]_{ki} = h_{ki}^{(l)}$, $i = 1, 2, \ldots, N_T$, $k = 1, 2, \ldots, K$.

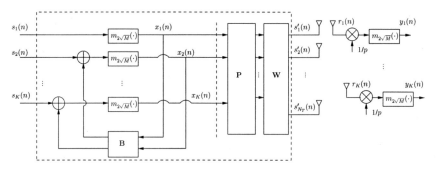

Fig. 9.16 Tomlinson-Harashima precoding applied to MIMO system

The MAI cancellation is facilitated by use of the QR decomposition of the channel matrix $\mathbf{H}^{(0)}$. Thus, we express $[\mathbf{H}^{(0)}]^H$ as

$$[\mathbf{H}^{(0)}]^H = \mathbf{QR}, \tag{9.48}$$

where $\mathbf{Q}$ is an $N_T \times K$ matrix, such that $\mathbf{QQ}^H = \mathbf{I}$, and $\mathbf{R}$ is a $K \times K$ upper triangular matrix with diagonal elements $\{r_{ii}\}$. Based on this decomposition of $[\mathbf{H}^{(0)}]^H$, the signal to be transmitted is precoded with the matrix transformation

$$\mathbf{W} = \mathbf{QA}, \tag{9.49}$$

where $\mathbf{A}$ is a $K \times K$ diagonal matrix with diagonal elements $1/r_{ii}$, $i = 1, 2, \ldots, K$. The $\{r_{ii}\}$ are real and positive [48]. The matrix $\mathbf{P} = p\mathbf{I}$ is a diagonal $K \times K$ matrix that is used simply for scaling the power of the transmitted signal and results in equal SNR for all users. Therefore, we have an effective channel matrix of the form

$$\begin{aligned} \mathbf{H}^{(0)}\mathbf{WP} &= [\mathbf{QR}]^H \mathbf{QAP} \\ &= p\mathbf{R}^H\mathbf{A}. \end{aligned} \tag{9.50}$$

We note that $\mathbf{R}^H\mathbf{A}$ is a $K \times K$ lower triangular matrix with unit diagonal elements. As a result, user k sees multiple-access interference from users $1, 2, \ldots, k-1$. We also note that the effective channel matrix $\mathbf{H}^{(0)}\mathbf{W} = \mathbf{R}^H\mathbf{A}$ will have full rank K, provided that $N_T \geq K$.

By reducing this channel matrix to a lower triangular matrix, we can now subtract the interference at the transmitter that each user would normally observe at their respective receivers. Thus, when the channel adds the same interference to the transmitted signal, the received signal at each receiver would be free of interference.

Taking advantage of the lower triangular matrix structure, successive interference cancellation is performed with the feedback filter defined by the matrix

$$\mathbf{B} = [\mathbf{I} - \mathbf{H}^{(0)}\mathbf{W}, \, -\mathbf{H}^{(1)}\mathbf{W}, \, -\mathbf{H}^{(2)}\mathbf{W}, \ldots, -\mathbf{H}^{(L-1)}\mathbf{W}], \quad (9.51)$$

where the matrix $(\mathbf{I} - \mathbf{H}^{(0)}\mathbf{W})$ is used to cancel the interference due to the other users that arises in the current symbol interval, and the terms $-\mathbf{H}^{(1)}\mathbf{W}, \, -\mathbf{H}^{(2)}\mathbf{W}, \ldots,$ $-\mathbf{H}^{(L-1)}\mathbf{W}$ are used to cancel the interference due to previous symbols.

To ensure that the subtraction of the interference terms do not result in an increase of transmitter power, we use the modulo operator, as in Tomlinson-Harashima precoding, to limit the range of the signal to the boundaries of the signal constellation. Thus, the output of the modulo operators for the nth symbol vector, as shown in Fig. 9.16 (for square QAM constellations) is

$$\mathbf{x}_n = \text{mod}_{2\sqrt{M}}[\mathbf{s}_n + \mathbf{B}\hat{\mathbf{x}}_n]$$
$$= \mathbf{s}_n + \mathbf{B}\hat{\mathbf{x}}_n - 2\sqrt{M}\mathbf{z}_n, \quad (9.52)$$

where the modulo operation is performed on each real and imaginary components of the vector $[\mathbf{s}_n + \mathbf{B}\hat{\mathbf{x}}_n]$, $\mathbf{x}_n$ is the $K \times 1$ vector at the output of the modulo operator, $\mathbf{s}_n$ is the $K \times 1$ data vector, $\hat{\mathbf{x}}_n$ is defined as

$$\hat{\mathbf{x}}_n = [\mathbf{x}_n^T, \, \mathbf{x}_{n-1}^T, \, \mathbf{x}_{n-2}^T, \ldots, \mathbf{x}_{n-(L-1)}^T]^T, \quad (9.53)$$

and $\mathbf{z}_n$ is a $K \times 1$ vector with complex-valued components that take on integer values, determined by the constraint that the real and imaginary components of $\mathbf{x}_n$ fall in the range of $[-\sqrt{M}, \sqrt{M})$. Therefore, the transmitted signal vector is expressed as

$$\mathbf{s}'_n = \mathbf{W}\mathbf{P}\mathbf{x}_n$$
$$= p\mathbf{W}\mathbf{x}_n \quad (9.54)$$

and the received signal vector is

$$\mathbf{r}_n = p\sum_{i=0}^{L-1}\mathbf{H}^{(i)}\mathbf{W}\mathbf{x}_{n-i} + \boldsymbol{\eta}_n, \quad (9.55)$$

and, hence,

$$\mathbf{P}^{-1}\mathbf{r}_n = \mathbf{x}_n + (\mathbf{H}^{(0)}\mathbf{W} - \mathbf{I})\mathbf{x}_n$$
$$+ \sum_{i=1}^{L-1}\mathbf{H}^{(i)}\mathbf{W}\mathbf{x}_{n-i} + \boldsymbol{\eta}'_n. \quad (9.56)$$

By substituting for $\mathbf{B}$ and $\mathbf{x}_n$ in (9.56), it follows that

$$\mathbf{P}^{-1}\mathbf{r}_n = \mathbf{s}_n + \boldsymbol{\eta}'_n - 2\sqrt{M}\mathbf{z}_n. \quad (9.57)$$

Consequently, the MAI and ISI are cancelled perfectly, resulting in the test statistics for the nth symbol vector as follows:

$$\mathbf{y}_n = \mathrm{mod}_{2\sqrt{M}}\left[\frac{1}{p}\mathbf{r}_n\right]. \tag{9.58}$$

9.4.2.1 Optimum Ordering of the Decentralized Receivers

The ordering of the K decentralized receivers affects the construction of the $K \times N_T$ channel matrix $\mathbf{H}^{(0)}$. There are $K!$ possible column permutations of $[\mathbf{H}^{(0)}]^H$ and, hence, there is one QR decomposition associated with each permutation. In turn, there are $K!$ transformation matrices $\mathbf{W} = \mathbf{QA}$, each of which requires a different transmit power. In order to minimize the total transmit power, it is necessary to search over all the column permutations. Such an exhaustive search procedure is computationally time-consuming, except for a small number of users. A method for simplifying the search for the optimum ordering is described in [19].

The error rate performance of the QR decomposition method described above has been evaluated in [1, 2]. Figure 9.17 illustrates the symbol error probability as a function of the SNR (total transmitted signal power over all antennas divided by N_0) for QPSK modulation, $L = 1, 2$ and $N_T = K = 2$. The Monte Carlo simulation results are also illustrated. The simulation results are obtained by transmitting 1,000 data symbols over each of 10,000 channel realizations.

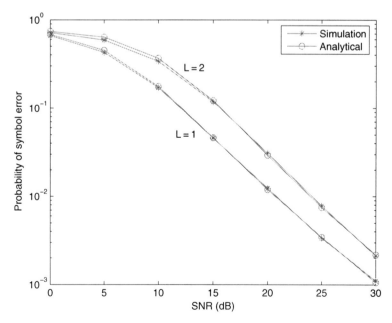

Fig. 9.17 Performance of optimal-ordered QR decomposition with $N_T = K = 2$ and $L = 1, 2$

Figure 9.18 shows analytical results of the symbol error rate performance for QPSK with $L = 1$ (flat fading), $K = 2$, and $N_T = 2, 3, 4$. We observe that the system performance improves with an increase in the number of transmit antennas, which reflects the benefit of spatial diversity.

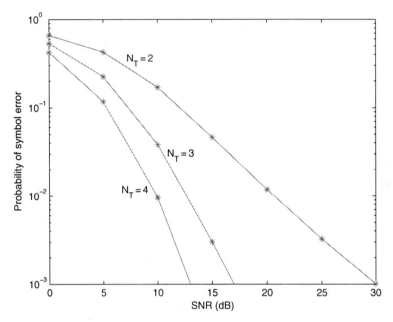

Fig. 9.18 Performance of optimal-ordered QR decomposition with $K = 2, L = 1$, and $N_T = 2, 3, 4$

Figure 9.19 shows simulation results comparing the error rate performance of the linear zero-forcing (ZF) and minimum MSE (MMSE) precoding methods with the QR decomposition method for QPSK modulation with $L = 1$ and $K = N_T = 4$. We observe that the performance of the QR decomposition method is better than that of the linear precoders at high SNR, but poorer at low SNR. It should be noted that the improvement in performance of the QR decomposition method at high SNR is obtained at a significantly higher computational complexity compared with the linear precoders.

9.4.3 Vector Precoding

Hochwald et al. [26] have described and evaluated the performance of a vector precoding method in which the data vector **s** to be transmitted to the K users is modified by the addition of a precoding vector with integer elements as illustrated in Fig. 9.20. That is, the modified signal vector is defined as

$$\mathbf{s}' = \mathbf{s} + \tau \hat{\mathbf{p}}, \tag{9.59}$$

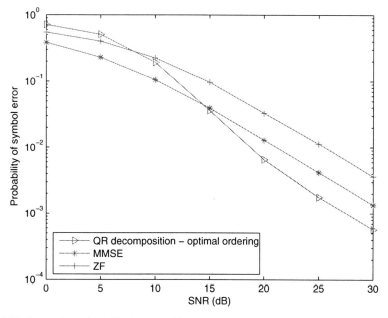

Fig. 9.19 Comparison of the QR decomposition and the linear precoders with $N_T = K = 4$

where the components of the signal vector are taken from a square QAM signal constellation, τ is a real positive number, and $\hat{\mathbf{p}}$ is a K-dimensional vector with complex-valued elements whose real and imaginary components are integers. Hence, for $N_T \geq K$, the transmitted signal vector is

$$\mathbf{x} = \mathbf{A}_T \left(\mathbf{s} + \tau \hat{\mathbf{p}} \right), \tag{9.60}$$

where the matrix $\mathbf{A}_T$ is given as

$$\mathbf{A}_T = \begin{cases} \gamma \mathbf{H} \left(\mathbf{H} \mathbf{H}^H \right)^{-1} & \text{zero-forcing criterion} \\ \gamma \mathbf{H}^H \left(\mathbf{H} \mathbf{H}^H + \beta \mathbf{I} \right)^{-1} & \text{MMSE criterion} \end{cases}. \tag{9.61}$$

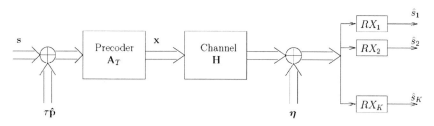

Fig. 9.20 Model of MIMO broadcast system employing vector precoding

The perturbation vector $\hat{\mathbf{p}}$ is chosen to minimize the power in the transmitted signal, i.e.,

$$\hat{\mathbf{p}} = \arg\min_{\mathbf{p}} \|\mathbf{A}_T (\mathbf{s} + \tau \mathbf{p})|^2 , \tag{9.62}$$

subject to the constraint that the real and imaginary components of each element of $\mathbf{p}$ are integers. Methods for solving the least-squares K-dimensional integer-lattice problem and the selection of the positive number τ are described by Hochwald et al. [26].

The received signal vector is

$$\mathbf{r} = \mathbf{H}\mathbf{A}_T \left(\mathbf{s} + \tau \hat{\mathbf{p}}\right) + \eta, \tag{9.63}$$

where η is the additive noise vector. Although the perturbation vector $\hat{\mathbf{p}}$ is not known to the receivers, the mth user assumes that its received signal has the form

$$r_m = \gamma \left(s_m + \tau p_m\right) + \eta'_m, \tag{9.64}$$

where η'_m includes the additive channel noise and the MAI from other users due to the non-zero scale factor β. Since each user knows γ and τ, the mth user performs the modulo operation on the real and imaginary components of r_m to remove p_m and passes the result to its decoder. It is demonstrated in Hochwald et al. [26] that this vector perturbation scheme achieves a signal diversity order comparable to the diversity achieved by the maximum-likelihood detector.

9.4.4 Lattice Reduction Method for Precoding

The lattice reduction method for precoding, described in this section, is similar to the Tomlinson-Harashima precoding for channels with known interference at the transmitter. As in the previous sections, the components of the signal vector to be transmitted are taken from a square QAM signal constellation and, thus, are elements of a lattice.

A lattice can be expressed in terms of its generator matrix $\mathbf{G}$ whose rows denote a basis for the lattice, i.e., all lattice points can be written as a linear combination of the rows of $\mathbf{G}$ with integer weighting coefficients. Any lattice Λ can have many generator matrices and many bases for representation of lattice points. In particular, if $\mathbf{F}$ is a square matrix with integer entries such that $\det(\mathbf{F}) = \pm 1$, then $\mathbf{F}^{-1}$ exists and its entries are all integers. Then, $\mathbf{G}' = \mathbf{G}\mathbf{F}$ is a generator of lattice Λ. The new generator matrix $\mathbf{G}'$ defines a new basis for the lattice Λ. A desirable property of the modified lattice basis is that it be an orthogonal, or close-to-orthogonal basis with the lowest basis vector norms. The process of finding such a basis for a lattice is called *lattice reduction*. Although lattice reduction in high dimensions is an NP-hard problem, a polynomial-time suboptimal lattice reduction algorithm described by Lenstra et al. [29], called the LLL algorithm, gives very good results.

The lattice reduction algorithm developed by Lenstra et al. [29] was designed for real lattices. Gan and Mow [20] have generalized the LLL algorithm to lattices in complex dimensions. Similar to real lattices, if $\mathbf{G}$ is a generator matrix of a complex lattice and $\mathbf{G}' = \mathbf{GF}$, where $\mathbf{F}$ is a square matrix with complex elements having integer real and imaginary parts, such that $\det(\mathbf{F}) = \pm 1$ or $\det(\mathbf{F}) = \pm j$, then $\mathbf{G}'$ is also a basis for the lattice generated by $\mathbf{G}$.

We may apply lattice reduction to the columns of $\mathbf{H}\left(\mathbf{HH}^H\right)^{-1}$ for the zero-forcing criterion or to the columns of $\mathbf{H}^H\left(\mathbf{HH}^H + \beta\mathbf{I}\right)^{-1}$ for the MMSE criterion. Thus, we obtain the matrix $\mathbf{F}_{ZF}$ or $\mathbf{F}_{MSE}$. For simplicity, let us consider the lattice reduction of the matrix $\mathbf{H}^+ = \mathbf{H}^H\left(\mathbf{HH}^H\right)^{-1}$, which yields the matrices $\mathbf{F}$ and $\mathbf{F}^{-1}$. The precoding matrix $\mathbf{A}_T$ is given as

$$\mathbf{A}_T = \mathbf{H}^+\mathbf{F}. \tag{9.65}$$

The transmitted signal vector is $\mathbf{x} = \mathbf{A}_T\mathbf{s}'$, where $\mathbf{s}' = \mathrm{mod}_{2\sqrt{M}}[\mathbf{Fs}]$, and the elements of the K-dimensional signal vector $\mathbf{s}$ are selected from an M-ary QAM signal constellation. The received signal vector is

$$
\begin{aligned}
\mathbf{y} &= \mathbf{Hx} + \boldsymbol{\eta} \\
&= \mathbf{HA}_T\mathbf{s}' + \boldsymbol{\eta} \\
&= \mathbf{Fs}' + \boldsymbol{\eta} \\
&= \mathrm{mod}_{2\sqrt{M}}[\mathbf{s}] + \boldsymbol{\eta}.
\end{aligned}
\tag{9.66}
$$

Thus, each receiver recovers its information symbol without interference from other users.

The lattice reduction technique has been shown to have a performance comparable to that achieved by the vector precoding method described previously. In fact, the signal diversity order achieved by the lattice reduction technique is comparable to the signal diversity order obtained by maximum-likelihood detection, but the lattice reduction technique has a much smaller computational complexity. The interested reader is referred to the papers by Windpassinger et al. [55] and Taherzadeh et al. [46].

9.5 Conclusions

The goal of this chapter was to present an overview of techniques to combat interference in wireless communication systems. We did this by considering typical situations where interference suppression and/or mitigation is necessary to ensure satisfactory system operation, and these examples included both uplink and downlink scenarios. The key waveform designs were single-carrier direct sequence, multicarrier direct sequence, and OFDM, and included both SISO and MIMO systems. The fundamental conclusion to be drawn from the material presented in this chapter is that many communication systems cannot function in the absence of explicit interference suppression/mitigation procedures, over and above the various modulation and error correction coding techniques that are employed in their design.

References

1. Amihood, P., Masry, E., Milstein, L.B., Proakis, J.G.: Analysis of a MISO pre-BLAST DFE technique for decentralized receivers. In: Proceedings of 40th Asilomar Conference on Signals, Systems and Computers, pp. 1587–1592. Pacific Grove, CA (2006).
2. Amihood, P., Masry, E., Milstein, L.B., Proakis, J.G.: Performance analysis of a pre-BLAST DFE technique for MISO channels with decentralized receivers. IEEE Trans. Commun. **55**, 1385–1396 (2007).
3. Batra, A., Lingam, S., Balakrishnan, J.: Multi-band OFDM: A cognitive radio for UWB. In: Conference Proceedings of IEEE International Symposium on Circuits and Systems (ISCAS), pp. 4094–4097. Island of Kos, Greece (2006).
4. Batra, A., Zeidler, J.R.: Narrowband interference mitigation in OFDM systems. In: Proceedings of the IEEE Military Communications Conference (MILCOM). San Diego, CA (2008).
5. Batra, A., Zeidler, J.R., Beex, A.A.: A two-stage approach to improving the convergence of least-mean square decision-feedback equalizers in the presence of severe narrowband interference. EURASIP J. Adv. Signal Process. (2008). Article ID 390102.
6. Baum, C.W., Pursley, M.B.: Bayesian methods for erasure insertion in frequency-hop communication systems with partial-band interference. IEEE Trans. Commun. **40**, 1231–1238 (1992).
7. Bingham, J.A.C.: Multicarrier modulation for data transmission : An idea whose time has come. IEEE Commun. Mag. **28**, 5–14 (1990).
8. Carron, G., Ness, R., Deneire, L., der Perre, L.V., Engles, M.: Comparison of two modulation techniques using frequency domain processing for in-house networks. IEEE Trans. Consumer Electron. **47**, 63–72 (2001).
9. Chong, L.L., Milstein, L.B.: Comparing DS-CDMA and multicarrier CDMA with imperfect channel estimation. In: Proceedings of 11th IEEE Workshop on Statistical Signal Processing, pp. 385–388. Singapore (2001).
10. Chouly, A., Brajal, A., Jourdan, S.: Orthogonal multicarrier techniques applied to direct sequence spread spectrum CDMA systems. In: Proceedings of the IEEE Global Telecommunications (Globecom) Conference, pp. 1723–1728. Houston (1993).
11. Cimini, L.J.: Analysis and simulation of a digital mobile channel using orthogonal frequency division multiplexing. IEEE Trans. Commun. **COM-33**, 665–675 (1985).
12. Coulson, A.J.: Bit error rate performance of OFDM in narrowband interference with excision filtering. IEEE Trans. Wireless Commun. **5**, 2484–2492 (2006).
13. Da Silva, V.M., Sousa, E.S.: Performance of orthogonal CDMA codes for quasi-synchronous communication systems. In: Proceedings of the IEEE International Conference on Universal Personal Communications (ICUPC), pp. 995–999. Ottawa (1993).
14. Darsena, D.: Successive narrowband interference cancellation for OFDM systems. IEEE Commun. Lett. **11**, 73–75 (2007).
15. Fazel, K., Kaiser, S.: Multi-carrier and Spread Spectrum Systems. John Wiley & Sons, Inc., New York, NY, (2003).
16. Fazel, K., Papke, L.: On the performance of convolutionally-coded CDMA/OFDM for mobile communication systems. In: Proceedings of the IEEE Personal, Indoor, and Mobile Radio Communications (PIMRC) Conference, pp. 468–472. Yokahama (1993).
17. Fischer, R.F.H.: Precoding and Signal Shaping for Digital Transmission. John Wiley & Sons, New York, NY (2002).

18. Fischer, R.F.H., Windpassinger, C., Lampe, A., Huber, J.B.: Space-time transmission using Tomlinson-Harashima precoding. In: Proceedings 4th International ITG Conference on Source and Channel Coding, pp. 139–147. Berlin (2002).
19. Foschini, G.J., Golden, G.D., Valenzuela, R.A., Wolniansky, P.W.: Simplified processing for high spectral efficiency wireless communication employing multi-element arrays. IEEE J. Selected Areas Commun. **17**, 1841–1852 (1999).
20. Gan, Y.H., Mow, W.H.: Accelerated complex lattice reduction algorithm applied to MIMO detection. In: Proceedings of 48th IEEE Global Telecommunications Conference, pp. 2953–2957. St. Louis, MO (2005).
21. Gerakoulis, D., Salmi, P.: An interference suppressing OFDM system for wireless communication. In: IEEE International Conference on Communications (ICC), pp. 480–484. New York, NY (2002).
22. Gray, R.M.: Toeplitz and circulant matrices: A review. Foundations Trends Commun. Inform. Theory **2**, 155–239 (2006).
23. Hara, S., Prasad, R.: Overview of multicarrier CDMA. IEEE Commun. Mag. **35**, 126–133 (1997).
24. Harashima, H., Miyakawa, H.: Matched-transmission technique for channels with intersymbol interference. IEEE Trans. Commun. **COM-20**, 774–780 (1972).
25. Haykin, S.: Adaptive Filter Theory, Fourth edn. Prentice Hall, Upper Saddle River, NJ (2002).
26. Hochwald, B.M., Peel, C.B., Swindlehurst, A.L.: A vector-perturbation technique for near-capacity multiantenna multiuser communication – Part II: Perturbation. IEEE Trans. Commun. **53**, 537–544 (2005).
27. IEEE 802.11: Wireless LAN medium access control (MAC) physical layer (PHY) specifications, amendment 1: High-speed physical layer in the 5 GHz band (1999).
28. Kondo., S., Milstein, L.B.: Performance of multicarrier DS CDMA systems. IEEE Trans. Commun. **44**, 238–246 (1996).
29. Lenstra, A.K., Lenstra, H.W., Lovász, L.: Factoring polynomials with rational coefficients. Math. Ann. **261**, 515–534 (1982).
30. Li, L.M., Milstein, L.B.: Rejection of narrow-band interference in PN spread-spectrum systems using transversal filters. IEEE Trans. Commun. **30**, 925–928 (1982).
31. Li, L.M., Milstein, L.B.: Rejection of CW interference in QPSK systems using decision-feedback filters. IEEE Trans. Commun. **COM-31**, 473–483 (1983).
32. Ling, A.S., Milstein, L.B.: Trade-off between diversity and channel estimation errors in asynchronous MC-DS-CDMA and MC-CDMA. IEEE Trans. Commun. **56**, 584–597 (2008).
33. Milstein, L.B.: Interference Suppression to aid acquisition in direct-sequence spread-spectrum communications. IEEE Trans. Commun. **36**, 1200–1207 (1988).
34. Milstein, L.B., Schilling, D.L., Pickholtz, R.L., Erceg, V., Kullback, M., Kanterakis, E.G., Fishman, D.S., Biederman, W.H., Salerno, D.C.: On the feasibility of a CDMA overlay for personal communications networks. IEEE J. Selected Areas Commun. **10**, 655–668 (1992).
35. Mitola, J., Maguire, G.Q.: Cognitive radio: making software radios more personal. IEEE Personal Commun. Mag. **6**, 13–18 (1999).
36. Nassar, C.R., Natarajan, B., Wu, Z., Wiegandt, D., Zekavat, S.A., Shattil, S.: Multi-carrier Technologies for Wireless Communications. Kluwer Academic Publishers, Boston, MA (2002).
37. Nilsson, R., Sjöberg, F., LeBlanc, J.P.: A rank-reduced LMMSE canceller for narrowband interference suppression in OFDM-based systems. IEEE Trans. Commun. **51**, 2126–2140 (2003).
38. Peel, C.B., Hochwald, B.M., Swindlehurst, A.L.: A vector-perturbation technique for near-capacity multiantenna multiuser communication – Part I: Channel inversion and regularization. IEEE Trans. Commun. **53**, 195–202 (2005).
39. Proakis, J.G.: Digital Communications, Fourth edn. McGraw Hill, Boston, MA (2001).
40. Proakis, J.G., Manolakis, D.G.: Introduction to Digital Signal Processing. MacMillian, New York, NY (1998).
41. Redfern, A.J.: Receiver window design for multicarrier communication systems. IEEE J. Select. Areas Commun. **20**, 1029–1036 (2002).

42. Shensa, M.J.: Time constants and learning curves of LMS adaptive filters. Tech. Rep. 312, Naval Ocean Systems Center, San Diego, CA (1978).
43. da Silva, C.R.C.M.: Performance of spectra-encoded ultra-wideband communication systems in the presence of narrow-band interference. Ph.D. thesis, University of California, San Diego (2005).
44. da Silva, C.R.C.M., Milstein, L.B.: The effects of narrow-band interference on UWB communication systems with Imperfect channel estimation. IEEE J. Select. Areas Commun. **24**, 717–723 (2006).
45. Snow, C., Lampe, L., Schober, R.: Error rate analysis for coded multicarrier systems over quai-static fading channels. IEEE Trans. Commun. **55**, 1736–1746 (2007).
46. Taherzadeh, M., Mobasher, A., Khandani, A.: Communication over MIMO broadcast channels using lattice-basis reduction. IEEE Trans. Inform. Theory **53**, 4567–4582 (2007).
47. Tomlinson, M.: A New Automatic Equalizer Employing Modulo Arithmetic. IEE Electron. Lett. **7**, 138–139 (1971).
48. Tulino, A.M., Verdu, S.: Random Matrix Theory and Wireless Communications. Now Publishers Inc., Boston, MA (2004).
49. Wang, J., Milstein, L.B.: Adaptive LMS filters for cellular CDMA overlay situations. IEEE J. Select. Areas Commun. **14**, 1548–1559 (1996).
50. Wang, T., Proakis, J.G., Masry, E., Zeidler, J.R.: Performance degradation of OFDM systems due to Doppler spreading. IEEE Trans. Wireless Commun. **5**, 1422–1432 (2006).
51. Wang, T., Proakis, J.G., Zeidler, J.R.: Analysis of per-channel equalized filtered multitone modulations over time-varying fading channels. In: Proceedings of the IEEE Personal Indoor and Mobile Radio Communications (PIMRC) (2006).
52. Wang, T., Proakis, J.G., Zeidler, J.R.: Interference analysis of filtered multitone modulation over time-varying fading channels. IEEE Trans. Commun. **55**, 717–727 (2007).
53. Wei, P., Zeidler, J.R., Ku, W.H.: Adaptive interference suppression for CDMA overlay systems. IEEE J. Select. Areas Commun. **12**, 1510–1523 (1994).
54. Weinstein, S.B., Ebert, P.M.: Data transmission by frequency-division multiplexing using the discrete Fourier transform. IEEE Trans. Commun. Technol. **COM-19**, 628–634 (1971).
55. Windpassinger, C., Fischer, R.F.H., Huber, J.B.: Lattice-reduction-aided broadcast precoding. IEEE Trans. Commun. **52**, 2057–2060 (2004).
56. Windpassinger, C., Fischer, R.F.H., Vencel, T., Huber, J.B.: Precoding in multi-antenna and multi-user communications. IEEE Trans. Wireless Commun. **3**, 1305–1316 (2004).
57. Wu, Z., Nassar, C.R.: Narrowband interference rejection in OFDM via carrier interferometry spreading codes. IEEE Trans. Wireless Commun. **4**, 1491–1505 (2005).
58. Yee, N., Linnartz, J.P.: Wiener filtering of multi-carrier CDMA in a Rayleigh fading channel. In: Proceedings of the IEEE Personal, Indoor, and Mobile Radio Communications (PIMRC) Conference, pp. 1344–1347. The Hague (1994).
59. Yee, N., Linnartz, J.P., Fettweis, G.: Multi-carrier CDMA in indoor wireless radio. In: Proceedings of the IEEE Personal, Indoor, and Mobile Radio Communications (PIMRC) Conference, pp. 109–113. Yokahama (1993).
60. Zeidler, J.R.: Performance analysis of LMS adaptive prediction filters. Proc. IEEE **78**, 1781 - 1806 (1990).
61. Zeidler, J.R., Satorius, E.H., Chabries, D.M., Wexler, H.T.: Adaptive enhancement of multiple sinusoids in uncorrelated noise. IEEE Trans. Acoust. Speech, Signal Process. **ASSP-26**, 240 - - 254 (1978).

Chapter 10
Cognitive Radio: From Theory to Practical Network Engineering

Ekram Hossain, Long Le, Natasha Devroye, and Mai Vu

10.1 Introduction

Under utilization of radio spectrum in traditional wireless communication systems [29], along with the increasing spectrum demand from emerging wireless applications, is driving the development of new spectrum allocation policies for wireless communications. These new spectrum allocation policies, which will allow unlicensed users (i.e., secondary users) to access the radio spectrum when it is not occupied by licensed users (i.e., primary users) will be exploited by the *cognitive radio* (CR) technology. Cognitive radio will improve spectrum utilization in wireless communication systems while accommodating the increasing amount of services and applications in wireless networks. A cognitive radio transceiver is able to adapt to the dynamic radio environment and the network parameters to maximize the utilization of the limited radio resources while providing flexibility in wireless access [42]. The key features of a CR transceiver include awareness of the radio environment (in terms of spectrum usage, power spectral density of transmitted/received

Ekram Hossain
Department of Electrical and Computer Engineering, University of Manitoba, Winnipeg, Manitoba, Canada
e-mail: ekram@ee.umanitoba.ca

Long Le
Department of Aeronautics and Astronautics, Massachusetts Institute of Technology, Boston, Massachusetts, USA
e-mail: longble@mit.edu

Natasha Devroye
Department of Electrical and Computer Engineering, University of Illinois at Chicago, Chicago, Illinois, USA
e-mail: devroye@ece.uic.edu

Mai Vu
Division of Engineering and Applied Sciences, Harvard University, Montreal, Quebec, Canada
e-mail: mai.h.vu@mcgill.ca

V. Tarokh (ed.), *New Directions in Wireless Communications Research*,
DOI 10.1007/978-1-4419-0673-1_10,
© Springer Science+Business Media, LLC 2009

signals, wireless protocol signaling) and intelligence. This intelligence is achieved through learning for adaptive tuning of system parameters such as transmit power, carrier frequency, and modulation strategy (at the physical layer), and higher layer protocol parameters.

Implementation of a cognitive radio will be based on the concept of dynamic spectrum access (DSA). Through DSA, frequency spectrum can be shared among primary users and cognitive radio users (i.e., secondary users) in a dynamically changing radio environment. There are two major flavors of dynamic spectrum access: dynamic licensing (for dynamic exclusive use of radio spectrum) and dynamic sharing (for coexistence) [3, 116]. Dynamic sharing can be of two types: horizontal spectrum sharing and vertical spectrum sharing. In the former case, all users/nodes have equal regulatory status while in the latter case all users/nodes do not have equal regulatory status (i.e., there are primary users and secondary users) and secondary users opportunistically access the spectrum without negatively affecting the primary users' performance.

In this chapter, we focus on vertical spectrum sharing in a cognitive radio network. In particular, we outline the recent information-theoretic advances pertaining to the limits of such networks. Information theory provides an ideal framework as well as tools and metrics for analyzing the fundamental limits of communication. The limits obtained provide benchmarks for the operation of cognitive networks, allowing researchers and engineers to gauge the efficiency of any practical network and guide their design. Spectrum sensing is one of the major functions of a cognitive radio the goal of which is to determine the activity of licensed users by periodically observing signals on the target frequency bands. We discuss some theoretical results on the effect of side information (e.g., spatial locations of the users, transmission probability of primary users) on the cognitive sensing performance. Analysis of interference is required to design cognitive radio parameters so that the impact of interference to the primary users can be minimized. We provide examples of this interference analysis in a cognitive radio system.

To this end, we discuss the practical implementation aspects of vertical spectrum sharing employing either an interference control or an interference avoidance approach and discuss open research challenges. An interference avoidance approach requires spectrum sensing and secondary users are allowed to access a particular spectrum band only if primary users are not detected on that band by certain sensing technique [84, 98]. An interference control approach allows primary and secondary users to transmit simultaneously on the same frequency band. Transmission powers of secondary users, however, should be carefully controlled such that the total interference created by secondary users at each primary receiver be smaller than the maximum tolerable level. In fact, this maximum interference level corresponds to an interference temperature limit which is mandated by FCC and/or primary network operators.

The rest of the chapter is organized as follows. Section 10.2 focuses on the information-theoretic limit of communication in a cognitive radio channel shared by a primary transmitter–receiver pair and a secondary transmitter–receiver pair. Section 10.3 describes some specific results on the cognitive sensing performance with side information on the spatial locations of the users. Section 10.4 focuses on the impact of cognitive users on the primary users in terms of interference power. Sections 10.5 and 10.6 describe the modeling and engineering design approaches for the two spectrum access paradigms, namely the interference control and the interference avoidance paradigms, respectively. In the rest of the chapter, we will use the terms "cognitive user" and "secondary user" interchangeably.

10.2 Information-Theoretic Limits of Cognitive Networks

In this section, we emphasize and explore the impact of *cognition*, defined as extra information (or side information) the cognitive radio nodes have about their wireless environment, on the information-theoretic limits of communication.

10.2.1 Cognitive Behavior: Interference Avoidance, Control, and Mitigation

Cognitive networks should achieve better performance than standard homogeneous networks[1] as they are able to (1) exploit the nodes' cognitive abilities, i.e., sensing and adapting to their wireless environment and (2) often (but not necessarily) exploit new policies in secondary spectrum licensing scenarios in which the agile cognitive radios are permitted to share the spectrum with primary users. Naturally, the extent to which the performance of the network can be improved depends on what the cognitive radios know about their spectral environment, and consequently, how they adapt to this. Cognitive behavior, or how the secondary cognitive users employ the primary spectrum, may be grouped into three categories, as also done with slight variations in [21, 25, 27, 36], each of which exploits varying degrees of knowledge of the wireless environment at the secondary user(s):

- **Interference avoidance (spectrum interweave):** The primary and secondary signals may be thought of as being orthogonal to each other: they may access the spectrum in a time-division-multiple-access (TDMA) fashion, in a frequency-division-multiple-access (FDMA) fashion, or in any fashion that ensures that the primary and secondary signals do not interfere with each other. The cognition required by the secondary users to accomplish this is knowledge of the spectral gaps (in, for example, time, frequency) of the primary system. The secondary users may then fill in these spectral gaps.
- **Interference control (spectrum underlay):** The secondary users transmit over the same spectrum as the primary users, but do so in a way that the interference seen by the primary users from the cognitive users is controlled to an acceptable level, captured by primary QoS constraints.[2] The cognition required is knowledge of the "acceptable levels" of interference at primary users in a cognitive user's

---

[1] Networks in which no nodes are cognitive radios.

[2] What constitutes an acceptable level will be described later and it may vary from system to system.

transmission range as well as knowledge of the effect of the cognitive transmission at the primary receiver. This last assumption boils down, in classical wireless channels, to knowledge of the channel(s) between the cognitive transmitter(s) and the primary receiver(s).

- **Interference mitigation (spectrum overlay):** The secondary users transmit over the same spectrum as the primary users, but in addition to knowledge of the channels between primary and secondary users (nature), the cognitive nodes have additional information about the primary system and its operation. Examples are knowledge of the primary users' codebooks, allowing the secondary users to decode primary users' transmissions, or in certain cases even knowledge of the primary users' message.

We consider a simple channel in which a primary transmitter–receiver pair (white, $\mathcal{PT}_x$, $\mathcal{PR}_x$) and a cognitive transmitter–receiver pair (gray, $\mathcal{ST}_x$, $\mathcal{SR}_x$) share the same spectrum, shown in Fig. 10.1. For this simple channel we will derive fundamental limits on the communication possible under each type of cognitive behavior. One information-theoretic metric that lends itself well to illustrative purposes and is central to many studies is the capacity region of the channel. Under Gaussian noise, we will illustrate different examples of cognitive behavior and will build up to the right illustration in Fig. 10.1, which corresponds to the rates achieved under different levels of cognition.

The basic and natural conclusion is that, the higher the level of cognition at the cognitive terminals, the higher the achievable rates. However, increased cognition often translates into increased complexity. At what level of cognition future secondary spectrum licensing systems will operate will depend on the available side information and network design constraints.

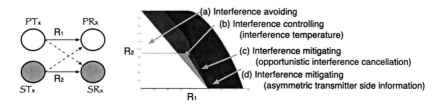

Fig. 10.1 The primary users (*white*) and secondary users (*gray*) wish to transmit over the same channel. *Solid lines* denote desired transmission, *dotted lines* denote interference. The achievable rate regions under four different cognitive assumptions and transmission schemes are shown on the right; (a)–(d) are in order of increasing cognitive abilities

10.2.2 Information-Theoretic Basics

A communication channel is modeled as a set of conditional probability density functions relating the inputs and outputs of the channel. Given this probabilistic characterization of the channel, the fundamental limits of communication may be expressed in terms of a number of metrics of which *capacity* is one of the most known and powerful. Capacity is defined as the supremum over all rates (expressed in bits/channel use) for which reliable communication may take place. While capacity is central to many information-theoretic studies, it is often challenging to determine. Inner bounds, or achievable rates, as well as outer bounds to the capacity may be more readily available. For more precise information-theoretic definitions we refer the reader to [17, 18, 110].

The additive white Gaussian noise (AWGN) channel with *quasi-static fading* is the example most used in this section. In the AWGN channel, the output Y is related to the input X according to $Y = hX + N$, where h is a fading coefficient (often modeled as a Gaussian random variable), and N is the noise which is $N \sim \mathcal{N}(0, 1)$. Under an average input power constraint $E[|X|^2] \leq P$, the well-known capacity is given by $C = \frac{1}{2} \log_2 \left(1 + |h|^2 P\right) = \frac{1}{2} \log_2 \left(1 + \text{SINR}\right) := \mathcal{C}(\text{SINR})$, where SINR is the received signal to interference plus noise ratio, and $\mathcal{C}(x) := \frac{1}{2} \log_2(1 + x)$.

We now proceed to analyzing three different classes of cognitive behavior.

10.2.3 Interference Avoidance: Spectrum Interweave

Secondary spectrum licensing and cognitive radio was arguably conceived with the goal and intent of implementing the interference-avoiding behavior [42, 79]. Cognition in this setting corresponds to the ability to accurately detect the presence of other wireless devices; the cognitive side information is knowledge of the spatial, temporal, and spectral gaps, or *white spaces* a particular cognitive Tx–Rx pair would experience. Cognitive radios would adjust their transmission to fill in the spectral (or spatial/temporal) void, as illustrated in Fig. 10.2, with the potential to drastically increase the spectral efficiency of wireless systems.

This type of behavior requires knowledge of the spectral white spaces. In a realistic system the secondary transmitter would spend some of its time sensing the channel to determine the presence of the primary user. As an illustrative example and idealization, we assume that knowledge of the spectral gaps is perfect: when primary communication is present the cognitive devices are able to precisely determine this presence, instantaneously. While such assumptions may be valid for the purpose of a theoretical study and provide outer bounds on what can be realistically achieved, practical methods for detecting primary signals have also been

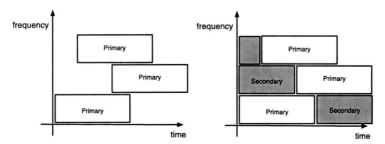

Fig. 10.2 Interference avoidance: a cognitive user senses the time/frequency "white spaces" and opportunistically transmits over the detected spaces

of great interest recently. A theoretical framework for determining the limits of
communication as a function of the sensed cognitive transmitter and receiver gaps
is formulated in [52,91]. Studies on how detection errors may affect the cognitive
and primary systems are found in [88,96,97]. Because current secondary spectrum
licensing proposals demand detection guarantees of primary users at extremely low
levels in harsh fading environments, a number of works have suggested improving
detection capabilities through allowing multiple cognitive radios to collaboratively
detect the primary transmissions [19,30,34,78].

Under our idealized assumptions, the rates R_1 of the primary Tx–Rx pair and R_2
of the cognitive Tx–Rx pair achieved through ideal white space filling are shown
as the inner white triangle in Fig. 10.1. When a single user transmits the entire
time in an interference-free environment, the intersection points on the axes are
attained. The convex hull of these two interference-free points may be achieved by
time sharing (TDMA fashion). Where on this line a system operates depends on
how often the primary user occupies the specific band. If the primary and secondary
power constraints are P_1 and P_2, respectively, then the white space filling rate region
may be described as follows:

White space filling region (a)

$$= \{(R_1, R_2)|0 \leq R_1 \leq t\mathcal{C}(P_1), 0 \leq R_2 \leq (1-t)\mathcal{C}(P_2), 0 \leq t \leq 1\}.$$

10.2.3.1 Interference Avoidance Through MIMO

In addition to detecting the spectral white spaces, interference at the primary user
may be avoided or controlled if the cognitive user is equipped with multiple antennas
and is able to place its transmit signal in the null space of the primary users receive
channel. In this scenario, the channel between the secondary-transmit antennas and
the primary-receive antennas must be known. Studies where the cognitive rates are
maximized subject to primary user communication guarantees (such as maximum
average interference power constraints) are considered in [50,51,111,113–115]. The
scenarios considered in these papers can be considered as an *interference-avoiding*
scheme if the tolerable interference at the primary receivers is set to zero, otherwise it falls under the *interference-controlled* paradigm we look at in the following
section.

10.2.4 Interference Control: Spectrum Underlay

When the interference caused by the secondary users on the primary users is permitted to be below a certain level (according to QoS constraints), the more flexible
interference-controlled behavior emerges. We note that this type of interference-
controlled behavior covers a large spectrum of cognitive behavior and we highlight
only three examples: an example of the resulting achievable rate region in small

networks, and throughput scaling laws in two different types of large spectrum underlay networks.

10.2.4.1 Underlay in Small Networks: Achievable Rates

A cognitive radio may simultaneously transmit with the primary user(s) while using its cognitive abilities to control the amount of harm it inflicts upon them. One common definition of harm involves the notion of *interference temperature*, a term first introduced by the FCC [63] to denote the average level of interference power seen at a primary receiver. In secondary spectrum licensing scenarios, the primary receiver's interference temperature should be kept at a level that will satisfy the primary user's desired quality of service. Provided the cognitive user knows (1) the maximal interference temperature for the surrounding primary receivers, (2) the current interference temperature level, and (3) how its own transmit power will translate to received power at the primary receiver, then the cognitive radio may adjust its own transmission power so as to satisfy any interference temperature constraint the primary user(s) may have. The work in [31, 35, 106, 109] considers the capacity of cognitive systems under various receive-power (or interference-temperature-like) constraints.

As an illustrative example, we consider a very simple interference-temperature-based cognitive transmission scheme. Assume in the channel model of Fig. 10.1 that each receiver treats the other user's signal as noise, a lower bound to what may be achieved using more sophisticated decoders [99]. The rate region obtained is shown as the light gray region (b) in Fig. 10.1. This region is obtained as follows: we assume the primary transmitter communicates using a Gaussian codebook of constant average power P_1. We assume the secondary transmitter allows its power to lie in the range $[0, P_2]$ for P_2 some maximal average power constraint. The rate region obtained may be expressed as follows:

Simultaneous-transmission rate region (b)

$$= \left\{ (R_1, R_2) | 0 \le R_1 \le C\left(\frac{P_1}{h_{21}^2 P_2^* + 1}\right), \right.$$

$$\left. 0 \le R_2 \le C\left(\frac{P_2^*}{h_{12}^2 P_1 + 1}\right), \ 0 \le P_2^* \le P_2 \right\}.$$

The actual value of P_2^* chosen by the cognitive radio depends on the interference-temperature or received-power constraints at the primary receiver.

10.2.4.2 Underlay in Large Networks: Scaling Laws

Information-theoretic limits of interference-controlled behavior has also been investigated for large networks, i.e., networks whose number of nodes $n \to \infty$.

We illustrate two types of networks: single-hop networks and multi-hop networks. In the former, secondary nodes transmit subject to outage-probability-like constraints on the primary network. In the latter, the multi-hop secondary network is permitted to operate as long as the scaling law of the primary network is kept the same as in the absence of the cognitive network.

Single-hop cognitive networks

In the planar network model considered in [103] multiple primary and secondary users coexist in a network of radius D (the number of nodes grows to ∞ as $D \rightarrow \infty$). Around each receiver, either primary or cognitive, a protected circle of radius $\epsilon_c > 0$ is assumed in which no interfering transmitter may operate. Other than the receiver-protected regions, the primary transmitter and receiver locations are arbitrary, subject to a minimum distance D_0 between any two primary transmitters. This scenario corresponds to a broadcast network, such as the TV or the cellular networks, in which the primary transmitters are base stations. The cognitive transmitters, on the other hand, are uniformly and randomly distributed with constant density λ. We assume that each cognitive receiver is within a D_{max} distance from its transmitter, the channel gains are path-loss-dependent only (no fading or shadowing) and that each user treats unwanted signals from all other users as noise.

The quality of service guarantee of the primary users is of the form

$$\Pr[\text{primary user's rate} \leq C_0] \leq \beta.$$

That is, the secondary users must transmit so as to guarantee that the probability that the primary users' rates fall below C_0 is less than a desired amount β. Some of the questions answered in [101, 103] that relate to this single-hop cognitive network setting may be summarized as follows:

• **What is the scaling law of the secondary network?** By showing that the average interference to the cognitive users remains bounded due to the finite transmission ranges D_{max} of the cognitive users and D_0 of the primary users, one can show that the lower and upper bounds to each user's average transmission rate are constant and thus the network throughput grows linearly with the number of users [103].

• **How should the network parameters be chosen to guarantee**

$$\Pr[\text{primary user's rate} \leq C_0] \leq \beta?$$

This interesting question is addressed in [45, 101] and is discussed in Section 10.4.2.

We now consider a cognitive network consisting of multiple primary and multiple cognitive users, where there is no restriction on the maximum cognitive Tx–Rx distance. We assume Tx–Rx pairs are selected randomly, as in a classical [38] standalone ad hoc network. Both types of users are ad hoc, randomly distributed according to the Poisson point processes with different densities. Here the quality of service guarantee to the primary users states that the scaling law of the primary ad hoc network does not diminish in the presence of the secondary network.

In [54] it is shown that, provided that the cognitive node density is higher than the primary node density, using multi-hop routing, *both* types of users, primary and cognitive, can achieve a throughput scaling as if the other type of users were not present. Specifically, the throughput of the m primary users scales as $\sqrt{m/\log m}$, and that of the n cognitive users as $\sqrt{n/\log n}$.

What is of particular interest in this result is that, to achieve these throughput scalings, the primary network need not change anything in its protocols; it is *oblivious* to the secondary network's presence. The cognitive users, on the other hand, rely on their higher density and a clever routing technique (in the form of *preservation regions* [54]) to avoid interfering with the primary users.

10.2.5 Interference Mitigation: Spectrum Overlay

In interference-mitigating cognitive behavior, the cognitive user transmits over the same spectrum as the primary user, but makes use of this additional cognition to mitigate (1) interference it causes to the primary receiver and (2) interference the cognitive receiver experiences from the primary transmitter. In order to mitigate interference, the cognitive nodes must have the primary system's *codebooks*. This will allow the cognitive transmitter and/or receiver to opportunistically decode the primary users' messages, which in turn may lead to gains for both the primary and secondary users, as we will see. We consider two types of interference-mitigating behavior:

1. **Opportunistic interference cancellation:** The cognitive nodes have the codebooks of the primary users. The cognitive receivers opportunistically decode the primary users' messages which they pull off of their received signal, increasing the secondary channel's transmission rate.
2. **Asymmetrically cooperating cognitive radio channels:** The cognitive nodes have the codebooks of the primary users, and the cognitive transmitter(s) has knowledge of the primary user's message. The cognitive transmitter may use this knowledge to carefully mitigate interference at the cognitive receiver as well as cooperate with the primary users in boosting its signal at its receiver.

10.2.5.1 Opportunistic Interference Cancellation

We assume the cognitive link has the same knowledge as in the interference-temperature case (b) and has some additional information about the primary link's communication: the primary user's *codebook*. Knowledge of primary codebook translates to being able to decode primary transmissions; next we suggest a scheme which exploits this extra knowledge.

In *opportunistic interference cancellation*, as first outlined in [85] the cognitive receiver opportunistically decodes the primary user's message, which it then subtracts off its received signal. This intuitively cleans up the channel for the cognitive pair's own transmission. The primary user is assumed to be oblivious to the cognitive user's operation, and so continues transmitting at power P_1 and rate R_1. When the rate of the primary user is low enough relative to the primary signal power at the cognitive receiver (or $R_1 \leq C\left(h_{12}^2 P_1\right)$) to be decoded by $\mathcal{SR}_x$, the channel $(\mathcal{PT}_x, \mathcal{ST}_x \rightarrow \mathcal{SR}_x)$ will form an information-theoretic multiple-access channel, whose capacity region is well known [17]. In this case, the cognitive receiver will first decode the primary's message, subtract it off its received signal, and proceed to decode its own. When the cognitive radio cannot decode the primary's message, the latter is treated as noise. The region (c) in Fig. 10.1 illustrates the gains opportunistic decoding may provide over the former two strategies.

10.2.5.2 Asymmetrically Cooperating Cognitive Radio Channels

We increase the cognition even further and assume the cognitive node(s) has the primary codebooks as well as the message to be transmitted by the primary sender(s). For simplicity of presentation we consider again the two-transmitter, two-receiver channel shown in Figs. 10.1 and 10.3. This additional knowledge allows for a form of *asymmetric cooperation* between the primary and cognitive transmitters. This asymmetric form of transmitter cooperation, first introduced in [22, 24], can be motivated in a cognitive setting in a number of ways. For example, if $\mathcal{ST}_x$ is geographically close to $\mathcal{PT}_x$ (relative to $\mathcal{PR}_x$), then the wireless channel $(\mathcal{PT}_x \rightarrow \mathcal{ST}_x)$ could be of much higher capacity than the channel $(\mathcal{PT}_x \rightarrow \mathcal{PR}_x)$. Thus, in a fraction of the transmission time, $\mathcal{ST}_x$ could listen to, and obtain the message transmitted by $\mathcal{PT}_x$. Other motivating scenarios may be automatic repeat request (ARQ) systems and heterogeneous sensor systems [21, 107].

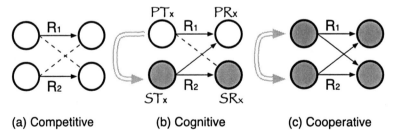

(a) Competitive   (b) Cognitive   (c) Cooperative

Fig. 10.3 Three types of behavior depending on the amount and type of side information at the secondary transmitter: (a) competitive: the secondary terminals have no additional side information, (b) cognitive: the secondary transmitter has knowledge of the primary user's message and codebook, and (c) cooperative: both transmitters know each others' messages. The *double line* denotes non-causal message knowledge

10.2.5.3 Background: Exploiting Transmitter Side Information

A key idea behind achieving high data rates in an environment where two senders share a common channel is interference cancellation or mitigation. Costa, in his famous paper "Writing on Dirty Paper" [16] applied the results of Gel'fand–Pinsker [33] to the AWGN channel, where he showed that in a channel with AWGN of power Q, input X, power constraint $E[|X|^2] \leq P$, and additive interference S of arbitrary power known non-causally to the *transmitter* but not the receiver,

$$Y = X + S + N, \quad E[|X|]^2 \leq P, \quad N \sim \mathcal{N}(0, Q)$$

the capacity is that of an interference-free channel, or

$$C = \max_{p(u|s)p(x|u,s)} I(U; Y) - I(U; S), \tag{10.1}$$

$$= \frac{1}{2} \log_2 \left(1 + \frac{P}{Q}\right). \tag{10.2}$$

This remarkable and surprising result has found its application in numerous domains including data storage [43, 66], watermarking/steganography [92], and most recently, *dirty-paper coding* has been shown to be the capacity-achieving technique in Gaussian MIMO broadcast channels [5, 105]. We now apply dirty-paper coding techniques to the Gaussian cognitive channel.

10.2.5.4 Bounds on the Capacity of Cognitive Radio Channels

Although in practice the primary message must be obtained causally, as a first step, numerous works have idealized the concept of message knowledge: whenever the cognitive node $\mathcal{ST}_x$ is able to hear and decode the message of the primary node $\mathcal{PT}_x$, it is assumed to have full a priori knowledge.[3] The one-way double arrow in Fig. 10.3 indicates that $\mathcal{ST}_x$ knows $\mathcal{PT}_x$'s message but not vice versa. This asymmetric transmitter cooperation present in the *cognitive* channel has elements in common with the *competitive* channel and the *cooperative* channels of Fig. 10.3, which may be explained as follows:

1. **Competitive behavior/channel:** The two transmitters transmit independent messages. There is no cooperation in sending the messages, and thus the two users *compete* for the channel. This is the same channel as the two-sender, two-receiver interference channel [7]. The largest to-date known general region for the interference channel is that described in [40] which has been stated more compactly in [12]. Many of the results on the cognitive channel, which contains an interference channel if the non-causal side information is ignored, use a similar rate-splitting approach to derive large rate regions [24, 56, 77].

---

[3] This assumption is often called the *genie assumption*, as these messages could have been given to the appropriate transmitters by a genie.

2. **Cognitive behavior/channel:** Asymmetric cooperation is possible between the transmitters. This asymmetric cooperation is a result of $\mathcal{ST}_x$ knowing $\mathcal{PT}_x$'s message but not vice versa.
3. **Cooperative behavior/channel:** The two transmitters know each others' messages (two-way double arrows) and can thus fully and symmetrically cooperate in their transmission. The channel pictured in Fig. 10.3(c) may be thought of as a two antenna sender, two single antenna receivers broadcast channel, where, in Gaussian MIMO channels, *dirty-paper coding* was recently shown to be capacity achieving [5, 105].

Cognitive behavior may be modeled as an interference channel with asymmetric, non-causal transmitter cooperation. This channel was first introduced and studied in [22, 24].[4] Since then, a flurry of results, including capacity results in specific scenarios of this channel have been obtained. When the interference to the primary user is weak ($h_{21} < 1$), rate region (d) has been shown to be the capacity region in Gaussian noise [58] and in related discrete memoryless channels [107]. In channels where interference at both receivers is strong both receivers may decode and cancel out the interference, or where the cognitive decoder wishes to decode both messages, capacity is also known [57, 70, 77]. However, the most general capacity region remains an open question for both the Gaussian noise and discrete memoryless channel cases.

When using an encoding strategy that properly exploits this asymmetric message knowledge at the transmitters, the region (d) in Fig. 10.1 is achievable in AWGN, and in the weak interference regime ($h_{21} < 1$ in AWGN) corresponds to the capacity region of this channel [58, 108]. The encoding strategy used assumes that both transmitters use random Gaussian codebooks. The primary transmitter continues to transmit its message of average power P_1. The secondary transmitter splits its transmit power P_2 into two portions, $P_2 = \eta P_2 + (1 - \eta) P_2$ for $0 \le \eta \le 1$. Part of its power, ηP_2, is spent in a *selfless* manner: on relaying the message of $\mathcal{PT}_x$ to $\mathcal{PR}_x$. The remainder of its power, $(1 - \eta) P_2$, is spent in a *selfish* manner on transmitting its own message using the interference-mitigating technique of *dirty-paper coding*. This strategy may be thought of as selfish, as power spent on dirty-paper coding may harm the primary receiver (and is indeed treated as noise at $\mathcal{PR}_x$). The rate region (d) may be expressed as follows [20, 58]:

$$\text{Asymmetric cooperation rate region (b)} \tag{10.3}$$

$$= \left\{ (R_1, R_2) | 0 \le R_1 \le C \left(\frac{(\sqrt{P_1} + h_{12}\sqrt{\eta P_2})^2}{h_{12}^2 (1 - \eta) P_2 + 1} \right), \tag{10.4} \right.$$

$$\left. 0 \le R_2 \le C \left((1 - \eta) P_2 \right), \ 0 \le \eta \le 1 \right\}. \tag{10.5}$$

By varying η, we can smoothly interpolate between strictly selfless behavior to strictly selfish behavior. Of particular interest from a secondary spectrum licensing

---

[4] It was first called the *cognitive radio channel* and is also known as the *interference channel with degraded message sets*.

perspective is the fact that the primary user's rate R_1 may be strictly increased with respect to all other three cases (i.e., the x-intercept is now to the right of all other three cases). That is, by having the secondary user possibly relay the primary's message in a selfless manner, the system essentially becomes a 2×1 multiple-input-single-output (MISO) system which sees all the associated capacity gains over non-cooperating transmitters or antennas. This increased gain could serve as a motivation for having the primary share its codebook and message with the secondary user.

While Fig. 10.1 shows the impact of increasing cognition (or side information at the cognitive nodes) on the achievable rate regions corresponding to protocols which make use of this side information, Fig. 10.4 shows the impact of transmitter cooperation. In this figure, the region achieved through asymmetric transmitter cooperation (*cognitive behavior*) is compared to the (1) Gaussian MIMO broadcast channel region (in which the two transmitters may cooperate, *cooperative behavior*, from [5, 105]), (2) the achievable rate region for the interference channel region obtained in [40] (the largest known to date for the Gaussian noise case, *competitive behavior*)[5], and (3) the time-sharing region where the two transmitters take turns using the channel (*interference-avoiding behavior*). We note that the framework for the Gaussian MIMO broadcast channel region may also be used to express an achievable rate region for the Gaussian asymmetrically cooperating channel [20].

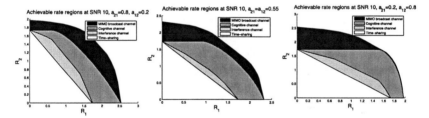

Fig. 10.4 Capacity region of the Gaussian 2×1 MIMO two-receiver broadcast channel (outer), cognitive channel (middle), achievable region of the interference channel (second smallest) and time-sharing (innermost) region for Gaussian noise powers $N_1 = N_2 = 1$, power constraints $P_1 = P_2 = 10$ at the two transmitters, and three different channel parameters h_{12}, h_{21}

While the above channel assumes non-causal message knowledge, a variety of two-phase half-duplex causal schemes have been presented in [24, 64], while a full-duplex rate region was studied in [4]. Many achievable rate regions are derived by having the primary transmitter exploit knowledge of the *exact* interference seen at the receivers (e.g., dirty-paper coding in AWGN channels). The performance of dirty-paper coding when this assumption breaks down has been studied in the context of a compound channel in [80] and in a channel in which the interference is partially known [37].

---

[5] The achievable rate region of [40] used in these figures (as the "interference channel" achievable region) assumes the same Gaussian input distribution as in [24] and is omitted for brevity.

Cognitive channels have also been explored in the context of multiple nodes and/or antennas. Extensions to channels in which both the primary and secondary networks form classical multiple-access channels have been considered in [11, 23]. Cognitive versions of the X channel [75] have been considered in [26, 53], while cognitive transmissions using multiple antennas without asymmetric transmitter cooperation have been considered in [115].

10.3 Cognitive Sensing with Side Information

Sensing is an inherent problem in a cognitive network that requires non-overlapping primary and secondary operations. Spectrum sensing has been pursued by a great number of researchers. We mention here only a specific result about the effect of side-information on cognitive sensing performance [44]. This side information can consist of spatial locations of the primary and cognitive receivers and *a priori* primary transmission probability. For sensing algorithms based on Bayesian energy detection, such side information affects the detection threshold and the resulting performance. Specifically, information on spatial locations can help stabilize the performance for a wide range of the primary activity factor. Highly skewed a priori primary transmission probability further helps improve the performance significantly.

In particular, consider a circular network with a single primary Tx–Rx pair and a single secondary Tx–Rx pair, as shown in Fig. 10.5. The primary receiver is at the center of the network, while both the primary and secondary transmitters are randomly and uniformly located within the disc. To the secondary transmitter, knowledge of the locations of the primary receiver (S_{tx}) and the secondary (its own) receiver (S_{rx}) are considered as side information.

For sensing based on Bayesian energy detection, the sensing threshold is chosen to minimize a total cost consisting of the interference caused from the secondary transmitters when the spectrum is in use and the transmission opportunity loss experienced by the secondary users not operating when the spectrum is idle. Figure 10.6[6] shows the sensing performance with various combinations of side information on the spatial locations. Comparisons with the standard *constant false alarm detector* (CFAR) [60] with $P_{FA} = 0.001$ and 0.01, without any side information, are also included. Spatial location information can improve the performance between 1.5 and 3 times, depending on the primary activity factor and the combination of information available. Figure 10.7[6] shows the performance with additional information on the primary a priori transmission probability ρ. When ρ is skewed ($\rho \neq 0.5$), then the knowledge of ρ further improves the detector performance dramatically.

While spectrum sensing is fundamental to the design of a cognitive radio network based on the interference avoidance approach, interference analysis is a fundamental

---

[6] The authors would like to thank Dr. Seung-Chul Hong for providing these figures.

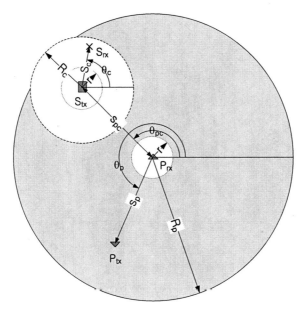

Fig. 10.5 Network configuration

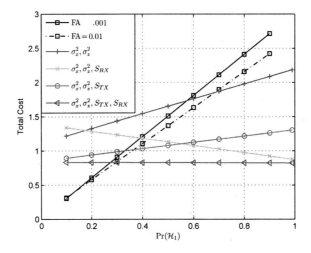

Fig. 10.6 Total cost comparisons without knowledge of ρ

part of cognitive radio design based on the interference control and mitigation paradigms. The next section deals with interference analysis in a cognitive radio network.

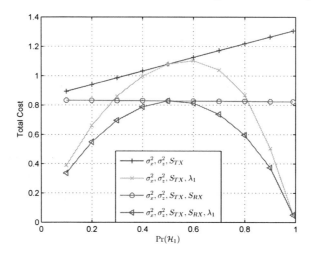

Fig. 10.7 Total cost comparisons with knowledge of ρ

10.4 Interference Analysis

Interference analysis has been studied by a number of authors (see, for example, [14, 31, 63, 106, 109]). The results can be used to design various network parameters to guarantee a certain performance to the primary users. Our objective here is to provide only an example of this interference analysis and its application in two different network settings: a network with beacons and a network with exclusive regions for the primary users.

Consider an extended, circular network in which the cognitive users are uniformly distributed with constant density λ. The network radius D increases with the number of cognitive users n. The interference generated by these cognitive users depends on their locations, which are random, and on the random channel fading. This leads to random interference. The average interference power to the worst-case primary users, which may be shown to be at the center of the circular network, can be computed as follows [102]:

$$E[I_n] = \frac{2\pi \lambda P}{(\alpha - 2)} \left(\frac{1}{\varepsilon^{\alpha-2}} - \frac{1}{D^{\alpha-2}} \right), \tag{10.6}$$

where α is the path-loss exponent, ε is a receiver-protected radius, and P is the cognitive transmit power. Provided that the path-loss exponent $\alpha > 2$, then the average interference is bounded, even with an infinite number of cognitive users ($n \to \infty$ or $D \to \infty$).

The average interference can be used to either limit the transmit power of the cognitive users or to design certain network parameters to limit the impact of interference on the primary users. Next, we discuss two examples of how the interference analysis can be applied to design network parameters.

10.4.1 A Network with Beacons

In a network with beacons, the primary users transmit a beacon before each transmission. This beacon is received by all users in the network. The cognitive users, upon detecting this beacon, will abstain from transmitting for the next duration. Such a mechanism is designed to avoid interference from the cognitive users to the primary users. In practice, however, because of channel fading, the cognitive users may sometime misdetect the beacon. They can then transmit concurrently with the primary users, creating interference. This interference depends on certain parameters, such as the beacon detection threshold, the distance between the primary transmitter and receiver, and the receiver-protected radius. By designing network parameters, such as the beacon detection threshold, one can control this interference to limit its impact on the primary users' performance.

Using a simple power detection threshold, the missing beacon probability can be shown to be

$$q = 1 - e^{-\gamma d^\alpha}, \tag{10.7}$$

where again α is the path loss exponent, γ denotes the ratio between power threshold and beacon transmit power (or the beacon detection threshold), and d is the distance from the cognitive user to the primary transmitter (the beacon transmitter). Given a certain activity factor of the cognitive users when missing the beacon, the generated interference can then be computed analytically [102]. Bounds on the interference can then help in the design of network parameters.

For example, the interference bound versus the beacon detection threshold can be plotted as in Fig. 10.8. This graph provides a specific rate at which the interference increases as the beacon threshold increases. The rate depends on other parameters such as α, D, ϵ, and P. The case when the cognitive transmitters are always transmitting (a beacon-less system) corresponds to $\gamma = \infty$.

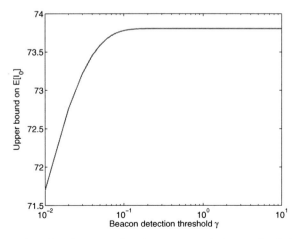

Fig. 10.8 An upper bound on the average interference versus the beacon threshold level

10.4.2 A Network with Primary Exclusive Regions

Another way of limiting the impact of cognitive users on primary users is to impose a certain distance from the primary user, within which the cognitive users cannot transmit. This configuration appears suitable to a broadcast network in which there is one primary transmitter communicating with multiple primary receivers. Examples include the TV network or the downlink in the cellular network. In such networks, primary receivers may be passive devices and are therefore hard to be detected by cognitive users, in contrast to the primary transmitter whose location can be easily inferred. Thus it may be reasonable to place an exclusive radius D_0 around the primary transmitter, within which no cognitive transmissions are allowed. Such a primary-exclusive region (PER) has been proposed for the upcoming spectrum sharing of the TV band [45, 76]. The cognitive transmitters are randomly and uniformly distributed outside the PER, within a network radius D from the primary transmitter. As the number of cognitive users increases, D increases. The network model is shown in Fig. 10.9.[7]

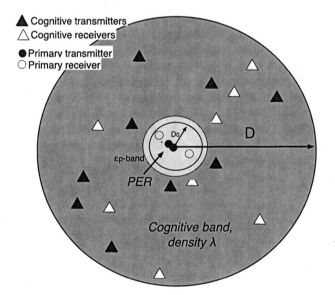

Fig. 10.9 A cognitive network consists of a single primary transmitter at the center of a *primary exclusive region* (PER) with radius D_0, which contains its intended receiver. Surrounding the PER is a protected band of width $\varepsilon > 0$. Outside the PER and the protected bands, n cognitive transmitters are distributed randomly and uniformly with density λ

Of interest is how to design the exclusive radius D_0, given other network parameters, to guarantee an outage performance to the primary users. This outage performance guarantees a certain data rate for a certain percentage of time for all primary

---

[7] We would like to thank Dr. Seung-Chul Hong for providing us with this figure.

receivers within the PER. The 'worst case' receiver is at the edge of the PER in a network with an infinite number of cognitive users ($D \to \infty$).

Using the interference power analysis (10.6), coupled with the outage constraint, an explicit relation between D_0 and other parameters, including the protected radius ε, the transmit power of the primary user P_0, and cognitive users P, can be established [101]. For example, the relation between D_0 and the primary transmitter power P_0 is shown in Fig. 10.10. The fourth-order increase in power here is in-line with the path-loss exponent $\alpha = 4$. The figure shows that a small increase in the receiver-protected radius ϵ can lead to a large reduction in the required primary transmit power P_0 to reach a receiver at a given radius D_0 while satisfying the given outage constraint.

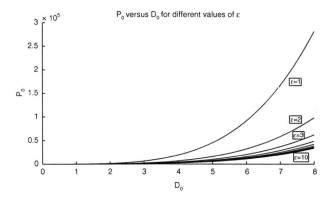

Fig. 10.10 Relationship between the primary transmit power P_0 and the exclusive region radius D_0

10.5 Practical Cognitive Network Engineering: Interference Control Approach

Interference control-based spectrum sharing allows simultaneous transmissions of primary and secondary users given the total interference constraints at primary receivers. These interference power constraints, in essence, require a sophisticated power control scheme for secondary transmitters. In order to meet interference constraints and QoS requirements for secondary users, channel gains among secondary users and from secondary users to primary receivers is usually required for proper power allocation. While collecting the channel gain information among secondary users are possible in many cases, obtaining channel gains from secondary transmitters to primary receivers is usually not trivial because primary networks may not assist secondary networks in measuring/estimating the channel gains. Hence, although in theory the interference control approach can be used in both centralized and distributed wireless networks, this spectrum access paradigm would be more applicable for networks with infrastructure such as cellular networks where channel state information can be readily obtained.

One typical example where the interference control approach can be employed for cellular-type networks is shown in Fig. 10.11. In this example, base stations (BSs) in the secondary network transmit in the downlink direction exploiting the licensed frequency bands used by primary users in the uplink direction. For this particular network setting, channel gains from secondary transmitters (i.e., secondary BS) to primary receivers (i.e., primary BS) can be estimated by the secondary networks using pilot signals transmitted from the primary BSs. Similar network setting was considered in [62, 68] where secondary users (i.e., cognitive radios) use an ad hoc mode for communication. In this section, we describe a typical spectrum-sharing model with QoS and fairness constraints for secondary users and interference constraints for primary users. For ease of exposition, we refer to the network setting in Fig. 10.11 in the model description where single-hop traffic flows are considered. We will discuss the scenario with multi-hop traffic flows later on.

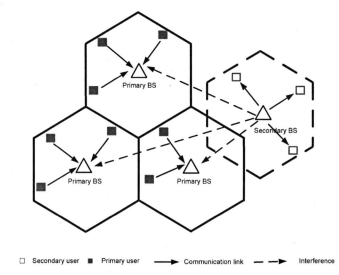

Fig. 10.11 Typical example of spectrum access using interference control paradigm

10.5.1 Single-Antenna Case

Engineering of wireless networks is in general much more challenging than engineering of the wireline counterpart. This is due to inherent transmission characteristics of a wireless channel with fading and shadowing. As a result, users who are assigned the same quantity of radio resources would achieve different throughput performances. Therefore, wireless network engineering should maintain certain fairness among different users such that users with unfavorable channel conditions still have satisfactory performance. In addition, most wireless applications have certain QoS requirements which can be usually described by different performance measures such as throughput, delay, delay jitter. These QoS requirements usually correspond to certain minimum transmission rates or signal to noise ratio for wireless users.

The problem of optimal spectrum sharing among secondary users can be formulated as an optimization problem with a suitable objective function and a set of constraints which capture user fairness, QoS constraints for secondary users, and interference constraints for primary users. Suppose there are n secondary users and m primary receivers. For the sake of brevity, the term secondary user here refers to a pair of secondary users who communicate with each other in an ad hoc mode or a secondary user communicates with the BS in a cellular setting.

Let R_i denote the achievable rate for secondary user i which depends on the amount of allocated power, bandwidth, noise, and interference it receives from other primary and secondary users. To engineer the cognitive radio network, we would choose a suitable objective function for the underlying spectrum-sharing optimization problem which could balance good overall network performance as well as fairness for the secondary users. In [82], one such objective function, which is parameterized by a parameter κ, was proposed as follows:

$$U(R_1, R_2, \ldots, R_n) = \sum_{i=1}^{n} f_\kappa(R_i), \qquad (10.8)$$

where $f_\kappa(x)$ is the utility function for one user which can be written as

$$f_\kappa(x) = \begin{cases} \ln(x) & \text{if } \kappa = 1 \\ \frac{x^{1-\kappa}}{1-\kappa} & \text{otherwise} \end{cases}. \qquad (10.9)$$

This general objective function can achieve different types of fairness depending on parameter κ. Specifically, for $\kappa = 0$ the total throughput is maximized, while $\kappa = 1$ achieves the proportional fairness for different users [61], $\kappa = 2$ achieves harmonic mean fairness, and $\kappa \to \infty$ provides max–min fairness. In general, the higher the value of κ the fairer the solution of the underlying optimization problem. Let h_{ij} denote the channel gain from the transmitter of secondary user j to primary receiver i, P_i denote transmission power of secondary transmitter i, and I_j denote the maximum tolerable interference level at primary receiver j. Suppose each secondary user i has a minimum QoS requirement described in terms of minimum rate B_i. The spectrum-sharing problem for secondary users under QoS and interference constraints can be formulated as follows:

$$\text{maximize} \quad U(R_1, R_2, \ldots, R_n)$$
$$\text{subject to}$$
$$R_i \geq B_i, \quad i = 1, 2, \ldots, n$$

$$\mu_j = \sum_{i=1}^{n} h_{j,i} P_i \le I_j, \quad j = 1, 2, \ldots, m,$$

where μ_j is the total interference created by secondary users at primary receiver j.

The optimization problem formulated above can be solved efficiently in cases where it is convex. For scenarios where the formulated problem is non-convex, a fast and suboptimal algorithm may be required. It is noted that the formulated optimization problem may not be feasible when its constraints are too stringent and/or the network load is too high. If this is the case, an admission control mechanism needs to be invoked to limit the number of admitted secondary users. Then, power allocation for the set of admitted secondary users can be performed. Using this framework, in [68], a solution approach was proposed for the joint admission control and power allocation problem for secondary users assuming code-division multiple access (CDMA) technology at the physical layer.

In order to obtain power allocation solutions for the aforementioned spectrum-sharing problems, channel gains among secondary users and from secondary transmitters to primary receivers should be estimated frequently. Unfortunately, instantaneous channel gains from the secondary transmitters to primary receivers may be not readily obtained. The secondary network, however, can estimate the corresponding channel gains in the reverse direction by exploiting pilot signals transmitted by primary receivers (i.e., primary BSs). Due to the reciprocal characteristic of the wireless channel, the secondary network can obtain required mean channel gains averaged over short-term fading. Let $\bar{h}_{ij}$ be the average channel gain from secondary transmitter I to primary receiver j which is estimated by exploiting the pilot signal. The interference constraints at primary receiver j can be written as

$$\bar{\mu}_j = \sum_{i=1}^{n} \bar{h}_{j,i} P_i \le \theta I_j, \quad j = 1, 2, \ldots, m, \tag{10.10}$$

where $\theta < 1$ is a conservative factor which should be chosen to make the interference constraints be violated with a small probability. Mathematically, we should have

$$\Pr[\mu_j > I_j \,|\, \bar{\mu}_j \le \theta I_j] \le \Gamma_0, \quad j = 1, 2, \ldots, m, \tag{10.11}$$

where Γ_0 is the desired interference constraint violation probability. Given information about fading channel statistics, the interference constraint violation probability written in (10.11) can be calculated. Hence, the power allocation solution can be obtained by determining the factor θ and the corresponding transmission powers for secondary transmitters [62]. In [71], a related spectrum-sharing problem was solved which maximizes throughput of secondary networks while sufficiently protecting primary users by maintaining a sufficiently high probability of detection. This paper, however, did not consider fairness among secondary users but tried to quantify the throughput and sensing tradeoff by finding an optimal sensing time.

10.5.2 Multiple Antenna Case

In scenarios where multiple antennas are available at secondary users and/or primary receivers, more care should be taken in performing power allocation for secondary users [71, 112, 115]. In general, the availability of multiple antennas at secondary transmitters and/or receivers provides potential multiplexing and diversity gains which would enhance performance of a cognitive radio network. In addition, if each primary receiver has multiple antennas, there are two different ways to impose interference constraints, namely one total interference power constraint over all receive antennas or a set of interference power constraints applied to each individual receive antenna. Also, if each primary transmitter has multiple antennas, power allocation among transmit antennas under total transmit power constraints and interference power constraints at primary receivers should be jointly considered.

In [112], joint beamforming and power allocation for a single-input multiple-output (SIMO) MAC of a cognitive radio network with multiple secondary transmitters and primary receivers each with one antenna was investigated. In this paper, solutions for sum-rate maximization problems with or without minimum SINR requirements for secondary users were proposed. In [115], power allocation for capacity maximization of a single pair of secondary users using MIMO was considered. Both the cases with one and multiple channels were investigated. Under this MIMO setting, optimal transmit power over transmit antennas was performed. It was shown that by exploiting multiple antennas, secondary users can balance between spatial multiplexing for their transmission while limiting interference to primary receivers. Note that these initial works on cognitive radio networks using multiple antennas consider either a MIMO setting for a single pair of secondary users or SIMO setting for multiple secondary users. Because sum-rate maximization usually favors users in good conditions, trade-off between throughput and fairness should be considered by maximizing a suitable utility function such as that in (10.8). Solving a spectrum-sharing problem with fairness consideration for secondary users and interference constraints for primary receivers in a multi-user MIMO cognitive radio network is still an open problem.

10.6 Practical Cognitive Network Engineering: Interference Avoidance Approach

In an interference avoidance approach, secondary users need to sense a frequency band of interest and transmit only if primary users are not detected on the chosen band. The interference avoidance approach is, therefore, more conservative than the interference control approach. However, no strict power control is required for this spectrum access paradigm. In this section, we discuss both scheduling and random access-based medium access control (MAC) techniques for this spectrum access approach. Both single-hop and multi-hop transmission scenarios are considered. Since the spectrum of interest to a secondary network is usually very broad, fast wideband spectrum sensing is very challenging. To solve this problem, the spectrum of interest can be divided into multiple narrow frequency bands where spectrum sensing can be done by cheap radio devices. For this reason, we only describe multi-channel MAC issues in the following.

10.6.1 Single-Hop Case

10.6.1.1 Scheduling-Based MAC

The scheduling-based MAC for single-hop flows would be mostly applicable for a point-to-multi-point network. This would be the case when multiple secondary users communicate with a base station (BS) or an access point (AP) using available licensed frequency bands [72, 73]. Given spectrum sensing results, a scheduler at a BS or an AP has full information about availability of all channels to make scheduling decisions. The scheduler can also opportunistically exploit fluctuations in channel quality of available channels due to fading to enhance throughput performance [72, 100]. As in a traditional scheduling problem, fairness among users should be taken into account in designing a scheduling algorithm. The key difference in a cognitive scheduling problem is that some channels may not be available for secondary users at some particular time. Therefore, statistical information regarding channel availability should be considered to maintain good long-term throughput and fairness performance for secondary users. Note that opportunistic scheduling considering multiple channels has been investigated in some recent works (e.g., in [65, 73]).

10.6.1.2 Random Access-Based MAC

A random access-based MAC protocol is needed when there is no network controller to coordinate spectrum allocation for multiple users. A typical application for the random access-based MAC is in ad hoc networks where a node can establish data connections with one or several neighboring nodes. In this case, a MAC protocol should perform the following functions:

- **Channel contention and reservation**: Each user with data to transmit needs to choose one or several available channels. The chosen channels should be available at both transmitter and receiver sides to avoid collision with primary users. Channel reservations should be informed to other neighboring users to avoid possible transmission collisions among secondary users.
- **Spectrum sensing**: Due to the presence of primary users, some channels may not be available for secondary users. Therefore, secondary users should sense their chosen channels to avoid collision with primary users.

In the following, we present some key design aspects and engineering approaches for a random access-based cognitive MAC protocol in single- and multi-user scenarios.

1. **Optimal MAC design for a single-cognitive-user scenario**: If there is a single pair of secondary users communicating with each other, there is no need to perform channel reservation and contention resolution. The key design problem for this setting is to exploit spectrum holes in all channels to optimize the secondary network performance. It is obvious that if the secondary user can sense all the channels quickly then it would simply find available channels and transmit data using these channels. In practice, sensing time is usually non-negligible and a user can only sense one or a small number of channels at once. Therefore, an optimal sensing and access strategy plays an important role in obtaining good network performance [8–10]. Such an optimal sensing and access strategy depends strongly on statistical properties of channel availability. In the following, we describe design approaches under two different assumptions about channel availability.

   - *Markovian assumption*: When data transmission of primary users shows correlation, availability of the channels can be modeled as a Markov chain [32, 117]. If the availability of one channel is independent of other channels, a two-state Markov chain can be used to model evolution between idle and busy states (i.e., available and not available for secondary users' transmission, respectively) of each channel.

     Figure 10.12 shows a transition diagram of a two-state Markov chain for one particular channel. Suppose there are L channels and time is divided into equal-sized time slots. Also, assume that a secondary user can sense L_1 channels at the beginning of each time slot and transmit data on available channels in the remaining interval of a time slot based on sensing outcomes. In the following, we will discuss the typical optimal spectrum and access problem where a secondary user wishes to find an optimal set of channels to sense in each time slot to achieve maximum long-term throughput performance.

     Specifically, the secondary user makes the decision in choosing the set of channels to sense in one particular time slot based on its decisions and sensing

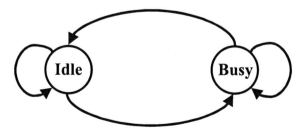

Fig. 10.12 Two-state Markov channel describing the availability of each channel

outcomes in all previous time slots. In [117], it has been shown that this
problem can be solved using the theory of partially observable markov decision process (POMDP) [89]. This is because by sensing only a subset of
channels in each time slot, the secondary user can obtain only partial information about availability of all the channels. According to the POMDP theory,
knowledge of the system state can be summarized in a belief vector [89].

Let $V = 2^L$ denote the number of system states each of which represents
idle/busy status on all L channels. Then, the belief vector can be denoted as
$\Omega(t) = [\omega_1(t), \omega_2(t), \ldots, \omega_V(t)]$ where $\omega_i(t)$ is the conditional probability
(given the decision and observation history) that the network is in state i at
the beginning of time slot t. In addition, the belief vector is the sufficient
statistic for the optimal sensing policy. In [89], it was shown that the optimal
solution of the POMDP problem can be found using a linear programming
approach. Given the optimal solution, the secondary user can find a set of
channels to sense in each time slot. It then updates the belief vector based on
the sensing outcomes which are used to find a solution for the next time slot.
Although the optimal solution for this opportunistic spectrum access problem
can be calculated, the computational complexity grows exponentially with
the number of channels. Therefore, a good and suboptimal spectrum sensing
policy is usually preferred [117].

- *Independence assumption*: In this case, the availability of each channel is assumed to be independent of time. Assume that each channel is either available
  or busy in each time slot and its transmission rate is chosen from a finite set
  of rates. Let T_s be the time slot interval, T_m be sensing and channel probing
  time. Here, sensing is used to verify availability of a particular channel and
  probing is used to find a current feasible rate on a channel. If a secondary user
  senses k channels, the normalized remaining time for data transmission is

$$c_k = 1 - k\frac{T_m}{T_s}. \tag{10.12}$$

We are interested in the optimal spectrum access problem where the secondary
user wishes to maximize its total transmission rate by adopting an optimal
sensing/probing strategy. Here, the more channels the secondary user senses,
the more likely it finds available channels with a cost of reducing the data
transmission time. Let p_i denote the probability that channel i is available for
the secondary user. If the secondary user knows p_i, it would sense channels
in the order of decreasing p_i. Otherwise, it can simply sense channels in a
random order. Without loss of generality, we number the channels in the order they are sensed by the secondary user. Assume that there are K possible
transmission rates on any channel each of which corresponds to one particular
modulation and coding scheme. Also, assume that the probability that rate k
is chosen on any channel is s_k. We further assume that the secondary user has
to make a decision after each sensing/probing regarding transmitting on the
current channel (if it is available) or continue sensing/probing other channels.
The operation of this optimal spectrum sensing/probing problem is shown in

Fig. 10.13. Note that the problem discussed here generalizes the opportunistic multi-band access problem in [86] to the cognitive radio context.

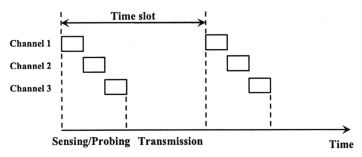

Fig. 10.13 Timing diagram for spectrum sensing/probing and access

Let r_k be the transmission rate of channel k, η_k be the achieved throughput after sensing/probing channel k, and Λ_k be the average throughput accumulated from the kth sensing/probing. Using optimal stopping theory [13], we have

$$\eta_k = \begin{cases} c_k \varphi_k r_k & \text{if } c_k \varphi_k r_k > \Lambda_{k+1} \\ \Lambda_{k+1} & \text{otherwise} \end{cases}, \tag{10.13}$$

where $\varphi_k = 1, 0$ represents the event that channel k is available or busy, respectively. Equation (10.13) can be interpreted as follows. If the obtained throughput due to the kth sensing/probing is larger than the expected throughput that would be achieved if the secondary user keeps sensing/probing further channels (i.e., $c_k \varphi_k r_k > \Lambda_{k+1}$), the secondary user would stop sensing/probing and transmit on the current channel. Note that the optimal sensing strategy is completely determined by Λ_k. We have

$$\Lambda_L = c_L \mathbf{E}\left[\varphi_L r_L\right] = c_L p_L \mathbf{E}\left[r_L\right] = c_L p_L \sum_{l=1}^{K} s_l r_l. \tag{10.14}$$

Other values of Λ_k ($k < L$) can be calculated using backward induction. Specifically, let us define $\Psi_k = \{l : c_k r_l > \Lambda_{k+1}\}$ and let $\bar{\Psi}_k$ be the complement of Ψ_k. Then, we have

$$\Lambda_k = c_k p_k \sum_{l \in \Psi_k} s_l r_l + \Lambda_{k+1} \left[(1 - p_k) + p_k \sum_{l \in \bar{\Psi}_k} s_l \right]. \tag{10.15}$$

The aforementioned optimal sensing/probing strategies are applied to scenarios with a single cognitive user. However, they can be used to develop a multi-user MAC protocol by taking a winner-take-all approach which can be described as follows. Secondary users who are currently backlogged perform contention on a control channel. The secondary user that wins the contention employs an optimal sensing/probing strategy to find good available channels for data transmission as discussed above. These contention and access operations are repeated in each fixed-size time slot. This method was employed in [55, 94] to develop multi-channel MAC protocols for cognitive radio networks. This design approach, however, has several limitations. On the one hand, for the discussed optimal control strategies a winning secondary user explores spectrum opportunities only in a subset of channels to limit the sensing/probing overhead. Therefore, spectrum holes in unexplored channels are wasted. On the other hand, a secondary user may find several available channels among explored channels as in the POMDP-based control strategy; however, its queue backlog may be smaller than the transmission capacity offered by these available channels. As a result of this, some valuable radio resources are wasted because user backlogs are not taken into consideration. Therefore, it is desirable that a cognitive MAC protocol should exploit transmission opportunities on all channels and avoid over-allocating capacity for secondary users considering users' queue backlogs.

2. **MAC protocol design for a multiple-cognitive-user scenario**: Design of a multi-channel MAC protocol is challenging due to the following reasons. First, as in a single-channel CSMA-based MAC protocol, the well-known hidden and exposed terminal problems still exist in a corresponding multi-channel MAC protocol. It is known that employment of RTS/CTS could not completely solve these problems. Although a dual busy tone approach, which employs two separate tones transmitted by a transmitter and a receiver in two narrow bands to protect RTS transmission and data reception, could remove the hidden terminal problem, this approach requires extra bandwidth and one more transceiver per user [41]. Second, different approaches for channel contention and reservations should be adopted depending on radio capabilities of cognitive devices. Therefore, there would be no universal MAC protocol that works well in all different scenarios. In general, users are expected to hear channel reservations made by other users to prevent possible collisions. Therefore, if each secondary user has only one transceiver, a user transmitting on one particular channel may not hear channel reservation negotiated on other channels. This problem is referred as a multi-channel hidden terminal problem [90]. Third, it is necessary to balance traffic load on different channels to reduce excessive contention overhead. Finally, a MAC protocol should provide fairness among different users.

Design of a multi-channel MAC protocol in general and for a cognitive network in particular strongly depends on radio capabilities of wireless/cognitive users, i.e., the number of radios each user has [81]. In the cognitive radio context, it also depends on dynamics of primary users and evacuation time requirements in case primary users return to a previously available channel. These aspects determine how fast secondary users can update a "spectrum map" and how often spectrum sensing/evacuation should be performed. There are some recent works on MAC protocol design for multi-channel and/or multi-user cognitive radio networks [15, 39, 46, 49, 55, 67, 74, 93, 94]. In the following, we discuss some important design principles and point out open research issues.

In general, there are two popular approaches to designing a multi-channel MAC protocol, namely common control channel and channel-hopping-based approaches [81]. In a common control channel approach, one chosen channel is used to exchange control information which determines data channels for contending users. This design approach usually works well under low traffic load but may have degraded performance under high traffic load due to congestion of the common control channel. In the channel-hopping-based approach, users hop through all channels by following a common or different hopping patterns. Two users who wish to communicate with each other must meet on one particular channel to perform channel reservations for data transmission. This design approach can resolve the congestion problem of the common control channel approach; however, its implementation is more complicated. In the following, we describe typical designs for scenarios where each cognitive user has two or one transceiver.

- *Each cognitive user has two transceivers*: We describe a typical MAC protocol design based on the common control channel approach for this setting here. For this design, each cognitive user employs one transceiver for channel contention and reservation and employs the other transceiver to transmit data on a chosen channel. Specifically, the control transceiver of any user is always tuned to a chosen control channel to transmit control messages and listen to channel reservations made by other users. With the dedicated control transceiver, each cognitive user always has up-to-date information about traffic load on each channel to make its own channel reservation. In addition to making channel reservations, each cognitive user should perform spectrum sensing frequently to have an up-to-date spectrum map. To avoid confusion between primary and secondary transmissions, spectrum sensing can be performed in pre-determined quiet periods during which secondary users shut off their transmissions to perform spectrum sensing.

  If the spectrum map changes slowly, cognitive users would have correct information about spectrum opportunities on all channels. When a cognitive user wants to transmit data to its neighbors, it transmits a CRTS message which contains a list of preferred channels to its intended receiver. The list of preferred channels consists of available channels learned from spectrum sensing with low secondary traffic load. The receiver upon receiving CRTS chooses one "best" channel in the received channel list and sends the chosen channel in a CCTS message to the transmitter. The transmitter upon receiving CCTS switches to the chosen channel for data transmission.

  Here, two different channel negotiation strategies can be adopted in designing a cognitive MAC protocol. In the first strategy, each available channel is only allocated for a single pair of cognitive users. Therefore, after a particular available channel is chosen by a pair of cognitive users, other cognitive users in their neighborhood should remove this channel from their available channel lists. A pair of secondary users should release their chosen channel after successfully transmitting a packet, i.e., they have to perform new channel negotiation/reservation through exchanging CRTS/CCTS messages before transmitting another packet. This channel negotiation strategy, however, creates a large amount of overhead due to CRTS/CCTS messages when the number of cognitive users is large. In the second channel negotiation strategy, multiple pairs of secondary users can choose the same channel. Hence, after exchanging CRTS/CCTS messages, cognitive users need to perform contention with other cognitive users choosing the same channel. Operation of this MAC protocol is illustrated in Fig. 10.14. Users reside on their chosen channels until they detect the presence of primary users or a pre-determined period T_{max} has been expired. T_{max} can be chosen to be equal to the required channel evacuation time (e.g., this value is 2s in 802.22 standard). This design could alleviate congestion on the control channel and reduce overhead. This is due to the reason that the traffic load on each channel is much lower than the total traffic load of all cognitive users, and cognitive users contend on the control channel less frequently.

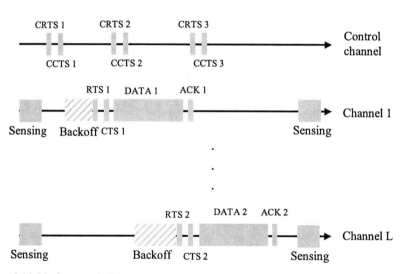

Fig. 10.14 MAC protocol with two transceivers per cognitive user

- *Each cognitive user has one transceiver*: In this case, the transceiver is used to exchange control information as well as transmit/receive data. To resolve the multi-channel hidden terminal problem mentioned above, a synchronous MAC protocol can be employed as proposed in [90]. Specifically, time is divided into periodic beacon intervals as illustrated in Fig. 10.15. Periodic beacon transmissions are used to synchronize all users. Cognitive users choose their channels by exchanging CRTS/CCTS messages during a channel reservation phase. If multiple cognitive users are allowed to choose the same channel, contention on each channel is resolved through exchanging RTS/CTS messages by the corresponding cognitive users. This design can potentially improve channel utilization when the beacon interval is large and several packets from different users can be transmitted in a data transmission phase. Moreover, sensing should be performed to find available channels by each cognitive user. In Fig. 10.15, sensing is performed at the beginning of each beacon interval based on which cognitive users reserve channels for a data transmission phase. In the case where a beacon interval is longer than a channel evacuation time, one or more sensing periods should be placed in the data transmission phase to protect primary users.

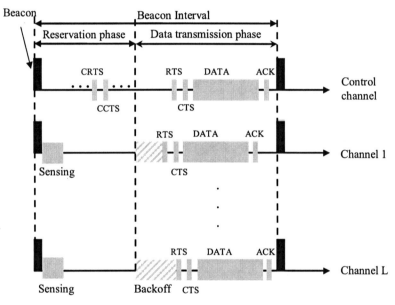

Fig. 10.15 MAC protocol with one transceiver per cognitive user

In the above MAC protocol, the control channel may be congested when the number of cognitive users is large. In addition, transmission time on all channels other than the control channel is wasted during the channel reservation phase. These problems can be resolved by channel-hopping-based MAC protocols [2, 81] where cognitive users hop through the channels by following a common or different hopping patterns. Cognitive users who want to communicate with each other wait until their partners hop to the same channel to exchange control information. For the case where users following different hopping sequences, each user learns hopping patterns of their neighbors by listening to corresponding broadcast seeds. When a user wants to communicate with its neighbor, it follows the intended neighbor's hopping pattern to exchange control information and negotiates data channels. In addition, sensing can be performed during pre-determined quiet periods to detect available channels.

Although some recent works have compared different multi-channel MAC protocols based on simple analysis and simulation [81], an accurate analytical model to quantify performance of multi-channel MAC protocols in wireless networks in general and in cognitive radio networks in particular is still an open problem. In addition, optimal control design for spectrum sensing and access was only done for a single-user scenario [8–10]. Developing an optimal MAC protocol for the multi-user scenario is an open and challenging problem. This is because throughput analysis needs to be performed for optimal design of a MAC protocol. Unfortunately, this is a complicated problem by itself. In fact, calculating throughput performance usually requires tracking detailed protocol evolutions which capture complicated relationships among many protocol parameters. Given the fact that a closed-form equation to calculate network throughput is usually difficult to obtain, finding an optimal design for channel sensing and negotiation operations of a MAC protocol is a very difficult task.

10.6.2 Multi-hop Case

In a multi-hop setting, cognitive users establish multi-hop transmissions with its peers where traffic is transmitted from sources to destinations through multiple communication links. Engineering multi-hop cognitive networks requires more care due to possible existence of different sets of primary users with different channel activity. Either an interference control or an interference avoidance approach can be employed to protect active primary users in the multi-hop setting. Employment of the interference control approach is, however, challenging because network-wide estimation of channel gains from secondary users to primary receivers may not be easy. Using the interference avoidance approach, if primary users are present on some channels, secondary users are simply prohibited to use these busy channels for their transmissions [87]. Detection of primary users using a certain spectrum sensing technique can be performed by each cognitive users to construct its local spectrum map.

In general, multi-hop communications can be established by scheduling or random access-based transmissions. It was known that CSMA/CA-based MAC protocols are quite inefficient for a multi-hop wireless network [69]. This is due to the hidden terminal problem and suboptimality of the backoff mechanism. Although implementation of a dual tone approach could remove the hidden terminal problem, more control bands and transceivers per user are required [41]. The scheduling-based approach could potentially achieve better throughput performance; however, a scheduling-based MAC usually requires centralized implementation where information about source-destination pairs, network topology, link conflict relationship, or channel gains needs to be gathered at a central control point to calculate optimal network configuration [1, 6, 59, 87]. A scheduling-based MAC is, therefore, more suitable to stationary wireless networks (e.g., wireless mesh networks) with slowly changing source-destination demands (e.g., in networks with aggregated traffic flows [59]).

The scheduling-based MAC design strongly depends on an interference model. In particular, the interference model determines sets of wireless links which can be activated simultaneously. There are two popular interference models, namely protocol model and physical model [38]. In the protocol model, interference relationship among wireless links is binary where wireless links outside interference range of one another can be transmitted simultaneously. Therefore, any two wireless links can be either conflicting with each other or can be activated simultaneously. This interference relationship is usually captured in a conflict graph to find an optimal network configuration [1] (e.g., optimal rate control, routing and scheduling for wireless networks). The physical model determines a set of active links by a corresponding set of constraints on minimum signal to noise plus interference ratio (SINR). In essence, a minimum SINR requirement for each wireless link is imposed to guarantee the desired bit error rate performance. Under the physical model, transmit power and channel gains among wireless links need to be gathered to determine sets of active links and corresponding optimal network configuration [59]. The physical model is more accurate but it is also more complicated than the protocol model. In general, the scheduling sub-problem in a cross-layer design problem is a bottleneck where sets of simultaneously activated wireless links must be determined such that desired end-to-end performance for all multi-hop flows can be achieved.

10.7 Conclusions

Results on the achievable rate of a cognitive radio link have been summarized for three different types of cognitive behavior, namely interference avoidance, interference control, and interference mitigation. In a spectrum underlay scenario with the interference-controlled behavior of the cognitive radio nodes, scaling laws of both single-hop and multi-hop cognitive networks have been described. Bounds on the capacity of a cognitive radio channel have been described for two types of interference-mitigating cognitive behavior, namely opportunistic interference cancellation (or decoding) and asymmetrical cooperation with primary transmitter. The general conclusion is, the higher the level of cognition (i.e., side information) at the cognitive radio nodes, the higher the maximum achievable rate for the cognitive radio channel. A similar conclusion holds for the cognitive sensing performance—specifically, information about spatial location of the primary and secondary receivers and primary user activity can improve the sensing performance significantly. To limit the impact of interference on primary users, network parameters have to be designed based on interference analysis. Examples of interference analysis have been provided for a beacon-enabled network and a network with primary exclusive regions.

Practical issues and potential approaches in design and engineering of channel access methods in a cognitive radio network have been described. With the interference control approach, a spectrum-sharing method has to ensure the rate (or SINR) and fairness constraints for secondary users as well as interference constraints for primary users. With the interference avoidance approach, spectrum sensing has to be performed efficiently so that the utilization of spectrum holes can be maximized and also the QoS requirements for the secondary users are met. Economic aspects of spectrum sharing (e.g., pricing), which have not been addressed in this chapter, will also need to be considered for practical design of cognitive radio systems [47, 48, 83, 95, 104]. Design and engineering of multi-user (single hop and multi-hop, single antenna and multiple antenna) cognitive radio networks is still in its infancy which deserves more research investigation.

References

1. M. Alicherry, R. Bhatia, and L. (E). Li, "Joint channel assignment and routing for throughput optimization in multi-radio wireless mesh networks," in *Proc. ACM Mobicom'05.*
2. P. Bahl, A. Chandra, and J. Dunagan, "SSCH: Slotted seeded channel hopping for capacity improvement in IEEE 802.11 ad hoc wireless networks," in *Proc. ACM Mobicom'04.*
3. M. M. Buddhikot, "Understanding dynamic spectrum access: Models, taxonomy and challenges," in *Proc. IEEE International Symposium on New Frontiers in Dynamic Spectrum Access Networks (DySPAN)*, April 2007, pp. 649–663.
4. S. H. Seyedmehdi, Y. Xin, and Y. Lian, "An achievable rate region for the causal cognitive radio," in *Proc. Allerton Conf. Commun. Control and Comp.*, Monticello, IL, Sept. 2007.
5. G. Caire and S. Shamai, "On the achievable throughput of a multi-antenna gaussian broadcast channel," *IEEE Trans. Inf. Theory*, vol. 49, no. 7, pp. 1691–1705, July 2003.
6. M. Cao, X. Wang, S. -J. Kim, and M. Madihian, "Multi-hop wireless backhaul networks: A cross-layer design paradigm," *IEEE J. Sel. Areas Commun.*, vol. 25, no. 4, pp. 738–748, May 2007.
7. A. Carleial, "Interference channels," *IEEE Trans. Inf. Theory*, vol. IT-24, no. 1, pp. 60–70, Jan. 1978.
8. N. B. Chang and M. Liu, "Optimal channel probing and transmission scheduling for opportunistic spectrum access," in *Proc. ACM Mobicom'07*, Sept. 2007.
9. N. B. Chang and M. Liu, "Competitive analysis of opportunistic spectrum access strategies," in *Proc. IEEE INFOCOM'08*, April 2008.
10. Y. Chen, Q. Zhao, and A. Swami, "Joint design and separation principle for opportunistic spectrum access in the presence of sensing errors," *IEEE Trans. Inf. Theory*, vol. 54, no. 5, pp. 2053–2071, May 2008.
11. P. Cheng, G. Yu, Z. Zhang, H.-H. Chen, and P. Qiu, "On the maximum sum-rate capacity of cognitive multiple access channel," http://arxiv.org/abs/cs.IT/0612024.
12. H. Chong, M. Motani, H. Garg, and H. E. Gamal, "On the han-kobayashi region for the interference channel," To appear in *IEEE Trans. Inf. Theory*, vol. 54, no. 7, pp. 3188–3195, July 2008.
13. Y. S. Chow, H. Robbins, and D. Siegmund, *Great Expectations: The Theory of Optimal Stopping*, Houghton Mifflin Company, Boston, USA, 1971.
14. T. Clancy, "Achievable capacity under the interference temperature model," *IEEE Conference on Computer Communications (INFOCOM)*, May 2007.
15. C. Cordeiro and K. Challapali, "C-MAC: A cognitive MAC protocol for multi-channel wireless networks," in *Proc. DySPAN'07*, April 2007.
16. M. Costa, "Writing on dirty paper," *IEEE Trans. Inf. Theory*, vol. IT-29, pp. 439–441, May 1983.
17. T. Cover and J. Thomas, *Elements of Information Theory*. New York: John Wiley & Sons, 1991.

18. I. Csiszár and J. Kőrner, *Information Theory: Coding Theorems for Discrete Memoryless System*. Akadémiai Kiadó, Budapest, Hungary, 1981.
19. C. da Silva, B. Choi, and K. Kim, "Distributed spectrum sensing for cognitive radio systems," in *Information Theory and Applications ITA Workshop*, Feb. 2007.
20. N. Devroye, "Information theoretic limits of cooperation and cognition in wireless networks," Ph.D. dissertation, Harvard, 2007.
21. N. Devroye, P. Mitran, M. Sharif, S. S. Ghassemzadeh, and V. Tarokh, "Information theoretic analysis of cognitive radio systems," in *Cognitive Wireless Communication Networks*, V. Bhargava and E. Hossain, Eds. Springer, 2007.
22. N. Devroye, P. Mitran, and V. Tarokh, "Achievable rates in cognitive radio channels," in *39th Annual Conference on Information Sciences and Systems (CISS)*, Mar. 2005.
23. N. Devroye, P. Mitran, and V. Tarokh, "Achievable rates in cognitive networks," in *2005 IEEE International Symposium on Information Theory*, Sept. 2005.
24. N. Devroye, P. Mitran, and V. Tarokh, "Achievable rates in cognitive radio channels," *IEEE Trans. Inf. Theory*, vol. 52, no. 5, pp. 1813–1827, May 2006.
25. N. Devroye, P. Mitran, and V. Tarokh, "Cognitive decomposition of wireless networks," in *Proc. CROWNCOM*, Mykonos Island, Greece, Mar. 2006.
26. N. Devroye and M. Sharif, "The multiplexing gain of MIMO X-channels with partial transmit side-information," in *IEEE International Symposium on Information Theory*, June 2007.
27. N. Devroye and V. Tarokh, "Fundamental limits of cognitive radio networks," in *Cognitive Wireless Networks: Concepts, Methodologies and Vision*, F. Fitzek and M. Katz, Eds. Springer, New York, USA, 2007.
28. R. Etkin, A. Parekh, and D. Tse, "Spectrum sharing for unlicensed bands," *IEEE J. Sel. Areas Commun.*, vol. 25, no. 3, pp. 517–528, Apr. 2007.
29. FCC, Spectrum Policy Task Force Report, No. 02-155, Nov. 2002.
30. G. Ganesan and Y. Li, "Cooperative spectrum sensing in cognitive radio networks," in *New Frontiers in Dynamic Spectrum Access Networks (DYSPAN)*, Nov. 2005.
31. M. Gastpar, "On capacity under receive and spatial spectrum-sharing constraints," *IEEE Trans. Inf. Theory*, vol. 53, pp. 471–487, Feb. 2007.
32. S. Geirhofer, L. Tong, and B. M. Sadler, "Dynamic spectrum access in the time domain: Modeling and exploiting whitespace," *IEEE Commun. Mag.*, vol. 45, no. 5, pp. 66–72, May 2007.
33. S. Gel'fand and M. Pinsker, "Coding for channels with random parameters," *Probl. Contr. Inf. Theory*, vol. 9, no. 1, pp. 19–31, 1980.
34. A. Ghasemi and E. Sousa, "Collaborative spectrum sensing for opportunistic access in fading environments," in *New Frontiers in Dynamic Spectrum Access Networks (DYSPAN)*, Nov. 2005.
35. A. Ghasemi and E. Sousa, "Capacity of fading channels under spectrum-sharing constraints," in *Proc. IEEE Int. Conf. Commun.*, Istanbul, Turkey, June 2006.
36. A. Goldsmith, S. Jafar, I. Maric, and S. Srinivasa, "Breaking spectrum gridlock with cognitive radios: An information theoretic perspective," *Proceedings of the IEEE*, vol. 97, no. 5, pp. 894–914, May 2009.
37. P. Grover and A. Sahai, "On the need for knowledge of the phase in exploiting known primary transmissions," 2nd IEEE Int. Symposium on New Frontiers in Dynamic Spectrum Access Networks, 2007 (DySPAN 2007), pp. 462–471, Dublin, Ireland, 17–20 April 2007.
38. P. Gupta and P. R. Kumar, "The capacity of wireless networks," *IEEE Trans. Inf. Theory*, vol. 46, no. 2, pp. 388–404, Mar. 2000.
39. B. Hamdaoui and K. G. Shin, "OS-MAC: An efficient MAC protocol for spectrum-agile wireless networks," *IEEE Trans. Mobile Comput.*, vol. 7, no. 8, pp. 915–930, Aug. 2008.
40. T. Han and K. Kobayashi, "A new achievable rate region for the interference channel," *IEEE Trans. Inf. Theory*, vol. IT-27, no. 1, pp. 49–60, 1981.
41. Z. J. Hass and J. Deng, "Dual busy tone multiple access (DBTMA) – A multiple access control scheme for ad hoc networks," *IEEE Trans. Commun.*, vol. 50, no. 6, pp. 975–985, Jun. 2002.

42. S. Haykin, "Cognitive radio: Brain-empowered wireless communications," *IEEE J. Sel. Areas Commun.*, vol. 23, no. 2, pp. 201–220, Feb. 2005.
43. C. Heegard and A. E. Gamal, "On the capacity of computer memories with defects," *IEEE Trans. Inf. Theory*, vol. 29, pp. 731–739, Sept. 1983.
44. S.-C. Hong, M. Vu, and V. Tarokh, "Cognitive sensing based on side information," *Sarnoff Conf.*, Apr. 2008.
45. N. Hoven and A. Sahai, "Power scaling for cognitive radio," *Int Conf. Wireless Netw., Comm. Mobile Comput.*, vol. 1, pp. 250–255, Jun. 2005.
46. A. C. C. Hsu, D. S. L. Wei, and C.-C. J. Kuo, "A cognitive MAC protocol using statistical channel allocation for wireless ad-hoc networks," in *Proc. IEEE WCNC'07*.
47. J. Huang, R. Berry, and M. Honig,"Distributed interference compensation for wireless networks," *IEEE J. Sel. Areas Commun.*, vol. 24, no. 5, pp. 1074–1084, May 2006.
48. J. Huang, R. Berry, and M. Honig, "Auction-based spectrum sharing," *ACM/Kluwer J. Mobile Netw. Appl. (MONET)*, vol. 11, no. 3, pp. 405–418, Jun. 2006.
49. S. Huang, X. Liu, and Z. Ding, "Opportunistic spectrum access in cognitive radio networks," in *Proc. IEEE INFOCOM 2008*, Apr. 2008.
50. T. Hunter, A. Hedayat, M. Janani, and A. Nostratinia, "Coded cooperation with space-time transmission and iterative decoding," in *WNCG Wireless*, Oct. 2003.
51. M. Islam, Y.-C. Liang, and A. T. Hoang, "Joint power control and beamforming for secondary spectrum sharing," in *Proc. of IEEE 66th Vehicular Technology Conference (VTC)*, Oct. 2007.
52. S. A. Jafar and S. Srinivasa, "Capacity limits of cognitive radio with distributed and dynamic spectral activity," *IEEE J. Sel. Areas Commun.*, vol. 25, pp. 529–537, Apr. 2007.
53. S. Jafar and S. Shamai, "Degrees of freedom region for the MIMO X channel," *IEEE Trans. Inf. Theory*, vol. 54, no. 1, pp. 151–170, Jan. 2008.
54. S.-W. Jeon, N. Devroye, M. Vu, S.-Y. Chung, and V. Tarokh, "Cognitive networks achieve throughput scaling of a homogeneous network," *IEEE Trans. Info. Theory*, submitted Mar 2008.
55. J. Jia, Q. Zhang, and X. Shen, "HC-MAC: A hardware-constrained cognitive MAC for efficient spectrum management," *IEEE J. Sel. Areas Commun.*, vol. 26, no. 1, pp. 106–117, Jan. 2008.
56. J. Jiang and X. Xin, "On the achievable rate regions for interference channels with degraded message sets," submitted to *IEEE Trans. Inf. Theory*, vol. 54, no. 10, pp. 4707–4712, Apr. 2007.
57. J. Jiang, X. Xin, and H. Garg, "Interference channels with common information," submitted to *IEEE Trans. Inf. Theory*, vol. 54, no. 1, pp. 171–187, Jan. 2008.
58. A. Jovicic and P. Viswanath, "Cognitive radio: An information-theoretic perspective," *Proc. IEEE ISIT 2006*, Seattle, USA, July 9–14, 2006.
59. A. Karnik, A. Iyer, and C. Rosenberg, "Throughput-optimal configuration of fixed wireless networks," *IEEE/ACM Trans. Networking*, vol. 16, no. 5, pp. 1161–1174, Oct. 2008.
60. S. M. Kay, *Fundamentals of Statistical Signal Processing, Volume 2: Detection Theory*. Prentice Hall, New Jersey, USA, 1998.
61. F. P. Kelly, A. Maulloo, and D. Tan, "Rate control for communication networks: Shadowing prices, proportional fairness, and stability," *J. Oper. Res. Soc.*, vol. 49, no. 3, pp. 237–252, Mar. 1998.
62. D. I. Kim, L. Le, and E. Hossain, "Joint rate and power allocation for cognitive radios in dynamic spectrum access environment," *IEEE Trans. Wireless Commun.*, vol. 7, no. 12, pp. 5517–5527, Dec. 2008.
63. P. J. Kolodzy, "Interference temperature: A metric for dynamic spectrum utilization," *Int. J. Netw. Manag.*, vol. 16, no. 2, pp. 103–113, Mar. 2006.
64. O. Koyluoglu and H. Gamal, "On power control and frequency re-use in the two-user cognitive channel," submitted to *IEEE Trans. Wireless Commun.*, 2007.
65. S. Kulkarni and C. Rosenberg, "Opportunistic scheduling: Generalizations to include multiple constraints, multiple Interfaces, and short term fairness," *ACM/Kluwer Wireless Netw. J. (WINET)*, vol. 11, no. 5, pp. 557–569, Sept. 2005.

66. A. Kusnetsov and B. Tsybakov, "Coding in a memory with defective cells," *Prob. Pered. Inform.*, vol. 10, pp. 52–60, Apr.–Jun. 1974.
67. L. Le and E. Hossain, "OSA-MAC: A MAC protocol for opportunistic spectrum access in cognitive radio networks," in *Proc. IEEE WCNC'08*, Mar.–Apr. 2008.
68. L. Le and E. Hossain, "Resource allocation for spectrum underlay in cognitive radio networks," *IEEE Trans. Wireless Commun.*, vol. 7, no. 12.
69. J. Li, C. Blake, D. S. J. D. Couto, H. I. Lee, and R. Morris, "Capacity of ad hoc wireless networks," in Proc. *ACM Mobicom'01*.
70. Y. Liang, A. Somekh-Baruch, H. V. Poor, S. S. (Shitz), and S. Verdú, "Capacity of cognitive interference channels with and without secrecy," *IEEE Trans. Inf. Theory*, vol. 55, no. 2, pp. 604–619, Feb. 2009.
71. Y. C. Liang, Y. Zeng, E. C. Y. Peh, and A. T. Hoang, "Sensing-throughput tradeoff for cognitive radio networks," *IEEE Trans. Wireless Commun.*, vol. 7, no. 4, pp. 1326–1337, Apr. 2008.
72. X. Liu, E. K. P. Chong, and N. B. Shroff, "Opportunistic transmission scheduling with resource-sharing constraints in wireless networks," *IEEE J. Sel. Areas Commun.*, vol. 19, no. 10, pp. 2053–2065, Oct. 2001.
73. X. Luo and K. Kar, "Joint scheduling and power allocation in multi-channel access point networks under QoS constraints," in *Proc. IEEE ICC'08*, May 2008.
74. L. Ma, X. Han, and C. C. Shen, "Dynamic open spectrum sharing MAC protocol for wireless ad hoc networks," in *Proc. DySPAN'05*, Nov. 2005.
75. M. Maddah-Ali, A. Motahari, and A. Khandani, "Signaling over MIMO multi-base systems: combination of multi-access and broadcast schemes," in *2006 IEEE International Symposium on Information Theory*, Jul. 2006.
76. M. Marcus, "Unlicensed cognitive sharing of TV spectrum: The controversy at the federal communications commission," *IEEE Comm. Mag.*, vol. 43, no. 5, pp. 24–25, May 2005.
77. I. Maric, R. Yates, and G. Kramer, "Capacity of interference channels with partial transmitter cooperation," *IEEE Trans. Inf. Theory*, vol. 53, pp. 3536–3548, Oct. 2007.
78. S. M. Mishra, A. Sahai, and R. W. Brodersen, "Cooperative sensing among cognitive radios," in *Proc. IEEE Int. Conf. Commun.*, Istanbul, Turkey, Jun. 2006.
79. J. Mitola, "Cognitive radio," Ph.D. dissertation, Royal Institute of Technology (KTH), 2000.
80. P. Mitran, N. Devroye, and V. Tarokh, "On compound channels with side-information at the transmitter," *IEEE Trans. Inf. Theory*, vol. 52, no. 4, pp. 1745–1755, Apr. 2006.
81. J. Mo, H.-S. W. So, and J. Walrand, "Comparison of multichannel MAC protocols," *IEEE Trans. Mobile Comput.*, vol. 7, no. 1, pp. 50–65, Jan. 2008.
82. J. Mo and J. Walrand, "Fair end-to-end window-based congestion control," *IEEE/ACM Trans. Netw.*, vol. 8, no. 5, pp. 556–567, Oct. 2000.
83. D. Niyato and E. Hossain, "Competitive spectrum sharing in cognitive radio networks: A dynamic game approach," *IEEE Trans. Wireless Commun.*, vol. 7, no. 7, pp. 2651–2660, Jul. 2008.
84. M. Oner and F. Jondral, "On the extraction of the channel allocation information in spectrum pooling systems," *IEEE J. Sel. Areas Commun.*, vol. 25, no. 3, pp. 558–565, Apr. 2007.
85. P. Popovski, H. Yomo, K. Nishimori, R. D. Taranto, and R. Prasad, "Opportunistic interference cancellation in cognitive radio systems," in *Proc. 2nd IEEE International Symposium on Dynamic Spectrum Access Networks (DySPAN)*, Dublin, Ireland, Apr. 2007.
86. A. Sabharwal, A. Khoshnevis, and E. Knightly, "Opportunistic spectral usage: Bounds and a multi-band CSMA/CA protocol," *IEEE/ACM Trans. Netw.*, vol. 15, no. 3, pp. 533–545, Jun. 2007.
87. Y. Shi and Y. T. Hou, "A distributed optimization algorithm for multi-hop cognitive radio networks," in *Proc. INFOCOM'08*, Apr. 2008.
88. O. Simeone, Y. Bar-Ness, and U. Spagnolini, "Stable throughput of cognitive radios with and without relaying capability," *IEEE Trans. Commun.*, vol. 55, no. 12, pp. 2351–2360, Dec. 2007.
89. R. Smallwood and E. Sondik, "The optimal control of partially observable Markov processes over a finite horizon," *Oper. Res.*, vol. 21, no. 5, pp. 1071–1088, Sept.–Oct. 1973.

90. J. So and N. Vaidya, "Multi-channel MAC for ad hoc networks: Handling multi-channel hidden terminals using a single transceiver," in *Proc. ACM Mobihoc'04*.

91. S. Srinivasa, S. Jafar, and N. Jindal, "On the capacity of the cognitive tracking channel," *Proc. IEEE Int. Symp. Inf. Theory*, July 2006.

92. Y. Steinberg and N. Merhav, "Identification in the presence of side information with application to watermarking," *IEEE Trans. Inf. Theory*, vol. 47, pp. 1410–1422, May 2001.

93. H. Su and X. Zhang, "Channel-hopping based single transceiver MAC for cognitive radio networks," in *Proc. CISS'08*, Mar. 2008.

94. H. Su and X. Zhang, "Cross-layer based opportunistic MAC protocols for QoS provisionings over cognitive radio wireless networks," *IEEE J. Sel. Areas Communs.*, vol. 26, no. 1, pp. 118–129, Jan. 2008.

95. J. Sun, E. Modiano, and L. Zheng, "Wireless channel allocation using an auction algorithm," *IEEE J. Sel. Areas Commun.*, vol. 24, no. 5, pp. 1085–1096, May 2006.

96. R. Tandra and A. Sahai, "SNR walls for signal detection," *IEEE J. Sel. Areas Commun.*, vol. 2, no. 1, pp. 4–17, Feb. 2008.

97. A. Tkachenko, "Testbed design for cognitive radio spectrum sensing experiments," Ph.D. dissertation, Berkeley, 2007.

98. J. Unnikrishnan and V. V. Veeravalli, "Cooperative sensing for primary detection in cognitive radio," *IEEE J. Sel. Signal Process.*, vol. 2, no. 1, pp. 18–27, Feb. 2008.

99. S. Verdú, *Multiuser Detection*. Cambridge University Press, Cambridge, UK, 2003.

100. P. Viswanath, D. Tse, and R. Laroia, "Opportunistic beamforming using dumb antennas," *IEEE Trans. Inf. Theory*, vol. 48, no. 6, pp. 1277–1294, Jun., 2002.

101. M. Vu, N. Devroye, and V. Tarokh, "On the primary exclusive regions in cognitive networks," submitted to *IEEE Trans. Wireless Comm.*, to appear.

102. M. Vu, S. S. Ghassemzadeh, and V. Tarokh, "Interference in a cognitive network with beacon," *IEEE Wireless Comm. Netw. Conf.*, Mar. 2008.

103. M. Vu and V. Tarokh, "Scaling laws of single-hop cognitive networks," submitted to *IEEE Trans. Wireless Comm.*, to appear.

104. F. Wang, M. Krunz, and S. Cui, "Price-based spectrum management in cognitive radio network," *IEEE J. Sel. Signal Process.*, vol. 2, no. 1, pp. 74–87, Feb. 2008.

105. H. Weingarten, Y. Steinberg, and S. Shamai, "The capacity region of the Gaussian MIMO broadcast channel," submitted to *IEEE Trans. Inf. Theory*, vol. 52, no. 9, pp. 3936–3964, Sept. 2006.

106. W. Weng, T. Peng, and W. Wang, "Optimal power control under interference temperature constraints in cognitive radio network," in *Proc. of IEEE Wireless Communications and Networking Conference*, Hong Kong, Mar. 2007.

107. W. Wu, S. Vishwanath, and A. Arapostathis, "Capacity of a class of cognitive radio channels: Interference channels with degraded message sets," *IEEE Trans. Inf. Theory*, vol. 53, no. 11, pp. 4391–4399, Jun. 2007.

108. W. Wu, S. Vishwanath, and A. Aripostathis, "On the capacity of interference channel with degraded message sets," submitted to *IEEE Trans. Inf. Theory*, vol. 53, no. 11, pp. 4391–4399, 2007.

109. Y. Xing, C. Marthur, M. Haleem, R. Chandramouli, and K. Subbalakshmi, "Dynamic spectrum access with QoS and interference temperature constraints," *IEEE Trans. Mobile Comput.*, vol. 6, no. 4, pp. 423–433, Apr. 2007.

110. R. Yeung, *A First Course in Information Theory*. Springer, New York, USA, 2002.

111. S. Yiu and M. Vu, "Interference reduction by beamforming in cognitive networks," submitted to Proc. IEEE Global Telecommun. Conf., pp. 1–6, Nov. 30–Dec. 4.

112. L. Zhang, Y. -C. Liang, and Y. Xin, "Joint beamforming and power allocation for multiple access channels in cognitive radio networks, *IEEE J. Sel. Areas Commun.*, vol. 26, no. 1, pp. 38–51, Jan. 2008.

113. L. Zhang, Y.-C. Liang, Y. Xin, and H. V. Poor, "Robust cognitive beamforming with partial channel state information," http://front.math.ucdavis.edu/0711.4414.

114. L. Zhang, Y. Xin, and Y.-C. Liang, "Power allocation for multi-antenna multiple access channels in cognitive radio networks," in *Annual Conference on Information Sciences and Systems (CISS)*, Mar. 2007.
115. R. Zhang and Y. Liang, "Exploiting multi-antennas for opportunistic spectrum sharing in cognitive radio networks," http://front.math.ucdavis.edu/0711.4414.
116. Q. Zhao and B. M. Sadler, "A survey of dynamic spectrum access," *IEEE Signal Process. Mag.,* vol. 24, no. 3, pp. 79–89, May, 2007.
117. Q. Zhao, L. Tong, A. Swami, and Y. Chen, "Decentralized cognitive MAC for opportunistic spectrum access in ad hoc networks: A POMDP framework," *IEEE J. Sel. Areas Commun.,* vol. 25, no. 3, pp. 589–600, Apr. 2007.

Additional Readings

1. K. R. Chowdhury and I. F. Akyildiz, "Cognitive wireless mesh networks with dynamic spectrum access," *IEEE J. Sel. Areas Communs.*, vol. 26, no. 1, pp. 168–181, Jan. 2008.
2. M. Gandetto and C. Regazzoni, "Spectrum sensing: A distributed approach for cognitive terminals," *IEEE J. Sel. Areas Commun.*, vol. 25, no. 3, pp. 546–557, Apr. 2007.
3. S. Geirhofer, L. Tong, and B. M. Sadler, "Cognitive medium access: Constraining interference based on experimental models," *IEEE J. Sel. Areas Commun.*, vol. 26, no. 1, pp. 95–105, Jan. 2008.
4. D. Niyato, E. Hossain, and L. Le, "Competitive spectrum sharing and pricing in cognitive wireless mesh networks," in *Proc. IEEE WCNC'08*, Mar.–Apr. 2008.

Chapter 11
Coded Bidirectional Relaying in Wireless Networks

Petar Popovski and Toshiaki Koike-Akino

11.1 Introduction

The techniques for coded bidirectional or two-way relaying have received significant attention in the recent years [1–15]. The mechanisms for two-way relaying or, more general, multi-way relaying leverage on two conceptual blocks. The *first conceptual block* is the shared nature of the wireless communication medium. On the one hand, this implies that there is interference when multiple transmissions are occurring simultaneously. On the other hand, the *wireless broadcasting is "cheap"* in a sense that a single transmission can be received by multiple nodes. The *second conceptual block* is the idea of network coding. In short, the traditional design of communication networks observes the data flows as conventional commodity flows. Therefore, a routing node in the network essentially replicates the data packets from an incoming link to an outgoing link (or multiple links, in case of multicast). The network coding recognizes that a data flow is different from a physical commodity flow and generalizes the routing such that the data on a given outgoing link is a function of the data from two or more incoming links.

To see how these building blocks accrue into novel modes of wireless communication, consider the example on Fig. 11.1 [1]. The communication scenario is that the node A has packets destined to C and vice versa. However, in this example we assume that the capacity of the direct channel between A and C is zero and the communication between them must be done by using B as a relay node. The packet from the source node i to the destination j is denoted by $\mathbf{D}_{ij}$. For this example, all the

Petar Popovski
Department of Electronic Systems, Aalborg University, Niels Jernes Vej 12,
DK-9220 Aalborg, Denmark.
e-mail: petarp@es.aau.dk

Toshiaki Koike-Akino
School of Engineering and Applied Sciences, Harvard University,
33 Oxford Street, Cambridge, MA 02138, USA.
e-mail: koike@seas.harvard.edu

V. Tarokh (ed.), *New Directions in Wireless Communications Research*,
DOI 10.1007/978-1-4419-0673-1_11,
© Springer Science+Business Media, LLC 2009

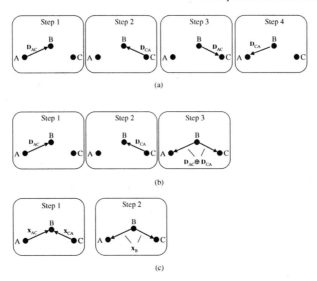

Fig. 11.1 (a) Uncoded bidirectional relaying; (b) example of coded bidirectional relaying with three steps (phases); and (c) example of coded bidirectional relaying with two steps

packets have identical sizes. The node B is neither source nor destination of the data traffic. The conventional (uncoded) relaying regards the transmission of $\mathbf{D}_{AC}$ and $\mathbf{D}_{CA}$ as two separate problems. Figure 11.1(a) shows that the uncoded relaying consumes four steps, each step having a duration of a time slot. On the other hand, the coded bidirectional relaying Fig. 11.1(b) considers the two transmissions as a single problem. In the first two slots the relay gathers the packets $\mathbf{D}_{AC}$ and $\mathbf{D}_{CA}$ from the respective source node. In Step 3 B broadcasts the packet $\mathbf{D}_B = \mathbf{D}_{AC} \oplus \mathbf{D}_{CA}$, where $\oplus$ is XOR operation. After receiving $\mathbf{D}_B$, the node A decodes $\mathbf{D}_{CA} = \mathbf{D}_B \oplus \mathbf{D}_{AC}$ by using its a priori knowledge of $\mathbf{D}_{AC}$. In an analogous way, C decodes $\mathbf{D}_{AC}$. Thus, to transfer the same amount of data, the coded two-way relaying requires only three time slots, which is an improvement of 33% over the uncoded relaying. This simple example clearly illustrates the two building blocks, mentioned above: (1) it utilizes the feature of "cheap wireless broadcast" to save one transmission slot and (2) the packets of different incoming communication flows are combined before being sent over the outgoing link (broadcast) of the relay node B.

In the example on Fig. 11.1(b) the relay node gathers the packets from A and C in a time-division manner. Alternatively, A and C can simultaneously transmit over the multiple-access channel [3], as shown in Fig. 11.1(c), where we use the notation $\mathbf{x}_i$ to denote the baseband signal transmitted by the node i. In Step 2, B transmits the signal $\mathbf{x}_B$, which is created as a function of the signal that B received in Step 1. This function does not necessarily assume that B is able to decode the individual signals. In the simplest case, B needs only to amplify the signal that it has received in Step 1 and broadcast it back to A and C in Step 2. Since A (C) a priori knows its contribution to the interfered signal received at B in Step 1, then it can utilize this information to reliably extract $\mathbf{x}_C$ ($\mathbf{x}_A$) from the signal $\mathbf{x}_B$. Ideally, if both A and C decode each other's signals correctly, the time to transmit the data is only two steps, which means throughput improvement over the uncoded relaying of 100%.

These illustrations of the coded two-way relaying suggest that the underlying ideas have a great potential to improve the performance of wireless networks. The chapter does not aim to cover all the aspects of the two-way relaying, which, for example, include multiple-antenna techniques, scenario-specific channel estimation, scheduling. Instead, the objective is to elaborate on several important ideas and insights brought by the two-way relaying scenario. We will see that already this simple three-node scenario gives rise to many novel techniques, such that the cases with more than three nodes are outside the scope.

The text is organized as follows. After the preliminaries in the next section, we first describe the bidirectional relaying techniques that require decoding at the relay. Section 11.4 is dedicated to the discussion of the techniques used when the relay does not decode the messages from the terminals. A class of such techniques is based on denoising (rather than decoding) and Section 11.5 outlines the special significance that the structured codes have for such denoising techniques. The next section departs form the information-theoretic setting to provide insight into the more practical aspects of the coded bidirectional relaying, by considering finite-length packets and practical modulation setting. The last section concludes the chapter.

11.2 Preliminaries

We assume that there are only two communication flows, $A \to C$ and $C \to A$, respectively. The relay B is neither a source nor a sink of any data in the system. All the nodes are half-duplex, such that a node can either transmit or receive at a given time. The rate at which node $i \in \{A, B, C\}$ transmits is denoted by R_i. During the transmission, the node i has n_i channel uses, such that the message that it sends has $n_i R_i$ bits. Let us denote by $N > n_i$ the total number of channel uses during the whole round of two-way relaying. If after N channel uses the message $W_{AC}(W_{CA})$ is received at $C(A)$ successfully, then the rate R_{AC} achieved from A to C and the rate R_{CA} achieved from C to A are given as follows:

$$R_{AC} = \frac{n_A R_A}{N}, \qquad R_{CA} = \frac{n_C R_C}{N}. \qquad (11.1)$$

We will be interested in determining the rate pair (R_{AC}, R_{CA}) and the sum (two-way) rate $R_{AC} + R_{CA}$. We will assume that in each round A and C transmit only fresh data, independent of any information exchange from the previous rounds.[1]

---

[1] This is analogous to what Shannon [16] refers as a scheme for capacity for the *restricted* two-way channel

The message sent from node i and destined for node j is denoted by W_{ij} and the corresponding binary representation is the vector $\mathbf{w}_{ij}$. The size (in bits) of the message is denoted by $|\mathbf{w}_{ij}|$. The codeword transmitted by node i is denoted by $\mathbf{x}_i$ and is a vector of dimension n_i, whose mth element is denoted by $x_i[m]$. The random variable that stands for a symbol sent (received) by node i is denoted by $X_i(Y_i)$. The received vector at node j is $\mathbf{y}_j$. We mainly consider Gaussian channels, explicitly stating if the channel is discrete.

For Gaussian channels X_i, Y_j are complex numbers, unless stated otherwise (Section 11.5.2). If only one node $U \in \{A, B, C\}$ is transmitting, then the received symbol at the node $V \in \{A, B, C\} \setminus U$ is given by

$$Y_V = h_{UV} X_U + Z_V \tag{11.2}$$

or if we want to emphasize that it is the mth symbol

$$y_V[m] = h_{UV} x_U[m] + z_V[m], \tag{11.3}$$

where h_{UV} is the complex channel coefficient between U and V. $z_V[m]$ is the complex-additive white Gaussian noise $\mathcal{CN}(0, N_0)$. The transmitted symbols have $E\{x_U[m]\} = 0$ and a normalized power $E\{|x_U[m]|^2\} = 1$. If A and C transmit simultaneously, then B receives

$$Y_B = h_1 X_A + h_2 X_B + Z_B. \tag{11.4}$$

Each node uses the same transmission power, which makes the links symmetric:

$$h_{AC} = h_{CA} = h_0 \quad h_{AB} = h_{BA} = h_1 \quad h_{CB} = h_{BC} = h_2. \tag{11.5}$$

This assumption is certainly restrictive, as one can optimize the two-way transmissions by allocating appropriate power to the nodes, while keeping some global power constraint for all the transmitters; however, that discussion is outside the scope of this text. The bandwidth is normalized to 1 Hz, such that the time is measured in number of symbols (channel uses) and the signal-to-noise ratios (SNR) are given as follows:

$$\gamma_i = \frac{|h_i|^2}{N_0}, \qquad i = 0, 1, 2 \tag{11.6}$$

and a point-to-point link with SNR of γ can reliably transfer up to

$$C(\gamma) = \log_2(1 + \gamma) \text{ [bit/s]}. \tag{11.7}$$

Without loss of generality, we can assume that

$$\gamma_2 \geq \gamma_1 \tag{11.8}$$

and we also assume that the direct link is worse than both links, i.e., $\gamma_0 < \gamma_1$. We will see that there are schemes in which the signal received over the direct link can be used as a side information to improve the end-to-end rates.

11.3 Two-Way Relaying with Decoding at the Relay

Here we discuss the methods for two-way relaying in which the relay node decodes the messages W_{AC} and W_{CA}. We consider a three-step scheme called *decode-and-forward (DF)* and a two-step scheme termed *joint-decode-and-forward (JDF)*. Both schemes consist of an *uplink* phase and a *broadcast* phase. In the uplink phase B gathers data from A and C, while in the broadcast phase B transmits to A, C.

11.3.1 The Uplink Phase

The uplink phase of the three-step scheme consists of two steps. In Step 1 the node A transmits W_{AC} using n_A symbols and in Step 2 the node C transmits W_{CA} using n_C symbols. The rates R_A and R_C should be chosen:

$$R_A \leq C(\gamma_1), \qquad R_C \leq C(\gamma_2), \tag{11.9}$$

where $C(\cdot)$ is defined in (11.7). After the uplink phase, B successfully decodes the $n_A R_A$ bits sent by A and the $n_C R_C$ bits sent by C. The message that B needs to relay to C and A is denoted as W_{BC} and W_{BA}, respectively. In the simplest case, the direct link between A and C is considered to have zero capacity, such that after the uplink phase, A has still not any information about the packet sent by C, and vice versa. Hence, when $\gamma_0 = 0$, B needs to completely relay the messages W_{AC} and W_{CA}. However, in general $\gamma_0 > 0$, such that the direct link between A and C carries non-zero information. In this case, at each step of the three-step scheme there is broadcast transmission, since one node is transmitting and two are receiving. In Step 1, A broadcasts to B and C at a rate R_A, where $R_0 \leq R_A \leq C(\gamma_1)$. The node C receives some partial information from the broadcast of A and vice versa, which decreases the amount of data that need to be broadcasted by B. In this case, W_{BC} contains $n_A[R_A - R_0]$ bits, while W_{BA} has $n_C[R_C - R_0]$ bits, where $R_0 \leq C(\gamma_0)$. The relay can use random binning [17] to create the messages that need to be relayed. With such approach, W_{BC} is uniquely determined from W_{AC} and W_{BA} is uniquely determined from W_{CA}. The uplink transmission of the three-step scheme is illustrated in Fig. 11.3(a).

While the uplink phase of the three-step scheme consists of two broadcast transmissions, the uplink phase of the JDF scheme consists of two simultaneous transmissions over a multiple-access (MA) channel. In this case the number of channel uses by A and C is equal, $n_A = n_C$, and the rates R_A and R_C should be selected within the capacity region of the MA channel [17] with B as a receiver (see Fig. 11.2):

$$R_A \leq C(\gamma_1), \qquad R_C \leq C(\gamma_2),$$
$$R_A + R_C \leq C(\gamma_1 + \gamma_2). \tag{11.10}$$

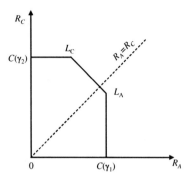

Fig. 11.2 Achievable rate region for the multiple-access channel used in the uplink phase

Due to the half-duplex restriction, in Step 1 both A and C cannot receive information over the direct link, such that the relayed messages cannot be reduced and $W_{BC} = W_{AC}$, $W_{BA} = W_{CA}$. The uplink transmission of the JDF scheme is shown in Fig. 11.3(b).

11.3.2 The Broadcast Phase

A feature of the two schemes described above is that the relay node B knows the messages W_{AC}, W_{CA} and has a complete freedom in combining them for the broadcast phase. We first describe a simple broadcast strategy that combines the data by using XOR. In relation to the quantity of data that B should relay to A, C, there are two different cases:

- $|\mathbf{w}_{BA}| \geq |\mathbf{w}_{BC}|$. The shorter packet $\mathbf{w}_{BC}$ is padded with $|\mathbf{w}_{BA}| - |\mathbf{w}_{BC}|$ zeros and the padded packet is denoted by $\mathbf{w}_{BC}^P$. Then B broadcasts the packet $\mathbf{w}_{BA} \oplus \mathbf{w}_{BC}^P$ at the transmission rate $R_B \leq C(\gamma_1)$, limited by the SNR of the weaker link.

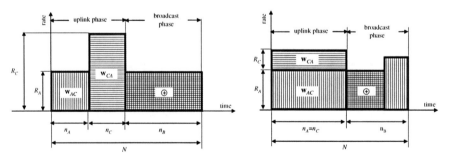

Fig. 11.3 Illustration of the uplink and the broadcast phase for the schemes with decoding at the relay: (a) three-step scheme and (b) two-step scheme

This is required since both A and C should correctly receive the XOR packet. This case of the broadcast phase is illustrated in Fig. 11.3(a).

- $|\mathbf{w}_{BA}| < |\mathbf{w}_{BC}|$. In this case we take the first $|\mathbf{w}_{BA}|$ bits of $\mathbf{w}_{BC}$ to create the packet $\mathbf{w}_{BC}^{(1)}$. The vector of the last $|\mathbf{w}_{BC}| - |\mathbf{w}_{BA}|$ bits of $\mathbf{w}_{BC}$ is denoted as $\mathbf{w}_{BC}^{(2)}$. The relay node transmits as follows:

  - The packet $\mathbf{w}_{BC}^{(1)} \oplus \mathbf{w}_{BA}$ at rate $R_B^{(1)} \leq C(\gamma_1)$
  - The packet $\mathbf{w}_{BC}^{(2)}$ at rate $R_B^{(2)} \leq C(\gamma_2)$, as only C needs to receive it

This simple broadcast is depicted in Fig. 11.3(b). Such a case can occur if during the uplink phase the rate pair (R_A, R_C) is selected to be, e.g., at the point L_A.

Figure 11.3 shows only two of the four possible combinations of the different options for the uplink phases and the broadcast phases. For example, also in the three-step scheme it can happen that $|\mathbf{w}_{BA}| < |\mathbf{w}_{BC}|$, such that the broadcast phase uses transmissions at two different rates.

11.3.3 Improved Broadcast Strategies

The simple broadcast strategy does not efficiently use the available degrees of freedom. In this section we describe two strategies that can enlarge the achievable rate region: (1) broadcast strategy based on superposition coding and (2) broadcast strategy with side information.

11.3.3.1 Broadcast with Superposition Coding

Superposition coding has been introduced as coding strategy that achieves the capacity region of degraded broadcast channels [17]. The Gaussian broadcast channels are degraded and here we explain the superposition strategy for the Gaussian case. We reuse the notations from the scenario above and consider the case in which B broadcasts to A and C. The codeword transmitted by B is

$$\mathbf{x}_B = \sqrt{1 - \theta}\mathbf{x}_{B,1} + \sqrt{\theta}\mathbf{x}_{B,2}, \qquad (11.11)$$

where θ is the power division coefficient. The codeword $\mathbf{x}_{B,1}$ should be decoded by both A and C, while the codeword $\mathbf{x}_{B,2}$ needs to be decoded only by C. When decoding $\mathbf{x}_{B,1}$, the codeword $\mathbf{x}_{B,2}$ should be treated as noise with power θ, such that its rate should satisfy

$$R_{B,1} \leq C\left(\frac{(1 - \theta)|h_{BA}|^2}{N_0 + \theta|h_{BA}|^2}\right) = C\left(\frac{(1 - \theta)\gamma_1}{1 + \theta\gamma_1}\right). \qquad (11.12)$$

C uses successive interference cancellation: after decoding $\mathbf{x}_{B,1}$, its contribution from the received signal is removed and $\mathbf{x}_{B,2}$ is decoded, such that its rate should satisfy

$$R_{B,2} \leq C(\theta\gamma_2). \qquad (11.13)$$

If B broadcasts during n_B channel uses, then A receives $n_B R_{B,1}$ bits, while C receives $n_B R_{B,1} + n_B R_{B,2}$ bits. This implies that this strategy is effective when the relay node B has more data destined for C than A, i.e., $|\mathbf{w}_{BC}| \geq |\mathbf{w}_{BA}|$. Recall the corresponding case from the simple broadcast strategy, in which we have split the broadcast data of B into two messages:

- The packet $\mathbf{w}_{BC}^{(1)} \oplus \mathbf{w}_{BA}$ is common for both A and C and is sent using the codeword $\mathbf{x}_{B,1}$, at a rate $R_{B,1}$.
- The packet $\mathbf{w}_{BC}^{(2)}$ is sent using the codeword $\mathbf{x}_{B,2}$, at a rate $R_{B,2}$.

Once the parameters of the uplink phase are fixed, one can pose the question which θ will result in most efficient broadcast? We can answer that by setting equalities in (11.12)–(11.13) and solving the balance equations for the data in the uplink and the broadcasts phases. For the three-step DF scheme the balance equations are given as

$$n_A(R_A - R_0) = n_B(R_{B,1} + R_{B,2}), \tag{11.14}$$

$$n_C(R_C - R_0) = n_B R_{B,1}. \tag{11.15}$$

The first and second balance equation is for $\mathbf{w}_{BC}$ and $\mathbf{w}_{BA}$, respectively. The only unknowns are n_B and θ, which can be found by solving the system of equations. For the JDF scheme the same equations are applied by setting $n_A = n_C$ and $R_0 = 0$.

11.3.3.2 Broadcast Strategy Optimized for Two-Way Relaying

In the previous broadcast strategies, the relay node "digitally" combines the data destined for the two different users by using the XOR operation. Therefore B can use codebooks and strategies that are used in the conventional broadcast scenario, where B is the source of information. On the other hand, B *is not the source* of information and each of the nodes A, C knows part of the data that B broadcasts, since it has itself sent the data in the uplink phase. This motivates to consider a different broadcast strategy, in which the codebooks at B are designed in order to account for the side information present at the nodes A, C, as introduced in [12].

Intuitive Description

Let us consider a very simplified scenario in which B has two bits $\mathbf{w}_{BA} = [c_1\ c_2]$ to be transmitted to A and four bits $\mathbf{w}_{BC} = [a_1\ a_2\ a_3\ a_4]$ to be transmitted to C. Furthermore, assume that A knows $\mathbf{w}_{BC}$ and C knows $\mathbf{w}_{BA}$ and this knowledge is used as a side information in the decoding process. We also make the following simplifying assumptions. The link $B - C$ has SNR γ_2 that is sufficient for C to reliably decode 16-QAM symbols sent by B. The link $B - A$ has a lower SNR and can reliably decode only QPSK symbols, but not constellations of a higher level. The term *reliable* as used here is not precise and does not reflect the probabilistic nature of the errors, but serves well for illustrating purposes. We pose the following question:

Can B use a single 16-QAM symbol to send both $\mathbf{w}_{BA}$ and $\mathbf{w}_{BC}$?

The trick is to observe that the 16-QAM constellation consists of four shifted QPSK cosets or subconstellations, see Fig. 11.4. A cannot reliably decode a 16-QAM symbol, but if it has a side information to which QPSK coset does the symbol belong, it can reliably decode it. Note that a "coset QPSK" has approximately 1.0 dB loss compared to the pure QPSK signal, but for the discussion here we assume that both can be received at A equally reliably. The mentioned decoding of cosets at A can be implemented as follows. The node B creates the following two-bit messages:

- The message $\mathbf{w}_{B,1} = [\, a_1 \; a_2 \,] \oplus [\, c_1 \; c_2 \,]$ and needs to be decoded by both A and C.
- The message $\mathbf{w}_{B,2} = [\, a_3 \; a_4 \,]$ needs to be decoded by C only.

To see how the transmitted 16-QAM symbol is selected for given $\mathbf{w}_{B,1}$, $\mathbf{w}_{B,2}$, we refer to Fig. 11.4. The node B uses $\mathbf{w}_{B,2}$ to determine which QPSK coset is being sent, while it uses $\mathbf{w}_{B,1}$ to determine the symbol within the coset. Now, since A knows $\mathbf{w}_{B,2}$ and thereby the coset a priori, it decodes the received signal by defining decision regions only for the coset of interest and ignoring the other 12 constellation points. After decoding $\mathbf{w}_{B,1}$, it retrieves the bits $[c_1 \; c_2]$ by using XOR. On the other hand, C decodes the transmission of B by using the full 16-QAM constellation, retrieves the 2-bit packets $\mathbf{w}_{B,1}$ and $\mathbf{w}_{B,2}$, and uses XOR to extract the bits $[a_1 \; a_2]$ from $\mathbf{w}_{B,1}$.

One may argue that we have again used digital combining with XOR. Nevertheless, here the decoding at A uses the side information in the "analog" part of the decoding, which is not the case with the usual XOR packet combining.

The result from [12] shows that a related broadcast strategy can be devised in information-theoretic sense by creating appropriate codebooks and decoding rules. Thus, the message from B can carry data to $A(C)$ at a rate $R_{BA}(R_{BC})$ where

$$R_{BA} \leq C(\gamma_1), \qquad R_{BC} \leq C(\gamma_2). \qquad (11.16)$$

With given data $\mathbf{w}_{BA}$ and $\mathbf{w}_{BC}$, this broadcast strategy is carried out as follows. Let us assume that B uses the maximal possible rates by putting the equality in (11.16). We first consider the case when

$$\frac{|\mathbf{w}_{BA}|}{C(\gamma_1)} \geq \frac{|\mathbf{w}_{BC}|}{C(\gamma_2)}. \qquad (11.17)$$

B divides $\mathbf{w}_{BA}$ into two messages $\mathbf{w}_{BA}^{(1)}$, $\mathbf{w}_{BA}^{(2)}$ such that

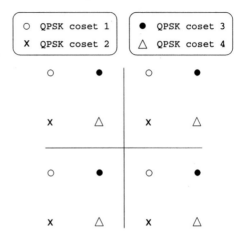

Fig. 11.4 Representation of 16-QAM with four QPSK subconstellations

- With $n_{B,1}$ channel uses and using the broadcast codebooks with side information it sends $\mathbf{w}_{BA}^{(1)}$ to A and $\mathbf{w}_{BC}$ to C, such that the following should be satisfied:

$$\frac{|\mathbf{w}_{BA}^{(1)}|}{C(\gamma_1)} = \frac{|\mathbf{w}_{BC}|}{C(\gamma_2)} \tag{11.18}$$

  from which the size of $\mathbf{w}_{BA}^{(1)}$ can be determined.
- With $n_{B,2}$ channel uses ordinary single-user codebook with rate $C(\gamma_1)$ to transmit $\mathbf{w}_{BA}^{(2)}$ to A.

From the above conditions the duration of the broadcast phase is found to be

$$n_B = n_{B,1} + n_{B,2} = \frac{|\mathbf{w}_{BA}|}{C(\gamma_1)}. \tag{11.19}$$

For the other case, when (11.17) is not satisfied, with a similar analysis it can be found that the duration of the broadcast phase is

$$n_B = \frac{|\mathbf{w}_{BC}|}{C(\gamma_2)}. \tag{11.20}$$

11.3.4 Numerical Illustration

Figures 11.5 and 11.6 compare the regions of achievable rate pairs (R_{AC}, R_{CA}) for different uplink/broadcast strategies of the schemes with decoding at the relay. Figure 11.5 compares the three-step DF scheme with the two-step JDF scheme when the simple broadcast strategy is used. The SNRs of the links are chosen such as to illustrate that none of the two achievable regions is completely contained in the

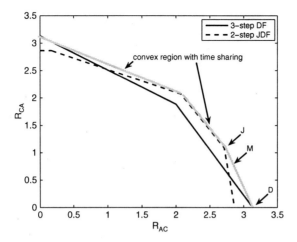

Fig. 11.5 Achievable rate regions with the three-step DF scheme and the two-step JDF scheme when the simple broadcast strategy is used. The parameters are $\gamma_0 = 0$ dB, $\gamma_1 = 15$ dB, $\gamma_2 = 20$ dB

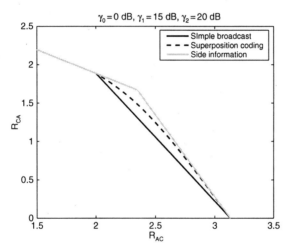

Fig. 11.6 Achievable rate regions with the three-step DF scheme when different broadcast strategies are used. The parameters are $\gamma_0 = 0$ dB, $\gamma_1 = 15$ dB, $\gamma_2 = 20$ dB

other. As shown in [7], the maximal two-way sum rate is achieved when $|\mathbf{w}_{BA}| = |\mathbf{w}_{BC}|$. The two strategies can be combined by time sharing [17] in order to achieve rate pairs that are not achievable by any single-strategy DF or JDF. The gray line shows the convex region of achievable rates when time sharing is used. For example, the rate pair at point M can be achieved by using the JDF strategy with rates at the point J for a fraction δ of time, while for the remaining fraction $(1 - \delta)$ of time using one-way transmission from A to C with B as helper (the rates at the point D).

Figure 11.6 compares the achievable rate region for the three-step DF scheme for the three different broadcast strategies. Note that the plot starts at a value $R_{AC} = 1.5$ in order to emphasize the differences between the strategies. As expected, the largest region is achieved by the broadcast strategy with side information. Similar results are obtained when JDF strategy is considered. In general, the benefits of the improved broadcast strategy are observable when the amount of data $|\mathbf{w}_{BC}|$ that need to be send over the stronger link are larger than $|\mathbf{w}_{BA}|$.

11.4 Two-Way Relaying Without Decoding at the Relay

The relay node B is not the intended destination of the data from the source and hence it is not necessary that it decodes the data. By leveraging on that observation, we can markedly increase the space of available communication strategies. In this section we discuss several techniques that do not require decoding at the relay. Note that the strategies without decoding at the relay are not novelty brought by the two-way relaying, and they have already been applied to one-way relaying, see, e.g. [18]. Nevertheless, the two-way relaying scenario has distinctive features that give rise to some completely novel strategies, such as denoise-and-forward (DNF), described further in the text.

We describe three strategies that have two steps, as in the JDF scheme: an uplink step over multiple-access channel, such that $n_A = n_C$, and a step of n_B channel uses for broadcast. As B does not obtain the messages W_{AC}, W_{CA}, it cannot combine them digitally by using the XOR operation. Instead, the signals from A and C are inherently combined through the multiple-access channel, such that the network coding is *analog* and done at the physical layer [4, 7, 11].

11.4.1 Amplify-and-Forward (AF)

Th strategy is amplify-and-forward (AF) [3, 6], in which the relay B amplifies and broadcasts the symbols of the received noisy signal $\mathbf{y}_B$ from the uplink phase, such that the vector transmitted in the broadcast phase is

$$\mathbf{x}_B = \beta \mathbf{y}_B = \beta(h_1 \mathbf{x}_A + h_2 \mathbf{x}_C + \mathbf{z}_B). \qquad (11.21)$$

The amplification factor β is chosen as

$$\beta = \sqrt{\frac{1}{|h_1|^2 + |h_2|^2 + N_0}} \qquad (11.22)$$

to make the average per-symbol transmitted energy at B equal to 1. For the AF scheme $n_B = n_A = n_C$. The mth symbol received by A is the amplified and additionally noised version of the mth symbol received by B in Step 1:

$$y_A[m] = \beta h_1 y_B[m] + z_A[m]$$
$$= \beta h_1^2 x_A[m] + \beta h_1 h_2 x_C[m] + \beta h_1 z_B[m] + z_A[m].$$

Assuming that A knows $x_A[m]$, h_1, h_2, and β, then it can remove the contribution of $\beta h_1^2 x_A[m]$ from $y_A[m]$ and obtain

$$r_A[m] = \beta h_1 h_2 x_C[m] + \beta h_1 z_B[m] + z_A[m], \tag{11.23}$$

which is a Gaussian channel for receiving $x_C[m]$ with SNR:

$$\gamma_{C \to A}^{(AF)} = \frac{\beta^2 |h_1|^2 |h_2|^2}{(\beta^2 |h_1|^2 + 1)N_0} = \frac{\gamma_1 \gamma_2}{2\gamma_1 + \gamma_2 + 1}. \tag{11.24}$$

This notation denotes that $\gamma_{C \to A}^{(AF)}$ is the SNR that determines the transmission rate R_C, such that $\mathbf{x}_C$ can be successfully decoded by A. Similarly, we can find the SNR which determines the rate R_A:

$$\gamma_{A \to C}^{(AF)} = \frac{\gamma_1 \gamma_2}{\gamma_1 + 2\gamma_2 + 1}. \tag{11.25}$$

The rate pair (R_A, R_C) used in Step 1 should satisfy

$$R_A \le C\left(\gamma_{A \to C}^{(AF)}\right), \qquad R_C \le C\left(\gamma_{C \to A}^{(AF)}\right) \tag{11.26}$$

and the achievable rate region is determined as

$$R_{AC} \le \frac{1}{2} C\left(\gamma_{A \to C}^{(AF)}\right), \qquad R_C \le \frac{1}{2} C\left(\gamma_{C \to A}^{(AF)}\right), \tag{11.27}$$

since the end-to-end rates are calculated over $2n_B$ channel uses.

11.4.2 Denoise-and-Forward (DNF)

Even if the relay does not decode the messages from A and C, it can still process the received signal $\mathbf{y}_B$ beyond mere amplification, while still not decoding the messages W_{AC}, W_{CA}. In general, we term those strategies denoise-and-forward (DNF) in order to emphasize that the noise is mitigated, but the signal is not decoded.

Motivating Example for Denoise-and-Forward (DNF)

Assume that the multiple-access channel at B is specified with $h_{AB} = h_{CB} = 1$:

$$Y_B = X_A + X_C + Z_B. \tag{11.28}$$

Let A, C use BPSK modulation $X_A, X_C \in \{-1, 1\}$. Assume at first that there is no noise $Z_B = 0$. Then the possible signals that B can observe are $\{-2, 0, 2\}$. If the received symbol is either 2 or -2 then B can infer that the signals sent by A and C are $(X_A, X_C) = (1, 1)$ and $(X_A, X_C) = (-1, -1)$, respectively. If B receives 0, then it has ambiguity whether the signals sent are $(X_A, X_C) = (1, -1)$ or $(X_A, X_C) = (-1, 1)$. However, e.g., if A sends 1 and learns that B has observed 0, then A can infer that $X_C = -1$. How much information does B need to send so that A and C can retrieve each other's symbols? One bit of information from B is sufficient: If B also uses BPSK modulation, then it can broadcast $X_B = -1$ when it observes $Y_B = -2$ or $Y_B = 2$ and it can broadcast $X_B = 1$ when $Y_B = 0$. One can easily check that, with the knowledge of X_A and X_B, A can infer X_C and, vice versa, knowing X_C and X_B, C can infer X_A.

  If the channel at B is noisy, then B needs to set appropriate decision regions for the symbols, see Fig. 11.7. The output of the decision process is a symbol from the set $\{-2, 0, 2\}$ and in the next step B applies *denoise mapping* in order to compress the ternary symbol from the decision process to a binary symbol $\{-1, 1\}$ that needs to be sent in the broadcast phase.

The example above outlines the main operations that need to be done by B to implement a DNF scheme: (1) decision process that quantizes the received signal and (2) mapping of the quantized signal to a message that is sent in the broadcast phase. In the example we have used *per-symbol denoising*, where the quantization and the mapping are done for each individual symbol. Such an approach is interesting when we consider non-information-theoretic analysis by considering finite-length packets and constellations, see Section 11.6.

For information-theoretic analysis, we need to consider *per-codeword denoising*. In that case, the decision/mapping at B is based on the received vector $\mathbf{y}_B$ during the $n_A = n_C$ channel uses of the multiple-access phase. The denoise mapping at B produces the message W_B and should be designed in such a way that by knowing

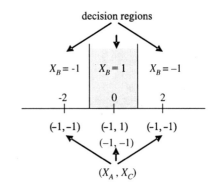

Fig. 11.7 Example of the decision regions and the denoise mapping when A, C use BPSK signaling and $h_{AB} = h_{CB} = 1$

W_{AC} and W_B, the node A can uniquely determine W_{CA} (analogous for C). More generally, the noisy versions of the broadcasted signal $\mathbf{x}_B$ are $\mathbf{y}_A$, $\mathbf{y}_C$, each being a vector of dimension n_B. After the broadcast phase, the node A should be able to decode W_{CA} from the observation $\mathbf{y}_A$ and using W_{AC} as a side information.

It is interesting to find the rate pair (R_{AC}, R_{CA}) that is on the outer bound of the achievable rate region. Assume that in the uplink phase, the transmission rates are $R_A = C(\gamma_1)$ and $R_C = C(\gamma_2)$. The relay B cannot decode W_{AC}, W_{CA} because the point $(R_A, R_C) = (C(\gamma_1), C(\gamma_2))$ is outside the capacity region of the multiple-access channel with B—this is shown by simply observing that $C(\gamma_1 + \gamma_2) < C(\gamma_1) + C(\gamma_2)$. Now assume that in the broadcast phase B uses $n_B = n_A = n_C$ to transmit $\mathbf{y}_B$ noiselessly, such that A, C can observe the exact output vector $\mathbf{y}_B$. Clearly, no DNF scheme can do better than this, since $\mathbf{y}_A$ and $\mathbf{y}_C$ can be represented as noisy copies of $\mathbf{y}_B$. If $\mathbf{x}_A$ is sent at rate R_A, then to be decodable at C it has to satisfy

$$R_A \le I(\mathbf{x}_A; \mathbf{y}_A | W_{CA}) \overset{(a)}{\le} I(\mathbf{x}_A; \mathbf{y}_B | W_{CA}) = n_A C(\gamma_1), \qquad (11.29)$$

where (a) follows from the data processing inequality [17]. However, C decodes the signal from A after $2n_A$ channel uses, such that the end-to-end rate from A to C is $\frac{1}{2}R_A = \frac{1}{2}C(\gamma_1)$. Making analogous observations for C, we can write that a point in the outer bound to the achievable region and the two-way sum rate is given by

$$(R_{AC}, R_{CA}) = \left(\frac{1}{2}C(\gamma_1), \frac{1}{2}C(\gamma_2) \right). \qquad (11.30)$$

The achievability of the outer bound with structured codes is discussed in Section 11.5.

11.4.3 Compress-and-Forward (CF)

An interesting variant of the DNF schemes is using the quantization/compression framework to determine the operations carried out at the relay node B and is termed compress-and-forward (CF) [15]. Figure 11.8 shows the data dependencies in the CF scheme.

After observing $\mathbf{y}_B$, B can obtain its quantized version $\hat{\mathbf{y}}_B$ as well as the message W_B, consisting of $n_B R_B$ bits, which is uniquely associated with $\hat{\mathbf{y}}_B$. We consider the decoding at C. The relay broadcasts $\mathbf{x}_B(W_B)$ and C receives $\mathbf{y}_C$. From the knowledge of $\mathbf{x}_C$ sent in the MA phase, C can create the set $\mathcal{W}^{MA}$ defined as

$$\mathcal{W}^{MA} = \{W_B | (\mathbf{x}_C, \hat{\mathbf{y}}_B(W_B)) \text{is jointly typical}\}. \qquad (11.31)$$

For formal definition of joint typicality, see [17]. Informally, we can say that from the knowledge of $\mathbf{x}_C$, the node C can create the set of candidate messages W_B that correspond to a set of quantized vectors $\{\hat{\mathbf{y}}_B\}$, such that the observed signal $\mathbf{y}_B$ is likely to be "close" to one of the vectors in that set $\{\hat{\mathbf{y}}_B\}$.

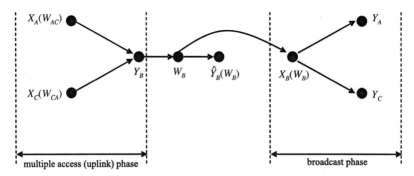

Fig. 11.8 Conceptual description of the compress-and-forward (CF) scheme

Next, after observing $\mathbf{y}_C$, C creates the set $\mathcal{W}^{BC}$ defined as

$$\mathcal{W}^{BC} = \{W_B | (\mathbf{x}_B(W_B), \mathbf{y}_C) \text{ is jointly typical}\}, \qquad (11.32)$$

which is the set of candidate messages that are likely to be sent by B in the broadcast phase. In a more conventional approach, one would select the transmission rate R_B so that W_B is decodable at C only from the observation $\mathbf{y}_B$. In that case $\mathcal{W}^{BC}$ should contain unique message W_B. The trick used here is W_B that does not need to be decodable only from $\mathbf{y}_C$. Instead, W_B can be found as the unique message that lies in the intersection $\mathcal{W}^{MA} \cap \mathcal{W}^{BC}$. Having found W_B, C can find the unique message W_{AC}.

An important element of the CF scheme is the quantization performed at B. For Gaussian channels, the parameter quantization in [15] is determined by using Gaussian test channel and the achievable rate regions are obtained by numerical optimization of the parameters of the test channel.

11.4.4 Numerical Illustration and Variations

In this section we illustrate the achievable rate regions for the schemes that do not require decoding at the relay. Figure 11.9 compares the achievable rate region for the AF scheme with the achievable rate region of the two-step JDF scheme (where the relay decodes). The point from the outer bound is also plotted as a reference. It is interesting to see that, for the chosen values of γ_1 and γ_2, the achievable region for AF contains points that are not achievable by decoding at the relay. From the results in [15] it can be seen that the CF scheme can bring a larger region as compared to AF, but not for all configurations of γ_1, γ_2.

In discussing the schemes without decoding at the relay, we have focused on the two-step schemes in which the uplink phase consists of transmissions over the multiple-access channel. The three-step schemes, which leverage on the direct link between A and C, can also utilize relaying which does not require decoding at the relay. For example, consider the possible way to use AF in the three-step scheme.

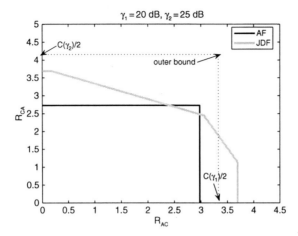

Fig. 11.9 Comparison of the achievable rate regions for the two-step schemes amplify-and-forward (AF) and joint decode-and-forward (JDF). The point from the outer bound is also plotted. The parameters are $\gamma_1 = 20$ dB and $\gamma_2 = 25$ dB

We can constrain $n_A = n_C = n_B$. B sums the n_A received symbols from the first step with the n_C symbols from the second step and amplifies the sum by respecting the transmit power constraint. Then C decodes by (a) combining the n_A symbols from Step 1 with the n_B symbols from Step 3 and (b) using its knowledge of the signal it has sent in Step 2. Another generalization of the schemes described here is the mixed forward [15], where the relay decodes the data in one direction (e.g., from A to C), while uses compression in the other direction (C to A).

11.5 Achieving the Two-Way Rates with Structured Codes

Using the arguments of random coding, it has been seen that it is hard, if possible at all, to achieve the maximal two-way sum rates, i.e., the outer bound point from Fig. 11.9. In this section we discuss the usage of structured codes in the DNF schemes, by which B can do reliable decisions about the combination of the codewords sent by A and C without actually decoding the messages W_{AC}, W_{CA}. We first discuss the case of discrete channel, where the structured codes are parity-check codes. For the Gaussian case, the structured codes are based on lattices.

11.5.1 Parity-Check Codes for Binary Symmetric Channels

In this section the input/output variables X_i, Y_j and the corresponding vectors $\mathbf{x}_i, \mathbf{y}_j$ are binary. The output of the multiple-access channel at the relay node B is defined as

$$Y_B = X_A \oplus X_C \oplus Z_B. \tag{11.33}$$

In the broadcast phase, the outputs at A and C are given as follows:

$$Y_A = X_B \oplus Z_A, \tag{11.34}$$
$$Y_C = X_B \oplus Z_C. \tag{11.35}$$

The noise variables $z_i, i \in \{A, B, C\}$ have Bernoulli distribution, with error probability $P(Z_i = 1) = p$. If B knows X_C a priori, then the capacity of the channel from A to B is

$$C_{AB|C} = 1 - H(p) = C_{CB|A}, \tag{11.36}$$

where analogously $C_{CB|A}$ is the capacity from C to B, if B knows X_A a priori. Let the source nodes select the transmission rates as

$$R_A = R_C = 1 - H(p). \tag{11.37}$$

We now use the coding theorem for the binary symmetric channel (BSC) for parity-check codes, see [19, Section 6.2]. This theorem states that the channel capacity of the binary symmetric channel can be approached arbitrarily closely by using the so-called coset codes. A (n, k) coset code is defined as a code with 2^k codewords of block length $n > k$ and each message has k bits. Each message is represented by a $1 \times k$ vector $\mathbf{w}$ and the corresponding codeword $\mathbf{x} \in \{0, 1\}^{[1 \times n]}$ is obtained as

$$\mathbf{x} = \mathbf{w}\mathbf{G} \oplus \mathbf{c}, \tag{11.38}$$

where $\mathbf{G}$ is a $k \times n$ binary matrix and c is $1 \times n$ binary vector. The theorem states that a code that is constructed with random selection of the bit entries in $\mathbf{G}$ and $\mathbf{c}$ can achieve the capacity of BSC.

Let us assume that, for a given codeword length n, the rates are selected

$$R_A = R_C = \frac{k}{n} < 1 - H(p) \tag{11.39}$$

and $\mathbf{G}$ and $\mathbf{c}$ are chosen according to the discussion above and are known to all the nodes. When the messages $\mathbf{w}_A$ and $\mathbf{w}_C$ are transmitted, A and C select their codewords, respectively, as follows:

$$\mathbf{x}_A = \mathbf{w}_A \mathbf{G} \oplus \mathbf{c}, \qquad \mathbf{x}_C = \mathbf{w}_C \mathbf{G}. \tag{11.40}$$

The received vector at B is

$$\mathbf{y}_B = \mathbf{x}_A \oplus \mathbf{x}_C \oplus \mathbf{z}_B = (\mathbf{w}_A \oplus \mathbf{w}_C)\mathbf{G} \oplus \mathbf{c}. \tag{11.41}$$

Conceptually, B can treat the received vector as an output of a BSC, over which a *virtual* transmitter sends the message $\mathbf{w}' = \mathbf{w}_A \oplus \mathbf{w}_C$ by using the same coset code as used by A. Then B tries to reliably decode the message $\mathbf{w}'$, rather than the individual messages $\mathbf{w}_A$ and $\mathbf{w}_C$. $\mathbf{m}'$ is a message of k bits that, by assumption, can be reliably decoded over the BSC. In the broadcast phase, B sends the codeword $\mathbf{x}' = \mathbf{w}'\mathbf{G} \oplus \mathbf{c}$, which can be reliably decoded by both A and C.

We can conclude that when the two–way relaying system features binary symmetric channels, the rate pair $(R_{AC}, R_{CA}) = (1 - H(p), 1 - H(p))$ of the outer bound point is achievable.

11.5.2 Gaussian Channel

At the moment we do not have a rigorous proof that the outer bound point is achievable in Gaussian channels and the discussion in this section should be taken as a possible way forward toward proving such achievability. Note that several works [10] have considered the usage of lattice codes for two–way relaying in Gaussian channels, but there is no known technique that certainly achieves the outer bound point.

In order to simplify the discussion, in this section we assume that the wireless channels are *real* Gaussian channels, such that the received signals are

$$Y_B = X_A + X_C + Z_B, \qquad (11.42)$$
$$Y_A = X_B + Z_A, \qquad (11.43)$$
$$Y_C = X_B + Z_C, \qquad (11.44)$$

where for the transmitter i the average power is $E[X_i^2] \le 1$ and the noise power is $E[Z_i^2] = \sigma^2$. We slightly abuse the notation used so far by defining

$$C(\gamma_1) = \frac{1}{2} \log_2(1 + \gamma_1) \qquad (11.45)$$

to correspond to the capacity of a real Gaussian channel with SNR γ_1. We assume that the transmission rates of A and C are chosen:

$$R_A = R(C) = C(\gamma_1) = C\left(\frac{1}{\sigma^2}\right). \qquad (11.46)$$

A lattice Λ is discrete subgroup of the Euclidean space $\mathbb{R}^n$ with the ordinary vector addition operation. If $\lambda_1, \lambda_2 \in \Lambda$, then $(\lambda_1 - \lambda_2) \in \Lambda$ and $(\lambda_1 + \lambda_2) \in \Lambda$. The lattices introduce algebraic structure for the codewords in the continuous channel. For detailed discussion on the lattices and their usage in achieving the capacity of Gaussian channel, the reader is directed to [20] and the references therein.

For our discussion, we use $n = n_A = n_B$, such that the transmitted codewords belong to $\mathbb{R}^n$. We use construction A to generate the lattice, as described in [20]. These are the steps to generate the lattices for A and C:

1. Pick a prime number p, such that the rate R and the number of channel uses n satisfy
$$p = 2^{nR}. \qquad (11.47)$$

2. Draw a generating vector $\mathbf{g} = (g_1, g_2, \ldots, g_n)$, where g_i is an integer uniformly picked from the set $\{0, 1, 2, \ldots, p - 1\}$.

3. Define the discrete codebook

$$\mathcal{C} = \{\mathbf{c} \in \mathbb{Z}_p^n : \mathbf{c} = (\mathbf{g} \cdot q) \bmod p \quad q = 0, 1, \ldots, p - 1\}. \tag{11.48}$$

4. The node A selects a random vector $\mathbf{u}_A = \{u_{A1}, u_{A2}, \ldots, u_{An}\}$, where each u_{Ai} is independently generated as a real number uniformly distributed in $[0, p)$. The codewords transmitted by A are obtained as follows: For each $c \in \mathcal{C}$, the corresponding codeword $\mathbf{x}_A(\mathbf{c})$ is obtained as

$$\mathbf{x}_A(\mathbf{c}) = \beta(p^{-1}[\mathbf{c} + \mathbf{u}_A]_p - 0.5), \tag{11.49}$$

where the notation $[v]_p$ denotes the per-component modulo-p operation:

$$[v]_p = \begin{cases} v & \text{if } v < p \\ v - p & \text{otherwise} \end{cases} \tag{11.50}$$

and the coefficient β is a normalization coefficient that ensures that the average power of the transmitted codewords is P. Using the crypto-lemma [21] (or see Lemma 1 in [20]), one can show that the ith component of each vector is a uniformly distributed real number in the interval $[-\frac{\beta}{2}, \frac{\beta}{2})$. Hence, in this case β should be chosen to be $\frac{\beta^2}{12} = P$ in order to match the power constraint.

5. The node C uses identical procedure as A, except that it generates an independent vector $\mathbf{u}_C$.

Let us consider the properties of the sum of such generated codewords. Each pair of codewords $(\mathbf{x}_A, \mathbf{x}_C)$ can be uniquely associated with the pair (q_A, q_C) where $q_A, q_C \in \{0, 1, \ldots, p - 1\}$ are the integers used in generating the correspondent $(\mathbf{c}_A, \mathbf{c}_C)$ in (11.48), such that we can use the notation $(\mathbf{x}_A(q_A), \mathbf{x}_C(q_C))$ or simply (q_A, q_C) to denote a pair of codewords. The sum of the codeword pair (q_A, q_C) is

$$s_i(q_A, q_C) = \beta(p^{-1}[[g_i q_A]_p + u_{Ai}]_p - 0.5 + p^{-1}[[g_i q_C]_p + u_{Ci}]_p - 0.5)$$

$$\overset{(a)}{=} \beta p^{-1}\left(([g_i q_A + u_{Ai}]_p + [g_i q_C + u_{Ci}]_p) - p\right), \tag{11.51}$$

where (a) follows from the distributive property of the modulo operation. The important component of $s_i(q_A, q_C)$ is

$$\tau_i(q_A, q_C) = [g_i q_A + u_{Ai}]_p + [g_i q_C + u_{Ci}]_p, \tag{11.52}$$

which can be represented as follows:

$$\tau_i(q_A, q_C) = [g_i[q_A + q_C]_p + u_{Ai} + u_{Ci}]_p + \delta_i p \tag{11.53}$$

$$= t_i + \delta_i p, \tag{11.54}$$

where

$$\delta_i = \begin{cases} 0 & \text{if } \tau_i(q_A, q_C) < p \\ 1 & \text{otherwise} \end{cases}. \tag{11.55}$$

On the other hand, using the crypto-lemma, one can infer that t_i is a real number that is uniformly distributed in $[0, p)$. Hence, the ith component of sum of the pair of codewords (q_A, q_C) can be represented by

$$s_i(q_A, q_C) = \beta p^{-1}(t_i + \delta_i p - p) = \beta(p^{-1}t_i + \delta_i - 1) = \beta(p^{-1}t_i - 0.5 + \delta_i - 0.5),$$
(11.56)

where $p^{-1}t_i - 0.5$ is a random variable that is uniformly distributed in $[-0.5, 0.5)$ and $\delta_i - 0.5$ is a binary random variable that can have a value of either -0.5 or 0.5. The distribution of δ_i is dependent on t_i and can be determined by observing that the distribution of $p^{-1}t_i + \delta_i - 1$ is equal to the distribution obtained from a sum of two random numbers uniformly distributed in $[-0.5, 0.5]$, which implies

$$P(\delta_i = 0|t_i) = \frac{t_i}{p}, \qquad P(\delta_i = 1|t_i) = 1 - \frac{t_i}{p}.$$
(11.57)

In the sequel we will need the following definition.

Definition 11.1. The codeword pair $(\mathbf{x}_A(q_A), \mathbf{x}_C(q_C))$ belongs to the class m if

$$[q_A + q_C]_p = (q_A + q_C) \bmod p = m.$$
(11.58)

There are p different classes, each containing p codeword pairs. If two codeword pairs (q_{A1}, q_{C1}) and (q_{A2}, q_{C2}) belong to the same class, then one can easily show that for each component i

$$t_i(q_{A1}, q_{C1}) = t_i(q_{A2}, q_{C2}),$$
(11.59)

such that $s_i(q_{A1}, q_{C1})$ and $s_i(q_{A2}, q_{C2})$ are either equal or their difference is exactly $\frac{\beta}{2}$.

The relay B receives

$$\mathbf{y}_B = \mathbf{s}(q_A, q_C) + \mathbf{z}_B.$$
(11.60)

If from $\mathbf{y}_B$ B can determine to which of the $p = 2^{nR}$ classes of codewords does (q_A, q_C) belong, then in the broadcast phase B can use "conventional" random codebooks to send the nR bits that describe the class of the pair of observed codewords. If A knows the class, then it can uniquely determine q_C and thereby W_{AC}. The same is valid for C and in such case we would have achieved the point of the outer bound.

Having introduced this framework, we can think that the communication from A, C to B is done between a virtual transmitter V and B. The virtual transmitter generates p codewords $\{\mathbf{x}_V\}$ where the ith component of each codeword is uniformly and independently distributed in $\left(-\frac{\beta}{2}, \frac{\beta}{2}\right)$. Before transmitting the signal to B, V generates *binary self-noise* at each component of $\{\mathbf{x}_V\}$: for the ith component, V looks at x_{Vi} and uses the appropriate distribution for δ_i, see (11.56), to determine whether to add or subtract $\frac{\beta}{2}$.

In the absence of the binary self-noise, the codewords of $\{x_V\}$ can be reliably decoded at B, which follows from the manner in which the codebook is generated and the noisy channel coding theorem for Gaussian channels. *We conjecture that there are codebooks $\{x_V\}$ such that B can reliably decode x_V even in the presence of the binary self-noise, but the rigorous proof for this claim is not available at this time and is a subject of ongoing work.*

11.6 Signaling Constellations for Finite Packet Lengths

The discussion hitherto has been mostly focused on information-theoretic aspects of the bidirectional relaying, which assumed arbitrarily long codewords/packets and error probability that is asymptotically zero. In this section, we focus on the practical aspects of the two-way relaying by considering finite-length packets and finite constellations. We restrict ourselves to the most interesting case, namely the two-step DNF schemes that do not require decoding at the relay. The schemes use with per-symbol denoising, rather than per-codeword one, and we address the following questions: How should the relaying node B generate a denoised signal for the broadcast phase by observing the received signal at the uplink phase? How many constellation points are required to reliably forward two distinct messages to the destination nodes A and C? Considering the case in which both terminals use QPSK constellations in the uplink phase, we present one example of signaling strategies for reliable bidirectional relaying. The ensuing discussions tell us two interesting results: (1) we should use multiple network-coding rules adaptively optimized according to the channel state information at the relay B and (2) some specific channel conditions necessitate the use of unconventional 5-ary signal constellations, rather than the QPSK.

11.6.1 XOR Denoising

When A, C are using QPSK modulation, the combined signal $h_1 X_A + h_2 X_C$ can have up to 16 possible values, depending on the channel coefficients h_1, h_2. As it has been noted in Section 11.4.2, the relay needs to compress the observed signal Y_B before forwarding it in the broadcast phase. Letting $\mathcal{M}(\cdot)$ be the QPSK signal mapper, the 2-bit digital data of the ML estimate $\hat{X}_i$ can be written as $\hat{w}_i = \mathcal{M}^{-1}(\hat{X}_i) \in \{0, 1, 2, 3\}$. In the DNF scheme that uses XOR-based network coding, the relay B creates the QPSK signal for the broadcast phase in the following way. It maps the received signal within the ML region for $(\hat{w}_A, \hat{w}_C)$ into the QPSK constellation as $X_B = \mathcal{M}(w_B)$ where $w_B = \mathcal{D}(\hat{w}_A, \hat{w}_C) = \hat{w}_A \oplus \hat{w}_C$. The mapper $\mathcal{D}(\cdot)$ denotes the digital denoising function. Since a terminal node knows its own information, it can detect the desired data from the denoising signal. The XOR denoising works very well for some channel conditions, e.g., $\phi \simeq 0$ where we define

$$h_2/h_1 = \rho \exp(j\phi). \qquad (11.61)$$

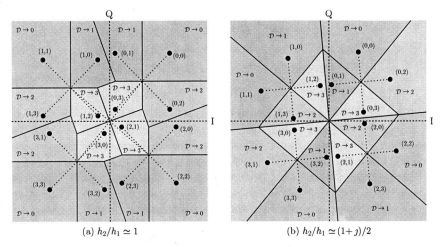

Fig. 11.10 Examples of received signal constellations at the relay B and XOR denoising (©[2008] IEEE. Reprinted with permission, IEEE Global Communication Conference)

Figure 11.10(a) shows an example of the received signal constellation Y_B at the uplink phase for $\rho \simeq 1$ and $\phi \simeq 0$. The four signal points at the center may be unreliable in the ML estimation because the distance between the points are short. This produces unreliable relaying when we adopt the JDF scheme at the relay. Meanwhile, as the DNF scheme maps such closest neighboring points into the same denoising signal, it offers a significant improvement in achievable throughput performance.

11.6.2 Adaptive Denoising with Quintary Cardinality

The denoising function $\mathcal{D}(\cdot)$ should be adaptively changed as a function of the channel state information h_1 and h_2 because the ML region highly depends on the channel condition. The XOR denoising function $\mathcal{D}(\hat{w}_A, \hat{w}_C) = \hat{w}_A \oplus \hat{w}_C$ does not work well for some specific channel conditions as illustrated in Fig. 11.10(b). This figure shows an example of the received signal y_B for $\rho \simeq 1/\sqrt{2}$ and $\phi = \pi/4$. As this figure shows, all the closest neighbor points are mapped into different denoising points. In particular, the distance between the ML points $(\hat{w}_A, \hat{w}_C) = (0, 1)$ and $(1, 2)$ is very short, which leads to an unreliable relaying performance.

It is very interesting to observe that for this case there does not exist a denoising function with 4-ary cardinality which can map all the closest neighbors into the same denoising point. The minimum achievable cardinality of the denoising function to avoid the distance shortening is 5, which implies that we need to use 5-ary modulation in the broadcast phase. For these channel conditions, the coding rule is given as

$$\mathcal{D}(\hat{w}_A, \hat{w}_C) = \begin{cases} 0 & \text{for } (\hat{W}_A, \hat{W}_C) \in \{(0,0), (1,1), (2,2), (3,3)\} \\ 1 & \text{for } (\hat{W}_A, \hat{W}_C) \in \{(0,3), (2,0), (3,1)\} \\ 2 & \text{for } (\hat{W}_A, \hat{W}_C) \in \{(0,1), (1,2), (2,3)\} \\ 3 & \text{for } (\hat{W}_A, \hat{W}_C) \in \{(3,2), (2,1), (1,0)\} \\ 4 & \text{for } (\hat{W}_A, \hat{W}_C) \in \{(1,3), (3,0), (0,2)\} \end{cases} \qquad (11.62)$$

The above denoising function performs well for $h_2/h_1 \simeq (1+j)/2$ or $(1-j)$. For the other channel conditions, we require three more 5-ary denoising functions and one more 4-ary denoising function to avoid all the possible distance shortening; in total there are six coding rules. For the 5-ary denoising, we should use some kind of 5-QAM signaling for broadcasting. Although it exhibits a slight loss in the Euclidean distances as compared to the QPSK modulation, the adaptive use of six denoising functions can bring a substantial benefit, as the results in the following section show.

11.6.3 End-to-End Throughput Performance

In Fig. 11.11, we show the performance comparisons in end-to-end throughput as a function of average SNR under Nakagami–Rice fading channels for a Ricean factor of 10 dB. We assume $E[|h_1|^2] = E[|h_2|^2]$ for simplicity. In this figure, we plot the curves for the conventional four-step relaying scheme, the three-step DF scheme with XOR network coding, and the two-step DNF scheme with XOR denoising and adaptive denoising. The direct link between A and C is assumed to have SNR

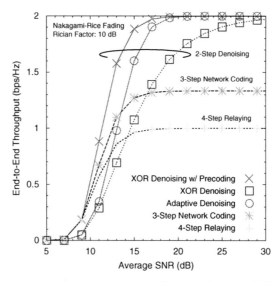

Fig. 11.11 End-to-end throughput performance as a function of average SNR in Nakagami–Rice fading channels for 10 dB Ricean factor (©[2008] IEEE. Reprinted with permission, IEEE Global Communication Conference)

of 0. Due to the time efficiency, the three-step network coding is superior to the four-step protocol with an improvement of 33%, as discussed before. The two-step DNF scheme further improves the throughput by a maximum of 100%. The XOR denoising can offer an excellent throughput performance if we can use the precoding technique for phase synchronization to achieve $\phi = 0$. However, it seems technically infeasible to obtain an accurate phase synchronization among the distributed terminals. Without such a precoding, the XOR denoising suffers from a serious performance degradation because of the distance shortening occurred in several channel conditions. This degradation can be significantly compensated by using adaptive network coding which allows the use of 5-ary modulations as well as 4-ary one.

11.7 Conclusions

The shared wireless medium fosters cooperative (relay-aided) communication modes among the proximate nodes. The two-way relaying scenarios lift such cooperative techniques to the next level, by leveraging on the recent ideas of network coding. We have shown that even the simple 3-node scenario with two-way relaying is a fertile ground for devising novel communication strategies. For a class of two-way relaying techniques, the relaying node decodes the messages before broadcasting them back to the terminals. For that class of techniques, the novel ideas are featured in the broadcast methods, since the relay has a large freedom in combining the two received messages. We have also elaborated on the techniques that do not require decoding at the relay, where the innovative techniques are combining several operations at the relay: detection, quantization, mapping, and recoding. In the information-theoretic setting, rather than using solely random codebooks, we have shown that the structured codes, such as the lattices, can have great utility if does not decode but only denoises the received signals. Besides the information-theoretic discussion, we have provided insights into the design of the two-way relaying schemes with practical modulation constellations.

The techniques for coded bidirectional relaying represent a prime example of the large freedom that a wireless network designer has in devising novel communication modes. This is in particular visible if we compare them with the protocol designs that are aligned to the layered protocol structure. The presented communication schemes somehow violate the layering, since (a) the intermediate nodes (network layer) do not decode the information before forwarding it further and (b) the adopted network-coding approach observes more than one flow simultaneously, while the layered architecture is single-flow-oriented. The approaches for coded bidirectional relaying leave largely open venues for future work: code designs, channel estimation, scheduling, retransmission protocols, etc. To name one more general topic— despite the apparent performance advantages in the simple three-node scenario—it is not straightforward to see how the techniques without decoding at the relay can scale to networks with many nodes.

References

1. Larsson, P., Johansson, N., and Sunell, K.-E.: Coded Bi-directional Relaying, *In Proc. 5th Scandinavian Workshop Ad Hoc Netw. (ADHOC'05)*, Stockholm, Sweden, May 2005.
2. Katti, S., Katabi, D., Hu, W., Rahul, H., and Médard, M.: The Importance of Being Opportunistic: Practical Network Coding for Wireless Environments, *In Proc. 43rd Annual Allerton Conf. Commun. Control Comput.*, Sep. 2005.
3. Popovski, P. and Yomo, H.: Bi-directional Amplification of Throughput in a Wireless Multi-Hop Network, *In Proc. IEEE 63rd Vehicular Technol. Conf. (VTC)*, Melbourne, Australia, May 2006.
4. Popovski, P. and Yomo, H.: The Anti-Packets Can Increase the Achievable Throughput of a Wireless Multi-hop Network, *In Proc. IEEE Int. Conf. Commun. (ICC)*, Istanbul, Turkey, Jun. 2006.
5. Hausl, C. and Hagenauer, J.: Iterative Network and Channel Decoding for the Two-Way Relay Channel, *In Proc. IEEE Int. Conf. Commun. (ICC)*, Istanbul, Turkey, Jun. 2006.
6. Rankov, B. and Wittneben, A.: Spectral Efficient Protocols for Half-Duplex Relay Channels, *IEEE J. Select. Areas Commun.*, vol. 25, pp. 379–389, Feb. 2007.
7. Popovski, P. and Yomo, H.: Physical Network Coding in Two-Way Wireless Relay Channels, *In Proc. IEEE Int. Conf. Commun. (ICC)*, Glasgow, Scotland, Jun. 2007.
8. Yomo, H. and Popovski, P.: Opportunistic Scheduling for Wireless Network Coding, *In Proc. IEEE Int. Conf. Commun. (ICC)*, Glasgow, Scotland, Jun. 2007.
9. Kim, S. J., Mitran, P., and Tarokh, V.: Performance bounds for bi-directional coded cooperation protocols, *IEEE Trans. Inform. Theory*, vol. 54, no. 11, pp. 5235–5241, 2008. http://arxiv.org/abs/cs/0703017.
10. Narayanan, K., Wilson, M. P., and Sprintson, A.: Joint Physical Layer Coding and Network Coding for Bi-directional Relaying, *In Proc. 2007 Allerton Conf. Commun. Control Comput.*, Monticello, IL, USA, 2007.
11. Katti, S., Gollakota, S., and Katabi, D.: Embracing Wireless Interference: Analog Network Coding, *In Proc. ACM SIGCOMM*, 2007.
12. Oechtering, T. J., Schnurr, C., Bjelakovic, I., and Boche, H.: Broadcast Capacity Region of Two-Phase Bidirectional Relaying, *IEEE Transactions on Information Theory*, vol. 54, no. 1, pp. 454–458, Jan. 2008.
13. Mitran, P.: The Diversity-Multiplexing Tradeoff for Independent Parallel MIMO Channels, *In Proc. IEEE Int. Symp. Inform. Theory (ISIT)*, Toronto, Canada, Jul. 2008.
14. Koike-Akino, T., Popovski, P., and Tarokh, V.: Denoising Maps and Constellations for Wireless Network Coding in Two-Way Relaying Systems, *In Proc. IEEE Globecom*, 2008, New Orleans, LA, Dec. 2008.
15. Kim, S. J., Devroye, N., Mitran, P., and Tarokh, V.: Achievable Rate Regions for Bi-directional Relaying, http://arxiv.org/abs/0808.0954.
16. Shannon, C. E: Two–Way Communications Channels, *In Proc. 4th Berkeley Symp. Math. Stat. Prob.*, Chicago, IL, Jun. 1961.
17. Cover, T. M., and Thomas, J.A: Elements of Information Theory, John Wiley & Sons Inc., New York, N. Y., 1991.
18. Laneman, J. N., Tse, D. N. C., and Wornell, G. W.: Cooperative Diversity in Wireless Networks: Efficient Protocols and Outage Behavior, *IEEE Trans. Infor. Theory*, vol. 50, no. 12, pp. 3062–3080, Dec. 2004.
19. Gallager, R.: Information Theory and Reliable Communication, New York: Wiley, 1968.
20. Erez, U. and Zamir, R.: Achieving $(1/2)\log(1 + \text{SNR})$ on the AWGN Channel with Lattice Encoding and Decoding, *IEEE Trans. Inform. Theory*, vol. 50, pp. 2293–2314, Oct. 2004.
21. Forney Jr., G. D.: On the Role of MMSE Estimation in Approaching the Information-Theoretic Limits of Linear Gaussian Channels: Shannon meets Wiener, *In Proc. 2003 Allerton Conf. Commun. Control Comput.*, Monticello, IL, USA, 2003.

Chapter 12
Minimum Cost Subgraph Algorithms for Static and Dynamic Multicasts with Network Coding

Fang Zhao, Muriel Médard, Desmond Lun, and Asuman Ozdaglar

12.1 Introduction

Network coding, introduced by Ahlswede et al. in their pioneering work [1], has generated considerable research interest in recent years, and numerous subsequent papers, e.g., [2–6], have built upon this concept. One of the main advantages of network coding over traditional routed networks is in the area of multicast, where common information is transmitted from a source node to a set of terminal nodes. Ahlswede et al. showed in [1] that network coding can achieve the maximum multicast rate, which is not achievable by routing alone. When coding is used to perform multicast, the problem of establishing minimum cost multicast connection is equivalent to two effectively decoupled problems: one of determining the subgraph to code over and the other of determining the code to use over that subgraph. The latter problem has been studied extensively in [5, 7–9], and a variety of methods have been proposed, which include employing simple random linear coding at every node. Such random linear coding schemes are completely decentralized, requiring no coordination between nodes, and can operate under dynamic conditions [10]. These papers, however, all assume the availability of dedicated network resources.

In this chapter, we focus on the former problem, which is to find the min-cost subgraph that allows the given multicast connection to be established (with appropriate coding) over coded packet networks. This problem has been studied in [11, 12]. The analogous problem for routed network is the Steiner tree problem, which is known

Fang Zhao, Muriel Médard, and Asuman Ozdaglar
Massachusetts Institute of Technology, 77 Mass Avenue, Cambridge, MA 02139, USA
e-mail: {zhaof, medard, asuman}@mit.edu

Desmond Lun
The Broad Institute of MIT and Harvard, 7 Cambridge Center, Cambridge, MA 02142, USA
e-mail: dlun@broad.mit.edu

V. Tarokh (ed.), *New Directions in Wireless Communications Research*,
DOI 10.1007/978-1-4419-0673-1_12,
© Springer Science+Business Media, LLC 2009

to be NP complete [13,14]. When coding is allowed, the min-cost subgraph problem can be formulated as a linear programming (LP) problem, and in this chapter, we examine algorithms to solve it for both static and dynamic multicasts.

Min-cost Subgraph for Static Multicasts

By static multicast, we refer to the case where a connection is set up for the user of a multicast group whose membership stays constant throughout the connection duration. The network topology, on the other hand, is not necessarily static. Lun et al. showed in [12] that the min-cost subgraph problem can be solved in a decentralized manner by using the dual subgradient method. Since subgraph optimization and coding can be decoupled in the multicast problem with network coding, and both of them can be done in a decentralized manner [10, 12], we have a completely decentralized system for coded multicast. There is no coordination between nodes, each node only knows the costs of its incoming and outgoing links, and communicates with its neighbors. In Section 12.3, we give an overview of this dual subgradient method, and study its convergence performance both theoretically and numerically.

There has been much work on using subgradient methods to solve the Lagrangian dual of a convex constraint optimization problem. The convergence behavior of the subgradient method used on the dual problem is well understood under various step size rules. However, in practice, the main interest is in solving the primal problem, and recovering, from the dual iterations, feasible or near-feasible primal solutions that converge to the optimal solution. There are special cases where the primal solutions computed as a by-product of the dual iterations are feasible and do converge, such as in [15], but this is not the case in general. There are only a few papers studying the recovery of primal solutions from the dual iterations, for example [16–18]. Recently, Nedić and Ozdaglar also looked at the convergence rate of the primal solutions in [19].

In Section 12.3, we present two slightly different formulations of the min-cost subgraph problem for network coding, both of which have the same optimal solutions. These two formulations give rise to two different distributed algorithms. One of them gives us a theoretical bound on the convergence rate of the primal solution; however, its intermediate primal solutions are not always feasible. The second one, on the other hand, produces a feasible subgraph after each iteration, which allows us to start the multicast with minimum delay. We would like to point out that this is possible due to the special structure of the network coding problem, and it is not true in general for dual subgradient method. Thus, coding is central to this formulation of the problem, even though it is not appearing explicitly. More details on this are presented in Section 12.3.1.

We also introduce heuristics to improve the convergence performance of our algorithm, and through simulations, we show that the algorithm produces significant reduction in multicast energy as compared to the centralized routing algorithm just after a few iterations, and it converges to the optimal solution quickly.

One of the challenges of wireless networks, such as ad hoc networks and sensor networks, is variability of the network topology. Topology changes can be caused by mobility of users, sleeping or waking up of nodes, or the shadowing effect due to moving obstacles. Our algorithm is also put to test in a dynamic wireless network model, and we show that the subgradient method is robust to topology changes, and nodes are able to adjust their transmission power levels to move smoothly and quickly to a new optimal subgraph in a distributed manner.

Min-cost Subgraph for Dynamic Multicasts

In applications such as real-time video distribution and teleconferencing, users can join or leave the multicast group at any time during the session. In such cases, we need to adjust the multicast subgraph to cater for the needs of this dynamic group. Lun et al. gave a dynamic programming formulation of this problem in [20], which aims to deliver continuous service to the users. However, link and code rearrangement, which are defined later, can still occur under their formulation.

In the context of traditional routing networks, this problem corresponds to the dynamic Steiner tree (DST) problem [21]. In DST, it is important to limit the number of rearrangements as a connection evolves, because rearranging a large multipoint connection may be time-consuming and may require significant use of network resources in the form of CPU time. In addition, rearrangement of a connection may result in the blocking of some parts of the connection as rearrangement proceeds. Therefore, the DST problem comes in two flavors [21, 22]. One is the nonrearrangeable version, in which rearrangement of existing routes is not allowed. In the other version, rearrangement is allowed, but the cost of rearrangements is taken into consideration.

The situation is similar in networks with coding. When the membership of the multicast group changes, we want to minimize the disturbance to existing users in the group by limiting both *link rearrangements* and *code rearrangements*. A *link rearrangement* occurs when some links in the current multicast subgraph are removed causing alternate paths to be used to serve existing users (see Fig. 12.1(a) for an example). Like in the routing networks, owing to the change in the physical connection, this kind of rearrangement causes disruptions to the continuous service to the multicast group. The second kind of rearrangement, which we call *code rearrangement*, is more subtle. Code rearrangement occurs when new incoming links are added to existing nodes in the multicast subgraph. Figure 12.1(b) shows an example for code rearrangements. Since random coding is used by the intermediate nodes in the subgraph to perform network coding, when a node has an additional incoming link, it has to generate a new set of random parameters to mix the incoming streams. All receivers downstream, therefore, have to use these new parameters and recompute the inversion matrix to decode the data streams. This scenario does not involve any physical switching of paths for the existing terminals, but it still causes a minor disruption to the continuous service due to this reprocessing of network coding parameters. Note that the disruptions caused by code rearrangements are generally smaller than that caused by link rearrangements.

In Section 12.4 we present algorithms that adapt to the changing demand of the multicast group, and at the same time, minimize disturbances to existing users. We also compare their performances through simulation.

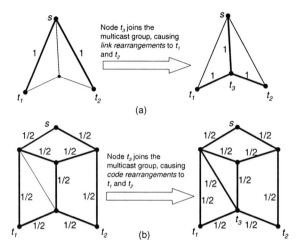

Node t_3 joins the multicast group, causing *link rearrangements* to t_1 and t_2

Node t_3 joins the multicast group, causing *code rearrangements* to t_1 and t_2

Fig. 12.1 (a) Example of an online step that causes link rearrangements to existing users and (b) example of an online step that causes code rearrangements to existing users. The multicast rate from source node s to terminal nodes $\{t_1, t_2, t_3\}$ is 1. The *thick lines* indicate links used in the multicast and the numbers against them indicate the rate of flow on them

12.2 Problem Formulation

In this section, we present the LP formulation of the min-cost subgraph problem in both wireline and wireless networks. We also derive the Lagrangian dual of these LP problems, which will be used in the distributed algorithms presented in Section 12.3.

12.2.1 Wireline Networks

We look at the problem of single multicast in wireline networks and model the network with a directed graph $G = (N, A)$, where N is the set of nodes and A is the set of links in the network. Each link $(i, j) \in A$ is associated with a non-negative number a_{ij}, which is the cost per unit flow on this link. We assume that the total cost of using a link is proportional to the flow, z_{ij}, on it. For the multicast, suppose we have a source node $s \in N$ producing data at a positive rate R that it wishes to transmit to a non-empty set of terminal nodes T in N.

It is shown in [1] that a subgraph z is capable of supporting a multicast connection of rate R from source s to T if and only if the min-cut from s to any $t \in T$ is greater than or equal to R. Hence, the problem of finding the min-cost subgraph can be formulated into the following LP problem [6]:

$$\text{minimize } f(z) = \sum_{(i,j)\in A} a_{ij} z_{ij}$$

$$\text{subject to } z_{ij} \geq x_{ij}^{(t)} \qquad\qquad \forall (i,j) \in A, t \in T,$$

$$\sum_{\{j|(i,j)\in A\}} x_{ij}^{(t)} - \sum_{\{j|(j,i)\in A\}} x_{ji}^{(t)} = \delta_i^{(t)} \ \forall i \in N, t \in T, \qquad (12.1)$$

$$x_{ij}^{(t)} \geq 0, \qquad\qquad \forall (i,j) \in A, t \in T,$$

where $x_{ij}^{(t)}$ corresponds to the virtual flow on link (i, j) for terminal t, z_{ij} is the actual flow on link (i, j) in the multicast subgraph, and

$$\delta_i^{(t)} = \begin{cases} R & \text{if } i = s, \\ -R & \text{if } i = t, \\ 0 & \text{otherwise.} \end{cases}$$

Although the decision variables, z_{ij}, are unbounded in the above formulation, it is easy to see that any optimal solution of (12.1), z^*, is bounded, i.e.,

$$0 \leq z_{ij}^* \leq b_{ij} \qquad \forall (i,j) \in A, \qquad (12.2)$$

for any $b_{ij} > R$. Thus, including the additional constraint (12.2) in (12.1) would not change the optimal solution set. However, it will affect its Lagrangian dual, and consequently, the algorithm for solving this problem. We will see later in Section 12.3.2 that this additional constraint can help us derive a theoretical bound on the convergence rate of our distributed algorithm.

To derive the Lagrangian dual problem for (12.1), we assign dual variable $p_{ij}^{(t)}$ to the constraint $z_{ij} \geq x_{ij}^{(t)}$ and leave the rest of the primal constraints in the dual objective function. In this way, z_{ij}'s do not appear in the dual problem, and this special structure of the network coding problem allows us to have a feasible primal solution after each dual iteration, as we will see in Section 12.3.1. The dual problem for (12.1) is given by

$$\text{maximize } q(p) = \sum_{t\in T} q^{(t)}(p^{(t)})$$

$$\text{subject to } \sum_{t\in T} p_{ij}^{(t)} = a_{ij} \qquad \forall (i,j) \in A, \qquad (12.3)$$

$$p_{ij}^{(t)} \geq 0 \qquad\qquad \forall (i,j) \in A, t \in T,$$

where

$$q^{(t)}(p^{(t)}) = \min_{x^{(t)}\in F_x^{(t)}} \sum_{(i,j)\in A} p_{ij}^{(t)} x_{ij}^{(t)} \qquad \forall t \in T, \qquad (12.4)$$

and $F_x^{(t)}$ is the bounded polyhedron of points $x^{(t)}$ satisfying the conservation of flow constraints

$$\sum_{\{j|(i,j)\in A\}} x_{ij}^{(t)} - \sum_{\{j|(j,i)\in A\}} x_{ji}^{(t)} = \delta_i^{(t)} \ \forall i \in N,$$

$$x_{ij}^{(t)} \geq 0 \qquad\qquad \forall (i,j) \in A.$$

Note that subproblem (12.4) is a standard shortest path problem with link costs $p_{ij}^{(t)}$, which can be solved using a multitude of distributed algorithms (e.g., distributed Bellman–Ford).

When constraint (12.2) is included in the primal problem, the dual problem becomes

$$\text{maximize } q(p) = \sum_{t\in T} q^{(t)}(p^{(t)}) + \sum_{(i,j)\in A} r_{ij}(p_{ij}) \qquad (12.5)$$

$$\text{subject to } p_{ij}^{(t)} \geq 0 \qquad\qquad \forall (i,j) \in A, t \in T,$$

where

$$r_{ij}(p_{ij}) = \min_{z\in F_z}(a_{ij} - \sum_{t\in T} p_{ij}^{(t)})z_{ij} \quad \forall (i,j) \in A,$$

and F_z is the bounded region of z given by

$$0 \leq z_{ij} \leq b_{ij} \quad \forall (i,j) \in A.$$

12.2.2 Wireless Networks

Under this model, we consider wireless networks where nodes are placed randomly within a 10×10 square with a radius of connectivity r. The energy required to transmit at unit rate to a distance d is taken to be d^2. Let f_{ij} be the cost function of link (i, j), and in our model, $f_{ij}(z_{ij}) = a_{ij}z_{ij}$ where $a_{ij} = d_{ij}^2$ is the energy required to send at unit rate over this link and z_{ij} is the rate of flow on this link. We justify this assumption in the cases where energy is the most significant constraint, but there are, for example, sufficient time or frequency slots to guarantee that no two transmissions ever interfere. This model is discussed in more depth in [6].

We consider wireless networks in which antennas are omnidirectional. When we transmit from node i to node j, we get transmission to all nodes whose distance from i is less than that from i to j "for free" — a phenomenon referred to as the "wireless multicast advantage" in [23]. If we impose an ordering $\preceq$ on the set of outgoing links from i, such that $(i, k) \preceq (i, j)$ if and only if $a_{ik} \leq a_{ij}$, we can then assume that we obtain a lossless broadcast link of unit rate from node i to all nodes k such that $(i, k) \preceq (i, j)$ for cost a_{ij}. Consider the example shown in Fig. 12.2, where there are three nodes within distance r from node i. If node i transmits with power a_{ij_3} to node j_3, the two nearer nodes, j_1 and j_2, also receive this information without additional cost. Thus, the situation here is quite different from the wireline case. Instead of picking links to transmit on in the wireline networks, the nodes in wireless networks pick power levels to transmit with, and this in turn determines their radius of coverage.

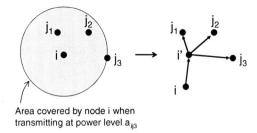

Area covered by node i when
transmitting at power level a_{ij_3}

Fig. 12.2 The "wireless multicast advantage" associated with omnidirectional antennas. The three destinations j_1, j_2, and j_3 can all be reached at the same time with cost a_{ij_3}, and this is equivalent to having three unit capacity links from i to j_1, j_2, and j_3. Here, we also included a virtual unit-capacity link (i, i') to impose the constraint that information transmitted on the three links must be the same

Similar to the wireline case, in the wireless network, the min-cost subgraph that can be used to perform the multicast with network coding is given by the following linear optimization problem:

$$\text{minimize } f(z) = \sum_{(i,j) \in A} a_{ij} z_{ij}$$

$$\text{subject to } \sum_{\{k|(i,k) \in A, (i,k) \succeq (i,j)\}} (z_{ik} - x_{ik}^{(t)}) \geq 0 \quad \forall (i, j) \in A', t \in T,$$

$$\sum_{\{j|(i,j) \in A\}} x_{ij}^{(t)} - \sum_{\{j|(j,i) \in A\}} x_{ji}^{(t)} = \sigma_i^{(t)} \quad \forall i \in N, t \in T, \quad (12.6)$$

$$x_{ij}^{(t)} \geq 0 \qquad\qquad\qquad \forall (i, j) \in A, t \in T,$$

Here, A' is a subset of A with the property that the constraint

$$\sum_{\{k|(i,k) \in A, (i,k) \succeq (i,j)\}} (z_{ik} - x_{ik}^{(t)}) \geq 0$$

is unique for all $(i, j) \in A'$.

The subgraph optimization scheme also uses the Lagrangian dual of (12.6) given below:

$$\text{maximize } q(p) = \sum_{t \in T} q^{(t)}(p^{(t)}) \qquad\qquad\qquad (12.7)$$

$$\text{subject to } \sum_{\{k|(i,k) \in A', (i,k) \preceq (i,j)\}} \sum_{t \in T} p_{ik}^{(t)} = a_{ij} \ \forall \ (i, j) \in A,$$

$$p_{ij}^{(t)} \geq 0 \qquad\qquad\qquad \forall \ (i, j) \in A', t \in T,$$

where

$$q^{(t)}(p^{(t)}) = \min_{x^{(t)} \in F_x^{(t)}} \sum_{(i,j) \in A} \left(\sum_{\{k|(i,k) \in A', (i,k) \preceq (i,j)\}} p_{ik}^{(t)} \right) x_{ij}^{(t)} \qquad (12.8)$$

and $F^{(t)}$ is the bounded polyhedron of points $x^{(t)}$ satisfying the conservation of flow constraints. Again, z_{ij}'s do not appear in the dual problem, so we can recover a feasible primal solution after each dual iteration, which is not possible for general subgradient methods.

To simplify the constraints in the dual problem (12.7), we can sort the outgoing links from node i in A' according to their costs. Note that in A', no two outgoing links from a node i are of the same cost. Consider the example in Fig. 12.2 where there are three outgoing links from i with $(i, j_1) \prec (i, j_2) \prec (i, j_3)$, the equality constraints in (12.7) with respect to these links become

$$\sum_{t \in T} p_{ij_1}^{(t)} = a_{ij_1},$$

$$\sum_{t \in T} p_{ij_1}^{(t)} + \sum_{t \in T} p_{ij_2}^{(t)} = a_{ij_2},$$

$$\sum_{t \in T} p_{ij_1}^{(t)} + \sum_{t \in T} p_{ij_2}^{(t)} + \sum_{t \in T} p_{ij_3}^{(t)} = a_{ij_3}.$$

These are equivalent to

$$\sum_{t \in T} p_{ij_1}^{(t)} = a_{ij_1},$$

$$\sum_{t \in T} p_{ij_2}^{(t)} = a_{ij_2} - a_{ij_1},$$

$$\sum_{t \in T} p_{ij_3}^{(t)} = a_{ij_3} - a_{ij_2}.$$

Therefore, if we define

$$s_{ij} = a_{ij} - \max_{\{k | (i,k) \in A', (i,k) \prec (i,j)\}} a_{ik}, \tag{12.9}$$

the dual problem (12.7) can be simplified to

$$\text{maximize} \sum_{t \in T} q^{(t)}(p^{(t)}) \tag{12.10}$$

$$\text{subject to} \sum_{t \in T} p_{ij}^{(t)} = s_{ij} \quad \forall (i, j) \in A,$$

$$p_{ij}^{(t)} \geq 0 \qquad \forall (i, j) \in A', t \in T.$$

12.3 Decentralized Min-cost Subgraph Algorithms for Static Multicast

We focus on static multicasts in this section. Section 12.3.1 gives an overview of the dual subgradient method for decentralized subgraph optimization. The convergence rate of this method is analyzed in Section 12.3.2. Various heuristics to improve the convergence performance of the canonical algorithm in both static and dynamic wireless networks are presented in Section 12.3.3, and Section 12.3.4 gives some numerical results.

12.3.1 Subgradient Method for Decentralized Subgraph Optimization

The subgraph optimization scheme in [6] tries to converge to the optimal primal solution by using a subgradient method on the dual problem. This algorithm is completely decentralized and each node has to know only the cost of its incoming and outgoing links, and exchange information with neighboring nodes. We first give an overview of the algorithm under the wireline network model in Section 12.3.1.1. Section 12.3.1.2 describes the extension of this algorithm to the wireless case.

12.3.1.1 Subgradient Method in Wireline Networks

In the following, we describe the distributed algorithms for solving (12.1) with and without constraint (12.2). We refer to the algorithm that solves problem (12.1) and its dual (12.3) as *Algorithm A*, and the algorithm for solving the primal with constraint (12.2) and dual (12.5) as *Algorithm B*. Most of the discussions and simulations in Section 12.3 are based on Algorithm A, since it has better convergence performance in practical settings. However, in Section 12.3.2, we use Algorithm B to derive a theoretical bound on the convergence rate of the primal solutions, which is not available for Algorithm A.

Algorithm A

1. **Initialize $p[0]$** — Before the first iteration, each node initializes $p[0]$.
2. **Compute $x[n]$** — In the nth iteration, use $p[n]$ as link costs, and run a distributed shortest path algorithm to determine $x[n]$.
3. **Update $p[n+1]$** — Update $p[n+1]$ using subgradient obtained through $x[n]$ values:

$$g_{ij}^{(t)}[n] = x_{ij}^{(t)}[n],$$

$$p[n+1] := \left[p[n] + \theta[n]g^{(t)}[n] \right]_P^+,$$

where $g[n]$ is the subgradient for $p[n]$, $\theta[n]$ is the step size for the nth iteration, and $[\cdot]_P^+$ denotes the projection onto the constraint set P in (12.3). This projection can be done in a distributed manner, and specifically $p_{ij}^{(t)}[n+1]$ is given by

$$p_{ij}^{(t)}[n+1] = \max\left(0, p_{ij}^{(t)}[n] + \theta[n]x_{ij}^{(t)}[n] + d_{ij}[n] \right), \qquad (12.11)$$

where $d_{ij}[n] \leq 0$ is a number computed based on the $p[n]$, $x[n]$, and $\theta[n]$ values
[6].

4. **Recover $\tilde{x}[n]$** — At the end of each iteration, nodes recover a primal solution,
$\tilde{x}[n]$, based on the dual computations. Let $\{\mu_l[n]\}_{l=1,\dots,n}$ be a sequence of convex
combination weights for each non-negative integer n, i.e., $\sum_{l=1}^{n} \mu_l[n] = 1$ and
$\mu_l[n] \geq 0$ for all $l = 1, \dots, n$. Further, let us define

$$\gamma_{ln} = \frac{\mu_l[n]}{\theta[l]}, \quad l = 1, \dots, n, \quad n = 0, 1, \dots,$$

and

$$\Delta\gamma_n^{\max} = \max_{l=2,\dots,n} \{\gamma_{ln} - \gamma_{(l-1)n}\}.$$

According to [16], if the step sizes $\{\theta[n]\}$ and the convex combination weights
$\{\mu_l[n]\}$ are chosen such that

a. $\gamma_{ln} \geq \gamma_{(l-1)n}$ for all $l = 2, \dots, n$, and $n = 0, 1, \dots,$

b. $\Delta\gamma_n^{\max} \to 0$ as $n \to \infty$, and

c. $\gamma_{1n} \to 0$ as $n \to \infty$ and $\gamma_{nn} \leq \delta$ for all $n = 0, 1, \dots,$ for some $\delta > 0$,

then we obtain an optimal solution to the primal problem (12.1) from any accumulation point of the sequence of primal iterates $\{\tilde{x}[n]\}$ given by

$$\tilde{x}_{ij}^{(t)}[n] = \sum_{l=1}^{n} \mu_l[n] x_{ij}^{(t)}[l], \quad n = 0, 1, \dots.$$

An example of a set of parameters that satisfy the above conditions are $\theta[n] = n^{-\alpha}$ for $n = 0, 1, 2, \dots$ where $0 < \alpha < 1$, and $\mu_l[n] = 1/n$ for $n = 1, 2, 3, \dots$
and $l = 1, \dots, n$.

5. **Determine $\tilde{z}[n]$** — Each node computes the $\tilde{z}_{ij}[n]$ values from the $\tilde{x}_{ij}^{(t)}[n]$ values.
   In order to minimize the cost, $\tilde{z}_{ij}[n] = \max_{t \in T} \tilde{x}_{ij}^{(t)}[n]$.
6. **Repeat** — Steps 2–5 are repeated until the primal solution has converged.

For details of this algorithm and related proofs, refer to [6].

Since the intermediate $\{\tilde{z}[n], \tilde{x}[n]\}$ values after each iteration are always feasible solutions to the primal problem, we do not have to wait till the primal solution
converges to start the multicast. Instead, the multicast can be started after the first
iteration, and we can shift the flows gradually through the iterations to operate on
a more cost-effective subgraph. Note that, in general, this is not true for dual subgradient methods, and it works out here due to the unique structure of the network
coding problem. Specifically, the flow variables, z_{ij}, are not involved in the flow
conservation constraints, and they do not appear in the dual iterations. This allows
us to pick feasible z-values after each dual iteration based on a set of feasible virtual
flows x.

If the boundedness constraint (12.2) is included in the primal problem, we have
Algorithm B for solving this new problem and its dual (12.5).

Algorithm B

1. **Initialize** $p[0]$.
2. **Compute** $x[n]$ **and** $z[n]$ — Computation of $x[n]$ is the same as that in Algorithm A. For $z[n]$, we have

$$
z_{ij}[n] = \begin{cases} 0 & \text{if } \displaystyle\sum_{t \in T} p_{ij}^{(t)} \leq a_{ij}, \\[2ex] b_{ij} & \text{if } \displaystyle\sum_{t \in T} p_{ij}^{(t)} \geq a_{ij}. \end{cases}
$$

3. **Update** $p[n+1]$ — Update $p[n+1]$ using subgradient obtained through $x[n]$ and $z[n]$ values:

$$
g_{ij}^{(t)}[n] = x_{ij}^{(t)}[n] - z_{ij}[n],
$$

$$
p_{ij}^{(t)}[n+1] = \max\left(0, p[n] + \theta[n]g_{ij}^{(t)}[n]\right).
$$

4. **Recover** $\tilde{z}[n]$ — Recovery of the primal solution $\tilde{z}[n]$ is done by taking a convex combination of all past $z[n]$ values, similar to the recovery of $\tilde{x}[n]$ in Algorithm A:

$$
\tilde{z}_{ij}^{(t)}[n] = \sum_{l=1}^{n} \mu_l[n] z_{ij}^{(t)}[l], \quad n = 0, 1, \ldots.
$$

5. **Repeat** — Steps 2–4 are repeated until the primal solution has converged.

As we shall see in Section 12.3.2, Algorithm B provides us a theoretical bound on the primal convergence rate of the min-cost subgraph problem, which is not available for Algorithm A. However, a major drawback of Algorithm B compared to Algorithm A is that the $\tilde{z}[n]$ values are not always feasible; therefore, we cannot start the multicast right away as in the case of Algorithm A. This is very undesirable in practice and is one of the main reasons why we only focus on Algorithm A in our simulations.

12.3.1.2 Subgradient Method in Wireless Networks

The main steps in the distributed min-cost subgraph algorithm for wireless networks are the same as that in Algorithm A of the wireline case, except for steps 3 and 5, in which some modifications are required. The details of the changes are highlighted below.

In step 3, when updating $p[n+1]$, the subgradient for $p_{ij}^{(t)}[n]$ in the wireless case is given by

$$
g_{ij}^{(t)}[n] = \sum_{\{k|(i,k)\in A, (i,k)\succeq(i,j)\}} x_{ik}^{(t)}[n],
$$

and again, $p_{ij}[n+1]$ is the Euclidean projection of $p_{ij}[n] + \theta[n]g_{ij}[n]$ onto the feasible set P_{ij}.

In step 5, we compute $\tilde{z}[n]$ based on the recovered primal solution $\tilde{x}[n]$. Recall that in the primal problem (12.6) we have the constraints

$$\sum_{\{k|(i,k)\in A,(i,k)\succeq(i,j)\}} (\tilde{z}_{ik} - x_{ik}^{(t)}) \geq 0 \ \forall (i,j) \in A', t \in T. \tag{12.12}$$

Assume that the sorted list of outgoing links from node i in A' based on their costs is $\{(i, j_1), \ldots, (i, j_k)\}$, and start from the most expensive links (i, j_k), the above constraint becomes

$$\tilde{z}_{ij_k} - \tilde{x}_{ij_k}^{(t)} \geq 0 \ \forall t \in T.$$

To minimize total cost, the optimal $\tilde{z}_{ij_k}$ value should be $\max_{t \in T} \tilde{x}_{ij_k}^{(t)}$. In cases where more than one outgoing links are of the same cost, we just need to make sure that the sum of the flows on these links satisfy constraint (12.12). The distribution of the total flow among these links can be done randomly without affecting the total cost. Once $\tilde{z}_{ij_k}$ value is determined, we can move on to the second most expensive link (i, j_{k-1}), whose constraint now becomes

$$\left(\tilde{z}_{ij_{k-1}} - \tilde{x}_{ij_{k-1}}^{(t)} \right) + \left(\tilde{z}_{ij_k} - \tilde{x}_{ij_k}^{(t)} \right) \geq 0 \ \forall t \in T,$$

and we have $\tilde{z}_{ij_{k-1}} = \max_{t \in T} (\tilde{x}_{ij_{k-1}}^{(t)} + \tilde{x}_{ij_k}^{(t)}) - z_{ij_k}$. By repeating the above process, we can obtain the optimal primal solution $\tilde{z}$ from $\tilde{x}$.

12.3.2 Convergence Rate Analysis

In this section, we study the convergence rates of our dual subgradient method presented in Section 12.3.1 in both the primal and the dual spaces. For clarity of presentation, we use the wireline model in this section, as its notations are much simpler than the wireless one. All results here can be easily extended to the wireless case.

12.3.2.1 Convergence Rate for the Dual Problem

The analysis and results in this section apply to both Algorithm A and Algorithm B. Here, we just present the analysis for Algorithm A, as the extension to Algorithm B is fairly straightforward.

With properly chosen step sizes, the standard subgradient method proposed in Section 12.3.1 converges to dual optimal solutions eventually [24], but it is difficult to analyze the convergence rate of the standard method. To this end, we consider the incremental subgradient method studied in [25]. The incremental subgradient method can be used here because the objective function in (12.3) is the sum of $|T|$ convex component functions, and the constraint set is non-empty, closed, and convex (see Chapter 2 of [25]). At each iteration, p is changed incrementally through a sequence of $|T|$ steps. Each step is a subgradient iteration for a single component function $q^{(t)}$. Thus, an iteration can be viewed as a cycle of $|T|$ subiterations. Denote the terminal nodes by $\{1, 2, \ldots, N_T\}$, where $N_T = |T|$. The vector $p[n+1]$ is obtained from $p[n]$ as follows:

$$\psi_0[n] := p[n],$$

$$\psi_i[n] := \left[\psi_{i-1}[n] + \theta[n] g^{(i)}[n]\right]_P^+,$$

$$p[n+1] := \psi_{N_T}[n].$$

We first prove two propositions that are useful for the convergence rate analysis.

Proposition 1 *Problem (12.3) satisfies the subgradient boundedness property, which means there exists a positive scalar C such that*

$$\|g\| \le C, \ \forall g \in \partial q^{(t)}(p[n]) \cup \partial q^{(t)}(\psi_{i-1,n}),$$

$$\forall i = 1, \ldots, N_T \quad \forall n.$$

Proof. This is true because $q^{(t)}$ is the pointwise minimum of a finite number of affine functions, and in this case, for every p, the set of subgradients $\partial q^{(t)}(p)$ is the convex hull of a finite number of vectors. Thus, the subgradients are bounded. $\square$

Proposition 2 *Let the optimal solution set be P^*, there exists a positive scalar μ such that*

$$q^* - q(p) \ge \mu(\text{dist}(p, P^*))^2 \quad \forall p \in P.$$

Proof. Problem (12.3) can be reformulated into a linear programming problem as follows:

$$\text{maximize } q'(v) = \sum_{t \in T} \sum_{i \in N} r_i^{(t)} \delta_i^{(t)} = R \sum_{t \in T} \left(r_s^{(t)} - r_t^{(t)}\right) \qquad (12.13)$$

$$\text{subject to } r_i^{(t)} - r_j^{(t)} \le p_{ij}^{(t)} \quad \forall (i, j) \in A, \ t \in T,$$

$$\sum_{t \in T} p_{ij}^{(t)} = a_{ij} \quad \forall (i, j) \in A,$$

$$p_{ij}^{(t)} \ge 0 \qquad \forall (i, j) \in A, t \in T.$$

The decision vector, v, is a concatenation of vectors p and r, and we denote the feasible set by V. For any feasible $p \in P$ from (12.3), there is a corresponding v in (12.13) with the same p-component and $q'(v) = q(p)$. Furthermore, for any feasible $v \in V$, we can extract a p vector from it that gives the same total cost in (12.3). Therefore, the two formulations (12.3) and (12.13) have the same optimal values, i.e., $q^* = q'^*$.

Since the set of solutions for a linear programming problem is a set of weak sharp minima [26], there exists a positive α such that

$$q'^* - q'(v) \ge \alpha(\text{dist}(v, V^*)) \qquad \forall v \in V.$$

So for any $p \in P$ in (12.3), we have

$$q^* - q(p) = q'^* - q'(v) \geq \alpha(\text{dist}(v, V^*)) \geq \alpha(\text{dist}(p, P^*)).$$

The last inequality comes from the fact that p/P^* is the projection of v/V^* on P, and the projection operation is non-expansive. Since P is a bounded polyhedron, the distance between any two points in P is bounded, i.e., $\text{dist}(p, p') \leq B$ for all $p, p' \in P$ for some positive B. Therefore,

$$q^* - q(p) \geq \frac{\alpha}{B}(\text{dist}(p, P^*))^2.$$

Let $\mu = \alpha/B$, and the proposition is proved. $\square$

With these propositions, we have the following result for constant step size.

Proposition 3 *For the sequence $\{p[n]\}$ generated by the incremental subgradient method with the step size $\theta[n]$ fixed to some positive constant θ, where $\theta \leq \frac{1}{2\mu}$, we have*

$$(\text{dist}(p[n + 1], P^*))^2 \leq (1 - 2\theta\mu)^{n+1}(\text{dist}(p[0], P^*))^2 + \frac{\theta|T|^2C^2}{2\mu} \qquad \forall n.$$
(12.14)

Proof. The proof for this proposition follows from Proposition 1.2 and the proof of Proposition 2.3 in [25]. Since the dual problem satisfies Proposition 1 (bounded subgradient), from Lemma 2.1 in [25], we have

$$||p[n+1] - r||^2 \leq ||p[n] - r||^2 - 2\theta(q(r) - q(p[n])) + \theta^2|T|^2C^2 \quad \forall r \in P, \quad \forall n.$$

Using this relation with $r = p^*$ for any optimal $p^* \in P^*$, we see that

$$||p[n+1] - p^*||^2 \leq ||p[n] - p^*||^2 - 2\theta(q^* - q(p[n])) + \theta^2|T|^2C^2 \quad \forall r \in P, \quad \forall n,$$
and by taking the minimum over all $p^* \in P^*$, we have (12.15)

$$(\text{dist}(p[n + 1], P^*))^2 \leq (\text{dist}(p[n], P^*))^2 - 2\theta(q^* - q(p[n])) + \theta^2|T|^2C^2$$
$$\leq (1 - 2\theta\mu)(\text{dist}(p[n], P^*))^2 + \theta^2|T|^2C^2 \quad \forall n,$$
(12.16)

where the last inequality comes from Proposition 2. From this relation, by induction, we can see that

$$(\text{dist}(p[n + 1], P^*))^2 \leq (1 - 2\theta\mu)^{(n+1)}(\text{dist}(p[0], P^*))^2$$
$$+ \theta^2|T|^2C^2 \sum_{i=0}^{n}(1 - 2\theta\mu)^i \quad \forall n,$$

which combined with

$$\sum_{i=0}^{n}(1 - 2\theta\mu)^i \leq \frac{1}{2\theta\mu},$$

yields the desired relation (12.14). $\square$

In summary, we have shown that the convergence rate for the incremental subgradient method on (12.3) is linear for a sufficiently small step size. However, only convergence to a neighborhood of the optimal solution set can be guaranteed, which is typical for constant step size rules. Moreover, our result also highlights the trade-off between the error and the convergence rate constant. The smaller the θ value, the smaller the size of the neighborhood, but on the other hand, we get a slower convergence.

12.3.2.2 Convergence Analysis for the Primal Problem

As mentioned in Section 12.3.1.1, it is advantageous to use Algorithm A in practice, as its primal solution is always feasible through the iterations. Unfortunately, due to the unboundedness of z_{ij} in formulation (12.1), it is very hard to derive its primal convergence rate. Therefore, in this section, we turn our focus to Algorithm B, for which we derive a bound on its convergence rate.

We first prove that our primal problem (12.1) with constraint (12.2) satisfies the Slater condition in Proposition 4, then present the main convergence rate result in Proposition 5.

Proposition 4 The Slater condition *There exists a vector* $\{\bar{z}, \bar{x}\}, \{\bar{z}, \bar{x}\} \in F$ *such that*

$$\bar{z}_{ij} > \bar{x}_{ij}^{(t)} \quad \forall (i, j) \in A, t \in T, \tag{12.17}$$

where $F = \{F_z, F_x\}$ *is the feasible set for the boundedness constraints for* z *and the conservation of flow constraints for* x.

Proof. For the virtual flows, $x_{ij}^{(t)}$, based on the conservation of flow constrains, there exists feasible solutions where $x_{ij}^{(t)} \leq R$ for all $(i, j) \in A$ and $t \in T$. Since the upper bound on z_{ij} is $b_{ij} > R$, we can always find a set of z that is strictly greater than the corresponding x. Therefore, our primal problem satisfies the Slater condition. $\square$

Proposition 5 *Let* $\{\bar{z}, \bar{x}\}$ *be a Slater vector satisfying (12.17), and C be the subgradient norm bound in Proposition 1, define*

$$B^* = \frac{2}{\gamma}(f(\bar{z}) - q^*) + \max\left\{\|p[0]\|, \frac{1}{\gamma}(f(\bar{z}) - q^*) + \frac{\theta C^2}{2\gamma} + \theta C\right\},$$

where $\gamma = \min_{\{i,j\} \in A, t \in T} \left(\bar{z}_{ij}^{(t)} - \bar{x}_{ij}^{(t)}\right)$. *If constant step size* θ *is used in the dual iterations, and simple averaging is used in the primal recovery, i.e.,* $\mu_l[n] = 1/n$ *for* $l = 1, 2, \ldots, n$, *then the primal cost after the nth iteration is bounded by*

$$f^* - \frac{1}{\gamma}[f(\bar{z}) - q^*]\frac{B^*}{n\theta} \leq f(\bar{z}[n]) \leq f^* + \frac{\|p[0]\|^2}{2n\theta} + \frac{\theta C^2}{2}. \tag{12.18}$$

Proof. We first derive the lower bound on $f(\tilde{z}[n])$ (the left-hand side of (12.18)). Recall that in Algorithm B, $g_{ij}^{(t)}[n] = g\left(z_{ij}[n], x_{ij}^{(t)}[n]\right) = x_{ij}^{(t)}[n] - z_{ij}[n]$, and

$$p[n+1] = \max(0, p[n] + \theta g[n]) \geq p[n] + \theta g[n],$$

we have

$$\theta g(z[n], x[n]) \leq p[n+1] - p[n] \qquad\qquad \forall n \geq 0.$$

Therefore, $\sum_{i=0}^{n-1} \theta g(z[i], x[i]) \leq p[n] - p[0] \leq p[n]$, where the last inequality follows from $p[0] \geq 0$. By the convexity of the function g, it follows that

$$g(\tilde{z}[n], \tilde{x}[n]) \leq \frac{1}{n}\sum_{i=0}^{n-1} g(z[i], x[i]) = \frac{1}{n\theta}\sum_{i=0}^{n-1}\theta g(z[i], x[i]) \leq \frac{p[n]}{n\theta}.$$

Because $p[n] \geq 0$, the positive elements in $g(\tilde{z}[n], \tilde{x}[n])$ satisfy $g(\tilde{z}[n], \tilde{x}[n])^+ \leq p[n]/n\theta$ for all $n \geq 0$. Let the amount of constraint violation of $(\tilde{z}[n], \tilde{x}[n])$ be $\|g(\tilde{z}[n], \tilde{x}[n])^+\|$, we have

$$\|g(\tilde{z}[n], \tilde{x}[n])^+\| \leq \frac{\|p[n]\|}{n\theta} \qquad \forall n \geq 1. \tag{12.19}$$

Given a dual optimal solution p^*, we have

$$q(p^*) = q^* \leq f(z) + (p^*)'g(z, x)$$

for any $x \in F_x$ and $z \in F_z$. Thus,

$$\begin{aligned} f(\tilde{z}[n]) &= f(\tilde{z}[n]) + (p^*)'g(\tilde{z}[n], \tilde{x}[n]) - (p^*)'g(\tilde{z}[n], \tilde{x}[n]) \\ &\geq q^* - (p^*)'g(\tilde{z}[n], \tilde{x}[n]). \end{aligned} \tag{12.20}$$

Because $p^* \geq 0$ and $g(\tilde{x}[n], \tilde{x}[n])^+ \geq g(\tilde{x}[n], \tilde{x}[n])$, we further have

$$-(p^*)'g(\tilde{z}[n], \tilde{x}[n]) \geq -(p^*)'g(\tilde{z}[n], \tilde{x}[n])^+ \geq -\|p^*\|\|g(\tilde{z}[n], \tilde{x}[n])^+\|. \tag{12.21}$$

From (12.19), (12.20), and (12.21), it follows that

$$f(\tilde{z}[n]) \geq q^* - \|p^*\|\frac{\|p[n]\|}{n\theta}. \tag{12.22}$$

Since our primal problem satisfies the Slater condition and the dual iterates have bounded subgradients, from Lemma 1 and 3 in [19], we have

$$\|p^*\| \leq \frac{1}{\gamma}(f(\bar{z}) - q^*) \quad\text{and}\quad \|p[n]\| \leq B^* \quad \forall n \geq 1,$$

where γ and B^* are defined in the above proposition. Substitute these bounds into (12.22), and we have the lower bound on $f(\tilde{z}[n])$

$$f(\tilde{z}[n]) \geq q^* - \frac{1}{\gamma}[f(\tilde{z}) - q^*]\frac{B^*}{n\theta}.$$

Next, we derive the upper bound on $f(\tilde{z}[n])$ (right-hand side of (12.18)). By the convexity of $f(z)$ and the definition of $(z[n], x[n])$ as a minimizer of the Lagrangian function $f(z) + p'g(z, x)$ over $x \in F_x$ and $z \in F_z$, we have

$$f(\tilde{z}[n]) \leq \frac{1}{n}\sum_{i=0}^{n-1} f(z[i])$$

$$= \frac{1}{n}\sum_{i=0}^{n-1}\left(f(z[i]) + p[i]'g(z[i], x[i])\right) - \frac{1}{n}\sum_{i=0}^{n-1}p[i]'g(z[i], x[i])$$

$$= \frac{1}{n}\sum_{i=0}^{n-1}q(p[i]) - \frac{1}{n}\sum_{i=0}^{n-1}p[i]'g(z[i], x[i])$$

$$\leq q^* - \frac{1}{n}\sum_{i=0}^{n-1}p[i]'g(z[i], x[i]).$$

$$(12.23)$$

Since $p[n+1] = \left[p[n] + \theta g[n]\right]^+$, by using the non-expansive property of projection and the fact that 0 is in the feasible region of the dual problem (12.5), we have

$$\|p[i+1]\|^2 \leq \|p[i]\|^2 + 2\theta p[i]'g[i] + \theta^2\|g[i]\|^2.$$

Since $g[i] = g(z[i], x[i])$, we further obtain

$$-p[i]'g(z[i], x[i]) \leq \frac{\|p[i]\|^2 - \|p[i+1]\|^2 + \theta^2\|g(z[i], x[i])\|^2}{2\theta}, \quad 0 \leq i \leq n-1.$$

By summing over $i = 0, 1, \ldots, n-1$, and combining with (12.23), we have

$$f(\tilde{z}[n]) \leq q^* + \frac{\|p[0]\|^2 - \|p[n]\|^2}{2n\theta} + \frac{\theta}{2n}\sum_{i=0}^{n-1}\|g(z[i], x[i])\|^2$$

$$\leq q^* + \frac{\|p[0]\|^2}{2n\theta} + \frac{\theta C^2}{2} \quad \forall n \geq 1.$$

$$(12.24)$$

By combining (12.22) and (12.24), we have the desired relation. □

This proposition shows that when using constant step size, the primal solutions converge to a neighborhood of the optimal solution with rate $O(1/n)$. Note that there is a trade-off between the size of the neighborhood and the convergence rate. If we want the primal solution to be close to the optimal one, we need to choose a small step size, but this would make the convergence rate very slow. This is a typical problem with using constant step size, and in our simulations, we avoid this situation by using diminishing step size.

12.3.3 Initialization and Primal Solution Recovery

In order to improve the convergence performance of the subgradient algorithm, we introduce some heuristics for steps 1 and 4 in the algorithm presented in Section 12.3.1.2. Specifically, we propose several methods for initializing the dual vector $p[0]$, and for recovering primal solutions $\{\tilde{x}[n]\}$, for both static and dynamic wireless networks.

12.3.3.1 Static Networks

We start with static networks, where the topology of the network is fixed throughout the multicast. We first introduce a naive way of initializing the dual variables.

- **Averaging method** — The simplest way to generate feasible initial values for the dual variables is to assign $p_{ij}^{(t)} = s_{ij}/N_T$ for all $t \in T$ and all $(i, j) \in A$. This method is useful in static networks since no prior information of the multicast problem is available at the nodes.

For the recovery of primal solution $\tilde{x}[n]$, we have the following two options.

- **Original primal recovery** — This is the recovery method presented in step 4 in Section 12.3.1 with simple averaging.
- **Modified primal recovery** — Using the original primal recovery method, we observed in simulations that the cost of the multicast starts at a high value, and then converges slowly to the optimal value through iterations. One reason for the slow convergence is that it is recovered by averaging $x[n]$ values from all the iterations. The effect of the first few high-cost iterations takes a large number of iterations later to dilute. A heuristic way to improve the convergence rate is to discard these "bad" primal solutions after some time, and just average over the most recent N_a number of iterations in primal solution recovery.

12.3.3.2 Dynamic Networks

As opposed to the static assumption, many wireless networks have topologies that are dynamic. Whenever a topology change occurs, we need to restart the distributed algorithm, as the subgraph used for multicast before the topology change might have become infeasible. In such cases, all the methods discussed in the previous section for dual variable initialization and primal solution recovery are still applicable. However, since the difference between the optimal solutions to the multicast problem before and after the changes is usually small, we should make use of the solutions $\hat{x}$ and $\hat{p}$ before the topology changes in the new iterations to improve the convergence rate. We propose additional methods to initialize p and update $\tilde{x}$ that make use of this old information.

For dual variable initialization, we present two additional heuristics.

- **Scaling method** — In this method, each node i scans through its set of outgoing links in A'. If a link (i, j) is an existing link in $\hat{A}'$ before the topology change, scale the $\{\hat{p}_{ij}^{(t)}\}$ values so that they satisfy the new dual constraints. Specifically, denoting $\sum_{t \in T} \hat{p}_{ij}^{(t)} = \hat{s}_{ij}$, we assign

$$p_{ij}^{(t)} = \hat{p}_{ij}^{(t)} \times \frac{s_{ij}}{\hat{s}_{ij}}.$$

  On the other hand, if link (i, j) is a new link after the topology change, we simply use

$$p_{ij}^{(t)} = s_{ij}/|T|.$$

- **Projection method** — In this method, we use an intermediate $\tilde{P}$ which is given by

$$\tilde{p}_{ij}^{(t)} = \begin{cases} \hat{p}_{ij}^{(t)} & \text{if } (i, j) \text{ is an old link,} \\ 0 & \text{if } (i, j) \text{ is a new link.} \end{cases}$$

  We can then project this $\tilde{P}$ onto the new feasible region of the dual problem using (12.11) to obtain an initial point P for the decentralized algorithm.

On the primal side, we observe that as long as no removed link was in the multicast subgraph used before the topology change, the old $\{\hat{x}[n]\}$ values from the previous iterations are still valid under the new topology. Thus, they can be used in the recovery of the current primal optimal solution. Based on this observation, we propose the following heuristics for primal solution recovery.

- **Look-back primal recovery** — When a topology change occurs, each node checks if any of its links used in the multicast is removed owing to topology change. If yes, it sends out a signal to all nodes, and $\{\tilde{x}[n]\}$ is computed based only on the new $\{x[n]\}$ values as in the original primal recovery method above. On the other hand, if no link is removed, the averaging is done over N_a iterations before and after the topology change. The assumption that nodes can be informed of the removal of an active link within one iteration is reasonable, since, in each iteration, distributed Bellman–Ford is used to compute $x[n]$ and sending such a signal to all nodes should take less time than running distributed Bellman–Ford.

12.3.4 Simulation Results

12.3.4.1 Static Networks

We use the wireless network model presented in Section 12.2.2 for our simulations, because wireless networks are a primal application for network coding. The random wireless networks are set up in a 10×10 square with a rate of connectivity $r = 3$. We run the distributed algorithm to determine the minimum energy subgraphs on these networks for multicast connections with unit rate. Here, the energy required to transmit at unit rate to a distance d is taken to be d^2. We assume that there is no collision or interference in the network. Simulation results showed that the standard subgradient method has a better convergence time as compared to the incremental subgradient method, thus, in this section, we only present results for the standard method for Algorithm A.

Figure 12.3 shows the average convergence performance for the proposed algorithms for networks with 30/50 nodes and 4/8 terminals in the multicast. The step sizes used in the subgradient method are $\theta[n] = n^{-\alpha}$ with $\alpha = 0.8$ for $n = 0, 1, \ldots$. For the modified primal recovery method, the parameter N_a is set to 30. As we can see, the two primal cost curves coincide for the first 30 iterations, and after that, the modified method converges to the optimal value faster than the original method.

To compare the performance of the proposed scheme to the cost of multicast when network coding is not used, we use the multicast incremental power (MIP) algorithm described in [23], which is a centralized heuristic algorithm to perform minimum-energy multicast in wireless networks. For the same setting, the average cost values for the multicast given by the MIP algorithm are also shown in Fig. 12.3. As can be seen, in both cases, even the initial high-cost values from our distributed algorithms are lower than that from the centralized MIP algorithm. Moreover, in fewer than 50 iterations, the cost of the multicast using modified primal recovery is within 5% higher than the optimal value. Therefore, in a small number of iterations, the decentralized subgraph optimization algorithms yield solutions to the multicast problem with energy significantly lower than that for multicast without network coding even if a centralized scheme is used.

To have a feel of how Algorithm B would have performed in the above scenario, we observe in Fig. 12.3(a) that after 200 iterations, the cost difference between Algorithm A and the optimal value is about 0.5. For algorithm B to arrive at a neighborhood of the optimal solutions of this size, the step size should be smaller than 0.016 even if we use a very small $C = 5$. With this step size, it would take Algorithm B thousands of iterations to arrive at where Algorithm B is in 200 steps. Therefore, for all our simulations, we will use Algorithm A only.

12.3.4.2 Dynamic Networks

To illustrate the performance of our algorithms in dynamic networks, we use random networks with mobile nodes. The mobility model used in our simulations is the random direction mobility model [27], where each node selects a random direction between $[0, 2\pi]$ and a random speed between [minspeed, maxspeed]. A node travels to the border of the simulation area in that direction, then randomly chooses another valid direction and speed, and continues the process. Note that our algorithms are applicable to all mobility models, and we have chosen this specific one for its simplicity.

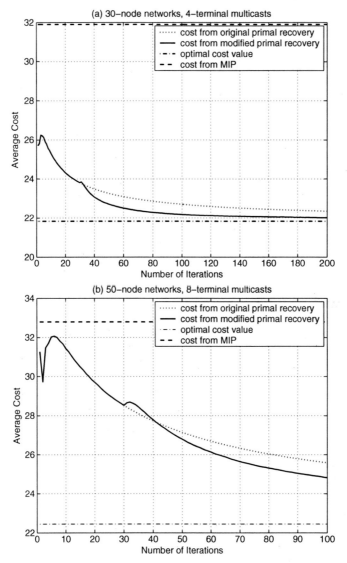

Fig. 12.3 Average cost of random 4/8-terminal multicasts in 30/50-node wireless networks, using the decentralized subgraph optimization algorithms and centralized MIP algorithm. For modified primal recovery method, $N_a = 30$

In our studies, we assume that the nodes are traveling at a speed that is slow relative to the node computation speed and link transmission rate. Under such assumptions, we consider the movement of the nodes in small discrete steps, and between each step, the set of links in the network and their costs are considered constant. We refer to the period between two discrete steps as a "static period," and let the number of subgraph optimization iterations performed within each static period be N_s.

We ran simulations for the various methods presented in Section 12.3.1. For dual variable initialization, we only present results based on the projection method, since it gives the best performance. Here, each node has a random speed in the interval $[0, 0.1]$ units/static period. We choose this range because the steps taken by the nodes with such speeds are relatively small compared to r, and our assumption that the network is static between steps is valid. Also, this is a relative speed of the nodes with respect to the static period, and we can vary N_s to simulate different actual speeds of the nodes.

To illustrate the typical performance of the subgraph optimization scheme in a mobile wireless network, Fig. 12.4 shows the costs for each iteration for an instance of the multicast problem. As expected, if we flush the memory of $\{\hat{x}[n]\}$ at the end of each static period, and start accumulation for the primal cost afresh, the cost of the multicast is very spiky. On the other hand, if old $\{\hat{x}[n]\}$ values are used when they are feasible, the primal cost is usually much smoother. Of course, if node movement renders the old $\{\hat{x}[n]\}$ values infeasible, we have no choice but to start afresh, and the curves for original primal recovery and look-back primal recovery coincide (as in the second static period in Fig. 12.4).

In Fig. 12.5, we show simulation results under different network and multicast settings, and for nodes with different speeds. The parameter N_a used in the modified primal recovery is set to 20. First, we compare the performance of the three options to recover primal solutions. Under the same settings, look-back primal recovery

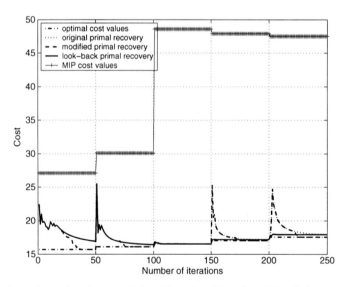

Fig. 12.4 Cost of a random four-terminal multicast in a 30-node mobile wireless network, with $N_s = 50$, under various algorithms. For the modified primal recovery method, we used $N_a = 20$ and, for the look-back primal recovery method, we used $N_a = 50$

gives the lowest average cost, followed by modified and original primal recovery. We also observe that when the same methods are used, the faster the node moves, the higher the average primal cost is, owing to the lack of time for the algorithm to converge. Also, a network with more nodes or a multicast with more terminals makes convergence of the decentralized algorithm slower, and thus results in higher average primal cost.

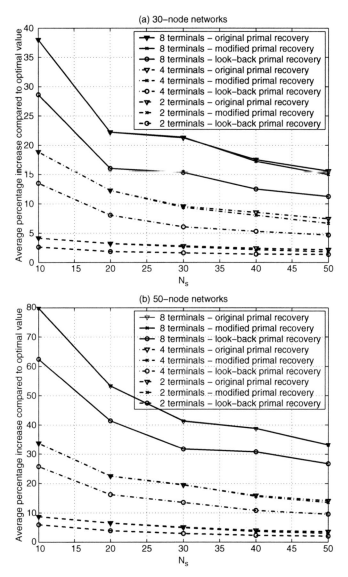

Fig. 12.5 Extra energy required for multicasts in mobile wireless networks using decentralized subgraph optimization scheme in terms of percentage of the optimal value

The simulation results have shown that the decentralized subgraph optimization scheme is robust in mobile wireless networks when the nodes are moving slowly relative to the computation and message exchange rate of the nodes. On average, it can track the changes in the optimal value closely, and, in most cases, requires lower energy for multicast than MIP even though the nodes are mobile and computation is done at each node in a distributed manner.

12.4 Min-cost Subgraph Algorithms for Dynamic Multicasts

For the dynamic multicast problem, there are two extreme cases. On one hand, we can simply find the new optimal subgraph whenever there is an update to the multicast group, and replace the existing subgraph with this new one. In this case, users in the group will experience a lot of disruptions, but the cost of the multicast is always kept minimal. On the other hand, we can enforce that no link or code rearrangement is allowed for all existing users throughout the multicast session. In this case, users enjoy uninterrupted services but, in general, the subgraph used will deviate further and further away from the optimal one. In this section, we present one algorithm to solve the nonrearrangeable version of the dynamic multicast problem, and three algorithms for the rearrangeable version. Simulation results show that one of the rearrangeable algorithms we propose, the α-scaled algorithm, can be used to strike a balance between cost and frequency of user disturbances in a distributed manner. Although we present our algorithms based on wireline networks, they can be easily extended to wireless networks.

12.4.1 Nonrearrangeable Algorithm

For simplicity, we assume that the rate of the multicast is lower than the capacity of the links, which is generally the case in current wireline networks. In a multicast session, a source node s transmits to a group of terminal nodes T, and the group changes over time. We refer to each change of the membership of the multicast group (either an addition or a removal of a terminal node) as an *online step*. The network model and problem formulation is the same as that in Section 12.2.1. This LP problem (12.1) can be solved by a number of methods, both centrally (e.g., Simplex method) and distributedly (e.g., subgradient method in Section 12.3.1). For the rest of this chapter, we denote any centralized/distributed algorithm that solves the LP problem as LP_{cent}/LP_{dist}, respectively.

For the dynamic multicast problem, the initial multicast subgraph is set up by solving (12.1). If we allow complete rearrangement, we can simply solve this problem again at each online step. However, to solve the nonrearrangeable version of the dynamic multicast problem, we have to prevent link and code rearrangements from happening. To meet the no-link rearrangement requirement, we basically need to make sure that the existing users still use the same path(s) for the multicast when the set T changes over time. This can be achieved by setting the cost of the links in the current subgraph G_c to zero. If the capacity of a link is larger than the rate used for the multicast, then the link is split into two virtual links, one with capacity equal to the rate used for the multicast and cost zero and the other with the remaining capacity and cost unchanged. For example, if link (i, j) has capacity $c_{ij} = 2$ and rate of flow $z_{ij} = 1$ for the multicast, then nodes i and j treat link (i, j) as two parallel links with capacities 1 each, and one of them has cost of 0, and the other one has cost of a_{ij}. After doing this, the current multicast subgraph becomes "free," and doing optimization on this new cost assignment will always lead to using the same path(s) to serve the existing users in the new subgraph. Therefore, link rearrangements are avoided.

One problem with the above method is that some links not necessary for the new terminal set might be included in the new subgraph after a removal of a terminal. This is because all links in the old subgraph are free, and some of these links might still be included in the solution to the LP problem even though they are not necessary in performing the multicast to the new terminal set. To solve this problem, instead of setting their costs to 0, we can set the cost of the used links to a small value ϵ, so that no extra link would be included in the optimal solution, and, at the same time, the used links are still almost free as compared to the other links.

As for code rearrangements, we want to prevent the usage of new links that go into existing nodes of the subgraph. To do that, each node in the subgraph can scan through its incoming links, and sets the cost of those unused links to a very large value, M. Again, if the capacity of an incoming link is not fully used in the current subgraph, we can split it into two parallel virtual links as above. These nodes then send the new costs of its incoming links to their corresponding tail nodes, and the new high costs can prevent these links from being used.

After making these changes to the link costs, when an online step occurs, we can simply run LP_{dist} again with the new costs, and obtain a feasible subgraph for the new multicast group without any link or code rearrangements. This algorithm is summarized in Fig. 12.6.

```
node i

nodeUsed = 0
for all (j,i) ∈ A
    if (j,i) ∈ G_c
        a_ji = ε
        nodeUsed = 1
    end
end
if nodeUsed = 1
    for all (j,i) ∈ A
        if (j,i) ∈ G_c    a_ji = M
    end
end
call LP_dist
```

Fig. 12.6 Nonrearrangeable algorithm

The above algorithm can be complicated due to the splitting of physical links into parallel virtual links. This requires more processing at the nodes and more co-ordination between the end nodes of the links. In addition, in the nonrearrangeable solution of the dynamic multicast problem, it is inevitable that the subgraph used would deviate further and further from the optimal subgraph. This is because the no-rearrangement requirement forces the subgraph we use as close as possible to the initial subgraph. Thus, when the multicast group changes further and further away from the original group over time, our multicast subgraph becomes more and more suboptimal.

To simplify this algorithm and keep the cost of the multicast low, we may need to make some compromise and allow some rearrangements. In the next section, we present three such heuristic algorithms.

12.4.2 Rearrangeable Algorithms

12.4.2.1 Algorithm for Minimizing Link Rearrangement (MLR)

One way to simplify the nonrearrangeable algorithm is to focus on eliminating link rearrangement only, and ignore code rearrangement. This can be easily done by setting the used link costs to a very small value ϵ after each online step as in the nonrearrangeable algorithm, and call LP_{dist} to solve the new LP problem. This algorithm, which we call the MLR (minimal link rearrangement) algorithm, is shown in Fig. 12.7.

The motivation for this algorithm comes from the observation that the complication of splitting links into used and unused portions arises when we have a non-tree subgraph, and things would be much simpler if we only have to deal with trees. This is because, in trees, each node only has one incoming link, and it has full information of the multicast. Thus, there is no worry about code rearrangement.

Notice that once the multicast subgraph becomes a tree, it will remain as a tree through the rest of the online steps. To see this, consider addition of a new node to the multicast group. Since the original subgraph G_c is considered "free" and each node in G_c has full information of the multicast, the new terminal node only needs to find the shortest path from any node in G_c to itself, and attach itself to the subgraph. As for the removal of a terminal, only a part of the tree may be removed, and the remaining graph should still be a tree. Since at every step, if G_c is not a tree, there is

```
node i

for all (j,i) ∈ A
    if (j,i) ∈ G_c
        a_ji = ε
    end
end
call LP_dist
```

Fig. 12.7 MLR algorithm

some positive probability that it will become a tree, and once it evolves into a tree, it will stay that way till the end of the multicast. Therefore, if we keep running the dynamic multicast session, the probability that we are dealing with trees goes to 1.

In addition, simulations on practical networks show that in more than 98% of the time, we do get the optimum Steiner tree at start-up. Therefore, we can focus on link rearrangements only and use the MRL algorithm. This algorithm still works if the initial subgraph is not a tree, the only difference is that we cannot guarantee that there will not be any rearrangements in such cases.

12.4.2.2 Algorithm for Limiting Multicast Cost (LMC)

If we use the MLR algorithm, it is expected that as time goes on, the subgraph used for multicast will move further and further away from the actual optimal subgraph for the current set of terminal nodes. As an alternative, we might want to introduce occasional rearrangements in order to keep the cost of the multicast close to optimal. We introduce the LMC algorithm, shown in Fig. 12.8, to do this. In this algorithm, the nodes run two programs in parallel, one of which generates the subgraph with no rearrangement using the algorithm presented above. We call this subgraph the *no-change subgraph* G_{nc}, and the cost of this subgraph C_{nc}. The other program keeps track of the optimal subgraph, G_{opt}, for the current set of multicast terminals, and the cost of G_{opt} is C_{opt}. At each step, the cost of the two subgraphs are compared, and if the cost of the no-change subgraph is higher than the optimal graph by a certain factor, β, we switch to the optimal subgraph. Using this method, we can control the trade-off between the frequency of disturbances to the users and the cost of the subgraph used for the multicast by changing the value of β. However, this method requires the nodes to keep track of two subgraphs, and centralized coordination is needed to compare the costs and make the nodes switch between two subgraphs simultaneously.

12.4.2.3 α-Scaled Algorithm

We now present a simple approximate algorithm that can trigger "auto-switching" between G_{nc} and G_{opt} in a distributed manner. Instead of assigning a very small cost to the used links as in the MLR algorithm, we can use a scaled value of the

```
call LP_cent to compute C_opt and G_opt
for all (j,i) ∈ A
    if (j,i) ∈ G_c
        a_ji = ε
    end
end
call LP_cent to compute C_nc and G_nc
if C_nc > C_opt × (1+β)
    use G_opt for the multicast
else
    use G_nc for the multicast
end
```

Fig. 12.8 LMC algorithm

original cost, i.e., for an existing link (i, j) in the subgraph, we use αa_{ij} as its cost in the future computations as long as it stays in the subgraph, where α is a scaling factor between 0 and 1. If $\alpha = 0$, it is the same as the MLR algorithm; and if $\alpha = 1$, we will be using the optimal subgraph every time. We refer to this algorithm as the α-scaled algorithm, and it is shown in Fig. 12.9.

To see why this heuristic works and how the constants α and β are related, consider the case of removal of a terminal node. The LMC algorithm compares the values of C_{nc} and $(1 + \beta)C_{opt}$ and picks the lower of the two. Since G_{opt} may overlap with the existing subgraph from before the online step, we assume the cost of this overlapping part of the subgraph is C_{ol} and the cost of the rest of the optimal subgraph is C_{others}. Thus, the comparison is equivalent to

$$\frac{1}{1 + \beta} \times C_{nc} \gtrless C_{ol} + C_{others}. \tag{12.25}$$

On the other hand, in the α-scaled algorithm, we are effectively choosing the lower cost between these two:

$$\alpha \times C_{nc} \gtrless \alpha \times C_{ol} + C_{others}. \tag{12.26}$$

If we set α to $1/(1 + \beta)$, we can see that (12.25) and (12.26) are very similar except the first term on the right-hand side. By scaling the existing link costs by α, we can satisfy the requirement that the cost of the subgraph used never goes over $(1 + \beta)C_{opt}$, but owing to the scaling factor α on C_{ol}, the approximate algorithm switches to the optimal subgraph more often than required by β. Using similar analysis, we have the same results for the case of addition of a terminal.

Thus, using an appropriate α to scale the costs of the used links, the optimization can trigger auto-switching between the two subgraphs, thus keeping the cost of the multicast low. In addition, we can make α a time-varying variable. In general, when a link is first added into the subgraph, it is likely that it will remain there for a while. However, the probability that the link remains in the optimal subgraph decreases with the online steps. To capture this characteristic, we can use a lower value for α for the first few online steps after a new link is added, and increase α gradually later on. Also, in a practical network, it may not be desirable to make back-to-back changes to the link connections, i.e., addition of a link to the multicast subgraph followed by an immediate removal of it in the next step. We can reduce the occurrence of such events by setting the α of new links to 0 for a few steps before raising it to the normal value of $1/(1 + \beta)$.

```
node i

for all (j,i) ∈ A
    if (j,i) ∈ G_c
        a_{ji} = α × a_{ji}
    end
end
call LP_dist
```

Fig. 12.9 α-Scaled algorithm

12.4.3 Simulation Results

We first present simulation results for the MLR algorithm. The network topologies used in the simulations are obtained from the Rocketfuel project [28]. In each simulation, we start with a multicast from a random source to a set of 10 random terminals. Subsequently, in each online step, we first randomly decide whether there is an addition or removal of terminal, and then randomly select a terminal to add/remove based on that decision. Figures 12.10 and 12.11 show the average increase of cost of the no-change subgraph compared to the cost of the optimal subgraph in terms of percentage of C_{opt}. The network topology used for Figs. 12.10 and 12.11 are backbones for Exodus (United States) and EBONE (Europe), respectively. As expected, the extra cost of the no-change subgraph grows approximately linearly with the online steps. In addition to the average curve, we also show the data points for each instance of the simulation in both Figs. 12.10 and 12.11. Note that there are cases when the cost of the no-change subgraph is as much as 60% higher than the optimal cost after 20 steps.

This undesirable phenomenon motivates the usage of the α-scaled algorithm. Figure 12.12 shows the simulation results for using the α-scaled algorithm on the network used in Fig. 12.10. Here, we aim to control the cost of the subgraph used to within $\beta = 30\%$ away from that of the optimal subgraph; thus, we use $\alpha = 0.75$. The average curve in Fig. 12.10 for the MLR algorithm is also shown here for comparison. The α-scaled algorithm provides lower cost for the multicasts as compared to the MLR algorithm. More importantly, the cost difference between the

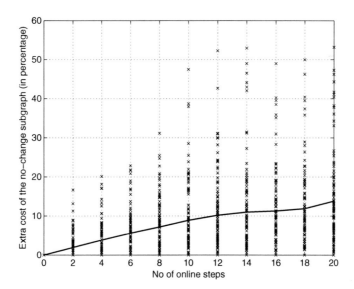

Fig. 12.10 Extra cost of the multicast subgraph generated by the MLR algorithm in terms of percentage of C_{opt} for the Exodus network. We are showing both the individual data points for each trial and the average curve

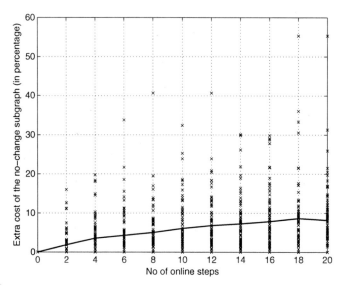

Fig. 12.11 Extra cost of the multicast subgraph generated by the MLR algorithm for the EBONE network in terms of percentage of C_{opt}

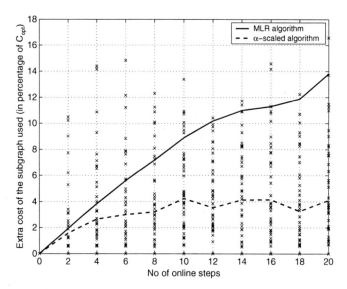

Fig. 12.12 Extra cost of the multicast subgraph generated by the α-scaled algorithm with $\alpha =$ 0.75 and the MLR algorithm in terms of percentage of C_{opt}, on the Exodus network. We are also showing the individual data points for each trial of the α-scaled algorithm

subgraph used and C_{opt} for the α-scaled algorithm is roughly constant after a while, and it does not grow over time. Of course, there is a price for this gain, which is the occasional switching from the no-change subgraph to the optimal subgraph. In this case, the average switching probability is 11.7%, which means, out of a hundred online steps, there are about 12 times when the existing users might experience disturbances to their transmissions. Furthermore, if we look closely at the data points for individual instances, we can see that, actually, none of the instance has gone over 20% higher than the optimal one. This is consistent with our discussion in Section 12.3 about the values of α and β. Therefore, if we want to have $\beta = 30\%$, we can use a lower value for α.

Finally, Fig. 12.13 shows the simulation results for the same network setup with different α values. As we can see, the higher the α value, the lower the average cost of the subgraph. At the same time, higher α values lead to higher switching rate. We observed that when α is equal to 0.5, the cost of the subgraph used is kept at around 9% higher than the optimal cost, whereas the switching probability is only 2.05%. Therefore, by selecting the α-value properly, we can keep the cost of the multicast close to optimal during the multicast session while causing few disturbances to the existing users.

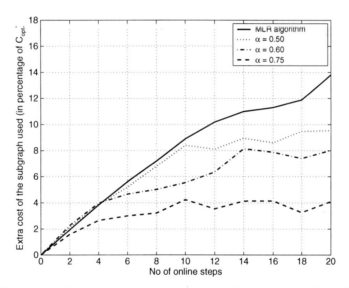

Fig. 12.13 Extra cost of the multicast subgraph generated by the α-scaled algorithm with various α-values, on the Exodus network

12.5 Conclusions

Subgraph optimization is an important problem in performing multicast with network coding. In this chapter, we studied the algorithms that solve this optimization problem for both static and dynamic multicasts. For static multicast, we presented distributed subgradient algorithms to find the min-cost subgraph and examined their convergence rate. Using the special structure of the network coding problem, we showed that with appropriately chosen step sizes, the dual problem converges to a neighborhood of the optimal solution linearly. Since the physical flow variables are decoupled from the dual iterations, we can obtain a feasible primal solution in each iteration. The convergence rate of the primal solutions is $O(1/n)$. We also proposed various heuristics for dual variable initialization and primal solution recovery to further improve the convergence performance. Simulation results show that the subgradient method produces significant reductions in multicast energy compared to centralized routing algorithms after just a few iterations. Moreover, the algorithm is robust to changes in the network and can converge to new optimal solutions quickly as long as the rate of change in the network is slow compared to the speed of computation and transmission.

For dynamic multicasts, in order to characterize the disturbances to users caused by the changes in the multicast subgraph, we introduced the concepts of link rearrangement and code rearrangement. We proposed both nonrearrangeable and rearrangeable algorithms for the dynamic multicast problem, and used simulation results to show that the α-scaled algorithm can effectively bound the growth of the multicast cost without causing too many disturbances to existing users.

Acknowledgments This work was supported by the BAE Systems National Security Solutions Inc. under grant "Control Based Mobile Ad-Hoc Networking Program (CBMANET)" (060786).

References

1. R. Ahlswede, N. Cai, S.-Y. R. Li, and R. W. Yeung, "Network information flow," *IEEE Trans. Inform. Theory*, vol. 46, no. 4, pp. 1204–1216, July 2000.
2. S.-Y. R. Li, R. W. Yeung, and N. Cai, "Linear network coding," *IEEE Trans. Inform. Theory*, vol. 49, no. 2, pp. 371–381, February 2003.
3. R. Koetter and M. Médard, "An algebraic approach to network coding," *IEEE/ACM Trans. Netw.*, vol. 11, no. 5, pp. 782–795, October 2003.
4. S. Jaggi, P. Sanders, P. A. Chou, M. Effros, S. Egner, K. Jain, and L. M. G. M. Tolhuizen, "Polynomial time algorithms for multicast network code construction," *IEEE Trans. Inform. Theory*, vol. 5, no. 6, pp. 1973–1982, June 2005.
5. T. Ho, R. Koetter, M. Médard, M. Effros, J. Shi, and D. Karger, "A random linear network coding approach to multicast," *IEEE Trans. Inform. Theory*, vol. 52, no. 10, pp. 4413–4430, October 2006.
6. D. S. Lun, N. Ratnakar, M. Médard, R. Koetter, D. R. K. T. Ho, E. Ahmed, and F. Zhao, "Minimum-cost multicast over coded packet networks," *IEEE Trans. Inform. Theory*, vol. 52, no. 6, pp. 2608–2623, June 2006.
7. T. Ho, M. Médard, J. Shi, M. Effros, and D. R. Karger, "On randomized network coding," in *Proc. of the 41th Annual Allerton Conference on Communication, Control, and Computing*, October 2003.
8. T. Ho, R. Koetter, M. Médard, D. R. Karger, and M. Effros, "The benefits of coding over routing in a randomized setting," in *Proc. 2003 IEEE International Symposium on Information Theory (ISIT'03)*, Yokohama, Japan, June–July 2003.
9. P. A. Chou, Y. Wu, and K. Jain, "Practical network coding," in *Proc. of the 41th Annual Allerton Conference on Communication, Control, and Computing*, October 2003.

10. T. Ho, B. Leong, M. Médard, R. Koetter, Y.-H. Chang, and M. Effros, "On the utility of network coding in dynamic environments," in *Proc. 2004 International Workshop on Wireless Ad-hoc Networks (IWWAN'04)*, 2004.

11. Y. Wu, P. A. Chou, and S.-Y. Kung, "Minimum-energy multicast in mobile ad hoc networks using network coding," *IEEE Trans. Commun.*, vol. 53, no. 11, pp. 1906–1918, November 2005.

12. D. S. Lun, N. Ratnakar, R. Koetter, M. Médard, E. Ahmed, and H. Lee, "Achieving minimum cost multicast: A decentralized approach based on network coding," in *Proc. IEEE Infocom*, vol. 3, March 2005, pp. 1607–1617.

13. K. Bharath-Kumar and J. M. Jaffe, "Routing to multiple destinations in computer networks," *IEEE Trans. Commun.*, vol. 31, no. 3, pp. 343–351, March 1983.

14. B. M. Waxman, "Routing of multicast connections," *IEEE J. Select. Areas Commun.*, vol. 6, no. 9, pp. 1617–1622, December 1988.

15. M. Chiang, "Nonconvex optimization of communication systems," in *Advances in Mechanics and Mathematics, Special Volume on Strang's 70th Birthday*, D. Gao and H. Sherali, Eds. Springer, New York, NY, U.S.A., 2008.

16. H. D. Sherali and G. Choi, "Recovery of primal solutions when using subgradient optimization methods to solve lagrangian duals of linear programs," *Oper. Res. Lett.*, vol. 19, pp. 105–113, 1996.

17. T. Larsson, M. Patriksson, and A. Strömberg, "Ergodic primal convergence in dual subgradient schemes for convex programming," *Math. Program.*, vol. 86, pp. 283–312, 1999.

18. K. C. Kiwiel, T. Larsson, and P. O. Lindberg, "Lagrangian relaxation via ballstep subgradient methods," *Math. Oper. Res.*, vol. 32, no. 3, pp. 669–686, August 2007.

19. A. Nedić and A. Ozdaglar, "Approximate primal solutions and rate analysis for dual subgradient methods," MIT LIDS, Tech. Rep., 2007.

20. D. S. Lun, M. Médard, and D. R. Karger, "On the dynamic multicast problem for coded networks," in *Proc. of WINMEE, RAWNET and NETCOD 2005 Workshops*, April 2005.

21. M. Imase and B. M. Waxman, "Dynamic steiner tree problem," *SIAM J. Discrete Math.*, vol. 4, no. 3, pp. 369–384, August 1991.

22. S. Raghavan, G. Manimaran, and C. S. R. Murthy, "A rearrangeable algorithm for the construction delay-constrained dynamic multicast trees," *IEEE/ACM Trans. Netw.*, vol. 7, no. 4, pp. 514–529, August 1999.

23. J. E. Wieselthier, G. D. Nguyen, and A. Ephremides, "Energy-efficient broadcast and multicast trees in wireless networks," *Mobile Netw. Appl.*, vol. 7, pp. 481–492, 2002.

24. D. P. Bertsekas, *Nonlinear Programming*. Athena Scientific, Nashua, NH, U.S.A, 1995.

25. A. Nedić, "Subgradient methods for convex minimization," Ph.D. dissertation, Massachusetts Institute of Technology, June 2002.

26. J. V. Burke and M. C. Ferris, "Weak sharp minima in mathematical programming," *SIAM J. Control Optim.*, vol. 31, no. 5, pp. 1340–1359, September 1993.

27. T. Camp, J. Boleng, and V. Davies, "A survey of mobility models for ad hoc network research," *Wireless Commun. Mob. Comput.*, vol. 2, no. 5, pp. 483–502, August 2002.

28. The rocketfuel project. [Online]. Available: www.cs.washington.edu/research/networking/rocketfuel

Chapter 13
Ultra Mobile Broadband (UMB)

Masoud Olfat

13.1 Introduction

UMB is designed to deliver mobile broadband wireless access systems that are
optimized for high spectral efficiency and short latencies using advanced technologies such as link adaptation, fast handoff, fast power control, inter-sector interference management, and multi-antenna transmission techniques. UMB is the
enhancement of CDMA2000 technologies (EVDO Rev-0, A, B, C) in the 3GPP2
standard community [1]. UMB was originally destined to support both FDD and
TDD modes, but later on more attention was paid to the development of the FDD
mode. As depicted in Fig. 13.1, although all of the previous technologies up to
EVDO-Rev B have been based upon code division multiple access (CDMA) air
interface, 3GPP2 standard community has decided to embark UMB based on orthogonal frequency division multiple access (OFDMA) to support higher capacity
traffics specially in urban environments. The same decision has been made by 3GPP
community to abandon CDMA in favor of OFDMA in their 3G LTE technology
[16]. WiMAX on the other hand has started its development based on OFDM and
OFDMA [14].

UMB standardization process started from July 2006, and the initial publication
was targeted for April 2007. It has been submitted to IEEE802.20 as a candidate for
FDD air interface technology.

The standard drafts define the technology into modules, where these modules
are called "protocols" and are grouped in "layers" and "planes". The standard draft
includes an overview document covering the physical, data link, and upper layers of
the air interface. It also includes separate physical layer and MAC layer documents,
security function documents, and several upper layer documents. All documents are

Masoud Olfat
Clearline,
593 Herndon Pkway, Herndon, VA 20171,
e-mail: masoud.olfat@clearline.com

V. Tarokh (ed.), *New Directions in Wireless Communications Research*,
DOI 10.1007/978-1-4419-0673-1_13,
© Springer Science+Business Media, LLC 2009

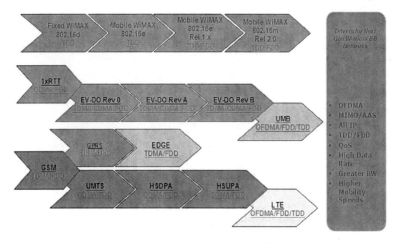

Fig. 13.1 Mobile broadband wireless technologies based upon OFDMA

stage 3, i.e., implementable requirements, containing specific language identifying mandatory and optional features on both the infrastructure and mobile equipments. This chapter outlines the highlights of UMB's physical, MAC, and upper layers very briefly.

At the moment, there are three front runner technologies for the 4G wireless broadband: 3GPP 3GLTE (long-term evolution), mobile WiMax, and 3GPP2 UMB. All three standards are offering low latency and high theoretical max speeds. AT&T and Verizon have already put their weights behind 3GLTE and Sprint Nextel (in process of creating a joint venture along with ClearWire, Intel, Google, Comcast, Time Warner Cable, and Bright House at the time of publishing this book) has embraced and already launched successfully mobile WiMax in one market in the United States [2]. Many other international operators have also announced their support for WiMAX (such as UQC in Japan or KT in Korea). However, according to ABI Research, and to my knowledge upto today, UMB's prospects are rather dim [3]. At the time of editing this book, no major operator has yet announced plans to trial or deploy UMB. Several of the major CDMA operators in the two primary markets are migrating to other technologies. Vendors cannot move forward with development unless their customers commit to trial this technology. Unlike UMB, WiMAX and LTE have ecosystems in place that offer support.

13.2 UMB Overall Architecture

UMB is consisted of several access networks (AN) that are playing the rolls of base stations and are providing two-way communications to a set of mobile or fixed subscribers called access terminals (AT). In TDD mode the ANs must be synchronized with each other, but in FDD mode they may or may not be synchronized. Each AN normally serves several ATs (point to multi-point). Although in many cases an AT communicates with only one sector of one AN, but specially in handoff areas, one AT might be in communication with more than one AN at a time. Each AN–AT pairing has its own separate protocol stack, referred to as a route, when one of the ANs is considered as serving AN and the rest are tunneled through the serving AN, without the serving AN needing to read or manage the content of the packets exchanged between the AT and the anchor AN. Both the serving AN and anchor AN are connected to Access GateWay as the first hop router (see Fig. 13.2). The tunneling system is designed in such a way that data on inter-AN interface can be tunneled directly to the AT. In other words, the communication can continue after a handoff without the need to transfer control state from the serving AN to target AN.

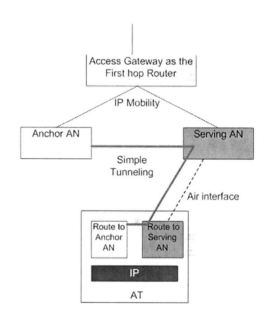

Fig. 13.2 Simple tunneling in UMB

UMB supports both unicast and broadcast/multicast modes. It allows enough flexibility to AN and AT to negotiate the configuration and their capabilities. It also allows data encryption through ciphering and integrity protection.

Unlike other earlier 3GPP2 technologies, and similar to WiMAX and 3GLTE, UMB supports a flat all IP architecture where all packets, voice, or data go through the same path. Flat networks provide faster responses, and therefore more applications can be offered. The flatter the network, the less hierarchy, and the higher overall reliability, and the lower OPEX of the network. In UMB, mobility is supported at the link layer and in the IP layer through mobile IP. Handoff to 1xEV-DO is also performed through mobile IP.

Figure 13.3 depicts the UMB protocol architecture and air interface reference model [4]. According to this depiction, the layers comprising the UMB protocol architecture are as follows:

- **Physical layer:** covering radio frequency, power, modulation and FEC coding, multiple antenna techniques, channelization, tone permutations, power control, etc. [5]
- **MAC layer:** that is a set of procedures to transmit and receive data through the physical layer over the air interface [6].
- **Radio link layer:** located below the MAC layer getting service from it, and includes QoS negotiation, multiplexing of higher layer packets, and support for various delivery modes [8].
- **Application layer:** including signaling, inter-route tunneling, support functionality, like RoHC for IP header suppression and EAP for authentication, and user plane support such as IP [9].
- **Connection control plane:** taking care of radio resource and connection management, as well as air link connection establishment and maintenance services [10].
- **Session control plane:** to provide protocol negotiation and protocol configuration services. [11].
- **Route control plane:** to create, maintain, delete, and in general manage routes [12].
- **Security functions:** include functions for key exchange, ciphering for encryption/decryption, digital signatures for and message integrity protection [7].

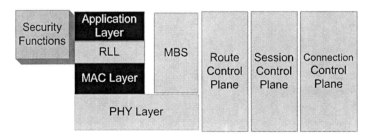

Fig. 13.3 UMB protocol architecture

UMB supports the channel bandwidths from 1.25 to 20 MHz, while multi-carrier operations and single-frequency reuse are also possible. Both downlink (forward link) and uplink (reverse link) are using OFDMA as mandatory PHY features, while CDMA is also mandatory for control channel in RL and an option for traffic in UL. In both directions rotational OFDM is also optional. Hybrid ARQ, link adaptation, and multiple antenna schemes are used to increase efficiency.

13.3 UMB Physical Layer

This section outlines the highlights of UMB PHY layer only for the FDD mode. As mentioned in Section 13.2, both DL and UL in UMB are based upon OFDMA. UMB PHY layer supports scalable bandwidth from 1.25 to 20 MHz with fine granularity.

OFDMA symbol structure: UMB supports scalable OFDMA in a sense that by changing the bandwidth the FFT size is changed such that the tone spacing is not decreased specially in smaller channel sizes. The concept of scalable OFDMA is explained in detail in Chapter 14. However, not all subcarriers can be modulated. A number of left and right tones are used as guard tones to protect data tones from out-of-band emissions. Any subcarrier which is not a guard subcarrier is defined to be a "used subcarrier". Table 13.1 defines the OFDM symbol numerology. Note that the actual bandwidth divided among tones is different and multiplied by a factor similar to undersampling. In this case, the chip rate defines the bandwidth, for example, if the bandwidth is 2.5 MHz, chip rate identified by $1/T_{CHIP}$ is equal to 2.4576 Mcps. Moreover, a multiplicative factor defined as N_{CP} determines the cyclic prefix duration. This factor is determined autonomously by the AT using the TDM 1 pilot waveform. N_{CP} could be $1, 2, 3$, or 4. The cyclic prefix is calculated based on the following equation. Based on this equation if $N_{CP} = 1$ the CP is $1/16$ of OFDM symbol duration, etc.:

Table 13.1 UMB OFDM symbol numerology

Parameters \ FFT size	128	256	512	1024	2048
Chip rate $1/T_{CHIP}$ (Mcps)	1.2288	2.4576	4.9152	9.8304	19.6608
Tone spacing $\frac{1}{T_{CHIP}N_{FFT}}$ (kHz)	9.6	9.6	9.6	9.6	9.6
Bandwidth (B) (MHz)	$B \leq 1.25$	$1.25 \leq B \leq 2.5$	$2.5 \leq B \leq 5$	$5 \leq B \leq 10$	$10 \leq B \leq 20$
Cyclic prefix duration, T_{CP} (μs)	6.51, 13.02, 19.53, or 26.04	6.51, 13.02, 19.53, or 26.04	6.51, 13.02, 19.53, or 26.04	6.51, 13.02, 19.53, or 26.04	6.51, 13.02, 19.53, or 26.04
Windowing guard, T_{WG} (μs)	3.26	3.26	3.26	3.26	3.26
OFDM symbol duration, T_s (μs)	113.93, 120.44, 126.95, or 133.46	113.93, 120.44, 126.95, or 133.46	113.93, 120.44, 126.95, or 133.46	113.93, 120.44, 126.95, or 133.46	113.93, 120.44, 126.95, or 133.46

$$T_{CP} = \frac{N_{CP}N_{FFT}T_{CHIP}}{16}, \quad N_{CP} = 1, 2, 3, 4. \qquad (13.1)$$

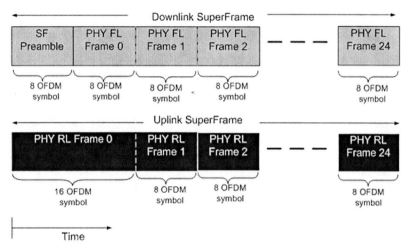

Fig. 13.4 UMB forward and reverse link superframe for full duplex FDD

Moreover, the windowing guard interval is a time domain function (mainly raised cosine function) used to shape the amplitude of the time domain OFDM signal calculated by

$$T_{WG} = \frac{N_{FFT} T_{CHIP}}{16}. \tag{13.2}$$

The OFDM symbol duration is then calculated by

$$T_s = N_{FFT} T_{CHIP} + T_{CP} + T_{WG}, \quad N_{CP} = 1, 2, 3, 4. \tag{13.3}$$

13.3.1 Superframe Structure

Figure 13.4 depicts the frame structure of UMB. As seen in this figure, in the forward link and reverse link, the units of transmission are called "superframe" or SF. An SF in the DL consists of a superframe preamble followed by 25 PHY frames. The SF preamble carries acquisition pilots and overhead channels and is used for initial system acquisition, as well as coarse channel estimation, broadcast information, and quick paging. The PHY frames are used for supporting data scheduling and resource allocation. The superframe preamble and each PHY frame in the FL and RL (except the first RL frame) consist of eight OFDMA symbols.

Two different FDD configurations are supported in UMB. One is full duplex FDD, when there is no gap between the PHY frames in FL and RL. In this case, both AN and AT can transmit and receive at the same time. In the second scenario (half duplex FDD), AN is still capable of transmitting and receiving at the same time of course on two different channel bandwidths with ample separation, while the mobile station does not include a duplexer and therefore can only transmit or receive at any moment of time. So, the scheduler at the AN must be careful not to schedule an AT to receive and transmit at the same time. This has two implications on the superframe structure as depicted in Fig 13.5. First, we need to apply a gap between the PHY frames to allow the device antenna to ramp up or down and switch from transmit to receive or vice versa. The second impact is that during the FL preamble, which is a broadcast message, no AT can send any RL message, so that portion of time is unused in the RL. The advantage is that the devices are simpler to implement and therefore less complex and less costly for end users. The disadvantage is the throughput hit caused by wasting the FL broadcasting time in the RL. The value of this guard time in UMB is calculated based on the following equation and is calculated for different N_{FFT} values to be equal to 78.13 μs:

$$T_g = 3N_{FFT}T_{CHIP}/4. \tag{13.4}$$

The duration of each superframe depends on two factors. The first factor is whether we are using half duplex or full duplex and the second factor is the value of cyclic prefix (N_{CP}). Assuming $T_g = 0$ for full FDD, and using Table 13.1, the following formula provides the values for superframe duration. Table 13.2 provides the values of these durations:

$$T_{SUPERFRAME} = 26 * 8 * T_s + 25 * T_g, \quad N_{CP} = 1, 2, 3, 4. \tag{13.5}$$

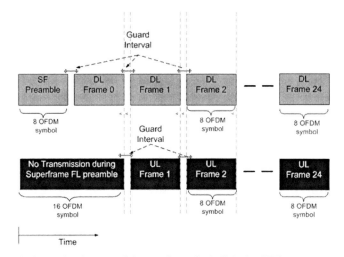

Fig. 13.5 UMB forward and reverse link superframe for half duplex FDD

13.3.1.1 FL Subcarrier Allocations

Unlike WiMAX, where the lowest possible data allocation is a slot in the downlink both distributed and adjacent permutation, in UMB the fundamental assignment

Table 13.2 UMB superframe duration for full and half duplex FDD in ms

N_{CP}	1	2	3	4
PHY frame duration	0.9114	0.9635	1.0156	1.01677
Superframe for full duplex	25.73	27.08	28.44	29.79
Superframe for half duplex	23.70	25.05	26.41	27.76

resource unit is a subcarrier. This is called a hop port. A hop port is a static resource that maps to a unique physical subcarrier, where the mapping changes over time. This concept is called "hopping." The concept of hop port differs depends on the permutation mode as described here [5].

In the forward links the subcarriers can be allocated to the users in two different modes. This classification is equivalent to both WiMAX and 3GLTE that support distributed and adjacent tone permutations. Similarly, UMB supports two modes of tone permutations, and subcarrier assignments to users. One is called distributed resource channel (DRCH) and the second one is called block resource channel (BRCH). Figure 13.6 illustrates the two tone permutation modes in UMB.

DRCH is equivalent to distributed tone permutation, and the subcarriers are scattered across entire bandwidth. A particular hopping pattern maps assigned hop ports to physical subcarriers in the frequency domain. The hop pattern changes from one OFDM symbol to the next. The channel and interference estimation is based on broadband common pilot. In this case, a tile assigned to a user is composed of 16 hop ports (tones) that are regularly spaced over a subzone (or the whole bandwidth, if the whole bandwidth is used for DRCH). The mapping changes every two OFDM symbols.

The BRCH is defined based upon the concept of tiles. Sixteen subcarriers of continuous hop ports in the frequency domain are grouped into a hop-port block. A tile consists of a hop-port block over a PHY frame of eight OFDM symbols. However, the set of tiles scattered across entire bandwidth. Some hop patterns map the hop ports to tiles in frequency domain, which cannot change for duration of one PHY frame. The hopping pattern can change from one PHY frame to another one. The hopping patterns among multiple sectors are independent from each other. This hopping pattern changes the mapping of hop ports to physical subcarriers in a pseudo-random manner from frame to frame, and therefore provides frequency and interference diversity. In the case of adjacent tone permutation, a tile is the minimum resource allocation unit for data transmission. The hopping pattern is identified by using a channel tree to specify sets of hop-port blocks. The desired channel and interference estimation over a tile is performed using dedicated pilots on those tiles.

However, UMB allows these two modes to coexist in the same PHY frame in two different approaches, in order to provide frequency diversity and frequency-selective

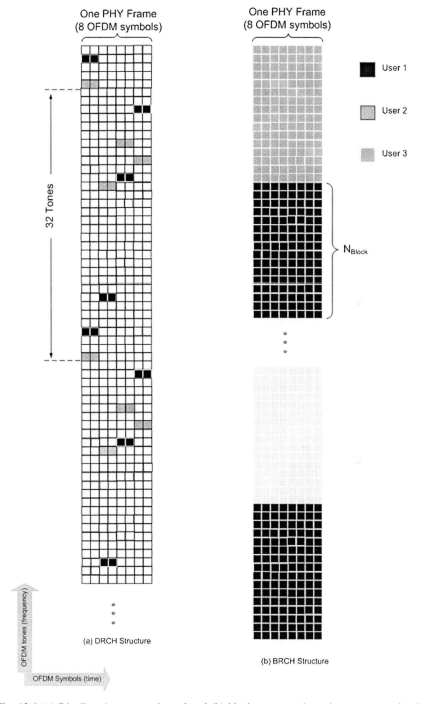

Fig. 13.6 (a) Distributed resource channel and (b) block resource channel tone permutation in UMB forward link

transmission simultaneously. The choice of resource multiplexing mode is based on the parameter called ResourceChannelMuxMode of the overhead messages protocol (see Fig. 13.7).

- **Multiplexing Mode 1:** In the first multiplexing mode, the underline permutation is DRCH, and DRCH punctures BRCH. The overhead channel indicates how many DRCHs are used so that ATs know the puncturing pattern.
- **Multiplexing Mode 2:** As seen in Fig. 13.7, in this case, DRCH and BRCH are only used on different subzones (subzone is defined as a set of subcarriers or hop ports of given size, usually 64 or 128). Overhead channel indicates which subzones are in DRCH and which subbands are in BRCH.

13.3.2 UMB FL Channelization

In the forward link, each superframe can be categorized into one of the following four channels: Preamble, control channel, pilot or data channel. UMB provides efficient channelization for encoding and decoding processing time, diversity, providing access to ATs, and scalability across various deployment bandwidths. This channelization scheme provides an acceptable reliability in FL and RL [13]. In the following, we describe the preamble functionality along with the channelization procedure in both the FL and RL in a little more detail.

Preamble: Figure 13.8 shows the structure of UMB FL preamble. The preamble like other frames consists of eight symbols. The first OFDMA symbol carries the Forward Primary BroadCast Control CHannel (F-PBCCH). F-PBCCH carries deployment-wide static parameters. The next four OFDMA symbols carry the Forward Secondary Broadcast Control CHannel (F-SBCCH) and the Forward Quick Paging CHannel (F-QPCH) in alternate superframes. In other words in even superframes F-SBCCH is sent over symbols 1–4 and in odd superframes F-QPCH is sent on those symbols (Fig. 13.8). The F-SBCCH carries sufficient information to enable the AT to demodulate FL data from the PHY frames. It gives information on FL hopping patterns, pilot structure, control channel structure, transmit antennas, multiplexing modes, etc. The last three symbols are TDM pilots, which are used for initial acquisition. These three OFDM symbols are denoted as TDM pilot 1, TDM pilot 2, and TDM pilot 3, respectively. TDM pilots 2 and 3 are additionally used to transmit the other sector interference channel as well. The TDM pilot 1 OFDM symbol forms the Forward ACQuisition CHannel (F-ACQCH), and the TDM Pilots 2 and 3 OFDM symbols form the Forward Other Sector Interference CHannel (F-OSICH). TDM1 is used for initial timing acquisition and coarse frequency offset recovery. TDM1 also carries information assisting in system determination by the AT. F-OSICH carries assisting RL interference (power) control for OFDMA data channels. It is conveyed as different phases on the two symbols. TDM 2 and TDM 3 include Walsh codes in the time domain, and then converted to the frequency domain using a DFT pre-coding before OFDM modulation.

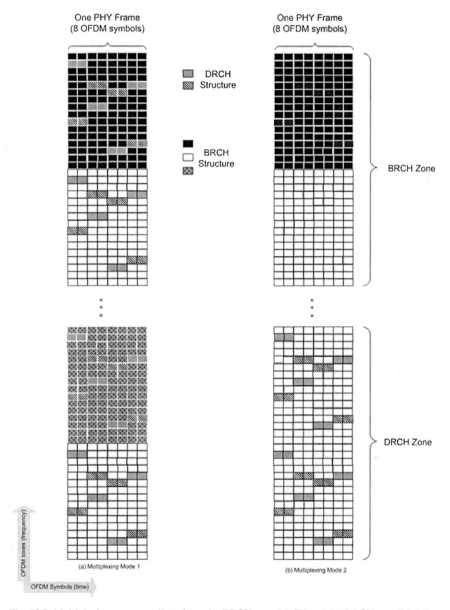

Fig. 13.7 Multiplexing resource allocations: (a) BRCH over DRCH and (b) DRCH over BRCH

Data channel: Forward data channel (F-DCH) is the primary channel for user data transmission. Cyclic redundancy checking (CRC) is appended to the data channel to validate the integrity of the packets. FEC coding and modulation as well as interleaving is also applied to this channel. Each mobile device is assigned a particular data channel either in distributed or adjacent mode. The location of each data

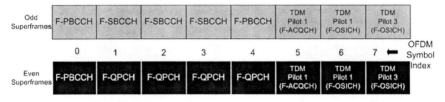

Fig. 13.8 UMB forward superframe preamble structure

channel assignment in frequency and time domain is communicated to the mobile station using a control channel called forward shared control channel (F-SCCH).

Pilot Channels: There are five types of pilot channels in UMB, namely forward common pilot channel (F-CPICH), forward dedicated pilot channel (F-DPICH), forward channel quality indicator pilot channel (F-CQIPICH), forward beacon pilot channel (F-BPCH), forward cell null channel (F-CNCH), and forward preamble pilot channel (F-PPICH).

- F-DPICH is present only in multiplexing mode 2, and transmitted in the BRCH subzones. It is present in each tile of assigned data transmission for channel estimation and interference estimation.
- F-CPICH is present in both multiplexing modes. No other pilots are used in multiplexing mode 1. Since F-DPICH is only used in the BRCH in multiplexing mode 2, F-CPICH is present only in DRCH subzones.
- F-CQIPICH is used by the AT to estimate the channel quality indicator (CQI) value and the best pre-coding matrix in case of closed-loop MIMO. It is sent only in case of multiplexing mode 2 and has a low overhead. F-CQIPICH is transmitted once every eight PHY Frames.
- F-BPICH is used to indicate the presence of the AN (base stations) on other channel bandwidths. It acts as out-of-band pilots to enable multi-carrier handoffs. A sector may transmit a sequence of periodic beacons (F-BPICH) on one or more channel carriers. The modulation of F-BPICH depends on the number of carriers, the ID of the carrier the beacon is sent over (2 bits), and the ID of the preferred sector (10 bits). A mobile station is able to decode beacons containing the ID of channel bandwidth carried over by its sector and sent by other sectors operating on other carrier. Therefore, the AT can identify neighboring sectors without tuning at those sectors and reading their signals. There are two different coding schemes for beacons, beacon codes A and B. Beacon code A is used when the number of used tones ($N_{FFT} - N_{GUARD}$) is greater than or equal to 422, otherwise Beacon code B is used. In each case Reed–Solomon coding is used to encode F-BPICH. The Reed–Solomon codes cause minimal interference to data traffic.

A beacon normally occupies an OFDM symbol where the whole power is transmitted is concentrated on a single tone, and zero energy on other tones (see Fig. 13.9). Beacon pilots are reserved every two superframes, and on those superframes, a number of OFDM symbols are reserved (the number is equal to the number of carriers). In this case, those symbols are dedicated to beacons, and no other channels are carried over those symbols. The actual position of the beacon symbol in the PHY frame is determined by a hash function of the sector ID. Each sector repeatedly transmits the same beacon code sequence on each carrier [15].

- F-CNCH defines subcarriers that are blanked by all the sectors in a cell. The main use of these pilots is measuring out-of-cell interference levels, because no other signal is sent over these channels by any sector.
- F-PPICH is used for coherent demodulation of F-OSICH. It is present only in the first two symbols of the preamble.

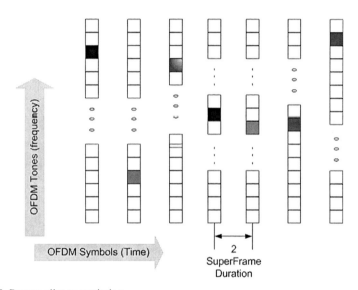

Fig. 13.9 Beacon pilot transmission

So, for each multiplexing mode, we can conclude the following:

1. For multiplexing mode 1, F-CPICH is present in every FL PHY frame and is distributed all over the bandwidths through the used subcarriers. F-DPICH and F-CQIPICH are absent in this mode.
2. For multiplexing mode 2, as explained before, there are two distinct subzones within a PHY frame, one is DRCH and the other is BRCH. F-CPICH is transmitted over the DRCH subzone for channel estimation of those zones in every FL PHY frame. For MIMO, the set of F-CPICH pilot tones is orthogonal for different antennas (Fig. 13.10). In BRCH subzones, F-DPICH is transmitted in every FL PHY for channel estimation of those zones. It is also used to estimate the second-order statistics of channel and interference of each BRCH tile. F-CQIPICH is transmitted in one out of every eight FL PHY frames and is used by the access terminal to measure FL channel quality and to support MIMO precoding for closed-loop MIMO (for concept see, for example, [17]).

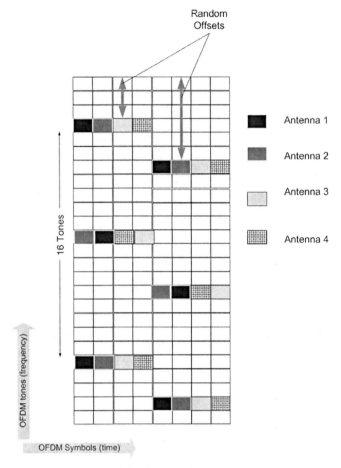

Fig. 13.10 Orthogonal forward link F-CPICH pilots for MIMO applications

Control Channels: The control channels in UMB are used for scheduling and managing the resources in the forward and reverse link. The control channels are all multiplexed together in one segment called forward link control segment (FLCS). There is no restriction in the location of FLCS ports. They can be located either over BRCH or DRCH. There should not be any overlap among pilot channels and control channels. FLCS is present in all FL PHY frames.

The control channels used in forward link are as follows:

1. The forward shared control channel (F-SCCH) is the most important control channel carrying the information about the data allocation opportunities for each AT. It carries information for the forward data channel transmission for one AT as well as group resource allocation. The assignment is mainly an encoded and CRC-protected blocks for data.

2. The forward reverse activity bit channel (F-RABCH) is a 1-bit indication, which shows whether AT should send any traffic on the reverse link CDMA segment or not.

3. The forward start of packet channel (F-SPCH) is used to indicate whether a persistent assignment is still valid. Persistent scheduling is referred to cases where an assignment is given to a mobile device for a period of time and is not expected to change for that duration. It is valid till it is explicitly de-assigned, or lost due to packet decoding failure, or expired due to recent overlapping assignment. This is unlike non-persistent scheduling, which is valid only for one packet transmission. It is also used to de-assign a previously assigned persistent scheduling, by indicating a value of 0 for the assignment (no packet is sent).

4. The forward fast other sector interference channel (F-FOSICH) is used for fast indications of interference levels over a given subzone in other sectors. If this channel is enabled, F-FOSICH is broadcast every PHY FL frame so the ATs are aware of the interference imposed by other sectors. Note that here, the entire bandwidth is divided evenly into several subzones. For example in case of 10 MHz bandwidth, the size of each subzone is 1.23 MHz. One might ask about the difference of this control signal with F-OSICH transmitted over the superframe preamble. The answer is that F-FOSICH carries the other sector interference indication in a faster rate but with a less coverage than F-OSICH.

5. The forward power control channel (F-PCCH) is used for reverse link closed-loop power control and is composed of 1-bit indicator, which commands the device to increase or decrease the RL CDMA control channel transmit power. Typical control rate is around 140 Hz.

6. The forward pilot quality indicator channel (F-PQICH) is a 4-bit quantized value representing RL pilot strength for each AT. It helps the mobile devices to choose the reverse link serving sector, when the AT needs to make decision during hand-off process. It is used for power control during RL control and data channels. The frequency of transmitting this control information is 67 Hz (one every 15 ms).

7. The forward interference over thermal channel (F-IOTCH) shows the interference level on a given subzone. The base station (AN) broadcasts interference over thermal (IoT) value in a subzone based on the amount of interference it observes on that subzone. If enabled, this is transmitted every FL PHY frame. F-IOTCH is used for dynamic interference control.

8. The forward acknowledgment channel (F-ACKCH) is used when the HARQ is used in the reverse link. It is used to acknowledge RL HARQ transmissions, sent in every PHY frame to acknowledge the associated RL PHY frame.

The FLCS shall operate on all DRCH resources if multiplex mode 1 is used, and can operate on either all DRCH resources or all BRCH resources if multiplex mode 2 is used.

FLCS is partitioned into two pieces: one common segment (CS) and zero or more link assignment block (LAB) segments. All of the above-mentioned control channels except F-SCCH are located only in the common segment. The LAB segments, if present, contain only the F-SCCH. However, if LAB does exist, F-SCCH could be present in CS.

Within CS, the control channels if present must follow the following order: F-ACKCH, F-SPCH, F-RABCH, F-PQICH, F-FOSICH, F-IOTCH, and F-PCCH. If LAB is not present and F-SCCH is present on CS, it is sent as the last control channel on CS. These channels are allocated and transmitted in FLCS common segment in a way to achieve the third-order diversity.

All control channels shall be multiplexed together onto a set of $N_{FLCS-BLOCKS}$ hop-port blocks, when

$$N_{FLCS-BLOCKS} = N_{FLCS-COMMON-BLOCKS} + 3 \times N_{FLCS-LAB-SEGMENTS},$$

where $N_{FLCS-COMMON-BLOCKS}$ and $N_{FLCS-LAB-SEGMENTS}$ are provided by overhead messages protocol.

If LAB segments exist, it sends data allocation to a user or a group of users, indicating the physical resources assigned as well as the modulation and coding to be used on that transmission. It also supports both persistent and nonpersistent assignments. Each LAB segment contains three tiles, and each F-SCCH will be populated onto one LAB segment (therefore, the total number of tiles belonging to LAB segments is a multiple of 3).

Table 13.3 summarizes all FL channels used in UMB.

13.3.3 Reverse Link in UMB

UMB's reverse link (RL) supports a hybrid OFDMA and CDMA air interface. The control channel on the RL is transmitted using power-controlled CDMA, and data traffic transmitted optionally either by OFDMA for high-rate transmission or CDMA for low-rate transmission. If both modes are used, they could be multiplexed in both time and frequency domains. The amount of resources allocated to each PHY mode can be configured by the operator, through the infrastructure RAN.

It is important to note that by CDMA, we mean multi-carrier CDMA, when the tones of an OFDM transmitter are CDMA modulated. In UMB, within one OFDM symbol, both CDMA and non-CDMA tones could be modulated by one IFFT module (see Fig. 13.11). As observed from this figure, the CDMA-modulated tones are DFT pre-coded. The advantage of pre-coding for CDMA tones is statistical multiplexing of bursty control and traffic signals, as well as reduced latency for control signaling.

CDMA in UMB is used for two reasons. First, every AT must use CDMA for transmitting RL control channels. Also, ATs can optionally use CDMA for sending low-rate, bursty, delay-sensitive data traffics such as VoIP. In general power control is an important factor for achieving acceptable SINR at the receiver. As a result, UMB has mandated the use of power control along with CDMA mode in RL for both control channel and traffic channel. As described in Section 13.3.2, F-RABCH is used to send acknowledgment for RL HARQ. This means fast power control and HARQ are supported for the CDMA data channel as well.

Table 13.3 All FL channels used in UMB superframe

Forward link control channels transmitted in superframe preambles		
Channel name	Description	Used for
F-PBCCH	Primary BroadCast Control CHannel	Carry deployment specific information
F-SBCCH	Secondary BroadCast Control CHannel	Carry sector-specific information for odd symbols
F-QPC	Quick Paging CHannel	Carry sector-specific information for even symols
F-ACQCH	Acquisition CHannel	Initial timing acquisition and coarse frequency offset recovery
F-OSICH	Other-Sector-Interference CHannel	Carry assisting RL interference
Forward link data channels		
Channel name	Description	Used for
F-DCH	Data CHannel	Carry user data
Forward link pilot channels		
Channel name	Description	Used for
F-DPICH	Dedicated Pilot CHannel	Channel estimation for Mux mode 2
F-CPICH	Common Pilot CHannel	Channel estimation in DRCH and both Mux modes
F-CQIPICH	Channel Quality Indicator PIlot CHannel	Estimate CQI for pre-coding matrix of closed-loop MIMO
F-BPICH	Beacon Pilot CHannel	Indicate the presence of AN on other channel BWs
F-CNCH	Cell Null CHannel	Measuring out-of-cell interference
F-PPICH	Preamble Pilot CHannel	Coherent demodulation of F-OSICH
Forward link control channels		
Channel name	Description	Used for
F-SCCH	Shared Control CHannel	Carries information about the data assignment for each AT.

(Continued)

Table 13.3 (Continued)

Channel name	Description	Used for
Forward link control channels		
F-RABCH	Reverse Activity Bit CHannel	Reverse link CDMA segment indication
F-SPCH	Start of Packet CHannel	Indicate persistent assignment
F-FOSICH	Fast Other Sector Interference CHannel	Interference levels over a given subzone indicator
F-PCCH	Power Control CHannel	Reverse link closed-loop power control
F-PQICH	Pilot Quality Indicator CHannel	RL pilot strength indicator
F-IOTCH	Interference Over Thermal CHannel	Interference level on a given subzone
F-ACKCH	Acknowledgment CHannel	RL HARQ acknowledgment

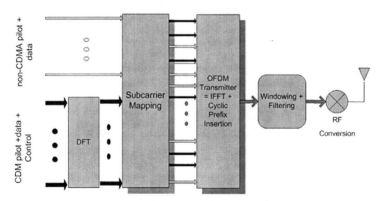

Fig. 13.11 Hybrid OFDMA/CDMA in UMB reverse link

Regular OFDMA transmission is mainly used for data packets and is used for single antenna, as well as quasi-orthogonal multiple antenna (QORL) and layered superposed OFDMA (LS-OFDMA) data transmission. The transmission of OFDMA is tile based, similar to the BRCH channelization on the forward link. Pilot patterns are defined that provide enough pilot resources for multiple users to share the same time–frequency resources to support QORL, SDMA, and LS-OFDMA, and also to provide sufficient channel estimation performance for time- and frequency-selective channels.

Normally the AN will indicate which flows are allowed on the CDMA traffic segment only, OFDMA traffic segment only, or both.

13.3.3.1 Reverse Link Physical Channels in UMB

Similar to forward link, there are three types of physical channels in reverse link, namely data channels, pilot channels, and control channels. However, we can categorize these physical channels in a different way, CDMA and OFDMA categorization. In other words, we can say that each RL PHY frame can be divided into a CDMA segment and an OFDMA segment.

The RL OFDMA segment consists of the following RL channels: reverse dedicated pilot channel (R-DPICH) the RL OFDMA pilot channel, reverse OFDMA dedicated control channel (R-ODCCH), and reverse acknowledgment channel (R-ACKCH) the OFDMA control channels, and Reverse OFDMA Data CHannel (R-ODCH) the OFDMA RL data channel.

The RL CDMA segment also consists of the following physical channels: reverse pilot channel (R-PICH), and reverse auxiliary pilot channel (R-AUXPICH) the pilot channels, reverse access channel (R-ACH), reverse CDMA dedicated control channel (R-CDCCH) the RL CDMA control channels, and finally reverse CDMA data channel (R-CDCH) the CDMA RL data channel.

Table 13.4 summarizes the reverse link physical channels and their definitions.

Reverse Link CDMA Physical Channel: The CDMA segment carries mandatorily the CDMA control channels R-ACH and R-CDCCH, the pilot channels R-PICH. It also optionally carries the data channel R-CDCH, and in that case the pilot channel R-AUXPICH can be carried optionally. Transmissions from different ATs in the CDMA segment are multiplexed in a CDMA fashion. In other words those waveforms are not orthogonal with respect to each other. The waveforms corresponding to the different channels that are carried on the CDMA segment are first generated in the time domain. The time domain waveforms of the different channels are then added together and the resulting waveform is converted to the frequency domain using a discrete Fourier transform (DFT) operation. The resulting frequency domain sequence is then mapped to the subcarriers of an OFDM symbol that are assigned to the CDMA segment for this AT.

1. **R-ACH** is a CDMA RL control channel used by AT for initial access, and for transition out of semi-connected state, or to handoff between sectors at the same or at different frequencies. The time domain sequence used in this channel is a Walsh sequence with a size of 1024 multiplied elementwise by a scrambling complex of the same size, and then time interleaved. The last 128 elements are selected (truncated), and the truncated sequence is multiplexed and passed through the DFT pre-coding.

2. **R-CDCCH** is a CDMA RL control channel carrying one or more of the following logical channels: channel quality indicator (CQI), the request channel by AT, the power amplifier headroom channel for power control purpose,

Table 13.4 All RL channels used in UMB

Reverse link CDMA segment physical			
Channel name	Description	Type	Used for
R-CDCCH	CDMA Dedicated Control CHannel	Control	Carry CQI, AT request channel, PA headroom
R-ACH	Access Channel	Control	Initial access and handoff
R-PICH	PIlot CHannel	Pilot	Helps RL power control and quality measurement
R-AUXPICH	AUXiliary PIlot CHannel	Pilot	Assist CDMA RL data transmission
R-CDCH	CDMA Data CHannel	Data	Carry CDMA RL data traffic
Reverse link CDMA segment physical			
Channel name	Description	Type	Used for
R-ACKCH	ACKnowledgment CHannel	Control	Acknowledges FL PHY Frames
R-ODCCH	OFDMA Dedicated Control CHannel	Control	Multiplexes several logical channels periodically
R-DPICH	Dedicated PIlot CHannel	Pilot	Channel estimation RL data and control transmission
R-ODCH	OFDMA Data CHannel	Data	RL OFDMA data traffic

and the power spectral density indication channel. The time domain sequence used in this channel is a Walsh sequence with a size of 1024 multiplied elementwise by a scrambling complex of the same size, and then time interleaved. The scrambled sequence is multiplexed and passed through the DFT pre-coding.

3. **R-PICH** is a CDMA RL pilot channel that helps the base station to support reverse link power control and reverse link quality measurement. The time domain sequence used in this channel is a Walsh sequence with a size of 1024 multiplied elementwise by a scrambling complex of the same size, and then time

interleaved. The scrambled sequence is multiplexed and passed through the DFT pre-coding.

4. **R-AUXPICH** is another CDMA RL pilot channel that is transmitted in every subsegment containing a CDMA data channel transmission. It is used to assist the AN estimate of the RL channel when the AT transmits CDMA traffic in RL. Moreover, this channel also carries information about the rate and HARQ transmission index of the R-CDCH transmission in the same subsegment. Transmission on R-AUXPICH is always aligned with the transmission on R-CDCH. The time domain sequence used in this channel is a Walsh sequence with a size of 1024 multiplied elementwise by a scrambling complex of the same size, and then time interleaved. The scrambled sequence is multiplexed and passed through the DFT pre-coding.

5. **R-CDCH** is used for CDMA data transmission in the reverse link. It supports a limited set of transmission formats. The packet formats on these channels are optimized for VoIP normally with EVRC vocoders, requiring three packet formats. Other types of flows may be transmitted on this segment subject to packet format limitation. The CDMA flow mapping is determined by AT using a distributed AT-centric CDMA MAC. As mentioned before, the traffic flow on these channels are DFT precoded, to provide statistical multiplexing (Fig. 13.11). The auxiliary pilots are transmitted in frames carrying CDCH transmissions. They occupy the same bandwidth as the data transmission. R-AUXPICH can also be used for control channel demodulation on frames where data are transmitted. The control subsegment hops over traffic sub-segments. R-CDCH always encodes using Rate-1/5 turbo coder or Rate-1/3 convolutional encoder.

The CDMA control segments are normally defined in such a way to occupy the whole channel bandwidth. In other words the CDMA control segment samples the wideband channel, and therefore allows AN to obtain knowledge about the SINR about the whole band in order to create an acceptable transmit power to achievable SINR mapping that is required for efficient scheduling.

The R-CDCCH is a common control segment shared by all users and is typically used to carry a periodic control channel.

Reverse Link OFDMA Physical Channel: The OFDMA segment carries one data channel R-ODCH, one pilot channel R-DPICH, and two control channels R-ACKCH and R-ODCCH. The transmission of OFDMA is tile based, similar to the BRCH channelization on the forward link. As described before, each hop-port block in UMB consists of 16 contiguous hop-ports, which are mapped by the hopping permutation to a contiguous set of subcarriers. Also, the set of subcarriers corresponding to a hop-port block does not change over one PHY frame. Therefore, the set of resources (over time and frequency) can be divided into units of tiles, where a tile is a contiguous 16×8 rectangle of hop-ports (16 in frequency and 8 in time) which are mapped to a contiguous 16×8 rectangle of subcarriers. Each tile on RL can be assigned to the CDMA segment, to the OFDMA segment, or may be left blank.

A pilot pattern is defined that provides enough pilot resources for multiple users to share the same time–frequency resources to support SDMA and LS-OFDMA, and also to provide sufficient channel estimation performance for time- and frequency-selective channels.

1. **R-DPICH** is a RL pilot OFDMA channel that provides dedicated pilots for the reverse OFDMA dedicated control channel (R-ODCCH) and reverse OFDMA data channel (O-ODCH) in order to allow the base station to perform channel estimation. R-DPICH shall be present only in tiles assigned to the OFDMA segment, not the ones assigned to CDMA segments. With the OFDMA segment, the tile may be assigned to the R-ODCH, the data channel, or R-ODCCH, the control channel. The location of these pilots in each tile is different for each of these cases and also depends on several parameters such as R-DPICH format index (could be 0 or 1), energy per modulation symbol, and code offset (an integer between 0 and 2). Figure 13.12 depicts the location of these pilots for different formats.
2. **R-ACKCH** is an OFDMA RL control used to acknowledge FL PHY frames transmitted on the forward data channel. In UMB the forward link serving sector (FLSS) and reverse link serving sector (RLSS) are not necessarily the same. R-ACKCH acknowledges the FLSS. The R-ACK channel is modulated on the OFDMA tones and punctures the RL data. The AT is assigned four subtiles to transmit its R-ACKCH. An algorithm ensures that every subtile either gets assigned to only one AT or, if it is assigned to more than one AT, the ATs use different exponential sequences. Figure 13.13 shows the modulation process and subtile mapping for R-ACKCH and Fig. 13.14 shows an example of the R-ACKCH resource assignments to five ATs, where there are five partial tiles, each divided into four subtiles.
3. **R-ODCCH** is another OFDMA RL control channel that multiplexes several logical channels in a periodic manner. Those logical channels are as follows:

   - *r-reqch: requests reverse link resources*
   - *r-bfch: feedback channel in support of forward link pre-coding and SDMA*
   - *r-sfch: feedback channel in support of forward link subband scheduling*
   - *r-cqich: forward link channel quality indicator channel*
   - *r-mqich: MIMO quality indicator channel*
   - *r-psdch: power spectral density (PSD) indicator channel*
   - *r-pahch: power amplifier (PA) headroom indicator*

4. **R-ODCH** is used for OFDMA data transmission in reverse link. It consists of either a data packet or an erasure sequence, both of which can span one or more RL PHY frames. This channel is power controlled in order to maintain desired other cell interference within a threshold. These channels are scheduled within tiles similar to FL BRCH channelization. It supports QORL and LS-OFDMA using orthogonal overlapped pilot sequences over each contiguous pilot cluster.

Table 13.4 summarizes the reverse link physical channels and their definitions.

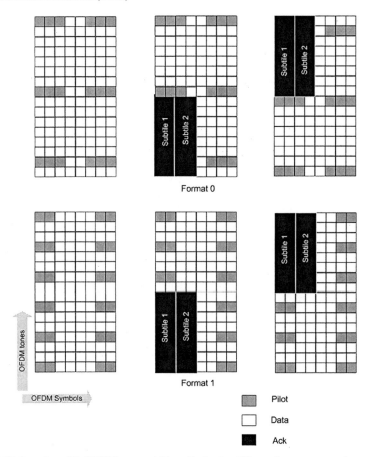

Fig. 13.12 Location of R-DPICH tones within a tile for the different formats

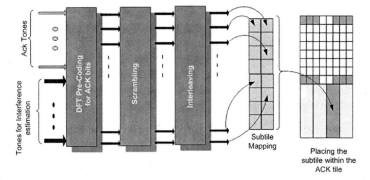

Fig. 13.13 R-ACKCH modulation and subtile mapping

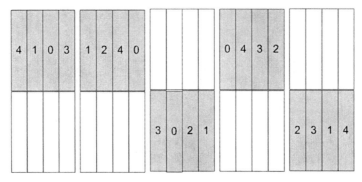

Fig. 13.14 R-ACKCH resource assignment to five ATs

13.4 UMB MAC Layer

UMB MAC layer is designed to provide reliable transmission between AN and multiple ATs. UMB MAC is composed of several components and entities mentioned below:

1. **Basic Packet Consolidation Protocol** provides the procedures and messages to carry multiple upper layer payloads in one MAC layer packet.
2. **Basic Superframe Preamble MAC Protocol** provides the procedures and messages for the operation of channels transmitted on the superframe preamble.
3. **Basic Access Channel MAC Protocol** provides the procedures and messages for the operation of the reverse access channel.
4. **Basic Forward Link Control Segment MAC Protocol** provides the procedures and messages for the operation of the channels transmitted on the forward link control segment.
5. **Basic Forward Traffic Channel MAC Protocol** provides the procedures and messages for the operation of the forward data channel.
6. **Basic Reverse Control Channel MAC Protocol** provides the procedures and messages for the operation of all the reverse link control channels except the reverse access channel.
7. **Basic Reverse Traffic Channel MAC Protocol** provides the procedures and messages for the operation of the reverse data channels.
8. **Basic RL QoS MAC Protocol** provides the procedures and messages for the access network to determine the QoS-related behavior of the access terminal.

Data Encapsulation: In FL the MAC layer of AN receives route packets, consolidates zero or more Route packets into a consolidated packet using the Packet Consolidation Protocol, invokes the forward traffic channel MAC to add layer-related headers and trailers, and forwards the resulting packet to the physical layer for transmission on the F-DCH.

In RL, the MAC layer of AT receives route packets, consolidates zero or more route packets into a consolidated packet using the Packet Consolidation Protocol

(PCP), invokes the reverse traffic channel MAC to add layer-related headers and trailers, and forwards the resulting packet to the physical layer for transmission on the R-ODCH or R-CDCH.

Figure 13.15 shows the basic data encapsulation process. On reception in each direction, the inverse functions are performed, and the received packets are delivered to the route protocol.

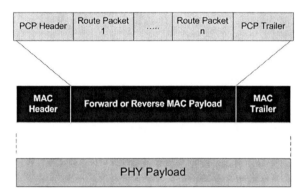

Fig. 13.15 UMB data encapsulation process

MAC Basic Access Channel Protocols: The basic access channel MAC protocol provides the procedures and messages required for an AT to transmit, and for an AN to receive, the access probe. An access probe may be used for initial access or handoff within an active set. The AT transmits a number of sequences of "access probes" until either a grant is received (within five physical frames) or until the AT gives up. The AN responds to an access probe with an access grant over the forward link control segment (FLCS) MAC protocol. The AN normally has at least one instance of this protocol for each sector in the network.

This protocol can be in one of the following two states at the AT:

- *Inactive State: In this state, the protocol waits for an activate command. This state occurs when the AT has not acquired an AN.*
- *Active State: In this state, the AT may transmit on the access channel.*

The AN can only be in the active state, when it monitors the access channel. Figure 13.16 shows the state diagram for basic access channel protocols.

The power level of the probes is set at the minimum originally and incrementally they increase with no limit, controlled by the AN. The timing between the probes and also the number of probes in each sequence as well as the maximum number of probe sequences are also controlled by AN. However, AT defines some of the parameters such as access type (initial access or handoff), AT access class, reason for access (paged or AT-initiated), selectivity of the quick paging format, QoS class of the intended service.

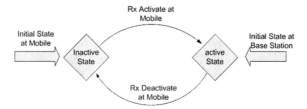

Fig. 13.16 State diagram for basic access channel MAC protocol

HARQ Interlace Structure for the Basic Physical Layer: HARQ is supported in both forward and reverse links. An HARQ interlacing structure consists of the timing relationship between different transmissions of the same MAC packet and of the acknowledgments, ACK or NACK, for these transmissions. Multiple HARQ interlacing structures are possible for both the forward and the reverse links. The HARQ interlacing structure can be different for different MAC packets and is determined by the assignment received on the F-SCCH. This structure could differ based upon relationship between the timings of the assignments, data transmission, ACK transmission, data retransmissions, etc. UMB supports synchronous HARQ on both FL and RL, i.e., synchronous re-transmissions and synchronous ACK/NACKs, thus reduces assignment overhead by eliminating explicit assignments for re-transmissions. Multiple HARQ interlace structures are possible for both FL and RL.

(1) Eight- and Six-Interlace Structures Without Extended Information: Support for eight interlaces is mandatory and with six interlaces is optional.

For FL, when a PHY frame is sent in FL, its HARQ retransmission is sent after eight/six PHY frames. This is true for repetition of HARQ retransmissions. Data transmission corresponding to an FL assignment that arrives in FL PHY frame k begins in FL PHY frame k. The acknowledgments are on RL PHY frame $k + 5/k + 3$. HARQ re-transmissions associated with the MAC packet that starts in PHY frame k occur in PHY frames $k + 8n/k + 6n$, $n = 1, \ldots$. Figure 13.17 shows examples of the timing relationship between FL packet transmissions and the associated acknowledgment transmissions for the eight-interlace structure. The multiple interlace structures allow ATs to have different decoding processing delays. Figures 13.17 and 13.18 show the eight- and six-interlace structures without extended information, respectively.

(2) Interlace Structure with Extended Transmissions: Support of this mode is optional for ATs. When the AT is close to cell edge or its power or link budget is limited, the extended transmission on FL is more appropriate. In this case, each HARQ transmission of a data packet spans three FL PHY frames. Data transmission corresponding to an FL assignment that arrives in FL PHY frame k begins in FL PHY frame k and the first transmission spans FL PHY frames $k, k + 1$, and $k + 2$ and is acknowledged on RL PHY frame $k + 5$. HARQ re-transmissions associated with this MAC packet occur in FL PHY frames $k + 8n, k + 8n + 1$, and $k + 8n + 2$ where n is the transmission index (see Fig. 13.19).

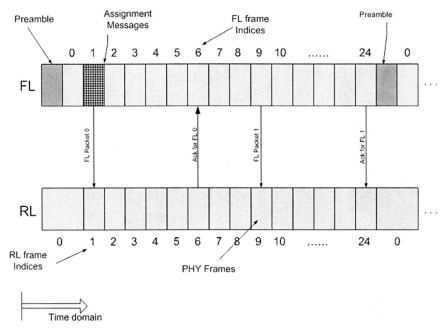

Fig. 13.17 FL eight interlace structure without extended transmissions

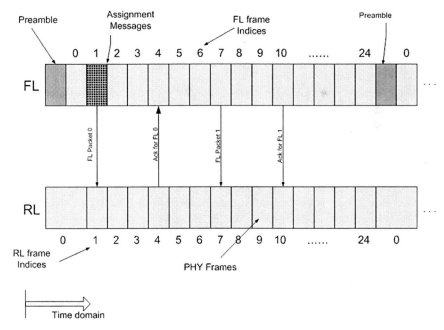

Fig. 13.18 FL six-interlace structure without extended transmissions

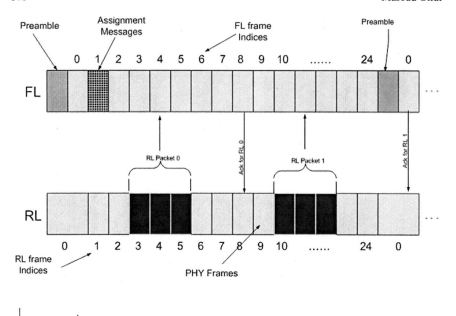

Fig. 13.19 FL eight-interlace structure with extended transmissions

UMB supports eight and six interlaces without extended information with two frames as well. These modes are used by the AN only in conjunction with group resource transmission, which we suffice by only showing a figure of eight-frame case; see Fig. 13.20

Basic Forward Link Control Segment MAC Protocol: The basic forward link control segment (FLCS) MAC protocol provides the procedures and messages required for MAC layer signaling. This includes AN transmissions on the F-SCCH, F-PCCH, F-PQICH, F-FOSICH, F-IOTCH, F-RABCH, F-ACKCH, and F-SPCH physical layer channels. For this protocol as well, MAC assumed two states for the AT, namely inactive and active states.

This protocol provides access grants, resource allocations, H-ARQ, and PHY layer support (pilots, power control, transmission diversity, timing offset) on forward channels that together form the FLCS. The resource allocation is established by the access network and communicated to the access terminal via assignment blocks, which include information such as MACID, identity of the allocated groups of hop-ports, persistence of allocation (e.g., permanent or temporary), packet format, H-ARQ interlace to use, physical layer support parameters.

Basic Forward Traffic Channel MAC Protocol: The basic forward traffic channel MAC protocol provides the procedures and messages required for an AN to transmit, and an AT to receive, the forward traffic channel.

Transmission on the forward traffic channel is multiplexed in time and frequency. An assignment on the forward traffic channel shall be specified by a set of hop-ports

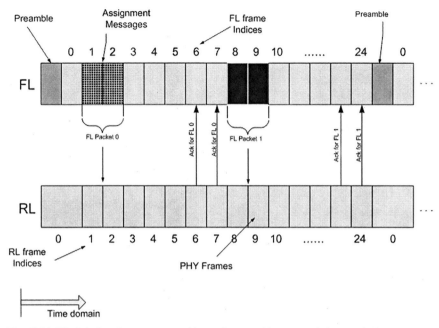

Fig. 13.20 FL eight-interlace structure with two frames without extended transmissions

(frequency) and a set of PHY frames in time. Each hop-port is specified by a hop-port index. Packets transmitted over the forward traffic channel are transmitted over the F-DCH physical layer channel that are assigned to ATs via assignment blocks that are sent over the F-SCCH. The duration of an assignment of hop-ports may or may not be pre-specified. There are four types of assignments: Assignments whose durations are pre-specified are known as non-persistent assignments and assignments whose durations are not pre-specified are known as persistent assignments. Non-persistent assignments may be unicast or they may be broadcast. Residual assignments (the set of hop-ports assigned for a particular access terminal for a particular set of PHY frames) have pre-specified durations, but are not included in the category of non-persistent assignments. They are opportunistically assigned when are not used by other valid ATs. Group resource allocations are also possible using this protocol through group assignment messages and group resource assignment (GRA) bit map over F-DCH. This is mainly used for applications such as VoIP that are mainly predictable in nature. GRA allows to statistically multiplex the resource allocation for several such applications and show each of those applications using a bit in the F-DCH bitmap, and therefore minimize the allocation overhead. Assignments can be extended or reduced and additional H-ARQ repetitions can be requested or redirected. Figure 13.21 shows the GRA bitmap structure.

Basic Reverse Control Channel MAC Protocol: The reverse control channel MAC defines the procedures for transmissions on the reverse pilot channel (R-PICH), the reverse CDMA dedicated control channel (R-CDCCH), the reverse

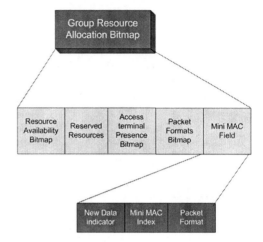

Fig. 13.21 GRA bitmap
structure

OFDMA dedicated control channel (R-ODCCH), and the reverse acknowledgment channel (R-ACKCH). The R-CDCCH and R-ODCCH channels carry several logical channels outlined in Section 13.3.3.1. The R-ODCCH carries the reverse MIMO quality indicator channel (r-mqich), the reverse channel quality indicator channel (r-cqich), the reverse request channel (r-reqch), the reverse subband feedback channel (r-sfch), and the reverse beam feedback channel (r-bfch). The R-PICH is a pilot channel. The R-CDCCH and R-ODCCH are CDMA and OFDMA dedicated control channels, respectively. The r-cqich is used by the access terminal to transmit the FL-quantized channel quality from different sectors to the access network. The r-reqch is used by the access terminal to request reverse link resources. The r-pahch is used by the access terminal to report its available transmit power. The r-psdch is used by the access terminal to report the path-loss differential between two sectors. The r-mqich carries MIMO channel quality information. The r-sfch is a feedback channel that is used by the access terminal to transmit the quantized FL channel quality measured for a subband in the FL serving sector (FLSS). The r-bfch is a feedback channel that is used by the access terminal to report a preferred beam index as well as supplemental channel quality information to enable SDMA transmission. SDMA transmission is defined in the PHY protocol. The R-ACKCH is used by the AT to acknowledge the MAC packets transmitted on the forward link.

Table 13.5 summarizes the control information sent on the reverse link.

Basic Reverse Traffic Channel MAC Protocol: The basic reverse traffic channel MAC protocol provides the procedures and messages required for an AT to transmit and for an access network to receive the reverse traffic channel. The AN maintains an instance of this protocol for every access terminal. Similar to other protocols, the AT might be in any of active or inactive states.

Transmission on the reverse traffic channel is multiplexed in time and frequency. An assignment on the reverse traffic channel shall be specified by a set of hop-ports and a set of PHY frames. Each hop-port is specified by a hop-port index. The duration of an assignment of hop-ports may or may not be pre-specified. Assignments

Table 13.5 Control information transmitted on the reversed link

Physical channel	Logical channel		Content
R-PICH		–	Reverse CDMA pilot channel
R-CDCCH	r-cqich	Channel quality indicator	4 bits+ 1 bit (desired FLSS flag)
	r-reqch	Request (RL resources)	6 bits (measure backlog)
	r-pahch	PA headroom	6 bits (code (carrier/thermal) power)
	r-psdch	Power spectral density	4 bits (code Tx/Rx power per sector)
R-ODCCH	r-mqich	MIMO channel quality indicator	Single codeword: 7 bits multiple codeword. 4 bits per layer
	r-cqich	Channel quality indicator	7 bits
	r-reqch	Request(RL resources)	6 bits (measure backlog)
	r-sfch	Subband feedback	4 bits (subband id) +4 bits (CQI delta)
	r-bfch	Beamforming feedback	6 bits (beam index) + 2 bits (CQI delta)
R-ACKCH		–	'1' - ACK (FL packet with good CRC)

whose durations are pre-specified are known as non-persistent assignments and assignments whose durations are not pre-specified are known as persistent assignments. The set of hop-ports assigned for a given set of PHY frames whose duration is not pre-specified is referred to as the "reverse link access terminal assignment" or RL-ATA. RL-ATAs are assigned via RLABs that have the persistent bit set to 1, and may be altered for a specified length of time via an RLAB with the persistent bit set to 0. An AT can have multiple RL-ATAs, as long as they do not overlap in time. The set of hop-ports assigned for a given set of PHY frames for a particular access terminal whose duration is pre-specified is referred to as the reverse link non-persistent access terminal assignment or RL-NP-ATA. The RL-NP-ATA is assigned via RLABs that have the persistent bit set to 0, which are denoted as NP-RLABs. An access terminal can have multiple RL-NP-ATAs, as long as they do not overlap in time. Sets of hop-ports in assignment blocks received from the FLCS MAC protocol are specified using the channel tree. Packets transmitted over the reverse traffic channel are transmitted over the R-ODCH physical layer channel. Access terminals are assigned R-ODCH resources (RL-ATAs, RL-NP-ATAs) via assignment blocks (RLABs, NP-RLABs) that are sent over the F-SCCH.

If RL-ATAs and RL-NP-ATAs coexist, no two RL-ATAs/RL-NP-ATAs can overlap in time. Moreover, an AT shall not have a non-empty RL-ATA that overlaps in time with a non-empty RL-NP-ATA. All hop-ports in the RL-ATA/RL-NP-ATA for an access terminal for a given PHY frame shall be combined for transmission over the physical layer channel (R-ODCH).

If the protocol is in the inactive state, the access terminal and the access network wait for an activate command.

Assignments can be extended or reduced and additional H-ARQ repetitions can be requested or redirected. In addition some control information can be sent on the data channels in band, via 8-bit control blocks (e.g., power amplifier headroom, request for RL resources, *C/I* changes reports).

RL QOS MAC Protocol: The RL QoS MAC protocol negotiates one set of attributes per stream defined in stream layer. These attributes allows the access terminal to make reverse link scheduling decisions, policing, shaping, and request generation. The attributes in this protocol may only be negotiated by the access network. The RL QoS MAC protocol maintains two token buckets for each stream: One of them is a bucket that helps scheduling of the packets belonging to a stream in the reverse link. The second one helps knowing when to declare overflow and begin dropping packets). The other one is used for policing and shaping of the traffic flow for a stream and may be used for scheduling decision on the reverse link. The access terminal uses these buckets to perform scheduling decisions and requests generation along with other attributes defined in this protocol. The bucket sizes (containing the number of octets available for transmission) are incremented based on the arrival rate of the stream and decremented based on the number of actually transmitted packets.

Handoff: Similar to WiMAX and 3GLTE, each active AT in UMB must maintain an active set. An active set is defined as a set of sectors maintained in an AT that are assigned a MAC-ID to the AT. The AT is allowed to switch the forward link serving sector (FLSS) and reverse link serving sector (RLSS) within the active set with fast physical layer signaling. The sectors are added to the active set ahead of switching time, during which connection setup is completed with the new sector. This allows fast switching between sectors, since the target sector is ready to accept the AT's traffic when the AT switches to the sector.

The AT constantly monitors forward and reverse channel quality of sectors within its active set to determine the best sector to get access. AT uses three sets of pilots to obtain these information (see Section 13.3.2). They are F-ACQCH and F-CPICH for obtaining the sector with best FL quality and F-PQICH to monitor the sector with the best RL quality.

In UMB it is not necessary for the forward link and reverse link serving sectors (FLSS/RLSS) to be the same. The AT selects the strongest FL sector. However, it tries to choose a sector whose RL quality is close enough to the RL control. To this end, AT uses the F-PQICH report transmitted by AN. It selects the strongest RL sector with adequate SINR to meet QoS requirements. It uses interference over thermal (IoT) reports by AN and uses power headroom and required power based on QoS class. During handoff indication, the AT indicates FL preference by sending R-CQICH to the target with desired FL serving sector.

When the RL sector is selected, AT uses r-reqch to indicate its RL preference to the target AN. When the target AN sends data or control assignment to the AT, handoff process is completed. When the AT switches from one sector to another, it uses the route associated with the new serving sector to transfer packets.

Because the target sector is set up to be the serving sector when it is added to route set, and because no connection state info is exchanged between sectors, the inter-sector interface is vastly simplified, enabling fast switching of layer 1 FLSS and RLSS entities during a handoff.

After the AT switches from the source sector to the target sector in a layer 1 handoff, it is likely that there are remnants or fragments of data packets in the route to the source sector that still need to be delivered to the AT. A tunneling mechanism ensures zero loss of packets during handoff.

Once a sector is in the route set, the sector will need to exchange messages with the AT to maintain the connected state, even though the current route is not serving. Layer 2 communication facilitates this process. This is in contrast with traditional networks, where the devices had to translate the protocols between the sectors, if it had to communicate with the new sector. In the UMB network design, the protocols between the AT and a sector in the route set are independent of the delivery service provided by the serving sector. The inter-sector interfaces are kept simple. A layer 2 communication is established between a sector and the AT as soon as the sector is added to the route set. A relationship is established between the new sector and the AT using the personality's defining protocols and attributes. All exchanges between the AT and a sector in the route set are communicated through a layer 2 tunnel between that sector and the serving sector. This is also called blind tunneling, where the serving sector blindly delivers the packets to the AT without interpreting the content. For example, transactions between the AT and the sector related to reporting radio measurements, QoS requests, and grants.

Power Control in UMB: UMB supports two power control modes: One is open-loop power control and the other one is closed-loop power control. The RL control channel power is controlled using fast closed-loop power control, which is used to set the transmit power level on the reverse link control channels that are transmitted periodically. Two main physical channels are used for power control in UMB. One is reverse pilot channel (R-PICH) which is an unmodulated DFT-precoded CDMA signal used to assist the AN for RL power control reference and RL quality measurement. The other one is F-PCCH forward power control channel, which carries reverse link power control commands. It also carries commands for closed-loop control of the reverse link control channel transmit power.

The RL-CDCH power control has the traffic channel power level set at an offset relative to R-PICH. The offset is based on traffic channel performance. The

RL-ODCH power control is enabled by setting the traffic channel power spectral density (PSD) level set at an offset relative to the control channel PSD level. This offset is adjusted based on interference indications received from neighboring sectors. The maximum traffic PSD offset is limited by inter-carrier interference.

The FLCS MAC protocol determines the number of the forward power control channel reports to be transmitted on each FL PHY frame, each having an index starting from 0. For each index, the FLCS MAC protocol determines a 1-bit value to be transmitted, the MACID of the AT to which this report is targeted and the power density at which it is sent.

There are two forward link physical channels that are not directly transmitted for power control, but since they carry interference information, they can be used for power control purposes. They are forward other sector interference channel, F-OSICH, that carries an other sector interference indication. This channel is transmitted over the superframe preamble. The other one is forward fast other sector interference channel, F-FOSICH, that carries an other sector interference indication transmitted at a faster rate but with less coverage than the F-OSICH. This channel is transmitted in the control segment of PHY frames.

F-FOSICH is used to indicate interference levels in a given reverse link hop-port subzone to the AT in other sectors. The FLCS MAC protocol determines the number of the F-FOSICH reports to be transmitted on each FL PHY frame, each having an index starting from zero. For each index, the FLCS MAC protocol determines a 4-bit value to be transmitted and the power density at which it is sent.

The frequency of power-level adjustment in closed-loop power control is 1 ms (one PHY frame).

13.5 Other PHY/MAC-layer features in UMB

Similar to WiMAX and 3GLTE, UMB has utilized most of advanced wireless technologies. It is a broadband wireless mobile technology supporting scalable OFDMA from 1.25 to 20 MHz. As described before, the frame structure is based upon superframes composed from preambles and 25 PHY frames in the FL. Each PHY frame is about 1 ms in the time domain, and this provides the chance for short latency for delay-sensitive functionalities. The system supports adaptive coding and modulation with resource-adaptive synchronous H-ARQ and turbo coding. Modulation schemes supported by the technology are QPSK, 16QAM, 8PSK, and 64QAM. Two base coding rates are supported: Rate 1/3 convolutional code for block lengths no greater than 128 and rate 1/5 turbo code for block lengths greater than 128. Puncturing or repetition is used at all code rates to other desired code rates.

HARQ is supported in both FL and RL. The HARQ re-transmission latency is eight (mandatory) and six (optional) PHY frames corresponding to 6–8 ms on both RL and FL. The system allows dividing the whole bandwidth to multiple subzones, and therefore the ATs and the AN have the chance to send the channel quality measures (CQI) at those subzones. This in turn provides the scheduler the opportunity to provide multi-user diversity gains for latency-sensitive traffic. In order to support HARQ, channel interleaver based on bit-reversal is used, which provides almost-regular puncture patterns and acceptable interleaver distance properties. At high-throughput levels, the HARQ transmissions use lower order modulations to avoid repetition of coded bits. This is called modulation step-down.

UMB utilizes the multiple antenna transmission in the FL (MIMO) to dramatically increase spectral efficiency at high SNRs by exploiting scattering. The OFDMA forward link features MIMO and open-loop transmit diversity, with support for single codeword MIMO with closed-loop rate and rank adaptation (SCW), multi-codeword (layered) MIMO with per-layer rate adaptation (MCW), and space–time transmit diversity (STTD). Frequency diversity and frequency-selective transmission are used in RL and FL to provide trade-offs for different channel and traffic mixes. Both open-loop and closed-loop MIMO are supported. Closed-loop MIMO and multi-user MIMO are supported by using precoding and SDMA.

UMB has exploited the features of 3GPP2 networks and has added several features such as group resource assignment (GRA) and persistent assignment, and dynamic sharing between control and data channel to improve VoIP capacity.

At the RL a hybrid OFDMA/CDMA transmission is supported. The CDMA transmission is DFT pre-coded, which allows for statistical multiplexing of various control channels, as well as optional low-rate delay-sensitive traffic. The reverse link supports quasi-orthogonal transmissions. The orthogonal transmissions are based on OFDMA, while non-orthogonal transmissions are based on layer-superposed OFDMA (LS-OFDMA) and SDMA. The precoded CDMA reverse link segment allows for statistical multiplexing of various control channels.

UMB supports L1 and L2 mobility by providing handoff. The system allows handoff support with FL softer handoff group selection, which improves edge user performance. Independent switching of FL and RL serving sectors is also allowed. In other words, it is not mandatory to have the same serving sector for FL and RL. To facilitate handoff a detailed neighbor sector/cell information is also provided.

UMB supports a trade-off between throughput and fairness through power control. The power control process in UMB is distributed, which is based on other cell interference indications.

Interference management through fractional frequency reuse is also supported. This results in improved coverage and edge user performance. Dynamic fractional frequency reuse can be used to optimize bandwidth utilization.

Rotational OFDM transmissions are optionally supported. In this mode, a set of complex modulation before being applied to IFFT block, in an OFDM transmitter, is rotated by a rotational matrix R_D, where D is the number of those modulation symbols. This is used only in DRCH mode and is optional for both AT and AN.

The technology provides special processing for broadcast and multicast services (BCMCS) data and optional support of supercast of unicast data on broadcast channels. Broadcast services use a single frequency network (SFN) transmission. Broadcast sub-bands are defined to be a set of 128 contiguous tones over one interlace, where at least one subband in each of those interlaces need to carry RL control signaling. Supercast is a service, wherein broadcast transmissions are overlapped with unicast transmissions (for users with high SINRs) in the same time and frequency resources. The unicast users that are scheduled on top of the broadcast layer decode the BCMCS signal first, and then decode the unicast signal after interference cancellation.

Security features are provided through privacy, integrity, and prevention of man-in-the-middle attacks, as much as possible. UMB provides encryption/decryption of packets at the radio-link protocol (RLP) level for OTA transmission/reception. The ciphering is tightly coupled with RLP, which is used for both control plane and user plane. Encryption/decryption is based on well-known advanced encryption system (AES or Rijndael) that uses 128-bit-long ciphering key. During key exchange protocol pairwise master keys (PMK) are pre-established between AN and AT, where up to three keys can be active at a time. Authentication in UMB is also based on extensive authentication protocol or EAP.

13.6 Conclusions

In this chapter, we briefly described some of the technical features of ultra mobile broadband (UMB) system, which is the next generation of 3GPP2 technologies base upon for providing high spectral efficiencies to mobile devices. To this end, 3GPP2 community has switched its attention from CDMA to MIMO-OFDMA solutions.

As standardized OFDM technology, UMB has a broader channel, better capacity, and throughput than its 3G predecessors. It represents a flexible platform for the seamless integration of voice, data, and multimedia. To this end, UMB has developed and exploited the state-of-the-art technologies to achieve these objectives. It also provides a deliberate channelization scheme to optimize encoding/decoding processing time, diversity, and access latency and memory requirement for the access at the terminals. It is scalable across deployment bandwidth to achieve acceptable reliability wireless communication. The channelization design is a general paradigm of all OFDMA systems.

References

1. 3GPP2 In: "3rd Generation Partnership Project 2," 2008, Available via http://www.3gpp2.org/
2. XOHM deployment, In: "XOHM, Intel and WiMAX Partners Celebrate New 4G Broadband Era in Baltimore" October 2008, Available via http://www.XOHM.com/
3. ABI Research In: "Ultra-Mobile Broadband (UMB) Might Be Dead on Arrival" 2008, Available via http://www.abiresearch.com
4. 3GPP2 C.S0084 − 000 − 0, In: "Overview for Ultra Mobile Broadband (UMB) Air Interface Specification," Ver. 3.0, August 2008, Available via http://www.3gpp2.org/Public/html/specs/tsgc.cfm

5. 3GPP2 *C.S0084* − 001 − 0, In: "Physical Layer for Ultra Mobile Broadband (UMB) Air Interface Specification," Ver. 3.0, August 2008, Available via `http://www.3gpp2.org/Public/html/specs/tsgc.cfm`
6. 3GPP2 *C.S0084* − 001 − 0, In: "Medium Access Layer for Ultra Mobile Broadband (UMB) Air Interface Specification," Ver. 1.0, April 2007, Available via `http://www.3gpp2.org/Public/html/specs/tsgc.cfm`
7. 3GPP2 *C.S0084* − 005 − 0, In: "Security Functions for Ultra Mobile Broadband (UMB) Air Interface Specification," Ver. 3.0, August 2008, Available via `http://www.3gpp2.org/Public/html/specs/tsgc.cfm`
8. 3GPP2 *C.S0084* − 003 − 0, In: "Radio Link Layer for Ultra Mobile Broadband (UMB) Air Interface Specification," Ver. 3.0, August 2008, Available via `http://www.3gpp2.org/Public/html/specs/tsgc.cfm`
9. 3GPP2 *C.S0084* − 004 − 0, In: "Application Layer for Ultra Mobile Broadband (UMB) Air Interface Specification," Ver. 3.0, August 2008, Available via `http://www.3gpp2.org/Public/html/specs/tsgc.cfm`
10. 3GPP2 *C.S0084*−006−0, In: "Connection Control Plane for Ultra Mobile Broadband (UMB) Air Interface Specification," Ver. 3.0, August 2008, Available via `http://www.3gpp2.org/Public/html/specs/tsgc.cfm`
11. 3GPP2 *C.S0084*−007−0, In: "Session Control Plane for Ultra Mobile Broadband (UMB) Air Interface Specification," Ver. 3.0, August 2008, Available via `http://www.3gpp2.org/Public/html/specs/tsgc.cfm`
12. 3GPP2 *C.S0084*−008−0, In: "Session Control Plane for Ultra Mobile Broadband (UMB) Air Interface Specification," Ver. 2.0, April 2007, Available via `http://www.3gpp2.org/Public/html/specs/tsgc.cfm`
13. M. M. Wang, M. Dong, In: "Channelization in Ultra Mobile Broadband Communication Systems: The Forward Link," August 2008, Proceeding of the International Wireless Communications and Mobile Computing Conference (IWCMC'08)
14. IEEE802.16-2005, In: "Part 16: Local and Metropolitan Area Networks–Air Interface for Fixed and Mobile Broadband Wireless Access Systems Amendement 2: Physical and Medium Access Control Layers for Combined Fixed and Mobile Operation in Licensed Bands and Corrigendum 1," February 2006, Available via `http://ieee802.org/16/`
15. M. Wang, T. Ji, J. Borran, and T. Richardson In: "Interference Management and Handoff Techniques in Ultra Mobile Broadband Communication Systems", The 10th International Symposium on Spread Techniques and Applications, August, 2008.
16. 3GPP TS 36.201, Evolved Universal Terrestrial Radio Access (E-UTRA); Long Term Evolution (LTE) Physical Layer; General Description, September 2008, Available via `http://www.3gpp.org/ftp/Specs/html-info/36-series.htm`
17. D. J. Love and R. W. Heath, Jr., In: "Multimode precoding for MIMO wireless systems," IEEE Trans. Signal Process., vol. 53, no. 1, pp. 3674–3687, October 2005.

Chapter 14
Mobile WiMAX

14.1 Introduction

Mobile WiMAX is a broadband wireless solution that enables convergence of
mobile and fixed broadband networks through a common wide-area broadband radio access technology and flexible network architecture. Mobile WiMAX ultimate
goal is to "mobilize the Internet [1]." The mobile WiMAX air interface is based on
the IEEE 802.16 Air Interface Standard [2]. Original WiMAX air interface adopted
multiple physical layer access mechanisms, while mobile WiMAX has selected
orthogonal frequency division multiple access (OFDMA) for improved multipath
performance in non-line-of-sight environments. Scalable OFDMA (SOFDMA) is
introduced in the IEEE 802.16e amendment to support scalable channel bandwidths
operation from 1.25 to 20 MHz [3]. The Technical Working Group (TWG) in
WiMAX forum has developed the mobile WiMAX system profiles [5]. The mobile WiMAX system profile defines the mandatory and optional features of the
IEEE standard that are necessary to build a mobile WiMAX-compliant air interface that can be certified by the WiMAX forum. The mobile WiMAX system profile enables mobile systems to be configured based on a common base feature set.
The mobile WiMAX system profiles are designed to ensure baseline functionality on terminals and base stations that are fully interoperable. Some elements of
the base station profiles are specified as optional to provide additional flexibility
for deployment based on specific deployment scenarios that may require different configurations that are either capacity optimized or coverage optimized. Mobile WiMAX Rel. 1.0 profile covers 5, 7, 8.75, and 10 MHz channel bandwidths
for licensed worldwide spectrum allocations in the 2.3, 2.5, and 3.5 GHz frequency
bands.

Masoud Olfat
593 Herndon Pkway, Herndon, VA 20171 USA.
e-mail: masoud.olfat@clearline.com

V. Tarokh (ed.), *New Directions in Wireless Communications Research*,
DOI 10.1007/978-1-4419-0673-1_14,
© Springer Science+Business Media, LLC 2009

14.2 Standardization Process

Just as the Wi-Fi wireless local area network (WLAN) was enabled by the IEEE 802.11 standard, the metropolitan area network (MAN) rocket was launched based on the IEEE 802.16 WirelessMAN standards. While IEEE802.16x laid out the specifications for standardization of WMAN, WiMAX forum, a global broadband wireless access (BWA) industry association, provides the quality control and certification to ensure successfully standardized deployment.

IEEE802.16 BWA started with line-of-sight (LOS) applications such as backhauling for cellular networks and providing wireless MAN to bring BWA to big buildings. For these applications, the standard community devised the wirelessMAN-SC (single carrier) as the physical layer solution. Here, the RF signal provided the communication between the two hubs located on top of big buildings that could see each other, and each hub was connected through wires to other nodes inside the building. Later on the applications expanded to include backhauling for wired and wireless LAN, and wireless MANs to bring BWA to homes and businesses as an alternative to DSL or cable access. These applications were not limited to LOS environments anymore and were deployed in multipath non-line-of-sight (N-LOS) morphologies. Consequently, the standard community used orthogonal frequency division multiplexing (OFDM) and orthogonal division multiple access (OFDMA) to define two new physical (PHY) layer designs wirelessMAN-OFDM and wirelessMAN-OFDMA in addition to wirelessMAN-SCa (single carrier for NLOS). These four PHY layer designs along with a common multiple access control (MAC) based upon DOCSIS (Fig. 14.1) were approved in the year 2004 in a standard document IEEE802.16-2004 [2]. This standard document provides broadband wireless access for fixed and nomadic services in N-LOS and multipath environments.

Although both the base stations and modems were fixed, the elements comprising the fading channel were not necessarily fixed. Cars, people, and other moving objects created a varying channel, and therefore the standard needed to consider

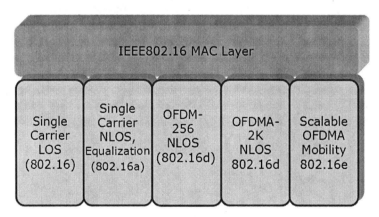

Fig. 14.1 IEEE802.16 common MAC and multiple PHY support

some elements of mobility in their design. Moreover, the main market explosion could have taken place when the technology addresses portability and mobility, bringing BWA directly to the end user. To this end, the IEEE802.16-2004 was amended by IEEE802.16-2005, which among many enhancements was mainly comprised of supporting three major features: (a) handover; (b) power control and idle/sleep mode for handling low-power mobile devices; and (c) new PHY mode named scalable-OFDMA [3].

Later on, the IEEE standard defined new projects to identify and resolve the bugs identified in both IEEE802.16-2004 and IEEE802.16-2005. These projects were called IEEE802.16-2006Cor2 (corrigendum) and four drafts of these documents were developed. The last draft was named IEEE802.16-2004Cor2/D4, which was amended to the main standard body IEEE802.16-2004, and its other amendment IEEE802.16-2005 to support the first full version of air interface mobile IEEE BWA.

In order to complete the support by the standard body, the IEEE802.16 committee defined the following working groups as well:

- 802.16f, 802.16g, and 802.16i network management
- 802.16h coexistence in license exempt frequency bands
- 802.16k bridging
- 802.16j multi-hop relay [7]

In 2007, a new project IEEE802.16Rev2 was embarked to combine IEEE802.16-2004 with all its amendments IEEE806.16-2005, IEEE802.16-2004Cor2/D3 and 4, 802.16f, 802.16g, into one standard document. Moreover, it aims to resolve some of the known shortcomings of the first IEEE802.16 release, such as high MAC overhead, and enhances the performance of the technology. By the time this book was under edition, the standard was on its last phase of finalization (sponsor ballot) and it is expected that by the end of 2008, the IEEE802.16-Rev2 standard is finalized [4].

14.2.1 WiMAX Forum

WiMAX (Worldwide Interoperability Microwave Access) is a non-profit organization formed to promote and certify conformance, compatibility, and interoperability of products based on IEEE 802.16 standards. This is the same role that Wi-Fi alliance is playing for IEEE 802.11 family of standards. It is worthwhile noting that IEEE802.16 is not the same as WiMAX. IEEE802.16 develops the technology specification, while WiMAX ensures conformance and interoperability of 802.16 products, and develops the network architecture for IEEE802.16 compliant equipments. Since the role of WiMAX forum could not be limited to certification and interoperability of IEEE802.16 compliant equipments, several working groups were created to achieve the following goals:

- To ensure a complete system solution: IEEE802.16 family of standard develops only the air interface technologies and at most the management elements. In order to define the end-to-end interoperable system solution the forum needs to develop core network architecture, and RAN design for 802.16 air interface. To this end, the forum has established the **Network Working Group (NWG)**.

- BWA technologies are deployed by some service operators. WiMAX forum has defined **Service Provider Working Group (SPWG)** to create requirement documents both for architecture and air interface profile.
- **Regulatory Working Group (RWG)** creates awareness for WiMAX spectrum requirements, responds to government consultations, and completes ETSI nomadic requirements documents.
- In order to facilitate global roaming among WiMAX operators around the world and harmonize the requirements, WiMAX forum has established **Global Roaming Working Group (GRWG)**.
- **Marketing Working Group (MWG)** to promote the WiMAX brands and the standards as basis for worldwide interoperability of BWA systems.
- **Application Working Group (AWG)** develops proof-of-concept (POC) demonstrations, applications and services guidelines documents, services access reference guide, application notes, and system simulation methodology.
- In order to ensure interoperability of multi-vendor products, **Technical Working Group (TWG)** defines the system profiles, PICS (protocol implementation conformance statement), Test Suite Structure and Test Protocols (TSS&TP).
- **Certification Working Group (CWG)** defines the certification process, certification test program, and certification test manual.

IEEE802.16 standard includes several mandatory as well as optional features. System profile, is a list of mandatory and optional features chosen from IEEE802.16 draft; the WiMAX vendors must implement to be WiMAX certified and interoperable [5].

The existing system profile is called Release 1.0 and is based upon IEEE802.16-2005. Figure 14.2 depicts the concept of system profiles. However, depending on the

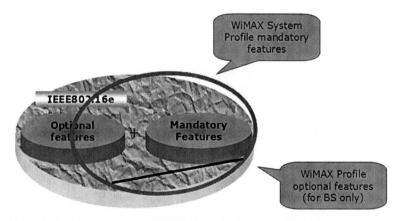

Fig. 14.2 Relationship between the IEEE802.16 standard draft and WiMAX system profile

used spectrum, the CWG has launched two different system profiles, system profile Release 1.0 Wave I and Wave II.

Certification waves define a subset of features identified in the mobile WiMAX profile used to certify products for a limited period of time. Certification Wave I includes any band class except band class 3, defined for products in the 2.5 GHz band (2.496–2.69 GHz).

Certification Wave 2 includes all band classes. The main reason for this differentiation is to exclude some of the features mandated by the WiMAX profile from some band classes like 2.3 GHz in Korea in the earlier time frame. WiMAX Wave I was mainly deployed in Korea early 2007 and was also referred to as WiBro (wireless broadband) by Korean Telecom (KT) and will be replaced by Wave II soon. The main features included in Wave II but not included in Wave I are MIMO, beamforming, some of optimized handoff features, some of optimized idle mode features, some of sleep mode features, IPv6 convergence sublayer, AMC permutation, pilot CINR measurement, some of QoS flows, and multicast/broadcast services.

WiMAX forum TWG is currently developing the system profile Release 1.5 to select the list of mandatory and optional features from IEEE802.16Rev2 standard draft, and also develop certification program for IEEE802.16Rev2 compliant equipments. Moreover, TWG is planning the same procedure for Release 2.0 profile based on IEEE802.16m standard draft (Table 14.1).

Table 14.1 WiMAX spectrum and channel bandwidths

Channel BW	1.25 MHz	5 MHz	7.0 MHz	8.75 MHz	10 MHz	20 MHz
FFT size	128	512	1024	1024	1024	2048
2.3–2.4 GHz		TDD		TDD	TDD	
2.305–2.32, 2.345–2.36 GHz		TDD			TDD	
2.496–2.69 GHz		TDD			TDD	
3.3–3.4 GHz		TDD	TDD		TDD	
3.4–3.8 GHz		TDD	TDD		TDD	

14.3 WiMAX Network Architecture

IEEE802.16 standard community developed the air interface specification, while the end-to-end network system architecture has been developed by WiMAX Networking Working Group. WiMAX NWG has adopted a three-stage standard development process similar to that followed in 3GPP standard body.

In the first stage, the used-case scenarios and service requirements are defined. This stage was mainly developed in the SPWG. Five used scenarios originally defined in SPWG, namely fixed, nomadic, portable, simple mobility, and full mobility. The main design guidelines outlined by SPWG and NWG stage 1 documents were to make WiMAX architecture aligned with wireline broadband networks such as cable networks, and at the same time support specific characteristics such as mobility, roaming, interworking. They required the architecture to (a) be functionally decomposable, (b) be modular and flexible for deployment, (c) support different usage models and scenarios, (d) decouple access and connectivity services, (e) support a variety of business models allowing different service providers, access providers, and application providers to work with each other, (f) allow loosely and tightly coupled interworking with existing wireless and wireline networks (such as 3GPP and 3GPP2), and finally (g) the procedures (security, management, QoS, provisioning, etc.) rely on IETF RFCs as much as possible.

Stage 2 developed the architecture that meets these used cases and requirement. In the following, we will outline some aspects of this network architecture very briefly without going into the details. In stage 2, the profiles A, B, and C were introduced. Later on, NWG removed profile A from its architecture definition [9].

In stage 3 the details of the protocols associated with each of these architectures are developed.

The first NWG Release 1.0 specification (stages 2 and 3) was published on the first quarter of 2007. As we edit this section, the development of WiMAX equipment and the first commercial deployment of mobile WiMAX are based upon NWG Release 1.0 version 1.2.2 stage 3 documents. Release 1.0 version 1.3 specification is expected to be the last revision of Release 1.0. Next release is called Release 1.5 whose specification is in development. The ASN profile C is the baseline in Rel 1.5 (profile A has been removed). Seventeen subteams were actively working on Release 1.5 activities. The main features considered for Rel. 1.5 are PCC/QoS, IMS Interworking, MCBCS, simple IP, PMIPv6, handover data integrity, RoHC, 3G interworking, emergency services, lawful intercept, ethernet services/VLAN support, over the air activation (OTA), provisioning/activation (OMA-DM, TR69), normative R8, diameter support, location-based services (LBS), fixed nomadic, UICC WiMAX support. Release 1.5 is expected to be finalized by the end of 2008 or early 2009. Release 2.0 also may start late 2008 or in 2009.

14.3.1 Network Reference Models

Figure 14.3 shows the network reference model (NRM)[8]. This model is a logical representation of network architecture and depicts the functional entities of WiMAX architecture and the network reference points. As seen in the figure, the NRM divides the WiMAX end-to-end architecture into three logical entities: (a) mobile device used by subscriber connecting to the access network using R1 interface; (b) access service network (ASN) forming the radio network, comprising one or more base stations and one or more ASN gateways (ASN-GW), owned by a network access provider (NAP); and (c) connectivity service network (CSN) providing IP

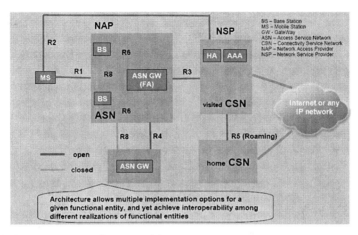

Fig. 14.3 WiMAX network reference model

connectivity and all IP core, and AAA (accounting, authorization, authentication) functionality, and owned by a network service provider (NSP). In the picture, home NSP is where the subscriber belongs and visited NSP is where the subscriber is being serviced.

14.3.2 ASN profiles

As seen in Fig. 14.3, the ASN is composed of one or more base stations and one or more ASN-GW and is responsible for providing basic connectivity, mobility-related functions, network discovery and selection (ND&S), AAA proxy, and radio resource management (RRM) functionalities. Depending on where each of these functions lies, whether at the BS or ASN-GW, and how the BS and ASN-GW are interconnected, different ASN profiles (profiles A, B, and C) define the WiMAX architectures. In profiles A and C, centralized ASN model is used, where the functions are split between BS and ASN-GW, and they are located in separate platforms through R6 interface. The major difference between profiles A and C is that in profile A the RRM functionalities are split, and radio resource assignment (RRA) is located in BS and radio resource control (RRC) lays in ASN-GW, while in profile C, the RRM is not split and is completely located at the BS. In 2007, NWG decided to drop profile A in favor of profile C.

In profile B, distributed ASN solution is used where the BS and ASN-GW functionalities are implemented in a single platform.

As seen in Fig. 14.3, the entities are interconnected through logical interfaces R1–R8. Each of these interfaces is composed of a set of protocols that define the communication between two logical entities. In the following, we define each of these interfaces or reference points:

- **Reference Point R1** consists of the MAC and PHY air interface protocols and procedures between MS and ASN (base station) as defined by IEEE802.16-2005 or later versions, and additional management protocols and parameters. This interface incorporates a physical connection as well.
- **Reference Point R2** consists of accounting, authentication, authorization, and IP host configuration management protocols and procedures between the MS and CSN or core network (AAA and HA servers). This interface is only a logical interface and does not reflect a physical interface between MS and CSN.
- **Reference Point R3** consists of the set of control plane protocols between the ASN and the CSN to support AAA, policy enforcement, and mobility management capabilities. It also encompasses the bearer plane methods (e.g., tunneling) to transfer user data between the ASN and the CSN.
- **Reference Point R4** includes a group of functions required to enable mobility of MS between multiple ASNs (ASN-GWs).
- **Reference Point R5** enables Internet working between the CSN operated by the home NSP and visited NSP.
- **Reference point R6** consists of the set of protocols and data plane for communication between the BS and the ASN-GW.
- **Reference Point R7** is an optional set of control protocols, e.g., for AAA and policy coordination in the ASN gateway as well as other protocols for coordination between the two groups of functions identified in R6.
- **Reference Point R8** consists of the set of control message flows (based on 802.16g) and optional data plane between the base stations for fast handover for an MS.

Considering these definitions, for profile B the interfaces R1, R2, R3, and R4 are open (standardized) and must be interoperable, while R6 could be proprietary by a particular vendor. However, for profile C, interfaces R1, R2, R6, R8, R4, and R3 are all open and standardized, and must be interoperable.

In the following, we will outline the functional elements of NWG stage 3 in both Release 1.0 version 1.3 and Release 1.5 very briefly.

Network discovery and selection: supporting both manual and automatic selection of appropriate networks by user's preference, if multiple networks are deployed in an environment. This includes NAP discovery, NSP discovery, NSP enumeration, and ASN attachment.

DHCP IP address assignment: to allocate dynamic point of attachment IP addresses to devices, or allocating IP addresses to ASN through AAA by home CSN. IPv6 addressing is also supported by including IPv6 access router functionality inside the ASN.

Quality of service architecture: IEEE802.16-2005 has provisioned an extensive model for supporting an application-based QoS for five different service classes. However, this model would not be completed without defining functional components in the network architecture to provision, create, modify, or delete different service flows. WiMAX forum QoS network architecture supports both static and dynamic service flow provisioning. In static mode, the service flows are provisioned by the base station, and the device cannot create new one or modify their parameters, whereas in dynamic mode it can.

14.3.3 Mobility Management

The WiMAX mobility management architecture was designed to achieve several goals. Among them, the following could impact the network architecture design:

- support seamless handoff at high mobility speeds with low packet loss and minimized handoff latency,
- minimize the number of control signaling required to perform handoff,
- separate control and data plane during handoff,
- support both mobile IP versions 4 and 6 (IPv4 and IPv6)-based mobility management,
- accommodate mobiles with multiple IP addresses and simultaneous IPv4 and IPv6 connections,
- support interworking and handoff to networks using other technologies,
- support roaming between NSPs,
- support static and dynamic home address configuration.

Mobility management is normally triggered in three scenarios: (1) when the MS moves from one BS to another, or (2) when an MS transitions from active mode in one ASN-GW to idle mode at a different ASN-GW, or (3) transferring the foreign agent (FA) for an MS from the serving FA to a new FA.

To support these requirements, WiMAX has incorporated both mobile IP version 4 (IPv4) and version 6 (IPv6). In IPv4, when MS moves between two BSs connected to the same foreign agent, FA (located in ASN-GW), it does not need to send any update. When MS moves between two BSs connected to different FAs, MS sends registration message, and the home agent (HA) located at the CSN forwards packets to the new FA care of address (CoA). In other words, anchored FA relays packets between HA and the new FA .

To this end, the WiMAX network supports two types of mobility: ASN-anchored mobility and CSN-anchored mobility.

- ASN-anchored mobility or micro-mobility or intra-ASN mobility (Fig. 14.4) is when the anchor foreign agent (FA) is maintained. In this mode, no care of address (CoA) update is required, when moving from one BS to another BS connected to different FAs. In this case, the mobility is transparent at the CSN. The handoff is performed at R6 and R4 only (R4 and R6 handover). It is also possible to keep the layer 3 connection to the same BS (anchor BS) through the handover and have data traverse from the anchor BS to the serving BS throughout the session.
- CSN-anchored mobility management or macro-mobility or inter-ASN mobility (Fig. 14.5) is when the MS changes to a new anchor FA. It involves transitions between FAs. MS needs re-registration with the home agent (HA). The handoff

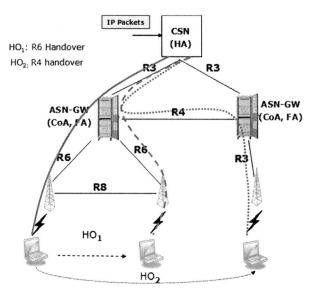

Fig. 14.4 WiMAX ASN-anchored mobility in two scenarios: R6 and R4 handover

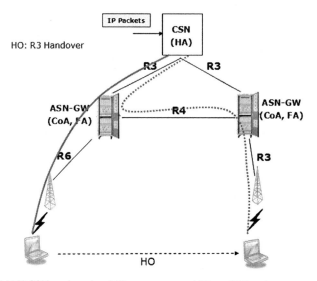

Fig. 14.5 WiMAX CSN-anchored mobility or macro-mobility or R3 handover

is performed at the R3 interface (R3 handover), with tunneling over R4 to transfer undelivered packets. The new FA and CSN exchange signaling messages to establish data forwarding path. There are two modes of mobile IP:

- client MIP or CMIP: when all MSs are mobile IP enabled;
- proxy MIP or PMIP: when the ASN-GW acts on behalf of client. In this case, the ASN-GW is instanced in ASN on behalf of the client that is not MIP capable or aware.

14.4 Physical Layer

As described in previous section, IEEE802.16 supports five different PHY modes (Fig. 14.1). WiMAX has adopted OFDM mode with 256 FFT size as the mandatory mode for fixed services operating in frequencies between 2 and 11 GHz. The channel bandwidths of operation supported and certified in this mode are 3.5, 7, 14, 4.375, and 8.75 MHz.

WirelessMAN-OFDMA was used as one of the major PHY layer modes in IEEE802.16-2004. In this mode, the number of tones was 2048 (2K FFT size) regardless of channel bandwidth (starting from 1.25 to 20 MHz). However, in OFDM systems the smaller the BW, the smaller the size of each tone (tone spacing) and therefore more sensitivity to the impairments caused by mobility. This phenomenon can be observed by looking at the following equation, which describes the OFDM time-domain signal in the presence of carrier frequency error ([11]):

$$y_i = x_i h_i e^{(j\pi\delta_f T_{FFT})} sinc(\delta_f T_{FFT}) + \sum_{i'=-FFT/2}^{FFT/2-1}$$

$$\times x_{i'} h_{i'} \frac{1}{T_{FFT}} \int_{u=0}^{T_{FFT}} e^{-j2\pi(\frac{i-i'}{T_{FFT}}-\delta_f)u} du + n_i. \qquad (14.1)$$

We have assumed the equation for the first symbol ($k = 0$) and ignored the phase noise. In this expression, y_i is the ith time-domain output of the IFFT block of an OFDM transmitter, x_i is the signal at the ith subcarrier in the frequency domain, and h_i and n_i are the channel and AWGN noise realization at the ith subcarrier. This equation is only valid if the frequency offset (δ_f) is smaller than the half of OFDM spacing. For larger offsets, the tones in frequency domain get shifted by one or more positions. The first term on the right-hand side of this equation corresponds to the attenuated time-shifted version of the transmitted symbol on the ith symbol, which can be reconstructed using time synchronization and one-tap equalization. The second term is the inter-carrier interference caused by frequency offset, which monotonically increases by the value of offset (up to half of tone spacing). Consequently, the ICI term can be seen as an additional noise term and can thus be represented as a degradation of SNR. Frequency offsets of up to 2% of the tone spacing are negligible, and in some cases 5–10% can also be tolerated.

When there is relative motion between the transmitter and receiver, a Doppler shift of the RF carrier results and introduces frequency error. Also, any error in the oscillators either at the transmitter or at the receiver can result in residual frequency error.

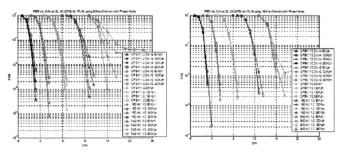

Fig. 14.6 Comparing the impact of tone spacing on the PER vs. SNR on different mobility speeds (a) 5MHz channel bandwidth with 2K FFT resulting in 2.5kHz tone spacing (b) 10MHz channel bandwidth with 2K FFT resulting in 5kHz tone spacing.

Figure 14.6 compares the PER vs C/N curve for different MCSs (modulation coding schemes) and different mobility speeds for an OFDM system with the same number of tones (2K), one with 5 MHz channel bandwidth and the other one with 10 MHz channel bandwidth. In the first figure, the tone spacing is half of that of the second figure. It is observed that specifically for higher modulation schemes (higher throughputs), as the speed increases (Doppler shift increases) the PER vs C/N degrades more rapidly.

Consequently, for mobile applications, the standard decided to keep the tone spacing constant if the channel bandwidth is one of the values of 1.25, 5, 10, or 20 MHz, by selecting 128, 512, 1024, and 2048 FFT sizes, respectively. The FFT size of 256 is not selected intentionally to avoid mixing up with fixed WiMAX profile (OFDM-256). For other channel bandwidths (such as 3.5, 7, 8.75 MHz) the tone spacing is defined in the system profile correspondingly. The mobile devices can discover the FFT size and bandwidth of operation either through scanning and initial network entry to the network or by inspecting the broadcasting message including the neighbor advertisement by the serving base station facilitating the handoff. Note that, due to fixed tone spacing, many base band parameters, such as OFDM symbol duration, cyclic shift, OFDM symbol duration, frame duration, remain unchanged. Consequently, re-implementation of many unnecessary BW-dependent parameters is eliminated. This is the principle of designing scalable OFDMA (S-OFDMA) used for mobile WiMAX [3]

It is worthwhile noting that in OFDM systems, cyclic prefix is used to provide for immunity against multipath fading, maintaining orthogonality among tones (by converting the linear convolution to cyclic convolution in time domain), and increasing the tolerance of the receiver to symbol time synchronization error. The values allowed for cyclic prefix (CP) by IEEE802.16-2005 are 1/16, 1/8, 1/4, and 1/2 of the OFDM symbol duration. In this case, the value of CP is decided by the base station (BS), and the mobile station (MS) recognizes the value of CP at the network entry process. However, mobile WiMAX system profile has selected the CP to be 1/8 of the symbol duration.

Mobile WiMAX duplexing mode: IEEE802.16 in principle is designed to support both time division duplexing (TDD) and frequency division duplexing (FDD) modes of operation. For FDD, both full FDD and half-duplex (HFDD) modes are envisioned. However, the IEEE802.16-2005 and mobile WiMAX Release 1.0 are mainly optimized for TDD mode.

14.4.1 S-OFDMA Frame Structure

A band class within the WiMAX system profile defines a particular frequency range (could be multiple subclasses within that range) as supported by the base station along with the channel bandwidths in that range, and channel rosters (center frequency separations). For example, the XOHM network [1] uses a 2.496–2.690 GHz spectrum with a 10 MHz channel bandwidth. This is called band class 3A.

In this section, we will use band class $3A$, as an example to describe WiMAX frame structure; however, the technology supports other band classes.

In order to increase the efficiency of the OFDM system, and to allow overlapping between adjacent channels without imposing interference as a result of out-of-band emission of the adjacent channel, oversampling is used in OFDM systems.Oversampling is also used for other reasons, such as decreasing peak to average power ration (PAPR) [15]. In general, if the oversampling ratio is n, the OFDM sampling frequency and the tone spacing are calculated as follows:

$$ToneSpacing = F = \frac{SamplingFrequency = F_s = floor(n.BW/8000) * 8000}{FFT}.$$
(14.2)

For band class $3A$, the BW is 10 MHz, the oversampling ratio is 1.12, and the FFT size is 1024, and so the tone spacing is 10.94 kHz. As mentioned, the CP interval is selected to be $1/8$ of the useful symbol duration, and therefore Table 14.2 could be easily derived for the PHY layer parameters for the channel BWs that are multiple of 1.25 MHz.

IEEE802.16 allows different values for the size of TDD S-OFDMA frame duration. The choice of frame duration is a trade-off between the delay and latency performances (such as handoff delay) and overhead performances. WiMAX has selected a value of 5 ms for the frame size. As seen in Fig. 14.11, the S-OFDMA frame is mainly composed of four parts, a downlink (DL) subframe, an uplink (UL) subframe, and two gaps between DL and UL, called TTG; transmit/receive transition gap (TTG), and receive/transmit transition gap (RTG). These are the time gaps between DL and UL portion (TTG) or UL and DL portion (RTG), when BTS switches transmission to receipt (TTG) or vice versa (RTG), and MS switches otherwise. It is needed for the BTS to turn around. These gaps are needed for the BS to ramp down or up, the BS TX/RX antenna to switch to actuate, and BS switch to receive or transmit.

Each of these values must be greater than 50 ns (SSRTG as defined by the standard, the maximum time needed for device to turn around), and TTG must include the round trip delay. The system profile has selected sum of RTG and TTG to be

Table 14.2 WiMAX OFDMA parameters

Parameter	Value			
Channel bandwidth (MHz)	1.25	5	10	20
Sampling frequency, F_s (MHz)	1.4	5.6	11.2	22.2
Sample time at the receiver, $(1/F_s)$ (ns)	714	178.5	89.25	44.625
FFT size	128	512	1024	2048
Tone spacing (KHz)	10.94			
Useful symbol time (1/tone spacing) (µs)	91.43			
Guard time (symbol time/8) (µs)	11.43			
OFDMA symbol time (symbol+guard) (µs)	102.94			

around 163.7 ns. If we subtract this value from 5 ms, the rest is divided into 47 OFDMA symbol durations. These symbol durations could be allocated arbitrarily to DL or UL subframes. However, we will see that there will be rules for allocation of these symbols to DL and UL.

14.4.2 Subchannel Permutation

In an OFDMA system, tones are grouped into subchannels. In the DL or UL, one or more subchannels carry the data for a user. Depending on the usage model, channel model, morphology, or the multiple antenna technology used various forms of tone permutations can provide better gains. In principal, there are two approaches to divide tones into subchannels. One is distributed and the other one is adjacent permutation. In distributed mode, tones that are allocated to a subchannel are selected from all over the frequency bandwidth and are not necessarily adjacent to each other. This mode is more appropriate for both mobile and fixed users and provides frequency diversity gain. In adjacent permutation tones allocated to a subchannel are all adjacent to each other in frequency domain. This case is more appropriate for adaptive coding and modulation, and provides better channel estimation capabilities, and since allows instantaneous feedback from the receiver to the transmitter, it is a better permutation mode for fixed, portable, and low-speed mobiles. Normally, for beamforming, and closed-loop MIMO techniques this mode is used. Figure 14.7 shows the difference between these modes.

IEEE802.16-2005 has defined several distributed permutation modes that differ mainly in terms of the number of pilots per subchannel, or the number of symbols per slot, etc. They are partial usage of subchannels (PUSC) (both segmented and with all subchannels), full usage of subchannels (FUSC), optional PUSC (OPUSC),

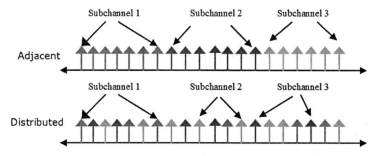

Fig. 14.7 Adjacent and distributed OFDMA permutation

optional FUSC (OFUSC), tile usage of subchannels 1 (TUSC1), and tile usage of subchannels 2 (TUSC2). The only defined adjacent permutation is called adaptive modulation coding (AMC). Some of these modes could only be used in DL and some could be used in both DL and UL. PUSC is the only mode that is mandatory by the standard, and the rest are optional. We will describe the PUSC and AMC (different modes) here briefly and refer the reader for more detailed description and also for the rest of permutations to the standard draft [2, 3].

PUSC was originally designed to allow virtual frequency reuse. In the networks where frequency reuse 1 ($n = 1$) is used, every sector might cause interference on adjacent sectors in both UL and DL. However, if frequency reuse of say 3 is used, the amount of interference is significantly reduced, and the sectors are not impacted by interference from adjacent sectors. However, this would require the availability of valuable spectrum and bandwidth on that area, as well as an accurate frequency planning. This is a luxury that is not available to all operators all around the world, due to the scarcity of spectrum and tight government regulations. By using PUSC, WiMAX allows a frequency reuse of 1 to be virtually mapped to a frequency reuse of 3. This removes the need for carrier management, while spectral mask is defined for only one carrier, and at the same time–frequency diversity is achieved.

Using PUSC in the DL, each symbol is divided into several clusters (depending on FFT size), each having 14 adjacent tones, the clusters are re-numbered and divided into six major groups, each group having either 24 or 16 clusters. In each cluster, first pilots are allocated, then the remaining tones are allocated to data subchannels.

These major groups are allocated to segments or sectors (by default, group 0 to sector 0, group 2 to sector 1, and group 4 to sector 2). If we have only three sectors, the rest of groups are given to these three sectors as well. In this case, each sector will get two major groups, which is one-third of the useful data tones, and the adjacent sectors use non-overlapping frequency segments that are virtually similar to frequency reuse of 3.

A DL PUSC slot includes two adjacent OFDM symbols and 24 data tones per symbol. As mentioned before, pilots had been already pre-allocated within each cluster.

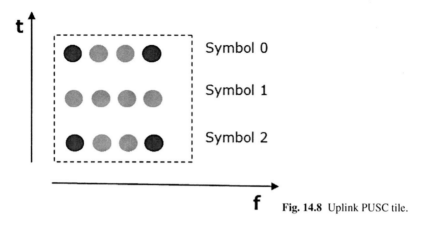

f   **Fig. 14.8** Uplink PUSC tile.

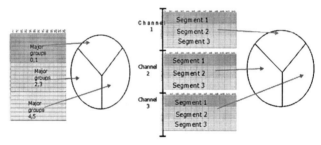

Fig. 14.9 PUSC permutation (a) with segmentation used for fractional frequency reuse and (b) all subchannels could be used for both reuses 1 and 3

In the UL, the concept of tiles are used, where a tile is a 4×4 structure constructed from 12 data and four pilot tones as depicted in Fig. 14.8. Each slot in UL PUSC is composed of six tiles, not necessarily adjacent in frequency domain. This results in 48 data tones and 24 pilot tones per slot.

Note that PUSC is particularly distinguished by the approach the tones are allocated to subchannels, and by the structure and density of pilot allocations. So, PUSC mode can be used as segmented PUSC (virtual frequency reuse or so-called "PUSC 1/3rd"), as well as when all major groups are allocated to all sectors. This is called "PUSC with all subchannels." Figure 14.9 shows these two configurations. If the operator is able to use three separate channel bandwidths for a real frequency reuse of 3, then depending on frequency planning, it might make sense to use PUSC with all subchannels, as segmented PUSC obviously reduces the system throughput since it reduces the effective bandwidth.

Tables 14.3 and 14.4 show the parameters used for DL and UL PUSC.

The only adjacent permutation used in WiMAX is AMC, which is mainly used for advanced antenna systems such as beamforming, or when link adaptation is

Table 14.3 PUSC DL parameters

Parameter	2K FFT	1K FFT	512 FFT	128 FFT
Num. of DC carriers	1	1	1	1
Num. of left guard tones	184	92	46	22
Num. of right guard tones	183	91	45	21
Num. of used tones	1681	841	421	85
Num. of tones per cluster	14	14	14	14
Num. of clusters	120	60	30	6
Num. of subchannels (SC)	60	30	15	3
Num. of data tones per SC per symb.	24	24	24	24

Table 14.4 PUSC UL parameters

Parameter	2K FFT	1K FFT	512 FFT	128 FFT
Num. of DC carriers	1	1	1	1
Num. of left guard tones	184	92	52	16
Num. of right guard tones	183	91	51	15
Num. of used tones	1681	841	409	97
Num. of subchannels (SC)	70	35	17	4
Num. of tones per SC	48	48	48	48
Num. of tiles	420	210	102	24
Num. of tones per tile per symb.	4	4	4	4
Num. of tiles per SC	6	6	6	6

essential. Normally when the mobile speed is low, the instantaneous feedback can be exploited, and therefore AMC could be considered as an appropriate permutation. AMC permutation is the same for UL and DL. In AMC, in each OFDM symbol pilots have fixed location in the frequency domain at each symbol.

AMC is defined based on the concept of bands, and hence the name of AMC band is applied to the AMC allocations. Each band consists of four consecutive rows of bins. Each bin is nothing but nine consecutive tones in frequency domain, with a pilot located in the middle of a bin.

AMC subchannels consist of six contiguous bins, either in time or frequency (with N bins by M symbols, and $N \times M = 6$) in the same band. This means there are three configurations for AMC permutations, 2×3, 3×2, and 1×6 (Fig. 14.10).

Each symbol is composed of 96 subchannels as opposed to 70 subchannels for PUSC. This is due to less number of pilots. So, in general AMC is capable of providing higher throughputs, less channel estimation reliability (due to less pilot density), and less frequency diversity, and is used more in low-speed environments.

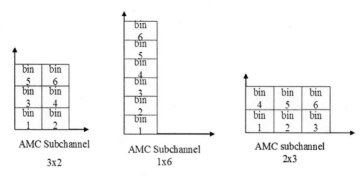

Fig. 14.10 AMC permutations, and different bin configurations

14.4.3 Frame Structure

For now, let us assume that there are 29 symbols allocated to DL and 18 symbols allocated to UL. In the DL subframe, the first symbol is called the preamble. Preamble is one OFDM symbol containing a known pseudo-noise (PN) sequence carrying a particular ID called IDCell (a number from 0 to 31), and segment ID (from 0 to 2), with different PAPR values. Total number of sequences are 114. It is used for initial timing and carrier synchronization, network scanning, and cell identification. It can be used for a coarse channel estimation, as well as signal to noise and interference (SINR) calculation.

The rest of the frames are divided into zones. A zone is nothing but a group of OFDMA symbols taking the same permutation scheme (such as PUSC or AMC). Within each zone, we have segments, which are consisted of one or more subchannels, through one or more adjacent symbols that create a single instance of MAC data unit.

If PUSC is used in both DL and UL, since each PUSC slot in DL and UL span over two and three OFDMA symbols consecutively the number of symbols in DL must be odd, and the number of symbols in UL must be a multiple of 3. As a result, having a total of 47 symbols, only a limited number of combinations are allowed for DL:UL separation, such as 41:6, 35:12, or 29:18.

The second symbol after preamble starts with a mandatory PUSC zone. It starts with frame control header (FCH), which is in the first and second OFDMA symbols after preamble consisted of one PUSC slot, using QPSK1/2 modulation with repetition 4. It includes the downlink frame prefix (DLFP), the burst profile of the DL-MAP following FCH (e.g., coding, repetition), used subchannel bitmap, and DL MAP length

Right after the FCH, a broadcast message called DL-MAP is included that is mainly aimed to include the map of DL allocation for all devices (and hence named DL MAP). Moreover, it defines all control information and contains synchronization field, could include allocation for other BS. The major part of the DL MAP is an information element (IE) called DL-MAP-IE that contains the allocations of all connections of all mobiles being allocated in this particular frame (the concept of connections per mobile and connection ID (CID) will be explained later). This IE for each connection per each mobile includes CID (unicast, multicast, or broadcast), downlink interval usage code (DIUC) to define the type of downlink burst type, OFDMA symbol offset (in unit of OFDMA symbols) at which the burst starts, subchannel index (the lowest index subchannel used for carrying the burst), power boost applied to the allocated data subcarriers, number of OFDMA symbols used to carry uplink burst, and repetition code used inside the allocation burst.

If the frame includes the UL subframe (which is the case for the majority of frames), the next message is called UL MAP that includes the UL-MAP defining UL allocation for all UL bursts transmitted by mobile stations. Like DL MAP, the main information element is called UL-MAP-IE that contains the CID, uplink interval usage code (UIUC) to define type of UL access and the burst type associated with the access, OFDMA symbol offset at which the burst starts, subchannel offset that is the lowest index subchannel used to carry the burst, number of OFDMA symbol used to carry the uplink burst, the number of subchannels with subsequent indices, the duration in units of OFDMA slots for the burst, and the repetition code used inside the allocated burst.

Note that there are other forms of DL MAPs and UL MAPs, such as compressed MAP, normal MAP, HARQ MAP, to optimize the size of DL and UL MAP to reduce the MAC overhead. Some of these MAPs are fit to specific situations, such as when HARQ is used. For the details of the structure of these MAPs and the usage case of these MAPs we refer the interested reader to [3]. DL MAP and UL MAP contain more information elements that define control messages [3]. The number of bits allocated to DIUC and UIUC is 4, and so a maximum of 16 DIUCs (UIUC) are defined. So, in order to increase these numbers DIUC=15 (UIUC=15) is called extended IE format to allow specific structures like MIMO.

In the UL, the first three symbols (or part of the first three symbols) are used for UL control signals. There are three major control signals sent in the UL, namely ranging used for synchronizing the power, time, and frequency, ARQ for sending the feedback to the BS transmitter, and finally the bandwidth request signals. The BW request and ranging allocations are mainly shared among mobiles and so the mobile stations contend over these resources.

The rest of DL and UL frames are divided into subchannels, and then allocated to users for data burst transmission (Fig. 14.11).

14.4.4 Channel Coding

Channel coding in mobile WiMAX includes randomization, forward error correction (FEC), bit interleaving, modulation, as well as repetition of orders 2, 4, and 6.

FEC adds redundancy to the signal in order to recover from errors without requiring retransmission. Mobile WiMAX includes several FEC schemes, out of which tail biting convolutional coding, in which the encoder memory is initialized with the

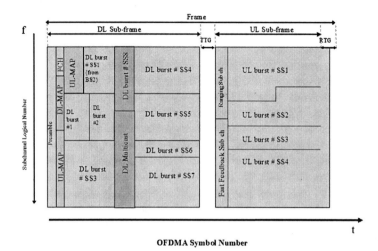

Fig. 14.11 WiMAX frame structure

last block being encoded is mandatory by the IEEE standard specifications. Other
FEC schemes defined by IEEE802.16e are as follows: (a) zero tailing convolutional
coding that appends a zero tail byte at the end of each bursts to initialize the viterbi
decoder, (b) convolutional turbo coding (CTC) used with HARQ to mitigate the effect of channel impairment and interference, (c) block turbo code (BTC), which is
the product of two simple component codes which are binary extended Hamming
codes or parity check codes, and (d) finally low density parity check (LDPC) that
provides competitive performance very close to Shannon capacity and reduced complexity. They are block codes, which are well suited for bursty packet switched data
services.

Although IEEE802.16-2005 identifies tail biting convolutional coding (CC) as
the only mandatory FEC mode, mobile WiMAX system profile has mandated all
equipments to use CTC, specially when HARQ is used. CC is mainly used at FCH
and DL and UL MAP, and CTC is used in the data region [5].

The modulation schemes used in this technology are QPSK, 16QAM, and
64QAM, with different data rates. Due to the limitation of the power and the error vector magnitudes (EVM) of the mobile devices, in the UL, 64QAM is only
allowed in the DL. Table 14.5 shows the data rates allowed for different modulation
schemes for different FEC modes.

Table 14.5 Coding rate for different modulation and FEC schemes in WiMAX

FEC / Modulation	CC-tail biting	BTC	CTC	LDPC
QPSK	1/2, 3/4	1/2, 3/4	1/2, 3/4	1/2, 2/3, 3/4, 5/6
16QAM	1/2, 3/4	1/2, 3/4	1/2, 3/4	1/2, 2/3, 3/4, 5/6
64QAM	1/2, 2/3, 3/4	1/2, 3/4	1/2, 2/3, 3/4, 5/6	1/2, 2/3, 3/4, 5/6

14.4.5 Multiple Antenna Modes in Mobile WiMAX

Mobile WiMAX is the first cellular technology that has materialized multiple antenna solutions, specially MIMO solutions in a large-scale cellular deployment. In this section, we avoid a detail description of the principles of multiple-antenna modes used in WiMAX, and just mention different modes supported by WiMAX technology, outline the highlight of some of them, and refer the interested reader for more implementation details to standard drafts.

In general, multiple antenna in wireless network could provide one or more of the following three gains; (a) providing SNR improvement through beamforming array gain, (b) link reliability and coding enhancement through diversity and coding gain, and (c) capacity improvement using multiplexing gain.

IEEE802.16 provides all three gains by allowing the following multiple antenna modes:

(a) Simple receive diversity by using multiple antenna at the receiver with sufficient spacing used for combating fading. The receiver uses either MRC or MMSE combining. Normally MRC has a significant gain over MMSE, but is more difficult to implement. Receive diversity can be done both in UL and DL, (b) diversity modes by space time/frequency coding again for combating fading. In this case transmit diversities of orders 2, 3, and 4 are supported by the standard, (c) beamforming by advanced antenna systems (AAS) to increases the coverage, and reduce the outage probability, (d) capacity enhancement by MIMO schemes, exploiting spatial multiplexing (SM) techniques, and (e) combination of beamforming/STFC or beamforming/MIMO by closed-loop MIMO mode, or using pre-coding mainly used for low-speed mobiles providing very high-throughput gains and coverage enhancements.

Space–time coding (STC) yields diversity gain without channel knowledge at the transmitter by coding across antennas (space) and across time and/or frequency [16]. This mode could provide capacity improvement as well as better spectral efficiency. The simplest STC method is Alamouti scheme (2 Tx,1 Rx antenna or 2 × 1)[12].

In spatial multiplexing (SM) transmit data stream is split into several parallel streams, which are transmitted from different antennas simultaneously along with signal processing at the receiver to detect the streams [13]. It provides capacity gain at no additional power or bandwidth consumption. Each stream has a lower data rate. The main computational complexity lies in the separation step. A major principle allowing the separation of streams in case of both SM and STC is the orthogonality of pilots between the two antennas. The pilot tones have to be allocated to the two antennas in such a way that signals transmitted from two antennas would not interfere on the pilot locations. Moreover, SM requires the correlation among two-transmit and two-receive antennas to be low (the correlation factor of the MIMO channel model to be low) and the SINR of the environment be high enough. Otherwise, SM is not an appropriate mode of operation.

For the same reason, WiMAX has used a clever approach named adaptive MIMO. In this approach, when a user is in a region where its SINR is above a threshold and the correlation factor of its channel matrix is low, or in a better word, its eigenvalue spread or its channel condition number (this is a number between 0 and 1) is low, SM is proposed, and as a result capacity gain is achieved. However, when the SINR is low, or the channel condition number is high, STC is used for that particular user (see Fig. 14.12) to increase the coverage.

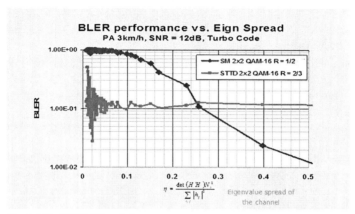

Fig. 14.12 MIMO DL switch

Note that the decision process could be made at the mobile station, and then fed back (or in a better word proposed) to the BS using some feedback mechanisms, or made at the BS directly, or a combination of these two. In each case, it is the BS that makes the final decision about the MIMO mode. Since the device has a better idea of the channel estimation and channel correlation, it can make a better judgment. However, the feedback duration is long, therefore relying only on MS feedback might not be appropriate. In any case, the decision process is not identified by the standard, and is implementation based. Standard has clearly defined the process by which mobile station informs the BS of its proposal, as well as BSs final decision of the MIMO mode per frame, and the provisions to make sure mobile and base stations are synchronized in terms MIMO mode in a frame-by-frame basis. These discussions are out of the scope of this chapter.

According to the number of transmit antennas, different MIMO patterns can be used for WiMAX. As mentioned above, IEEE has envisioned two, three, and four antennas at the BS. If we define the rate of transmission as the number of symbols transmitted per channel use, the standard supports rates 1, 2, and 3 using the matrices **A**, **B**, and **C**. These matrices for 2 and 4 antenna transmissions are defined as in Table 14.6. For rate 3, you may refer to the IEEE specifications [3].

Mobile WiMAX system profile has limited the DL MIMO to 2×2 mode [5]. So, only matrices **A** and **B** are used for mobile WiMAX release 1.0. Note that matrix **A** for two antennas is simply the Alamouti code used for STC and matrix **B** for SM transmission. Using Table 14.6 we can see that the following rates are achievable: For two antennas, matrix **A** achieves rate 2 and matrix **B** achieves rate 2. If three or

Table 14.6 MIMO matrices

	Matrix A	Matrix B	Matrix C
2 antennas	$\mathbf{A} = \begin{bmatrix} s_i & -s_{i+1}^* \\ s_{i+1} & s_i^* \end{bmatrix}$	$\mathbf{B} = \begin{bmatrix} s_i \\ s_{i+1} \end{bmatrix}$	
4 antennas	$\mathbf{A} = \begin{bmatrix} s_i & -s_{i+1}^* & 0 & 0 \\ s_{i+1} & s_i^* & 0 & 0 \\ 0 & 0 & s_{i+2} & -s_{i+3}^* \\ 0 & 0 & s_{i+3} & s_{i+2}^* \end{bmatrix}$	$\mathbf{B} = \begin{bmatrix} s_i & -s_{i+1}^* & s_{i+4} & -s_{i+6}^* \\ s_{i+1} & s_i^* & s_{i+5} & -s_{i+7}^* \\ s_{i+2} & -s_{i+3}^* & s_{i+6} & s_{i+4}^* \\ s_{i+3} & s_{i+2}^* & s_{i+7} & s_{i+5}^* \end{bmatrix}$	$\mathbf{C} = \begin{bmatrix} s_i \\ s_{i+1} \\ s_{i+2} \\ s_{i+3} \end{bmatrix}$

four antennas are used for transmission, rate 3 can be achieved by matrix $\mathbf{C}$, rate 2 with matrix $\mathbf{B}$, and rate 1 with matrix $\mathbf{A}$.

Beamforming is another mode of multiple antenna transmission in IEEE802.16, where multiple antenna elements transmit with different gain and phase to collectively form beams. By changing the gain and phase of the signal at each antenna we can steer the beam. It can maximize the transmitted energy toward the desired mobile, increase the coverage, provide the diversity gain, mitigate noise and interference, and improve signal quality (SINR). In addition to simple switched beams, WiMAX supports different types of transmit beamforming such as single-stream BF (using spatial channel, spatial correlation), where no particular antenna separation is required, and can be used in urban environment, and interference nulling, where all interferences toward a particular mobile are removed. The later mode requires small separation (half a wavelength antenna separation) and is mainly appropriate to morphologies without many paths. Other beamforming mode is multistream BF or SDMA where interference toward multiple mobiles at the same time is eliminated.

Regardless of beamforming type, in order to achieve beamforming gain, we need to support coherent signal processing at the receiver, and the transmitter requires instantaneous channel knowledge to identify the beamforming gains applied to the transmission signal. For that reason, beamforming is mainly appropriate for low-speed mobiles. WiMAX has envisioned different methods to provide the instantaneous channel knowledge to the BS. The channel knowledge could be obtained by estimating the channel in the UL direction at the BS and applying that to the DL using TDD channel reciprocity, or receiving the instantaneous feedback from mobile through either dedicated pilots (digital feedback) or channel sounding (analog feedback)[3].

Closed-loop MIMO and/or MIMO plus beamforming is also another mode of operation for providing high level of capacity and coverage gain for low-speed mobiles. This mode is enabled by using pre-coding [19], wherein a matrix W is multiplied by the transmission matrix (**A**, **B**, or **C**). BS uses channel quality indication fed back from mobile to calculate this matrix. Closed-loop MIMO can be implemented in different ways: (a) selection of a number of antennas among several antennas, (b) antenna grouping, (c) codebook method (CQICH feedback), and (d) using UL channel sounding (analog precoding).

A simpler mode of operation is also supported by WiMAX called cyclic delay diversity (CDD) or cyclic shift transmit diversity (CSTD), in which two or more antennas are employed at the base station, but those BSs are not transmitting at the same time. With CSTD, each antenna element in transmit array sends a circularly shifted version of the same OFDM time-domain symbol. For example if there are M transmit antennas at the base station and if antenna 1 sends an unshifted version of the OFDM symbol, then antenna m transmits the same OFDM symbol, but circularly shifted by $(m - 1)D$ time-domain samples. Note that each antenna adds a cyclic prefix after circularly shifting the OFDM symbol and thus the delay-spread protection offered by the cyclic prefix is unaffected by the CSTD. This mode has two advantages:

- WiMAX does not allow multiple streams over DL and UL MAP, and therefore STC is not allowed on those regions. As a result, DL/UL MAP could suffer from low SINR. Using CDD, this gain can be provided for the MAP area as well. Figure 14.13 taken from [10] provides some comparisons for MAP coverages.
- Using CDD, unlike STC and MIMO, the whole power is used by one antenna. As a result, not only space-time diversity is achieved (through delayed transmission and switching antenna elements), but also each antenna does not lose 3 dB due to power sharing.

For the UL, due to the limitation of device size, multiple antenna transmission specially from hand-held devices is problematic. First, MSs cannot afford to have

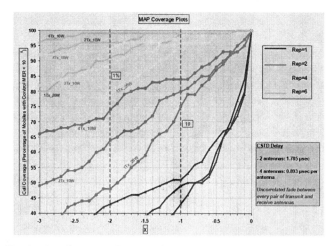

Fig. 14.13 Simulated performance of control channel coverage for ITU channel

two power amplifiers, which causes them to be bigger, more complex, and hence more expensive. Moreover, an efficient MIMO transmission requires considerable antenna separation, which is not apparently possible on mobile devices. Only a limited class of devices (e.g., CPEs) have these capabilities. For these reasons, IEEE has devised a new scheme called UL collaborative MIMO, in which each device only transmits from one antenna, but two devices are selected by the BS to collaborate and share a time–frequency resource for UL transmission. This is similar to sending two streams from two antennas with a difference that two antennas are on two separate devices, of course with significant separations. The only requirement is that pilots must be orthogonal for those two devices as depicted in Fig. 14.14. The processing at the base station is very similar to the processing of a two-branch MIMO transmission, with a minimal or zero impact on the MS complexity. BS scheduler needs to be modified slightly as well. Note that unlike DL MIMO, UL collaborative MIMO does not provide any enhancement on device throughput, but improves overall UL sector throughput.

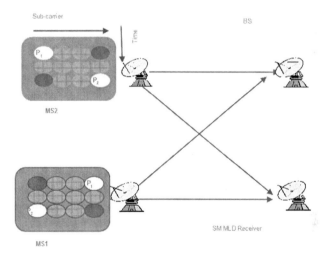

Fig. 14.14 UL collaborative MIMO

14.4.6 Power Control and Link Adaptation

Power control is used to adjust MS power so that UL signal strength received at BS falls within an acceptable range. The ideal level depends on assigned modulation/coding rates.

We need to perform power control at MS in order to combat propagation changes, near–far problem, fast fading (changes over frames), lognormal fading (changes over 100's of frames), power law changes (changes over 1000's of frames), as distance to BS changes, and interference changes.

There are two general types of power control in WiMAX, namely open-loop and closed-loop power control.

Closed-loop power control is mainly used at the network entry process and is shifted to open loop when the data session starts. It could be used during periodic ranging and BW request as well. In closed loop, MS adjusts power based on instruction from BS achieved using ranging messages, PMC-RSP message, power control IE, UL MAP fast tracking IE, and fast power control. All of these schemes have high overheads and require power adjustment every nth frame in order to track fast fading where significant signal level fluctuations occur every n frames.

In open-loop power control MS adjusts power based on estimate of propagation loss, noise plus interference supplied by the BS in the DL MAP, required SINR for the assigned modulation/coding/repetition rate, and offset from the BS. Since the TDD mode is assumed in WiMAX Rel 1.0, accurate path loss can be calculated using DL RSSI measurement of the preamble at the MS as a representative of UL propagation loss.

If any packets were received from MS in previous frame, fast power control can be achieved. Normally open-loop power control is performed at the data region and is faster than closed-loop power control.

Open-loop power control can be done in both passive and active modes. In passive UL open-loop power control, the mobile station does not add the correction term for MS-specific power offset, and the offset is not controlled by MS.

In active UL open-loop power control, the mobile station does add the correction term for MS-specific power offset. This term is called offset-SS-per-SS, and the offset is controlled by MS.

WiMAX forum system profile has decided to make closed-loop power control mandatory in the profile, and out of two modes for open-loop power control only passive open-loop power control is included in the system profile.

Link Adaptation: Efficient and accurate link adaptation is required to ensure WiMAX multi-rate performance. In CDMA systems, power control is key for multi-rate performance and in OFDMA systems link adaptation is the key.

IEEE80216e includes multiple types of report mechanisms and metrics for the support of DL link adaptation. Interoperability and system performance requires smart selection of these supported features.

Link adaptation mechanisms should address dynamic nature of interference knowing that loading is not constant, and the fact that interference is not reciprocal. WiMAX link adaptation mechanism is insensitive to changes in parameters, channel model, and considers mobility into account.

The link adaptation process is performed by sending SINR measured at the mobile receiver through feedback channels called channel quality indicator channel (CQICH). These SINR values are either averaged over the subchannels, over the PUSC slots, or AMC band.

They are reported through fast feedback allocation subheader (MAC layer dedicated packets), using either CQICH dedicated channels or CQICH enhanced allocation dedicated channels. Both support 4 and 6 bits, where the 6 bits case supports multiple CQICH and band AMC report.

How these reports are used at the BS is mainly implementation based, and not identified in the standard.

IEEE802.16 allows the mobile station to measure the SINR in three different regions:

- Preamble, with the advantage that it is sent in each and every frame, but it does not reflect the noise and interference in data zone because of preamble structure. It is sited specifically for reuse 1 and reuse 3 systems and not other reuse patterns (6 or more).
- Pilots, which reflect data zone interference between cells better than preamble. However, it does not reflect boosted data zone, AAS, changing load of data zones.
- Data, which are the best reflection of data zone interference level, and does not need the compensation of pilot/preamble boosting. However, it is prone to variations of interference level (due to change of loading, boosting, AAS), and it is more complicated to calculate and report.

Mobile WiMAX allows two types of feedbacks, one is physical SINR, or PCINR, where the actual measured and averaged SINR at the MS is fed back to the BS. At the BS the scheduler uses these values to determine the actual power level as well as the modulation and coding scheme (MCS) for the corresponding mobile station. The second type is called effective CINR or ECINR, where the MS, according to its own channel estimation and the water fall curve (BER vs. SINR curve), identifies the best MCS, and proposes the MCS value to the BS, rather than sending the actual CINR value. Even in this case, it is up to BSs scheduler to accept or reject this MCS value proposed by the MS.

HARQ: Automatic request (ARQ) in WiMAX can be done both in physical layer and in MAC layer. ARQ in physical layer is called hybrid ARQ or HARQ. It is faster and more reliable than MAC layer ARQ. HARQ mitigates the effect of channel impairment and interference, increases the data throughput by the multiple parallel data streams, enhances the robustness of transmitted bursts, efficiently resolves the problem of channel quality difference, and overcomes the adaptation error of the AMC in fading channel.

HARQ is supported in both DL and UL. It builds on the fast uplink feedback channels available in the OFDMA mode. HARQ is nothing but fast retransmission of packets in case of errors. Consequently it uses time diversity achieved by combining previously erroneously decoded packet and retransmitted packet. In case of a NACK (not acknowledgement), HARQ re-transmits more redundancy and receiver combines them with the previously sent data to achieve better SNR and coding gain against time-varying channel.

Two types of HARQ implementations are envisioned in IEEE802.16. One is chase combining (CC), where in case of NACK, transmitter sends the same copy that was sent previously. The MCS has to be the same for each packets, and therefore the implementation of this mode is very simple.

The second HARQ mode is called incremental redundancy (IR) HARQ, where in case of NACK, transmitter sends part of codeword that may be different from previous first transmission (subpackets). The subpacket should show a complementary property for better performance. IR HARQ shows more coding gain and better SNR, and the MCS can be adapted for each subpacket. It also provides more flexibility to adapt to the time variation of the channel. However, it requires more DL or UL MAP overhead, its implementation is more complex, and requires more memory at the receiver.

WiMAX system profile has mandated chase combining in Rel 1.0 only and is investigating the inclusion of incremental redundancy in the future versions.

14.5 Medium Access Control Layer

WiMAX medium access control layer resides above the PHY layer and is responsible for controlling and multiplexing different types of streams over the physical media, which is the wireless air interface in this case. IEEE802.16 MAC layer is developed starting from cable modem DOCSIS [18] and is built based upon the request/grant principles. As depicted in Fig. 14.15 the WiMAX MAC layer is consisted of three main sublayers and some service access points (SAP):

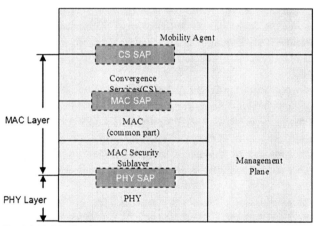

Fig. 14.15 MAC sublayers

The MAC service-specific convergence sublayer maps the transport layer specific external network traffic to the particular MAC connection ID. It is the interface to the upper layer entities through CS SAP. It also performs classification of service data units (SDU) coming from upper layers (layer 3) and header compression/suppression to increase the efficiency of the system.

The MAC common part (CPS) is the core MAC for access and performs all operations on the packets that are independent from upper layer (layer 3) and lower layer (PHY) such as BW allocation, and connection maintenance. It includes uplink scheduling, bandwidth request/grant, connection control, ARQ, ranging, and QoS control.

The privacy or security sublayer is located between CPS and PHY. It performs traffic encryption, authentication, and secure key exchange between the MS and the BS.

IEEE802.16 MAC has a point-to-multipoint (PMP) structure, where a central BS is sectorized and in each sector, independent set of antennas transmit multiple independent data stream simultaneously.

The mobile stations share the UL to the BS on a demand basis. Depending on the class of service utilized, the MS may be issued continuing rights to transmit, or the right to transmit may be granted by the BS after receipt of a request from the user. In addition to individually addressed messages, messages may also be sent on multicast connections in DL direction (control messages and video distribution are examples of multicast applications) as well as broadcast to all stations.

The MAC is connection oriented. In order to associate varying levels of QoS, all data communications to and from an MS are in the context of a transport connection. Service flows may be provisioned when an MS is entered in the network. Shortly after MS is registered, transport connections are associated with these service flows (one connection per service flow) to provide a reference against which to request bandwidth. Additionally, new transport connections may be established when a customer's service needs change. Each service flow defines the QoS parameters for the PDUs that are exchanged on the connection. All bandwidth requests and resource allocations are performed in the context of a service flow not for the MS itself. Bandwidth is granted by the BS to an MS as an aggregate of grants in response to per-connection requests from the MS.

Connections are identified by a 16-bit connection IDs (CID). At mobile initialization, two pairs of management connections, basic connections (UL and DL), and primary management connections (UL and DL) shall be established between the MS and the BS, and a third pair of management connections (secondary management, DL and UL) may be optionally generated. The basic connection is used by the BS MAC and MS MAC to exchange short, time-urgent MAC management messages. The primary management connection is used by the BS MAC and MS MAC to exchange longer, more delay-tolerant MAC management messages. The secondary management connection is used by the BS and MS to transfer delay-tolerant, standards-based messages (such as DHCP and TFTP). For data bearer services, the BS and the MS may initiate the service flows based upon the provisioning information.

Other than several CIDs, each MS is also uniquely identified by a 48-bit universal MAC address. It is used during the initial ranging process to establish the appropriate connections for a mobile. It is also used as part of the authentication process by which the BS and MS each verify the identity of the other.

The convergence sublayer at the top of the MAC layer has three major functions. These key functions are (1) classification of the higher layer PDUs into the appropriate CIDs, (2) delivery and receipt of resulting CS PDUs, and finally (3) payload header suppression/rebuilding (PHS). In order for the classifier to map higher layer packets to a CID, a set of "protocol-specific" matching criteria are needed, such as IPV4, or IPV6. Each classifier includes some parameters. For example IPv4 ethernet classifier that includes ethernet addresses, ethernet user priority field, IP addresses, and port numbers. The mapping also creates an association with the service flow of that connection, such as QoS characteristics.

In the CS sublayer also the repetitive portion of packet header is removed or compressed. This is done either by using payload header suppression (PHS) protocol or robust header suppression (ROHC) protocol. ROHC is a standard for how to compress IP, UDP, TCP, and RTP packet headers. It is designed mainly for links where packet loss is high, such as wireless links. It is ideal for streaming applications such as voice and video. It reduces headers to only 1 or 3 bytes. If a voice gross data rate with IPv4 and UDP header without header suppression is 48 kbps, with PHS it could become 25.6 kbps and the same thing with ROHC is about 19.2 kbps.

Scanning and Network Entry: The mobile station needs to find and connect to the network. The mobile station starts scanning the channel list based on some predefined protocols (IEEE802.16g defines these standard-based protocols, but a simple version could be based on NSP-ID or NAP-ID). The scanning means performing correlation between the known sequences at MS with the time-shifted version of the sequences carried over by the preamble of each BS and finding the ones with highest correlation values. There are 114 different sequences associated with 114 different BSID (a 48-bit ID whose first 24 bits represent the operator ID), and the MS can lock itself to any of those BSs.

The process of network entry as depicted in Fig. 14.16 follows the scanning, which includes (a) initial ranging (to be explained later), (b) 802.16 link initial-

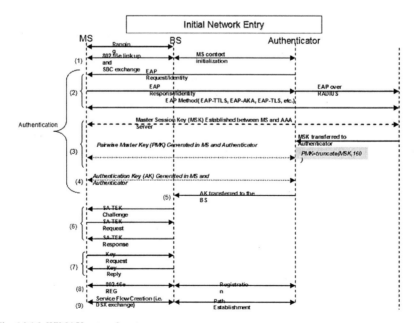

Fig. 14.16 WiMAX network entry process

ization (c) MS context initialization, (d) MS and BS capability negotiation (SBC negotiation), (e) authentication, (f) security association initialization and exchange, (g) registration, (h) Tx UL power report, (i) service flow initiation (DSX exchange), (j) and finally path establishment.

As observed from Fig. 14.16 some of these procedures are defined between MS and BS (over R1) and some between BS and ASN-GW (over R6) and some between MS and AAA (over R2).

By listening to the downlink and UL channel descriptor (DCD and UCD) mobile can obtain physical characteristics of the downlink, as well as the physical characteristics of the uplink.

Ranging: Ranging is a collection of processes to maintain the quality of the RF communication link, synchronize MS timing, frequency and power offsets, and to request power and/or downlink burst profile change. It is the first order of power control, and aligns timing to properly meet propagation delay. It also aligns frequency shift to compensate for synthesizer drift. There are essentially two types of ranging:

- **Initial Ranging:** Allows mobile to adjust transmission parameters at the network entry stage so that it can maintain uplink communications with the BS. MS starts the process by sending a randomly selected CDMA ranging code to the BS. If no response is received, the MS incrementally increases the power and tries again using a random backoff process.
- **Periodic Ranging:** Allows a registered in-session mobile to adjust transmission parameters periodically so that it can maintain uplink communications with the BS. This process is done periodically using a contention-based CDMA ranging process.

The contention codes (CDMA codes) are allocated to ranging and bandwidth request. There are 256 codes of length 144 bits each. BS can distinguish transmissions based on these different codes. This allows efficient use of resource with minimal overhead. The code allocations are announced in the UCD. Both initial and periodic ranging can be done using this process. The MS is anonymous to the BS in this mode.

Ranging can be done in a fast mode as well (fast ranging), when a dedicated area is allocated to the mobile for ranging process. This is mainly done during optimized handoff process.

Regardless of ranging process, UL-MAP indicates the allocated ranging subchannels for initial ranging, periodic ranging, and bandwidth requests. It is composed of one or more groups of six or eight adjacent subchannels.

Bandwidth Request: Bandwidth request in WiMAX is performed similar to ranging process. The MS sends the contention-based CDMA code for BW request. The BS responds using a broadcast message (CDMA-Alloc-IE) allocating opportunities for BW request. The MS that has sent the CDMA code uses the allocation to request the bandwidth either by using a stand-alone MAC bandwidth request header or in a piggybacked mode with a data transmission using MAC grant management subheader. Bandwidth requests may be either incremental or aggregate. However, the piggyback request must be incremental. The BS using its scheduling procedure and based upon the QoS of the service flow grants or rejects the request.

Registration: Registration is the last step of network entry and allows entry into the network. The registration request may include IP version, convergence sublayer capabilities, as well as ARQ and fragmentation parameters. Registration response include secondary management CID with delay-tolerant messages such as SMNP and DHCP.

WiMAX MAC supports both packing small SDUs and fragmentation of large SDUs, such that multiple small SDUs arriving from upper layer could be packed on a single MAC PDU, and at the same time one large SDU could be fragmented into multiple smaller MAC PDUs.

MAC ARQ: MAC layer ARQ is a mechanism for retransmission of lost blocks. It includes acknowledgments of successfully received packets as well as negative acknowledgment for failed packets. There are timers for receiving packets. ARQ feedback information can be sent as either stand-alone MAC message on basic CID or piggybacked on existing connection. MAC ARQ is optional, but when implemented, it is enabled on a per-unidirectional connection basis. The per-connection ARQ shall be specified and negotiated during connection creation.

WiMAX support different types of MAC ARQ. The three types are selective ACK, cummulative ACK, and selective with cummulative ACK.

14.5.1 Quality of Service

The service flows define the QoS parameters that are assigned to each connection. The QoS parameters could be maximum-sustained rate, minimum-reserved rate, jitter, latency, etc. MAC layer supports activating, changing, and deleting service flows. The establishment, modification, and deleting service flows are performed by using DSA-REQ (CID, SFID, QoS parameter set), DSC-REQ, DSD-REQ messages. They are referred as DSX process. The process of service flow initiation could be done either statically or dynamically. In dynamic service flow creation, each mobile based on its application requirement can create its service flow using DSA-REQ message, while in static mode, all MSs in a network are allocated the same service flow at the initiation by the network. Service flows are unidirectional and identified by a 32-bit service flow ID (SDIF).

WiMAX Rel 1.0 supports five different types of service flows. Each of these types are distinguished by a set of QoS parameters, and could be used for particular type of applications. They are distinguished by the way BS handles their scheduling support for data transport on a connection. Each connection is associated with a single scheduling service. A scheduling service is determined by a set of QoS parameters that quantify aspects of its behavior. These service flow types are as follows:

- **Unsolicited Grant Services (UGS)** to support real-time constant bit rate (CBR) applications such as T1, where the allocations are granted in an unsolicited periodically offers.The overhead and latency of MS request/grant is eliminated. It is mainly used for T1/E1 and VoIP without silence suppression.

- **Real-Time Poling Services (rtPS)** to support real-time variable size data packets on a periodic basis, where the allocations are granted for data packets per request. It offers unicast request opportunities to meet the flow's real-time needs and is appropriate for applications such as MPEG video.

- **Extended Real-Time Poling Services (ErtPS)** to support real-time variable bit rate in an unsolicited manner, where allocations are granted in an unsolicited offers for a duration, and the request/grants are updated in some intervals. It has less request/grant overhead than rtPS. It is mainly designed and more appropriate for voice over IP services with silence suppression.

- **Non-Real-Time Poling Services (nrtPS)** to support non-real-time variable size data packets for unicast request opportunities. It is mainly used for applications such as e-mail and FTP

- **Best Effort (BE)** to support best effort traffic. No minimum service level is required, and handled by space-available basis.

14.5.2 Power Saving Mode

One of the major problems with mobile devices is conserving battery powers. To that end, WiMAX has three states for each mobile station in the network. Active MSs are the ones that are actively transmitting or receiving traffic and are allocated CQICH channels. There are two non-active modes, namely sleep mode and idle mode.

In sleep mode the MS powers down in order to minimize the power usage, while stay connected, i.e., performing neighbor BSs ranging or handoff. In this mode, the MS keeps all allocations including UL CQICH, etc., but turns off the receive and TX power, and wakes up periodically (the period and length of sleep interval depends on power saving class and is already negotiated with BS) to perform the following: (a) periodic ranging, (b) handoff, (c) UL transmission if necessary, (d) checking the DL MAP to see if there is any DL traffic indication for the MS. If there is no traffic for the MS, it goes back to the sleep interval. The listening interval is always fixed, but the sleep interval could be different, and depends on the type of service flows and connections the MS carries and the power saving classes they belong to. Each power saving class is a group of connections having common demand properties, and could be of different service flows. The demand properties could be intervals between consequent allocations. WiMAX defines three power class types identified by their parameter sets, and procedure of activation/deactivation. For example, power class 1 is mainly used for BE service flows.

In idle mode, the mobile is powered down and is not connected to the BS anymore. In this mode, MS gives up all CQICH allocations, and the mobiles only become periodically available for DL broadcast traffic messaging without registration at a specific BS as the MS traverses an air link environment populated by multiple

BSs, typically over a large geographic area. Idle mode benefits MS by removing the active requirement for handoff, and all normal operation requirements. By restricting MS activity to scanning at discrete intervals, idle mode allows the MS to conserve power and operational resources.

Idle mode also benefits the network by eliminating air interface and network HO traffic from essentially inactive MS. Both MS and BS can initiate idle mode by deregistration process. In this mode, the mobile does not perform ranging, or hand off when it traverses the border of cells or sectors coverage. Both mobile station and BS can initiate return from this mode to active mode. MS goes back to active mode if it has traffic to send in the UL. However, the BS-initiated idle mode termination is performed by a process called "Paging." A mobile in idle mode, wakes up periodically to listen to paging channels. The paging controller normally resides somewhere above the BS, here at the ASN-GW. A paging group is a logical group of BSs offering a contiguous coverage region where the MS does not need to transmit in the UL, yet paged in the DL. All MSs in a paging group are paged by one entity. If mobile crosses the border of the paging zones, it has to perform the location update process.

There are two types of idle to active mode transition. In the regular transition mode, we assume that all MS context is erased, when the mobile is entered the idle mode, and therefore when it wants to go back to idle mode, it has to perform full network entry, including scanning new base stations, authentications. However, if the MS context is maintained at the ASN-GW (until a timer named resource holding timer expires), the MS does not need to perform full network entry, and mainly an initial ranging and registration is enough. This is called optimized idle mode re-entry. This re-entry mode significantly reduces the idle to active transition latency, but requires enough memory at the ASN-GW for each idle mobile.

In general, to compare the sleep and idle mode, idle mode provides a battery conservation and resource reservation, because in this mode, MS is not connected to the BS, do not perform ranging and handoff, and does not retain valuable UL CQICH allocations. Sleep mode MS is connected to BS, maintains UL control channel allocations, and performs handoff and ranging, and is less efficient in terms of power conservation and battery reservation.

However, sleep mode is more efficient in terms of transition to active mode latency, and therefore takes less time to initiate data transfer.

14.5.3 Multicast Broadcast Services

In addition to unicast services, WiMAX supports broadcast and multicast services (MBS). Broadcasting refers to unidirectional PMP service in which data are transmitted from a single source to all user terminals in the associated service area. Multicasting, on the other hand, refers to unidirectional PMP services in which data are transmitted from a single source to multiple user terminals that are subscribed to the service in the associated service area.

Each MBS connection is associated with specific QoS parameters and security parameters (secured both in MAC layer and in upper layer to prevent unauthorized access to MBS). MBS traffic is sent regardless of MS states.

WiMAX supports two types of MBS access, single-BS MBS, wherein MS is registered to only one BS for MBS, and multi-BS MBS, where more than one BS participate in transmitting multicast/broadcast data from service flow(s) (Fig. 14.17).

In single-BS MBS, each multicast MAC SDU is transmitted only once per BS channel, and MS is registered to BS. In multi-BS MBS all MSs need to register to MBS at the network level, not to any BS. It improves the reliability of reception through macro-space diversity. Each BS shall transmit the same PDU, using the same transmission mechanism (symbol, subchannel, modulation, etc.) at the same time. In this mode BS synchronization is needed. So, the CIDs should be the same for all BSs and MSs on the same channel.

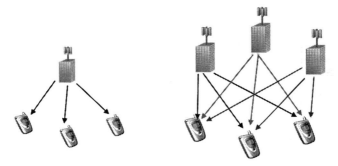

Fig. 14.17 (a) Single-BS MBS and (b) multi-BS MBS

To this end, for Multi-BS MBS, the concept of MBS zone is defined, which is a group of synchronized BSs that transmit same multicast or broadcast data using the same CID and SA, and the same transmission mechanism, such as symbol, subchannel, modulation. It is important to mention that MBS traffic is sent regardless of MS states.

MBS in WiMAX can be done using a dedicated carrier frequency to increase the broadcast or multicast throughput, or as a shared service with a unicast carrier frequency, where a permutation zone is carved somewhere in the DL subframe for MBS traffic.

14.5.4 Handoff

Handoff is a process by which a mobile station moves its connection from one BS to another BS or from one sector of a BS to another sector, when it crosses the border of cells or sectors, without interrupting or losing the connection to the network.

IEEE802.16 supports different types of handoff. They are hard handoff or break before make, soft handoff or make before break, fast base station switch or FBSS, and finally optimized hard handoff. Other modes such as drop handoff can be supported as well.

Since WiMAX profile has mandated the optimized hard handoff, here we focus our attention only on this mode, and the interested reader is referred to the standard draft [3].

There are three metrics defined in IEEE802.16 to trigger the scanning process. They are SINR, RSSI (received signal strength indicator), and round trip delay (RTD). When any of these values go below a predefined threshold (above the threshold in case of RTD), the mobile requests a scanning period from the BS, which is a time duration when the mobile is temporarily out of BSs DL and UL data connection to scan the neighboring BSs and sectors. The list of neighboring BSs are identified in a message called NBR-ADV or neighbor advertisement message which is broadcast by the BS in the DCD, and includes a list of neighboring BSs along with their physical characteristics, such as channel BW.

When the MS decides for the best alternative BS (in WiMAX sectors are also identified by BSID and so referred to by BS), it informs serving BS about its decision by a message called MOB-MSHO-REQ (Fig. 14.18). The serving BS negotiates with neighbor BS to set up ranging intervals. The serving BS responds by another message and at the same time initiates the context transfer through R6 and R4 (if it requires moving to another ASN-GW). The MS indicates that it is going to handover to its intended BS by sending the handoff indication message (MOB-HO-IND). At this time all DL and UL traffic are interrupted. MS synchronizes with the target BS and sends a ranging request which includes the BSID. This ranging

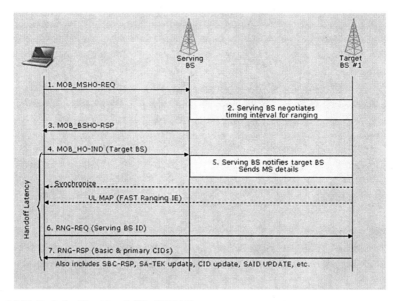

Fig. 14.18 Optimized hard handoff in WiMAX

process is performed using fast ranging, with a dedicated ranging allocation, and does not include CDMA allocation and contention. The target BS sends a ranging response which includes the operational parameters, such as, basic and primary CIDs.

The ranging response indicates which network entry steps can be omitted, i.e., SBC-REQ/RSP, PKMv2 authentication, TEK establishment, registration, network address acquisition, time of day acquisition, etc.

14.5.5 Security and Authentication in WiMAX

The privacy sublayer has two major functionalities; (1) privacy key management (PKMv2) including authentication and key exchange and (2) encapsulation protocol including data confidentiality and data integrity.

Privacy support in WiMAX covers both authentication and encryption and spans in four different areas:

1. Device authentication, to authenticate the device connecting to the network, and performed at the start of a session, using device credential and certificates and cryptographic keys. These keys are mainly loaded by either the manufacturer or the operator or service provider.
2. User authentication, to authenticate the subscriber, performed at the start of a session, and continued periodically and also at the time of network re-entry, using name and password of the user, set at the service initialization.
3. Message authentication or data integrity, during session using hash algorithms derived from secret keys stored at the initialization process.
4. Message encryption during a session using encryption algorithms derived from encryption keys.

The main authentication method used in WiMAX NWG and IEEE802.16-2005 is called extensible authentication protocol or EAP, which is used for both authentication and key distribution. EAP is a framework for authentication, derived from IETF RFC 3748, and negotiates pairwise master key (PMK) between ASN-GW and MS, and is used to generate encryption keys.

PKMv2 is defined in 802.16e-2005, and includes authentication/authorization protocol. It establishes a shared authorization key (AK). It is essentially a three-way handshake to set up the security associations (SA), and traffic encryption keys (TEK) between the MS and BS.

The communication between MS and BS is secured in WiMAX networks by encryption and the process of integrity protection. WiMAX uses advanced encryption standard (AES) for encryption. Using PKMv2, 128-bit keys are allowed in all modes.

PKMv2 derives keying material from the authentication key (AK), which is distributed to ASN by AAA when authentication is completed (Fig. 14.19). Traffic encryption key (TEK) is also randomly generated by the BS and transported to MS using key encryption key for encryption.

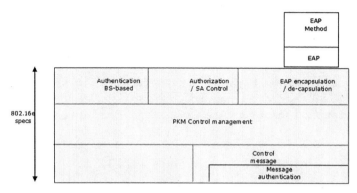

Fig. 14.19 Security and authentication in WiMAX

14.6 WiMAX Performance

Mobile WiMAX has been recently deployed in commercial networks and has proved in reality that it is an efficient broadband wireless technology for urban and dense urban environments. The average sector throughput and the average throughput delivered to the subscribers in the DL and UL are well beyond what the existing CDMA networks can deliver (about $2.5-3\times$). These values are obtained by distributing several users in the network with different mobility speeds, and averaging over time. However, in the following, we will briefly outline some of the conservative simulation results obtained using WiMAX Rel 1.0 parameters, published in NGMN reports [17], using the WiMAX vendor results.

Baseline simulation assumptions and configuration parameters for system-level performance evaluation of the reference mobile WiMAX system profile are listed below.

A DL:UL split of 29:18 OFDMA symbols was assumed in the TDD mode. Seven OFDMA symbols on DL and three OFDMA symbols on the UL were accounted for overhead, and this resulted in a roughly 3:2 ratio of 22:15 OFDMA symbols for transmission of data. Data spectral efficiency is used as a system performance metric and it is obtained by normalizing the average sector throughput on each link by the bandwidth used specifically for TDD split 29:18. Downlink performance for WiMAX has been evaluated assuming non-ideal channel estimation with realistic CQICH feedback and scheduling with 20 users per sector.

A 2×2 antenna configuration with adaptive MIMO switching between space–time coding (STC) matrix **A** and spatial multiplexing (SM) matrix **B** is assumed on the DL. A 1×2 antenna configuration with collaborative spatial multiplexing (CSM) is assumed on the UL. Frequency reuse of 1 is modeled in this simulation model. This results in additional interference. XOHM commercial network uses a frequency reuse of 3, and therefore the level of interference is significantly lower. The average sector throughputs for reuse 3 are 11 and 4 Mbps for DL and UL, respectively. Tables 14.7 and 14.8 summarize the simulation parameters and results of a WiMAX network, respectively.

Table 14.7 Simulation parameters

Parameter	Assumption/value
Cellular layout	Hexagonal grid, 19 cell sites, $n = 1$ reuse, 3 sectors per site
Site-to-site distance	1.5 km
Distance-dependent path loss	$L = 130.19 + 37.6 \log 10(R)$, R (km)
Shadowing standard deviation	8 dB
Penetration loss	10 dB
Carrier frequency	2.5 GHz
Channel model	Mixture of ITU PedB3 km/h (60%)– VehA30 km/h (30%)–vehA120 km/h (10%)
BS TX power	43 dBm (10 MHz bandwidth)
MS Tx power	23 dBm
Bs, MS height	BS (32 m), MS (1.5 m)

Table 14.8 Summary of WiMAX performance

Performance metric	Downlink	Uplink	Assumptions
Peak user data rate (bps/Hz)	6.07 bps/Hz (38 Mbps @ 10 MHz)	2.19 bps/Hz (8.4 Mbps @ 10 MHz)	DL: 2 × 2 MIMO, 64QAM5/6 UL: 1 × 2 MIMO, 16QAM3/4 29:18 DL:UL split
Average sector throughput (bps/Hz/Cell)	1.37 (8.45 Mbps @ 10 MHz)	0.75 (2.87 Mbps @ 10 MHz)	DL: 2 × 2 MIMO, UL: 1 × 2 UL C-MIMO,
Round trip delay (ms)	20–35		
Connection setup time (ms)	85–150		20% HARQ retransmission probability
Handoff interruption time (ms)	50–60		

14.7 Future Work Toward IMT-Advanced

In October 2007, WiMAX was approved as an ITU approved technology in IMT-2000 bands at 2.6 GHz. However, to address the requirements for systems beyond IMT-2000, IEEE802.16 community has embarked on a new work item early 2008 named IEEE802.16m [6]. This work extends the capabilities of IEEE802.16 air interface in order to meet the requirements for future wireless applications and services. These requirements are defined by ITU-R IMT-Advanced [13].These requirements are summarized as (a) supporting both TDD and FDD modes, (b) providing at least twice spectral efficiency, (c) supporting wider bands to enable higher data rates, and (d) supporting higher speed up to 350 km/h.

The IMT-Advanced compliant technologies may be available for commercial deployment beyond 2011. The WG is currently in process of defining the technical specification of the 802.16 m air interface and is expected to finalize the standard by mid-to-late 2009. The main subjects being addressed in 802.16 m are as follows:

- Higher peak rates and average sector throughput ($> 4\times$ improvement over mobile WiMAX)
- Lower access and handover latency (<10 ms, <30 ms)
- Fair distribution of quality of service ($> 4\times$ better cell edge throughput)
- Higher mobility (up to 500 km/h)
- Better coverage (at least 3 dB better link budget)
- Larger VoIP capacity
- Enhanced broadcast and multicast services
- Enhanced location-based services
- Deployment flexibility
- Higher order single- and multi-user MIMO techniques (4×4)
- Integrated relay
- Interference management
- Standards-based techniques for multi-radio coexistence
- Multicarrier (multi-channel) support
- Self-organization and optimization

14.7.1 Conclusions

WiMAX is nothing but the beginning of mobilizing the Internet. It allows the customers to hold the power and knowledge of the entire Internet in their hand anywhere, anytime, with lighting up any number of devices, offering an always-connected experience.

As standardized OFDM technology, which has been already commercially deployed, WiMAX has a broader channel, better capacity, and throughput than any existing technology near term. It represents a flexible platform for the seamless integration of voice, data, and multimedia. However, the IEEE802.16 standard community and WiMAX forum have embarked new releases to achieve higher throughputs and meet the growing need for robust, interactive, mobile multimedia applications.

To this end, WiMAX has developed and exploited the most state-of-the-art technologies to achieve these objectives. In this chapter, very briefly we outlined some of these technologies. However, this community is continuing to enhance the performance of the technology to meet the ever-growing need of subscribers to have access to higher throughput and better services and applications.

References

1. XOHM Deployment, In: "XOHM, Intel and WiMAX Partners Celebrate New 4G Broadband Era in Baltimore," Oct. 2008, Available via `http://www.XOHM.com/`.
2. IEEE802.16-2004, In: "Local and Metropolitan Area Networks – Part 16: Air Interface for Fixed Broadband Wireless Access Systems," Oct. 2005, Available via `http://ieee802.org/16/`.
3. IEEE802.16-2005, In: "Part 16: Local and Metropolitan Area Networks–Air Interface for Fixed and Mobile Broadband Wireless Access Systems Amendment 2: Physical and Medium Access Control Layers for Combined Fixed and Mobile Operation in Licensed Bands and Corrigendum 1," Feb. 2006, Available via `http://ieee802.org/16/`.
4. IEEE802.16Rev2/D6, In: "Local and Metropolitan Area Networks–Part 16: Air Interface for Fixed and Mobile Broadband Wireless Access," Jul. 2008, Available via `http://ieee802.org/16/`.
5. WiMAX Forum, In: "WiMAX Forum Mobile System Profile Release 1.0, rev. 1.5.0," Nov. 2007, Available via `http://www.wimaxforum.org/`.
6. IEEE802.16m Broadband Wireless Access Working Group, In: "IEEE802.16m Evaluation Methodology," Dec. 2007, Available via `http://ieee802.org/16/`.
7. IEEE802.16j, In: "IEEE Standard for Local and Metropolitan Area Networks, Part 16: Air Interface for Fixed and Mobile Broadband Wireless Access Systems: Multihop Relay Specification, P802.16j/D3," Feb. 2008, Available via `http://ieee802.org/16/`.
8. WiMAX Network Architecture, In: "WiMAX Forum Network Architecture, Stage 3: Detailed Protocols and Procedures, NWG MAINTENANCE IMPLEMENTATION REVIEW DRAFT Methodology," Release 1, Version 1.3.0, Sept. 2008, Available via `www.wimaxforum.org/`.
9. WiMAX Network Architecture, In: "WiMAX End-to-End Network Systems Architecture, (Stage 2: Architecture Tenets, Reference Model and Reference Points)," Aug. 2006, Available via `www.wimaxforum.org/`.
10. WiMAX Forum White Paper, In: "Mobile WiMAX V Part I: A Technical Overview and Performance Evaluation," Mar. 2006, Available via `www.wimaxforum.org/`.
11. R. Prasad, In: "OFDM for Wireless Communications Systems," P:120-130, Artech House Publishers, 2004.
12. S. M. Alamouti, In: "A simple transmit diversity technique for wireless communications," IEEE J. Sel. Areas Commun., vol. 16, no. 8, pp. 1451–1458, Oct. 1998.
13. G. J. Foschini, In: "Layered space-time architecture for wireless communication in a fading environment when using multiple antennas," Bell Labs Tech. J., vol. 1, no. 2, pp. 41–59, Autumn 1996.
14. ITU Global Standard for International Mobile Telecommunications 'IMT-Advanced', `http://www.itu.int/ITU-R`.
15. M. Olfat and K. J. R. Liu, In: "Low Peak to Average Power Ratio Cyclic Golay Sequences for OFDM Systems," IEEE International Conference on Communication, ICC, Paris 2004.
16. W. Su, Z. Safar, M. Olfat, and K. J. R. Liu, In: "Obtaining Full Diversity Space-Frequency Codes from Space-Time Codes via Mapping," IEEE Trans. Signal Process, special issue on Signal Processing for Multiple-Input Multiple-Output (MIMO) Wireless Communications Systems, vol. 51, no. 11, Nov. 2003.
17. NGMN Alliance, In: "NGMN TE WP1 Radio Performance Evaluation, Phase 2 Report," Feb. 2008, Available via `http://www.NGMN.org`.
18. DOCSIS, In: "DOCSIS : Data Over Cable Service Interface Specifications," Available via `http://http://www.docsis.org/`.
19. D. J. Love and R. W. Heath, Jr., In: "Multimode Precoding for MIMO Wireless Systems," IEEE Trans. Signal Process., vol. 53, no. 1, pp. 3674–3687, Oct. 2005.

Chapter 15
An Overview of 3GPP Long-Term Evolution Radio Access Network

Sassan Ahmadi

15.1 Introduction

The growing demand for mobile Internet and wireless multimedia applications such as Internet browsing, interactive gaming, mobile TV, video and audio streaming has motivated development of broadband wireless access technologies in recent years. As a result, the 3rd Generation Partnership Project (3GPP) initiated the work on the long-term evolution (LTE) in late 2004. LTE will ensure 3GPPs competitive edge over other cellular technologies. The evolved UMTS terrestrial radio access network (E-UTRAN) substantially improves end-user throughputs, sector capacity and reduces user-plane and control-plane latencies, bringing significantly improved user experience with full mobility. With the emergence of Internet Protocol (IP) as the protocol of choice for carrying all types of traffic, LTE is expected to provide support for IP-based traffic with end-to-end quality of service (QoS). Voice traffic will be supported mainly as voice over IP (VoIP) enabling better integration with other multimedia services. Initial deployments of LTE are expected by 2010 and commercial availability on a larger scale will likely happen a few years later.

Unlike its predecessors, which were developed within the framework of Release 99 UMTS architecture, 3GPP has specified the evolved packet core (EPC) architecture to support the E-UTRAN through reduction in the number of network elements and simplification of functionality but most importantly allowing for connections and handover to other fixed and wireless access technologies, providing the network operators the ability to deliver a seamless mobility experience. 3GPP has set aggressive performance requirements for LTE that rely on improved physical layer technologies such as orthogonal frequency division multiplexing (OFDM) and single-user and/or multi-user multiple-input multiple-output (MIMO) techniques,

Sassan Ahmadi
Intel Corporation, Mail-Stop: JF3-336,
2111 NE 25th Ave, Hillsboro, OR 97124, USA
e-mail: `sassan.ahmadi@intel.com`

V. Tarokh (ed.), *New Directions in Wireless Communications Research*,
DOI 10.1007/978-1-4419-0673-1_15,
© Springer Science+Business Media, LLC 2009

and streamlined Layer 2 protocols and functionalities. The main objectives of LTE
are to minimize the system and user equipment (UE) complexities, to allow flexible spectrum deployment in the existing or new frequency bands, and to enable
coexistence with other 3GPP radio access technologies.

In this chapter a comprehensive overview of 3GPP LTE radio access technology and E-UTRAN using a top-down systematic approach will be provided. All key
functional and performance features of LTE protocols will be discussed in a hierarchical manner starting from the overall network architecture and further focusing
on various elements of the radio access network. The latest link and system-level
performance results for LTE will be further provided.

15.1.1 Chronology of 3GPP Air Interface Technology Development

The 3GPP long-term evolution symbolizes the migration of the universal mobile
telecommunication system (UMTS) family of standards from systems that supported both circuit-switched and packet-switched voice/data communications to an
all-IP, packet-only system. The development of the LTE air interface is closely coupled with 3GPP system architecture evolution (SAE) project to define the overall system architecture and evolved packet core network. Figure 15.1 summarizes
the history of the 3GPP standard development in the past decade. It is shown
that the 3GPP standards have evolved toward higher performance and data rates,

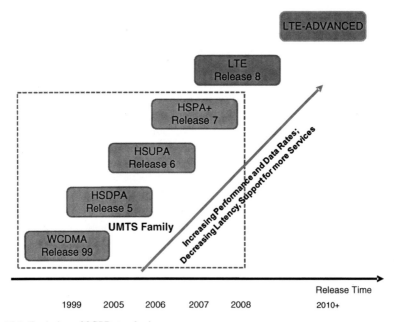

Fig. 15.1 Evolution of 3GPP standards

lower access latencies, and increasing capability to support emerging wireless applications. To achieve higher downlink and uplink data rates, UMTS operators today are upgrading their 3G networks with high-speed downlink packet access (HSDPA), which was specified in 3GPP Release 5, and high-speed uplink packet access (HSUPA), which was specified in 3GPP Release 6. The HSDPA and HSUPA, collectively known as HSPA, have been further upgraded in Releases 7 and 8 with additional features such as higher order modulations in the downlink and uplink, downlink open-loop MIMO, improvements of Layer 2 protocols, and continuous connectivity for packet data users. Release 8 of 3GPP air interfaces specifies LTE and SAE. In September 2007, the first set of LTE physical layer specifications was released. Completion of the rest of the specifications is expected by the end of 2008 followed by the user equipment conformance testing specifications. The 3GPP is currently working toward development of an advanced standard specification to address the IMT-Advanced requirements and services for the fourth generation of cellular systems [19].

15.1.2 3GPP LTE System Requirements

The evolved UMTS terrestrial radio access (E-UTRA) is expected to efficiently support mobile Internet as well as a variety of wireless applications such as HTTP, FTP, real-time and non-real-time video streaming, VoIP, interactive gaming. Therefore, LTE has been designed to provide very high data rate and low air-link access latency in order to satisfy the requirements for the existing and emerging applications. The bandwidth capability of an LTE-compliant UE is much higher than that of previous 3GPP releases, enabling much higher throughput and peak data rates in the downlink and uplink (i.e., 20 MHz for both transmission and reception in Release 8). This gives more flexibility to the service providers to adapt their services depending on the amount of available spectrum or the ability to start with limited spectrum for lower cost and later increase the spectrum for extra capacity.

A summary of LTE system requirements is provided in Table 15.1 [1]. Beyond these metrics, LTE has also targeted at minimizing complexity and power consumption and cost-effective migration from UMTS systems. Enhanced multicast services, enhanced support for end-to-end quality of service (QoS), and minimization of the number of options and redundant features in the architecture have also been taken into account.

The spectral efficiency in the LTE downlink (DL) is three to four times of that of HSDPA, while in the uplink (UL), it is two to three times of that of HSUPA assuming a loaded network in both cases. The handover procedure within LTE system is intended to minimize interruption time to less than that of circuit-switched handovers in 2G networks. Moreover the handovers to 2G/3G systems from LTE are designed to be seamless. Another distinct feature of LTE relative to the previous releases is the range of mobility that the system can support. The LTE system has been designed to provide optimal performance up to 15 km/h and to maintain connectivity

Table 15.1 3GPP LTE system requirements [1]

Metric	Requirement
Peak data rate	Downlink: 100 Mb/s (5 bps/Hz) within a 20 MHz downlink spectrum allocation; uplink: 50 Mb/s (2.5 bps/Hz) within a 20 MHz uplink spectrum allocation
Mobility	Optimized for low mobile speed from 0 to 15 km/h, higher mobile speed between 15 and 120 km/h should be supported with high performance. Mobility across the cellular network shall be maintained at speeds from 120 to 350 km/h (or even up to 500 km/h depending on the frequency band)
Control-plane latency	Transition time of less than 100 ms from camped state to active state, transition time of less than 50 ms between dormant state and active state
User-plane latency	Less than 5 ms in unload condition (i.e., single user with single data stream) for small IP packets
Average user throughput/MHz	Downlink: three to four times Release 6 HSDPA
	Uplink: two to three times Release 6 enhanced uplink
Spectrum efficiency (bits/sc/Hz/site) (loaded network)	Downlink: three to four times Release 6 HSDPA
	Uplink: two to three times Release 6 enhanced uplink
Coverage	Throughput, spectrum efficiency, and mobility targets should be met for 5 km cells, and with a slight degradation for 30 km cells. Cells range up to 100 km should not be precluded

with users that move up to 350 km/h (up to 500 km/h depending on the frequency band and deployment scenarios). The support of higher mobility motivated shorter transmission time intervals (TTI), faster feedback mechanisms and link adaptation compared to the legacy UMTS systems [1].

The user plane (U-plane) and control plane (C-plane) consists of a set of functions for processing and transmission of user data and control/signaling, respectively and will be described in the next sections.

15.2 Overall Network Architecture

The evolved UMTS terrestrial radio access network consists of evolved eNBs or equivalently E-UTRA base stations, providing the E-UTRA user-plane and control-plane protocol terminations toward the user equipment (see Fig. 15.2). Note that the radio network controller (RNC) functions have been included in the eNB to reduce the architectural complexity and further reduce the latency across the network. The eNBs are interconnected with each other by means of the X2 interface [6, 11]. The eNBs are also connected by means of the S1 interface to the EPC or more specif-

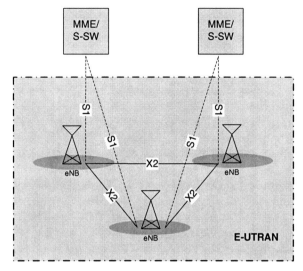

Fig. 15.2 E-UTRAN architecture ([6])

ically to the mobility management entity (MME) through the S1-MME reference point and to the serving gateway (S-GW) via the S1-U interface (see 3GPP TR 23.882 for more information). The S1 interface supports a multipoint connection among MMEs/serving gateways and eNBs. The E-UTRAN overall architecture is described in [6, 11].

Some general principles taken into consideration in the design of E-UTRAN architecture and the E-UTRAN interfaces are as follows [6]:

- Signaling and data transport networks are logically separated.
- E-UTRAN and EPC functions are fully separated from transport functions. Addressing schemes used in E-UTRAN and EPC are not associated with the addressing schemes of transport functions. The fact that some E-UTRAN or EPC functions reside in the same equipment as some transport functions does not make the transport functions part of the E-UTRAN or the EPC.
- The mobility for RRC connection is fully controlled by the E-UTRAN.
- The interfaces should be based on a logical model of the entity controlled through this interface.
- One physical network element can implement multiple logical nodes.

The eNB typically performs the following functions [6]:

- Radio resource management (RRM) that includes radio bearer control, radio admission control, connection management, dynamic allocation of resources to UEs in both uplink and downlink (i.e., scheduling).
- Header compression and encryption of user payloads.
- Selection of an MME at UE attachment when no routing to an MME can be determined from the information provided by the UE.
- Routing of U-plane data toward S-GW.

- Scheduling and transmission of paging messages (originated from the MME).
- Scheduling and transmission of broadcast information (originated from the MME).
- Measurement and reporting for support of mobility and scheduling.

The development of the SAE, whose objective was to set a framework for an evolution or migration of the 3GPP system to a higher data rate, lower latency, packet-optimized system that supports multiple RATs began in late 2004 and almost shortly after the initiation of the LTE project. The focus of this work was on the packet-switched domain with the assumption that voice services are supported in this domain.

The MME is the key control node for the LTE access network. It is responsible for idle mode UE tracking and paging procedure including retransmissions. The MME functions include non-access stratum (NAS) signaling and NAS signaling security, inter-core-network signaling for mobility between 3GPP access networks, idle mode UE accessibility (including control and execution of paging retransmission), tracking area list management (for UE in idle and active modes), packet data network (PDN) gateway and S-GW selection, MME selection for handovers with MME change, roaming, authentication of the users, and bearer management functions including dedicated bearer establishment [6, 11]. The NAS signaling terminates at the MME and it is also responsible for generation and allocation of temporary identities to the UEs. It verifies the authorization of the UE to camp on the service providers public land mobile network and enforces UE roaming restrictions.

The serving gateway (S-GW) routes and forwards user data packets, while also acting as the mobility anchor for the user plane during inter-eNB handovers and as the anchor for mobility between LTE and other 3GPP technologies. The S-GW functions include local mobility anchor point for inter-eNB handover, mobility anchoring for inter-3GPP mobility, E-UTRAN idle mode downlink packet buffering and initiation of network-triggered service request procedure, lawful interception, packet routing and forwarding, and transport-level packet marking in the uplink and the downlink [6, 11].

The packet data network gateway (P-GW) provides connectivity of the UE to external packet data networks by being the point of exit and entry of traffic for the UE. A UE may have simultaneous connectivity with more than one P-GW for accessing multiple packet data networks. The P-GW functions include per-user packet filtering, lawful interception, UE IP address allocation, and transport-level packet marking in the downlink [6, 11].

15.3 LTE Protocol Structure

In this section, we describe the functions of the different protocol layers and their location in the LTE protocol structure. Figures 15.3 and 15.4 show the user-plane and control-plane protocol stacks, respectively. In the C-plane, the NAS functional block is used for network attachment, authentication, setting up bearers, and mobility

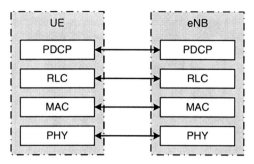

Fig. 15.3 User-plane protocol stack ([6])

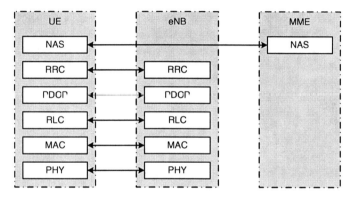

Fig. 15.4 Control-plane protocol stack ([6])

management. All NAS messages are ciphered and integrity protected by the MME and UE. The radio resource control (RRC) layer in the eNB makes handover decisions based on neighbor cell measurements reported by the UE, performs paging of the users over the air interface, broadcasts system information, controls UE measurement and reporting functions such as the periodicity of channel quality indicator (CQI) reports and further allocates cell-level temporary identifiers to active users. It also executes transfer of UE context from the serving eNB to the target eNB during handover and performs integrity protection of RRC messages. The RRC layer is responsible for setting up and maintenance of radio bearers. Note that RRC layer in 3GPP protocol hierarchy is considered as Layer 3 [6].

In the U-plane, the Packet Data Convergence Protocol (PDCP) layer is responsible for compressing or decompressing the headers of user-plane IP packets using robust header compression (RoHC) to enable efficient use of air interface resources [6, 9]. This layer also performs ciphering of both user-plane and control-plane traffic. Because the NAS messages are carried in RRC, they are effectively double ciphered and integrity protected, once at the MME and again at the eNB.

The radio link control (RLC) layer is used to format and transport traffic between the UE and the eNB [6, 8]. The RLC provides three different reliability modes for data transport, i.e., acknowledged mode (AM), unacknowledged mode (UM), and transparent mode (TM). The unacknowledged mode is suitable for transport of real-time services since such services are delay sensitive and cannot tolerate delay due to retransmissions. The acknowledged mode is appropriate for non-real-time services such as file transfers. The transparent mode is used when the size of packet data units is known in advance such as for broadcasting system configuration information. The RLC layer also provides sequential delivery of service data units to the upper layers and eliminates duplicate packets from being delivered to the upper layers. It may also segment the service data units. Furthermore, there are two levels of retransmissions for providing reliability, the hybrid automatic repeat request (HARQ) at the MAC layer and ARQ at the RLC layer. The ARQ is required to handle residual errors that are not corrected by HARQ that is kept simple by the use of a single bit error-feedback mechanism. An N-process stop-and-wait HARQ is employed that has asynchronous retransmissions in the DL and synchronous retransmissions in the UL. Synchronous HARQ means that the retransmissions of HARQ sub-packets occur at predefined periodic intervals. Hence, no explicit signaling is required to indicate to the receiver the retransmission schedule. Asynchronous HARQ offers the flexibility of scheduling retransmissions based on air interface conditions (i.e., scheduling gain).

The description of medium access control (MAC) and physical layers (PHY) will be provided in more detail in the next sections [2–5, 7].

15.4 Overview of the LTE Physical Layer

The main functional elements of the physical layer processing are described in the following sections [2–6].

15.4.1 Multiple Access Schemes

LTE uses asymmetric multiple access schemes in the downlink and uplink. The multiple access schemes for the LTE physical layer are based on orthogonal frequency division multiple access (OFDMA) with a cyclic prefix (CP) in the downlink and single-carrier frequency division multiple access (SC-FDMA) with CP in the uplink. OFDMA technique is particularly suited for frequency-selective channels and high data rate. It transforms a wideband frequency-selective channel into a set of parallel flat fading narrowband channels. This ideally allows the receiver to perform a less complex equalization process in frequency domain, i.e., single-tap frequency-domain equalization.

15.4.1.1 Downlink multiple access (OFDMA)

The downlink transmission scheme is based on conventional OFDMA using a cyclic prefix. A 10 ms radio frame is divided into ten equally sized subframes. Each subframe is further divided into two slots of 0.5 ms length. The basic transmission parameters in the downlink are specified in Table 15.2. The CP length is chosen to be longer than the maximum delay spread in the radio channel. For LTE, the normal CP length has been set to 4.69 μs, enabling the system to tolerate delay variations due to propagation over cells up to 1.4 km. Note that the insertion of CP increases the Layer 1 overhead and hence reducing the overall throughput. To provide enhanced multicast and broadcast services, LTE has the capability to transmit multicast and broadcast over a single frequency network (MBSFN), where a time-synchronized common waveform is transmitted from multiple eNBs, allowing macro-diversity combining of multi-cell transmissions at the UE. The cyclic prefix is utilized to cover the difference in the propagation delays, which makes the MBSFN transmission appear to the UE as a transmission from a single large cell. Transmission on a dedicated carrier for MBSFN with the possibility to use a longer CP with a subcarrier bandwidth of 7.5 kHz is supported as well as transmission of MBSFN on a carrier with both MBMS and unicast transmissions using time division multiplexing [3].

Table 15.2 LTE OFDMA parameters [3, 16]

Parameter		Value					
Channel bandwidth (MHz)		1.4	3	5	10	15	20
Number of resource blocks		6	15	25	50	75	100
Number of occupied subcarriers		72	180	300	600	900	1200
IDFT/FFT size		128	256	512	1024	1536	2048
Subcarrier spacing Δf (kHz)		15 (7.5)					
Sampling rate (MHz)		1.92	3.84	7.68	15.36	23.04	30.72
Samples/slot		960	1920	3840	7680	11520	15360
CP size (μs)	Normal CP (Δf=15 kHz)	5.21 (first symbol of the slot) 4.69 (other symbols of the slot) 7 symbols/slot					
	Extended CP (Δf=15 kHz)	16.67 6 symbols/slot					
	Extended CP (Δf=7.5 kHz)	33.33 3 symbols/slot					

15.4.2 Operating Frequencies and Bandwidths

E-UTRA is designed to operate in the frequency bands defined in Table 15.3. The requirements are defined for 1.4, 3, 5, 10, 15, and 20 MHz bandwidth with a specific configuration in terms of number of resource blocks (see Table 15.2).

Figure 15.5 illustrates the relationship between the total channel bandwidth and the transmission bandwidth, i.e., the number of resource blocks. The channel raster is 100 kHz which means the center frequency must be a multiple of 100 kHz. To support transmission in paired and unpaired spectrum, two duplexing schemes are

Table 15.3 LTE band classes [20]

E-UTRA band	Uplink (UL) eNB receive UE transmit F_{UL_low} - F_{UL_high}			Downlink (DL) eNB transmit UE receive F_{DL_low} - F_{DL_high}			UL–DL band separation F_{DL_low} - F_{UL_high}	Duplex mode
1	1920 MHz	-	1980 MHz	2110 MHz	-	2170 MHz	130 MHz	FDD
2	1850 MHz	-	1910 MHz	1930 MHz	-	1990 MHz	20 MHz	FDD
3	1710 MHz	-	1785 MHz	1805 MHz	-	1880 MHz	20 MHz	FDD
4	1710 MHz	-	1755 MHz	2110 MHz	-	2155 MHz	355 MHz	FDD
5	824 MHz	-	849 MHz	869 MHz	-	894MHz	20 MHz	FDD
6	830 MHz	-	840 MHz	875 MHz	-	885 MHz	35 MHz	FDD
7	2500 MHz	-	2570 MHz	2620 MHz	-	2690 MHz	50 MHz	FDD
8	880 MHz	-	915 MHz	925 MHz	-	960 MHz	10 MHz	FDD
9	1749.9 MHz	-	1784.9 MHz	1844.9 MHz	-	1879.9 MHz	60 MHz	FDD
10	1710 MHz	-	1770 MHz	2110 MHz	-	2170 MHz	340 MHz	FDD
11	1427.9 MHz	-	1452.9 MHz	1475.9 MHz	-	1500.9 MHz	23 MHz	FDD
12	[TBD]	-	[TBD]	[TBD]	-	[TBD]	[TBD]	FDD
13	777 MHz	-	787 MHz	746 MHz	-	756 MHz	21 MHz	FDD
14	788 MHz	-	798 MHz	758 MHz	-	768 MHz	20 MHz	FDD
⋮								
33	1900 MHz	-	1920 MHz	1900 MHz	-	1920 MHz	N/A	TDD
34	2010 MHz	-	2025 MHz	2010 MHz	-	2025 MHz	N/A	TDD
35	1850 MHz	-	1910 MHz	1850 MHz	-	1910 MHz	N/A	TDD
36	1930 MHz	-	1990 MHz	1930 MHz	-	1990 MHz	N/A	TDD
37	1910 MHz	-	1930 MHz	1910 MHz	-	1930 MHz	N/A	TDD
38	2570 MHz	-	2620 MHz	2570 MHz	-	2620 MHz	N/A	TDD
39	1880 MHz	-	1920 MHz	1880 MHz	-	1920 MHz	N/A	TDD
40	2300 MHz	-	2400 MHz	2300 MHz	-	2400 MHz	N/A	TDD

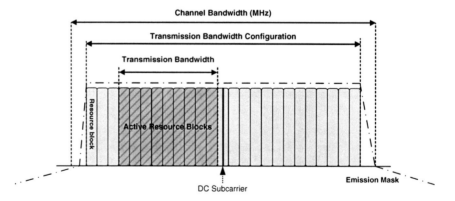

Fig. 15.5 Illustration of the relationship between channel bandwidth and transmission bandwidth (3GPP)

supported: frequency division duplex (FDD), allowing both full and half duplex terminal operation, and time division duplex (TDD). Table 15.3 shows the band classes where the LTE systems can be deployed.

15.4.2.1 Uplink multiple access (SC-FDMA)

The basic transmission scheme in the uplink is single-carrier transmission (SC-FDMA) with cyclic prefix to achieve uplink inter-user orthogonality and to enable efficient frequency domain equalization at the receiver side. The frequency domain generation of the SC-FDMA signal, also known as DFT-spread OFDM, is similar to OFDMA and is illustrated in Fig. 15.6. This allows for a relatively high degree of commonality with the downlink OFDMA baseband processing using the same parameters, e.g., clock frequency, subcarrier spacing, FFT/IFFT size. The use of SC-FDMA in the uplink is mainly due to relatively inferior peak-to-average power ratio (PAPR) properties of OFDMA that result in worse uplink coverage compared to SC-FDMA. The PAPR characteristics are important for cost-effective design of UE power amplifiers.

The subcarrier mapping determines which part of the spectrum is used for transmission by inserting a suitable number of zeros at the upper and/or lower end shown in Fig. 15.6. Between each DFT output sample $L - 1$ zeros are inserted. A mapping with $L = 1$ (as shown in Fig. 15.6) corresponds to localized transmissions, i.e., transmissions where the DFT outputs are mapped to consecutive subcarriers. There are two subcarrier mapping schemes (localized and distributed) that could be used in the uplink. However, LTE only specifies localized subcarrier mapping in the uplink.

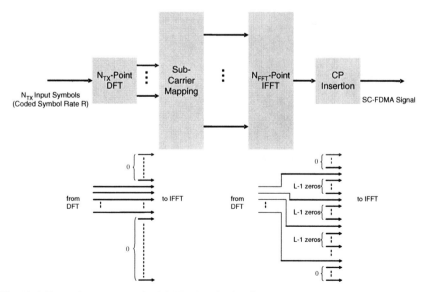

Fig. 15.6 Transmitter structure for SC-FDMA with localized and distributed subcarrier mapping schemes. Note that $N_{TX} < N_{FFT}$

15.4.2.2 Overview of the SC-FDMA concept

In order to demonstrate the similarities and differences between OFDMA and
SC-FDMA baseband processing, let us assume that one wishes to transmit a sequence of eight QPSK symbols as shown in Fig. 15.7 [15]. In the OFDMA case assuming $N_{TX} = 4$, four QPSK symbols would be processed in parallel, each of them
modulating its own subcarrier at the appropriate QPSK phase. After one OFDMA
symbol period, a guard period or cyclic prefix is inserted to mitigate the multi-path
effects. For SC-FDMA, each symbol is transmitted sequentially. With $N_{TX} = 4$,
there are four data symbols transmitted in one SC-FDMA symbol period. The higher
rate data symbols require four times the bandwidth and so each data symbol occupies 60 kHz of spectrum assuming a subcarrier spacing of 15 kHz. After four data
symbols, the CP is inserted. Note the OFDMA and SC-FDMA symbol periods are
the same [15, 18].

15.4.3 Frame Structure

Downlink and uplink transmissions are organized into radio frames with 10 ms duration. LTE supports two radio frame structures: Type 1, applicable to FDD duplex
scheme, and Type 2, applicable to TDD duplex scheme.

Frame structure Type 1 is illustrated in Fig. 15.8. Each 10 ms radio frame is
divided into 10 equally sized subframes. Each subframe consists of two equally
sized slots. For FDD, 10 subframes are available for downlink transmission and 10
subframes are available for uplink transmissions in each radio frame. Uplink and

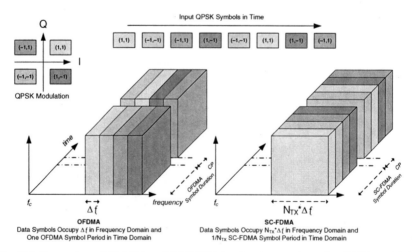

Fig. 15.7 Comparison of OFDMA and SC-FDMA baseband processing using QPSK modulation
with $N_{TX} = 4$ subcarriers. The input QPSK symbols, shown on top, are fed into the modulator
from left to right ([15])

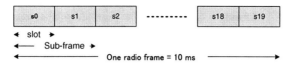

Fig. 15.8 Frame structure Type 1

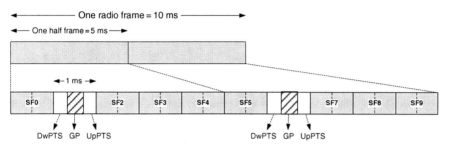

Fig. 15.9 Frame structure Type 2 (for 5 ms switch-point periodicity)

downlink transmissions are separated in the frequency domain. The transmission time interval (TTI) of the downlink/uplink is 1 ms [3,6].

Frame structure Type 2 is illustrated in Fig. 15.9. Each 10 ms radio frame consists of two half-frames of 5 ms each. Each half-frame consists of eight slots of 0.5 ms length and three special fields: downlink pilot time slot (DwPTS), guard period (GP), and uplink pilot time slot (UpPTS). The length of DwPTS and UpPTS is configurable subject to the total length of DwPTS, GP, and UpPTS being equal to 1ms (see [3] for more details). Both 5 and 10 ms switching-point periodicities are supported. The first subframe in all configurations and the sixth subframe in configuration with 5 ms switching-point periodicity consist of DwPTS, GP, and UpPTS. The sixth subframe in configuration with 10 ms switching-point periodicity consists of DwPTS only. All other subframes consist of two equally sized slots.

For TDD systems, the GP is reserved for downlink to uplink transition. Other subframes/fields are assigned for either downlink or uplink transmission as shown in Table 15.4. Uplink and downlink transmissions are separated in the time domain.

15.4.4 Physical Resource Blocks

The smallest time–frequency resource unit used for downlink/uplink transmission is called a resource element, defined as one subcarrier over one symbol [3]. For both TDD and FDD duplex schemes as well as in both downlink and uplink, a group of 12 subcarriers contiguous in frequency over one slot in time form a resource block

Table 15.4 Various permissible uplink/downlink configurations in frame structure Type 2 where "S" denotes the special subframe [3].

Configuration	Switch-point periodicity (ms)	Subframe number									
		0	1	2	3	4	5	6	7	8	9
0	5	DL	S	UL	UL	UL	DL	S	UL	UL	UL
1	5	DL	S	UL	UL	DL	DL	S	UL	UL	DL
2	5	DL	S	UL	DL	DL	DL	S	UL	DL	DL
3	10	DL	S	UL	UL	UL	DL	DL	DL	DL	DL
4	10	DL	S	UL	UL	DL	DL	DL	DL	DL	DL
5	10	DL	S	UL	DL	DL	DL	DL	DL	DL	DL
6	10	DL	S	UL	UL	UL	DL	S	UL	UL	DL

(RB) as shown in Fig. 15.10 (corresponding to one slot in the time domain and 180 kHz in the frequency domain). Transmissions are allocated in units of RB. One downlink/uplink slot using the normal CP length contains seven symbols. There are 6, 15, 25, 50, 75, and 100 RBs corresponding to 1.4, 3, 5, 10, 15, 20 MHz channel bandwidths, respectively [14–16]. Note that the resource block size is the same for all bandwidths.

15.4.5 Modulation and Coding

The baseband modulation schemes supported in the downlink and uplink of LTE are QPSK, 16QAM, and 64QAM. The channel coding scheme for transport blocks in LTE is turbo coding similar to UTRA, with a coding rate of $R = 1/3$, two eight-state constituent encoders and a contention-free quadratic permutation polynomial (QPP) turbo code internal interleaver [4]. Trellis termination is performed by taking the tail bits from the shift register feedback after all information bits are encoded. Tail bits are padded after the encoding of information bits. Before the turbo coding, transport blocks are segmented into octet-aligned segments with a maximum information block size of 6144 bits. Error detection is supported by the use of 24-bit CRC. The permissible modulation schemes for various physical channels in the downlink and uplink of LTE are shown in Table 15.5 [3].

15.4.6 Physical Channel Processing

Figure 15.11 illustrates different stages of LTE physical channel processing in the downlink and uplink. In the downlink, the coded bits in each of the codewords are scrambled for transmission on a physical channel. The scrambled bits are modulated to generate complex-valued modulation symbols that are later mapped

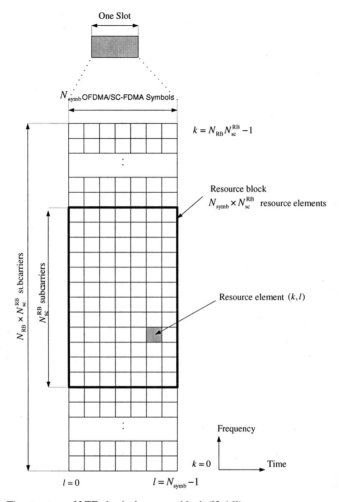

Fig. 15.10 The structure of LTE physical resource block ([3, 16])

to the complex-valued modulation symbols onto one or several transmission layers. The complex-valued modulation symbols on each layer are pre-coded for transmission and are further mapped to resource elements for each antenna port. The complex-valued time domain OFDMA signal for each antenna port is then generated following these stages [6].

In the uplink, the baseband signal is processed by scrambling the input-coded bits and then by modulation of scrambled bits to generate complex-valued symbols. The complex-valued modulation symbols are transform-precoded (DFT-based precoding) to generate complex-valued symbols that are later mapped to resource elements. The complex-valued time domain SC-FDMA signal for each antenna port is then generated.

Table 15.5 Physical channels and signals and their corresponding modulation schemes for the LTE downlink and uplink [3]

Downlink	Physical channels	Physical broadcast channel (PBCH)	QPSK
		Physical downlink control channel (PDCCH)	QPSK
		Physical downlink shared channel (PDSCH)	QPSK, 16QAM, 64QAM
		Physical multicast channel (PMCH)	QPSK, 16QAM, 64QAM
		Physical control format indicator channel (PCFICH)	QPSK
		Physical HARQ indicator channel (PHICH)	BPSK modulated on I and Q with the spreading factor 2 or 4 orthogonal codes
	Physical signals	Reference signals (RS)	Complex I+jQ pseudo-random sequence of length 31 Gold sequence derived from cell ID
		Primary synchronization signal	One of three Zadoff-Chu sequences
		Secondary synchronization signal	Two 31-bit BPSK M-sequence
Uplink	Physical channels	Physical uplink shared channel (PUSCH)	QPSK, 16QAM, 64QAM
		Physical uplink control channel (PUCCH)	BPSK, QPSK
		Physical random access channel (PRACH)	uth Root Zadoff-Chu sequence
	Physical signals	Demodulation reference signal (DRS) (narrowband)	Zadoff-Chu
		Sounding reference signal (SRS) (wideband)	Based on Zadoff-Chu

15.4.7 Reference Signals

Three types of downlink reference signals are defined: cell-specific reference signals associated with non-MBSFN transmission; MBSFN reference signals associated with MBSFN transmission, and UE-specific reference signals. There is one reference signal transmitted per downlink antenna port [3]. The cell-specific downlink reference signals consist of predetermined reference symbols that are inserted in the first and the third before last OFDMA symbol of each slot and are used for DL channel estimation [3, 6]. The exact sequence is derived from cell identifiers. The number of downlink antenna ports equals 1, 2, or 4. The reference signal sequence is derived from a pseudo-random sequence and results in a QPSK-type con-

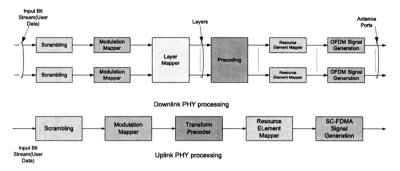

Fig. 15.11 Overview of physical channel processing ([3])

stellation. Cell-specific frequency shifts are applied when mapping the reference signal sequence to the subcarriers. The two-dimensional reference signal sequence is generated as the symbol-by-symbol product of a two-dimensional orthogonal sequence and a two-dimensional pseudo-random sequence. There are three different two-dimensional orthogonal sequences and 168 different two-dimensional pseudo-random sequences. Each cell identity corresponds to a unique combination of one orthogonal sequence and one pseudo-random sequence, thus allowing for 504 unique cell identities (i.e., 168 cell identity groups with three cell identities in each group).

The downlink MBSFN reference signals consist of known reference symbols inserted every other subcarrier in the third, seventh, and eleventh OFDMA symbols of subframe in case of 15 kHz subcarrier spacing and extended cyclic prefix (Fig. 15.12).

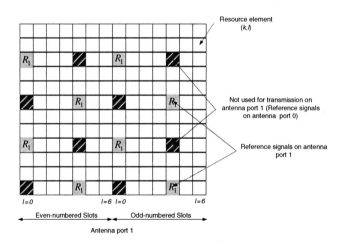

Fig. 15.12 Downlink reference signals for antenna port 1 when two antenna ports are used

The uplink reference signals, used for channel estimation for coherent demodulation, are transmitted in the fourth SC-FDMA symbol in each slot assuming normal CP. LTE uses dedicated reference signals in the uplink. The uplink reference signal sequence length equals the size (number of subcarriers) of the assigned resource. The uplink reference signals are based on Zadoff-chu sequences that are either truncated or cyclically extended to the desired length. There are two types of uplink reference signals: (1) the demodulation reference signal is used for channel estimation in the eNB receiver in order to demodulate control and data channels. It is located on the fourth symbol in each slot (for normal cyclic prefix) and spans the same bandwidth as the allocated uplink data and (2) the sounding reference signal provides uplink channel quality information as a basis for scheduling decisions in the base station. The UE sends a sounding reference signal in different parts of the bandwidths where no uplink data transmission is available. The sounding reference signal is transmitted in the last symbol of the subframe. The configuration of the sounding signal, e.g., bandwidth, duration, and periodicity, is given by higher layers. Figure 15.13 shows the reference signals corresponding to antenna port 1 in the downlink (assuming two antenna ports). The physical resources at the locations of the reference signals of another antenna port are not used for data transmission on the other antenna ports.

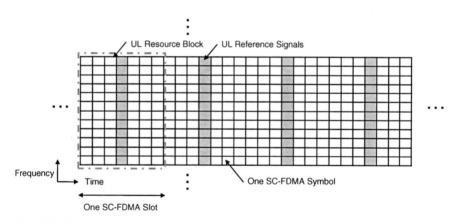

Fig. 15.13 Uplink physical resource blocks and reference signals

15.4.8 Physical Control Channels

As shown in Fig. 15.14, the physical downlink control channel (PDCCH) is located in the first n OFDMA symbols of each subframe where n is 3 and consists of transport format and resource allocation related to DL-SCH and PCH, and HARQ information related to DL-SCH as well as transport format, resource allocation, and HARQ information related to UL-SCH. The transmission of control signaling from

these groups is mutually independent. Multiple physical downlink control channels
are supported and a UE monitors a set of control channels [3, 6, 14, 15].

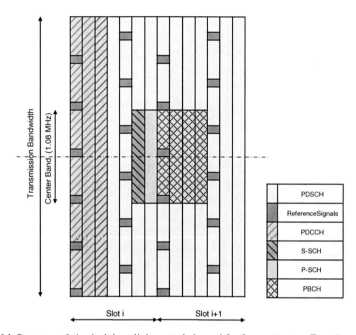

Fig. 15.14 Structure of physical downlink control channel for frame structure Type 1

Control channels are formed by aggregation of control channel elements (CCE);
each control channel element consists of a set of resource elements. Different code
rates for the control channels are realized by aggregating different numbers of control channel elements. QPSK modulation is used for all control channels.

There is an implicit relationship between the uplink resources used for dynamically scheduled data transmission, or the DL control channel used for assignment,
and the downlink ACK/NACK resource used for feedback. The physical uplink control channel (PUCCH) is mapped to a control channel resource in the uplink as
shown in Fig. 15.15. A control channel resource is defined by a code and two resource blocks, consecutive in time, with hopping at the slot boundary. Depending
on the presence or the absence of uplink timing synchronization, the uplink physical
control signaling can differ. In the case of time synchronization being present, the
out-of-band control signaling consists of channel quality indicator (CQI), acknowl-
edge/negative acknowledge (ACK/NACK), and scheduling request (SR). Note that
a UE only uses PUCCH when it does not have any data to transmit on PUSCH. If a
UE has data to transmit on PUSCH, it would multiplex the control information with
data on PUSCH.

By the use of uplink frequency hopping on PUSCH, frequency diversity effects
can be exploited and interference can be averaged. The UE derives the uplink resource allocation and frequency hopping information from the uplink scheduling

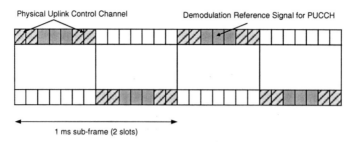

Fig. 15.15 Structure of physical uplink control channel for frame structure Type 1

grant. The downlink control information format 0 is used on PDCCH to convey the uplink scheduling grant. LTE supports both intra-subframe and inter-subframe frequency hopping. It is configured per cell by higher layers. In intra-subframe hopping, the UE hops to another frequency allocation from one slot to another within one subframe. In inter-subframe hopping, the frequency resource allocation changes from one subframe to another. The CQI reports are provided to the scheduler by the UE, measuring the current channel conditions. If MIMO transmission is used, the CQI includes necessary MIMO-related feedback.

The HARQ feedback in response to downlink data transmission consists of a single ACK/NACK bit per HARQ process. The PUCCH resources for SR and CQI reporting are assigned and can be revoked through RRC signaling. An SR is not necessarily assigned to UEs acquiring synchronization through the RACH (i.e., synchronized UEs may or may not have a dedicated SR channel). The PUCCH resources for SR and CQI are lost when the UE is no longer synchronized.

In LTE, uplink control signaling includes ACK/NACK, CQI, scheduling request indicator, and MIMO codeword feedback. When users have simultaneous uplink data and control transmission, control signaling is multiplexed with data prior to the DFT to preserve the single-carrier property in uplink transmission. In the absence of uplink data transmission, this control signaling is transmitted in a reserved frequency region on the band edge as shown in Fig. 15.15. Note that additional control regions may be defined as needed. Allocation of control channels with their small occupied bandwidth to band edge resource blocks reduces out-of-band emissions caused by data resource allocations on inner band resource blocks and maximizes the frequency diversity for control channel allocations while preserving the single-carrier property of the uplink waveform.

15.4.9 Physical Random Access Channel

The physical layer random access preamble, illustrated in Fig. 15.16, consists of a cyclic prefix of length TCP and a sequence part of length TSEQ. There are five random access preamble formats specified by the standard [3] where the parameter values depend on the frame structure and the random access configuration.

Higher layers control the preamble format. Each random access preamble occupies a bandwidth corresponding to six consecutive resource blocks for both frame structures. The transmission of a random access preamble, if triggered by the MAC layer, is restricted to certain time and frequency resources. These resources are enumerated in increasing order of the subframe number within the radio frame and the physical resource blocks in the frequency domain such that index 0 correspond to the lowest numbered physical resource block and subframe within the radio frame. The random access preambles are generated from Zadoff-Chu sequences with zero correlation zone. The network configures the set of preamble sequences the UE is allowed to use. There are 64 preambles available in each cell. The set of 64 preamble sequences in a cell is found by including first, in the order of increasing cyclic shift, all the available cyclic shifts of a root Zadoff-Chu sequence with the logical index RACH_ROOT_SEQUENCE, where RACH_ROOT_SEQUENCE is broadcasted as part of the system information [3, 17].

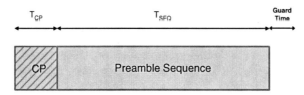

Fig. 15.16 Random access preamble format

The random access procedure in LTE consists of four steps. In step 1, the preamble is sent by UE. The time/frequency resource where the preamble is sent is associated with a random access radio network temporary identifier (RA-RNTI). In step 2, a random access response is generated by eNB and is sent on the downlink-shared channel. It is addressed to the UE using the temporary identifier and contains a timing advance value, an uplink grant, and a temporary identifier. Note that eNB may generate multiple random access responses for different UEs which can be concatenated inside one MAC protocol data unit. The preamble identifier is contained in the MAC sub-header of each random access response so that the UE can detect whether a random access response for the used preamble exists. In step 3, UE will send an RRC CONNECTION REQUEST message on the uplink common control channel, based on the uplink grant received in the previous step. In step 4, the eNB sends back a MAC PDU containing the uplink CCCH service data unit that was received in step 3. The message is sent on downlink-shared channel and addressed to the UE via the temporary identifier. When the received message matches the one sent in step 3, the contention resolution is considered successful [14, 17]; otherwise the procedure is restarted at step 1.

Paging is used for setting up network-initiated connection. A power-saving efficient paging procedure should allow the UE to sleep without having to process any information and only to briefly wake up at predefined intervals to monitor paging information (as part of control signaling in the beginning of each subframe) from the network [17].

15.4.10 Cell Search

Cell search is a procedure performed by a UE to acquire timing and frequency synchronization with a cell and to detect the cell ID of that cell. LTE uses a hierarchical cell search scheme similar to WCDMA. E-UTRA cell search is based on successful acquisition of the following downlink signals: (1) the primary and secondary synchronization signals that are transmitted twice per radio frame and (2) the downlink reference signals as well as based on the broadcast channel that carries system information such as system bandwidth, number of transmit antennas, and system frame number. The 504 available physical layer cell identities are grouped into 168 physical layer cell identity groups, each group containing three unique identities. The secondary synchronization signal carries the physical layer cell identity group and the primary synchronization signal carries the physical layer identity 0, 1, or 2 [6].

As shown in Fig. 15.14, the primary synchronization signal is transmitted on the sixth symbol of slots 0 and 10 of each Type 1 radio frame; it occupies 62 subcarriers, centered on the DC subcarrier. The primary synchronization signal is generated from a frequency domain Zadoff-Chu sequence. The secondary synchronization signal is transmitted on the fifth symbol of slots 0 and 10 of each radio frame. It occupies 62 subcarriers centered on the DC subcarrier. The second synchronization signal is an interleaved concatenation of two length-31 M-sequences. The concatenated sequence is scrambled with a scrambling sequence given by the primary synchronization signal [14, 17].

As shown in Fig. 15.14, PBCH is transmitted on symbols 0–3 of slot 1 and occupies 72 subcarriers centered on the DC subcarrier. The coded BCH transport block is mapped to four subframes within a 40 ms interval. The 40 ms timing is blindly detected, i.e., there is no explicit signaling to indicate 40 ms timing [6]. These channels are contained within the central 1.08 MHz (corresponding to six resource blocks) frequency band so that the system operation can be independent of the channel bandwidth.

15.4.11 Link Adaptation

Uplink link adaptation is used in order to improve data throughput in a fading channel. This technique varies the downlink modulation and coding scheme based on the channel conditions of each user. Different types of link adaptation are performed according to the channel conditions, the UE capability (such as the maximum transmission power and maximum transmission bandwidth), and the required QoS (such as the data rate, latency, and packet error rate). These link adaptation methods are as follows: (1) adaptive transmission bandwidth, (2) transmission power control, and (3) adaptive modulation and coding [6].

15.4.12 Multi-antenna Techniques in LTE

For the LTE downlink, a 2×2 configuration for MIMO is assumed as baseline configuration, i.e., two transmit antennas at the base station and two receive antennas at the terminal side [3, 6, 14, 17]. Configurations with four transmit or receive antennas are also supported in the specifications. Different downlink MIMO modes are supported in LTE, which can be adapted based on channel condition, traffic requirements, and UE capability. Those include transmit diversity, open-loop spatial multiplexing (no UE feedback), closed-loop spatial multiplexing (with UE feedback), multi-user MIMO (more than one UE is assigned to the same resource block), and closed-loop rank= 1 precoding.

In LTE spatial multiplexing, up to two codewords can be mapped onto different layers. One codeword represents an output from the channel coder. The number of layers available for transmission is equal to the rank of the channel matrix. Precoding in transmitter side is used to support spatial multiplexing. This is achieved by multiplying the signal with a precoding matrix prior to transmission. The optimum precoding matrix is selected from a predefined codebook which is known to both eNB and UE. The optimum precoding matrix is the one which maximizes the capacity.

The UE estimates the channel and selects the optimum precoding matrix. This feedback is provided to the eNB. Depending on the available bandwidth, this information is made available per resource block or group of resource blocks, since the optimum precoding matrix may vary between resource blocks. The network may configure a subset of the codebook that the UE is able to select from. In case of UEs with high velocity, the quality of the feedback may deteriorate. Thus, an open-loop spatial multiplexing mode is also supported which is based on predefined settings for spatial multiplexing and precoding. In case of four antenna ports, different precoders are assigned cyclically to the resource elements. The eNB will select the optimum MIMO mode and precoding configuration. The information is conveyed to the UE as part of the downlink control information on PDCCH.

In order for MIMO schemes to work properly, each UE has to report information about the channel to the base station. Several measurement and reporting schemes are available which are selected according to MIMO mode of operation and network choice. The reporting may include wideband or narrowband channel quality indicator (CQI) which is an indication of the downlink radio channel quality as experienced by this UE, precoding matrix indicator (PMI) which is an indication of the optimum precoding matrix to be used in the base station for a given radio condition, and rank indication (RI) which is the number of useful transmission layers when spatial multiplexing is used [3, 14, 17].

In case of transmit diversity mode, only one codeword can be transmitted. Each antenna transmits the same information stream, but with different coding. LTE employs space frequency block coding (SFBC) as transmit diversity scheme. A special precoding matrix is applied at transmitter side in the precoding stage in Fig. 15.11.

Cyclic delay diversity (CDD) is an additional type of diversity which can be used in conjunction with spatial multiplexing in LTE. An antenna-specific delay is applied to the signals transmitted from each antenna port. This effectively introduces artificial multi-path to the signal as seen by the receiver. As a special method of delay diversity, cyclic delay diversity applies a cyclic shift to the signals transmitted from each antenna port [14].

For the LTE uplink, multi-user MIMO (MU-MIMO) can be used. Multiple user terminals may transmit simultaneously on the same resource block. The scheme requires only one transmit antenna at UE side. The UEs sharing the same resource block have to apply mutually orthogonal pilot patterns. To take advantage of two or more transmit antennas, transmit antenna selection can be used. In this case, the UE has two transmit antennas but only one transmission chain. A switch will then choose the antenna that provides the best channel to the eNB [14, 17].

15.5 Overview of the LTE Layer 2

The Layer 2 functions in LTE are classified into the following categories: medium access control (MAC) functions, radio link control (RLC) functions, and Packet Data Convergence Protocol (PDCP) functions [6–10]. Figures 15.17 and 15.18 illustrate the structure of Layer 2 in LTE downlink and uplink. The service access point (SAP) for peer-to-peer communication is marked with circles at the interface between the sublayers. The SAP between the physical layer and the MAC sublayer provides the transport channels. The SAP between the MAC sublayer and the RLC sublayer provides the logical channels. The multiplexing of several logical channels (i.e., radio bearers) on the same transport channel (i.e., transport block) is performed by the MAC sublayer [6].

The services and functions provided by the MAC sublayer can be summarized as follows:

- Mapping between logical channels and transport channels.
- Multiplexing/de-multiplexing of RLC protocol data units corresponding to one or different radio bearers into/from transport blocks delivered to/from the physical layer on transport channels.
- Traffic volume measurement reporting.
- Error correction through HARQ.
- Priority handling between logical channels of one UE.
- Priority handling between UEs through dynamic scheduling.
- Transport format selection.

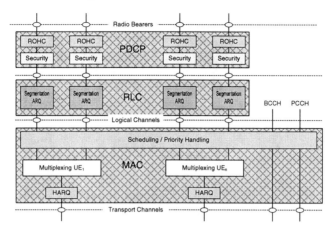

Fig. 15.17 LTE Layer 2 structure in the downlink ([6])

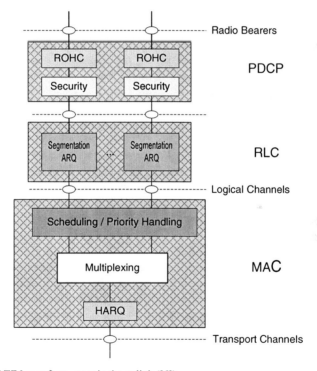

Fig. 15.18 LTE Layer 2 structure in the uplink ([6])

15.5.1 Logical and Transport Channels

The different logical and transport channels in LTE are illustrated in Figs. 15.19 and 15.20, respectively. Each logical channel type is defined by what type of information is transferred. The logical channels are generally classified into two groups:

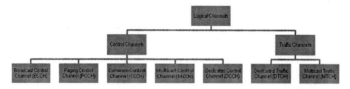

Fig. 15.19 Classification of LTE logical channels

Fig. 15.20 Classification of LTE transport channels

(1) control channels (for the transfer of control-plane information) and (2) traffic channels (for the transfer of user-plane information) as shown in Fig. 15.19 [6].

The control channels are exclusively used for transfer of control-plane information. The control channels supported by MAC can be classified as follows (see Fig. 15.19):

- Broadcast control channel (BCCH): A downlink channel for broadcasting system control information.
- Paging control channel (PCCH): A downlink channel that transfers paging information and system information change notifications. This channel is used for paging when the network does not know the location of the UE.
- Common control channel (CCCH): Channel for transmitting control information between UEs and eNBs. This channel is used for UEs having no RRC connection with the network.
- Multicast control channel (MCCH): A point-to-multipoint downlink channel used for transmitting MBMS control information from the network to the UE, for one or several MTCHs. This channel is only used by UEs that receive MBMS.
- Dedicated control channel (DCCH): A point-to-point bi-directional channel that transmits dedicated control information between a UE and the network. It is used by UEs that have RRC connection.

The traffic channels are exclusively used for the transfer of user-plane information. The traffic channels supported by MAC can be classified as follows (as shown in Fig. 15.20):

- Dedicated traffic channel (DTCH): A point-to-point bi-directional channel dedicated to a single UE for the transfer of user information.
- Multicast traffic channel (MTCH): A point-to-multipoint downlink channel for transmitting traffic data from the network to the UE. This channel is only used by UEs that receive MBMS.

The physical layer provides information transfer services to MAC and higher layers. The physical layer transport services are described by how and with what characteristics data are transferred over the radio interface. This should be clearly separated from the classification of what is transported, which relates to the concept of logical channels at MAC sublayer. As shown in Fig. 15.20, downlink transport channels can be classified as follows:

- Broadcast channel (BCH) that is characterized by fixed, predefined transport format and is required to be broadcast in the entire coverage area of the cell.
- Downlink shared channel (DL-SCH) that is characterized by support for HARQ; support for dynamic link adaptation by varying the modulation, coding, and transmit power; possibility for broadcast in the entire cell; possibility to use beamforming; support for both dynamic and semi-static resource allocation; support for UE discontinuous reception to enable power saving; and support for MBMS transmission.
- Paging channel (PCH) that is characterized by support for UE discontinuous reception in order to enable power saving, requirement for broadcast in the entire coverage area of the cell, mapped to physical resources which can be used dynamically also for traffic or other control channels.
- Multicast channel (MCH) which is characterized by requirement to be broadcast in the entire coverage area of the cell, support for macro-diversity combining of MBMS transmission on multiple cells, support for semi-static resource allocation. The uplink transport channels are classified as follows (see Fig. 15.20):
- Uplink shared channel (UL-SCH) which is characterized by possibility to use beamforming, support for dynamic link adaptation by varying the transmit power and modulation and coding schemes, support for HARQ, support for both dynamic and semi-static resource allocation.
- Random access channel (RACH) that is characterized by limited control information and collision risk.

The mapping of the logical channels to the transport channels in the downlink and uplink is shown in Fig. 15.21. As shown in Figs. 15.17 and 15.18, the main services and functions provided by the RLC sublayer include transfer of upper-layer PDUs supporting AM or UM, TM data transfer, error correction through ARQ (since CRC check is provided by the physical layer, no CRC is needed at RLC level), segmentation according to the size of the transport block, re-segmentation of PDUs that need to be retransmitted, concatenation of SDUs for the same radio bearer, in-sequence delivery of upper-layer PDUs except during handover, duplicate detection, and protocol error detection and recovery.

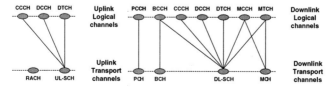

Fig. 15.21 Mapping of logical to transport channels in the downlink and uplink

15.5.2 ARQ and HARQ in LTE

E-UTRA provides ARQ and HARQ functionalities. The ARQ functionality provides error correction by retransmissions in acknowledged mode at Layer 2. The HARQ functionality ensures delivery between peer entities at Layer 1. The HARQ within the MAC sublayer is characterized by an N-process stop-and-wait protocol and re-transmission of transport blocks upon failure of earlier transmissions. The ACK/NACK transmission in FDD mode refers to the downlink packet that was received four subframes earlier. In TDD mode, the uplink ACK/NACK tim-ing depends on the uplink/downlink configuration. For TDD, the use of a single ACK/NACK response for multiple PDSCH transmissions is possible. A total of eight HARQ processes are supported [6, 17].

An asynchronous adaptive HARQ is utilized in the downlink. The uplink ACK/NACK signaling in response to downlink (re)transmissions is sent on PUCCH or PUSCH. The PDCCH signals the HARQ the process number and whether it is a transmission or retransmission. The retransmissions are always scheduled through PDCCH. A synchronous HARQ scheme is supported in the uplink. The maximum number of retransmissions can be configured per UE basis as opposed to per radio bearer. The downlink ACK/NACK signaling in response to uplink (re)transmissions is sent on PHICH. The ARQ functionality within the RLC sublayer is responsible for retransmission of RLC PDUs or RLC PDU segments. The ARQ retransmissions are based on RLC status reports and optionally based on HARQ/ARQ interactions. The polling for RLC status report is used when needed by RLC and status reports can be triggered by upper layers [6].

15.5.3 Packet Data Convergence Sublayer (PDCP)

Services and functions provided by the PDCP sublayer for the U-plane include header compression and decompression, transfer of user data between NAS and RLC layer, sequential delivery of upper-layer PDUs at handover for RLC AM, du-plicate detection of lower-layer SDUs at handover for RLC AM, retransmission of PDCP SDUs at handover for RLC AM, and ciphering.

Services and functions provided by the PDCP for the C-plane include ciphering and integrity protection and transfer of control-plane data where PDCP receives PDCP SDUs from RRC and forwards it to the RLC layer and vice versa.

15.6 Radio Resource Control Functions (RRC)

The main services and functions of the RRC sublayer include [6] the following:

- Broadcast of system information.
- Paging.

- Establishment, maintenance, and release of an RRC connection between the UE and E-UTRAN including allocation of temporary identifiers between UE and E-UTRAN and configuration of signaling radio bearer(s) for RRC connection.
- Security functions including key management.
- Establishment, configuration, maintenance, and release of point-to-point radio bearers.
- Mobility functions including UE measurement reporting and control of the reporting for inter-cell and inter-RAT mobility, handover, UE cell selection and reselection and control of cell selection and reselection, context transfer at handover.
- Establishment, configuration, maintenance, and release of radio bearers for MBMS services.
- QoS management functions.
- UE measurement reporting and control of the reporting.

The RRC consists of the following states:

- RRC_IDLE is a state where a UE-specific discontinuous reception (DRX) may be configured by upper layers. In the idle mode, the UE is saving power and does not inform the network of each cell change. The network knows the location of the UE to the granularity of a few cells, called the tracking area (TA). The UE monitors a paging channel to detect incoming traffic, performs neighboring cell measurements and cell selection/reselection, and acquires system information.
- RRC_CONNECTED is a state where transfer of unicast data to/from UE is performed and the UE may be configured with a UE-specific DRX or discontinuous transmission (DTX). The UE monitors control channels associated with the shared data channel to determine whether data are scheduled for it, provides channel quality and feedback information, performs neighboring cell measurements and measurement reporting, and acquires system information.

The main difference between MAC and RRC control lies in the signaling reliability. The signaling corresponding to state transitions and radio bearer configurations should be performed by RRC sublayer due to signaling reliability. The different characteristics of MAC and RRC control are summarized in Table 15.6.

Table 15.6 Summary of the differences between MAC and RRC control ([8])

	MAC Control		RRC Control
Control Entity	MAC		RRC
Signaling Type	PDCCH	MAC Control PDU	RRC Message
Signaling Reliability	$\sim 10^{-2}$ (no retransmission)	$\sim 10^{-3}$ (after HARQ)	$\sim 10^{-6}$ (after ARQ)
Control Latency	Very Short	Short	Longer
Extensibility	None	Limited	High
Security	No Integrity Protection No Ciphering	No Integrity Protection No Ciphering	Integrity Protected Ciphering

15.7 Mobility Management and Handover in LTE

In order to support mobility, an LTE-compliant UE may conduct the following measurements (physical layer measurements include reference signal received power and E-UTRA carrier received signal strength indicator [6]):

- Intra-frequency E-UTRAN measurements.
- Inter-frequency E-UTRAN measurements.
- Inter-RAT measurements.

Measurement commands are used by E-UTRAN to order the UE to start measurements, modify measurements, or stop measurements. The reporting criteria that are used include event-triggered reporting, periodic reporting, and event-triggered periodic reporting. In E-UTRAN RRC_CONNECTED state, network-controlled UE-assisted handovers are performed and various DRX cycles are supported. In E-UTRAN RRC_IDLE state, cell reselections are performed and DRX is supported. The intra-E-UTRAN-access mobility support for UEs in evolved packet system (EPS) connection management (ECM-CONNECTED) handles all necessary steps for relocation or handover procedures, such as processes that precede the final handover decision on the serving network, preparation of resources on the target network, commanding the UE to the new radio resources, and finally releasing resources on the serving network side. It contains mechanisms to transfer context data between eNBs and to update node relations on C-plane and U-plane. The UE makes measurements of attributes of the serving and neighbor cells to enable the process [6].

Depending on whether the UE needs transmission/reception gaps to perform the relevant measurements, measurements are classified as gap-assisted or non-gap assisted. A non-gap-assisted measurement is a measurement on a cell that does not require transmission/reception gaps to allow the measurement to be performed. A gap-assisted measurement is a measurement on a cell that does require transmission/reception gaps to allow the measurement to be performed. Gap patterns are configured and activated by the RRC functional block [6].

The handover procedure is performed without EPC involvement, i.e., preparation messages are directly exchanged between the eNBs. The release of the resources at the serving eNB (or source eNB) during the handover completion phase is triggered by the eNB. Figure 15.22 illustrates the handover procedure between eNBs within the same MME/S-GW. More information on the handover procedure can be found in [6].

After the downlink path is switched at the S-GW, downlink packets on the forwarding path and on the new direct path may arrive interchanged at the target eNB (see Fig. 15.22). The target eNB should first deliver all forwarded packets to the UE before delivering any of the packets received on the new direct path.

Upon handover, the serving eNB forwards in an orderly manner all downlink PDCP SDUs that have not been acknowledged by the UE to the target eNB. The serving eNB discards any remaining downlink RLC PDUs. Correspondingly, the

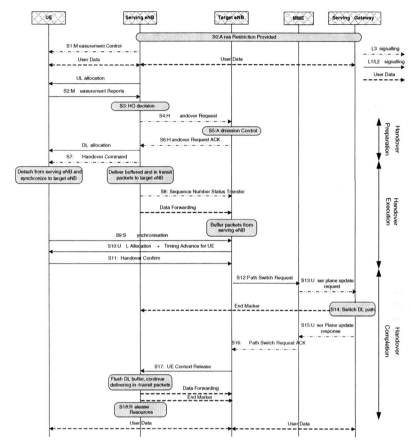

Fig. 15.22 Intra-MME/S-GW handover procedures ([6])

source eNB does not forward the downlink RLC context to the target eNB. Measurements to be performed by a UE for intra/inter-frequency mobility can be controlled by E-UTRAN, using broadcast or dedicated control/signaling. In RRC_IDLE state, a UE follows the measurement parameters defined for cell reselection. In RRC_CONNECTED state, a UE follows the measurement configurations specified by RRC managed by the E-UTRAN. Intra-frequency and inter-frequency neighbor cell measurements are defined as follows:

• Neighbor cell measurements performed by the UE are intra-frequency measurements when the current and target cell operate on the same carrier frequency.
• Neighbor cell measurements performed by the UE are inter-frequency measurements when the neighbor cell operates on a different carrier frequency compared to the current cell.

Table 15.7 Simulation parameters

Parameter	Assumption/value
Cellular layout	Hexagonal grid, 19 cell sites, 3 sectors per site
Inter-site distance (ISD)	500 m (cases 1 and 2), 1732 m (case 3)
Distance-dependent path loss	$L = I + 37.6 \log 10(R)$, R (km) I=128.1 at 2 GHz
Lognormal shadowing	Similar to UMTS 30.03, B 1.41.4 (TR 101.112)
Shadowing standard deviation	8 dB
Correlation distance of shadowing	50 m
Shadowing correlation	Between cells 0.5
	Between sectors 1.0
Penetration loss	10 dB (cases 1 and 3), 20 dB (case 2)
Carrier frequency	2.0 GHz
Channel model	Typical urban (TU)
UE speeds	3 km/h (cases 1 and 3), 30 km/h (case 2)
Total BS TX power	46 dBm (10 MHz bandwidth)
UE power class	24 dBm
Inter-cell interference modeling	UL: explicit modeling (all cells occupied by UEs)
Minimum distance between UE and cell	≥ 35 m
Traffic model	VoIP, full buffer
Number of users per cell	10

15.8 LTE Performance

Table 15.7 shows the simulation parameters for performance evaluation of
E-UTRA against the requirements [13]. Downlink performance for E-UTRA has
been evaluated assuming non-ideal channel estimation with realistic CQI feedback
and frequency-selective scheduling with 10 users per sector. For MIMO, explicit
frequency/spatial multi-cell interference is modeled. Control channel overhead is
also explicitly modeled ($n = 3$). The resource block is composed of 12 subcarriers
(180 kHz) by seven OFDMA symbols.

Uplink performance for E-UTRA has been evaluated assuming non-ideal channel estimation. Localized frequency division multiplexing is used and no uplink
sounding is assumed. Therefore, only long-term average channel quality and received interference levels are known at the base station. Although resource blocks
are multiplexed in the frequency domain, frequency-selective scheduling is not used
in the simulation.

VoIP performance is calculated based on AMR codec (12.2 kbps source rate)
with 50% voice activity factor assuming a two-state Markov model. Control channel
overhead is also explicitly modeled (i.e., $n = 3$ symbols are used for Layer 1/Layer
2 control signaling). Group scheduling in the uplink was also assumed.

15.9 Future Work Toward IMT-Advanced

To address the requirements for systems beyond IMT-2000 or IMT-Advanced specified by ITU-R [19], the 3GPP has embarked on a new work item to further extend

Table 15.8 Summary of LTE performance for case 1 (source 3GPP)

Performance metric	Downlink	Uplink	Assumptions
Peak user data rate (bps/Hz)	7.5 bps/Hz	3.75 bps/Hz	DL: 2 × 2 MIMO, 64QAM
			UL: 1 × 2 MIMO, 64QAM
	(150 Mbps @ 20 MHz)	(75 Mbps @ 20 MHz)	$R = 1$ (code rate)
Sector throughput (bps/Hz/cell)	1.63	0.86	
5% CDF cell edge User throughput (bps/Hz/cell)	0.05	0.03	DL: 2 × 2 MIMO UL: 1 × 2 MIMO
VoIP capacity (active users/MHz/cell)	68	44	
U-plane latency (ms)	5		
Connection setup time (ms)	50–80		20% HARQ retransmission probability
Handoff interruption time (ms)	12–18		

the capabilities of LTE air interface in order to meet the requirements for future wireless applications and services. The LTE-Advanced is expected to be part of 3GPP Release 10 and completed by the end of 2010. The IMT-Advanced compliant technologies may be available for commercial deployment beyond 2011. The requirements for LTE-Advanced are provided in [12]. The following are some of the unique features that are being considered as part of LTE-Advanced development:

- Multi-carrier support: RF carrier aggregation is considered for LTE-Advanced in order to support downlink transmission bandwidths larger than 20 MHz. In this case, two or more component RF carriers, each with a bandwidth up to 20 MHz, are aggregated. A terminal may simultaneously receive one or multiple component carriers depending on its capabilities.
- Spatial multiplexing: Extension of LTE downlink spatial multiplexing to eight layers is considered for LTE-Advanced. Uplink spatial multiplying up to four layers is under consideration.
- Coordinated multiple point transmission and reception: Coordinated multipoint transmission/reception is considered for LTE-Advanced as a tool to improve the coverage of high data rates, the cell-edge throughput and/or to increase system throughput. Downlink coordinated multipoint transmission implies dynamic coordination among multiple geographically separated transmission points (e.g., base stations). Examples of coordinated transmission schemes include coordinated scheduling and/or beamforming. This means that data may be simultaneously transmitted to a single UE from one of the transmission points. Furthermore, the scheduling is coordinated across multiple transmission points to control the inter-cell interference. Joint processing or transmission of data to

a single UE through coherent or non-coherent simultaneous transmission from multiple transmission points to improve the received signal quality. Coordinated transmission in the uplink multipoint reception implies reception of the transmitted signal at multiple, geographically separated points. Scheduling decisions can be coordinated among cells to control interference.

- Relaying functionality: Relaying is considered for LTE-Advanced to improve the coverage and throughput, group mobility, temporary network deployment and to provide coverage in new areas. The relay node is connected to the radio access network. The connection can be either in-band, where the network-to-relay link shares the same band with direct network-to-UE links, or out-of-band, where the network-to-relay link does not operate in the same band as the direct network-to-UE link.

15.10 Conclusions

3GPP has recently completed development of Release 8 specifications, specifying a new IP-based radio access network as the evolution of UTRAN. The 3GPP LTE RAN substantially improves end-user throughputs, sector capacity and reduces user-plane and control-plane latencies, bringing significantly improved user experience with full mobility. 3GPP has further specified the evolved packet core architecture to support the E-UTRAN through reduction in the number of network elements and simplification of functionality, providing the network operators the ability to deliver a seamless mobility experience.

In this chapter, a comprehensive overview of 3GPP LTE radio access technology and E-UTRAN was provided using a top-down systematic approach. All key functional elements and performance features of LTE protocols were described in a hierarchical manner starting from the overall network architecture and further focusing on various components of the radio access network.

References

3GPP Specifications used in this document can be found at http://www.3gpp.org/ftp/Specs/html-info/36-series.htm.

1. 3GPP TS 25.913, Requirements for Evolved UTRA (E-UTRA) and Evolved UTRAN (E-UTRAN), March 2006.
2. 3GPP TS 36.201, Evolved Universal Terrestrial Radio Access (E-UTRA); Long Term Evolution (LTE) Physical Layer; General Description, September 2008.
3. 3GPP TS 36.211, Evolved Universal Terrestrial Radio Access (E-UTRA); Physical Channels and Modulation, September 2008.
4. 3GPP TS 36.212, Evolved Universal Terrestrial Radio Access (E-UTRA); Multiplexing and Channel Coding, September 2008.
5. 3GPP TS 36.213, Evolved Universal Terrestrial Radio Access (E-UTRA); Physical Layer Procedures, September 2008.

6. 3GPP TS 36.300, Evolved Universal Terrestrial Radio Access (E-UTRA) and Evolved Universal Terrestrial Radio Access Network (E-UTRAN); Overall description; Stage 2, September 2008.
7. 3GPP TS 36.321, Evolved Universal Terrestrial Radio Access (E-UTRA); Medium Access Control (MAC) Protocol Specification, September 2008.
8. 3GPP TS 36.322, Evolved Universal Terrestrial Radio Access (E-UTRA); Radio Link Control (RLC) Protocol Specification, September 2008.
9. 3GPP TS 36.323, Evolved Universal Terrestrial Radio Access (E-UTRA); Packet Data Convergence Protocol (PDCP) Specification, September 2008.
10. 3GPP TS 36.331, Evolved Universal Terrestrial Radio Access (E-UTRA); Radio Resource Control (RRC); Protocol Specification, September 2008.
11. 3GPP TS 36.401, Evolved Universal Terrestrial Radio Access Network (E-UTRAN); Architecture Description, September 2008.
12. 3GPP TR 36.913, Requirements for Further Advancements for E-UTRA (LTE-Advanced), June 2008.
13. Long Term Evolution (LTE): Overview of LTE Air-Interface, Technical White Paper, Motorola, Inc.
14. UMTS Long Term Evolution (LTE) Technology Introduction, ROHDE & SCHWARZ, September 2008. (http://www.rohde-schwarz.com).
15. 3GPP Long Term Evolution: System Overview, Product Development, and Test Challenges, Agilent Technologies, May 2008. (http://www.agilent.com)
16. Hyung G. Myung, Technical Overview of 3GPP LTE, May 2008. (http://hgmyung.googlepages.com/scfdma).
17. E. Dahlman et al., 3G Evolution: HSPA and LTE for Mobile Broadband, 2nd Edition, Oxford, UK: Academic Press, October 2008.
18. Hyung G. Myung, Single Carrier FDMA, May 2008. (http://hgmyung.googlepages.com/scfdma).
19. ITU Global Standard for International Mobile Telecommunications IMT-Advanced, http://www.itu.int/ITU-R.
20. 3GPP TS 36.101, Evolved Universal Terrestrial Radio Access (E-UTRA); User Equipment (UE) radio transmission and reception, September 2008.

Index

3GPP
   physical layer, 438
3GPP LTE
   frame structure, 442
   overview, 430–464
   protocol structure, 436
   system requirements, 433

Access techniques
   multi-carrier based, 49
Adaptive denoising, 313
Amplify and forward, 302
Angle of arrival, 8
Aperiodic correlation, 71
ASN profiles, 395

Barker sequences, 71
Basic access channel protocol, 375
Bayesian energy detection, 264
Beam patterns
   array factor, 178
   average, 188
   circular arrays, 183
   distribution, 190
   distribution of maxima in sidelobes, 194
   fixed nodes, 177
   linear arrays, 180
Binary sequences
   merit factor, 71–76
Block modulated multicarrier systems
   interference mitigation, 223
Body area networks, 18–20
Broadcast MIMO systems
   interference suppression, 236

Capacity
   MIMO channel, 16

Cell search, 452
Channel models
   60 GHz and terahertz systems, 22
   body area networks, 18
   short range vehicular networks, 20
   spatial, 9
Channel sounder, 11
   directional, 17
Channel state estimation, 220
Coded multicarrier systems
   interference mitigation, 233
Codes
   Walsh-Hadamard, 227
Cognitive behavior, 255
Collaborative beamforming, 175, 197
   randomly distributed nodes, 185
Compress and forward (CF), 305
Convergence, 318
Cooperative path
   optimal, 207
   routing, 207
Cooperative wireless networks, 199–216
   system model, 202–205

Decentralized algorithms, 325
Decision feedback equalization, 129
Delay spread, 7
Denoise and forward (DNF), 303
Dirty paper channel, 151
Distribution
   Gaussian, 8
   Laplace, 9
   Rayleigh, 6
   Ricean, 6
   Weibull, 6
Doppler spectrum, 6
Dual subgradient method, 318

Equalization, 127

Factor graphs, 121, 124
FDE
  single carrier, 52
Flow priority, 203
Forward link subcarrier allocations, 357

Hadamard difference sets
  cyclic, 64
Hadamard transform, 67
Harmonic mean fairness, 271
Horizontal spectrum sharing, 252

IEEE 802.22, 280
Interference rejection, 217–248
Interference suppression, 218
Iterative algorithm
  convergence, 133
Iterative receivers, 119–136

Jacobi sequences, 75

Legendre sequences, 75
Linear precoding, 237
Low correlation quaternary, 78
LTE
  ARQ, 458
  HARQ, 458
  mobility management, 460
  multi-antenna techniques, 453
  overview of layer 2, 454–457
  performance, 462

MAC layer
  IEEE802.16, 390
MAP sysmbol detection, 119–121
Max-min fairness, 271
MC-CDMA, 227
Merit factor
  binary sequences, 71–76
MIMO
  asymptotic analysis, 149
  Capacity, 141
  Channel equalization, 167
  Closed loop receiver design for uplink, 160
  Closed loop transmitter design for downlink,
    161
  detection, 131
  fading channel, 147
  Gaussian, broadcast channel, 150
  Gaussian, capacity region, 142
  Limited feedback systems, 167
  System model, 140

Mobile WiMAX
  multiple antennas, 409
Modulation and coding, 444
Multi-antenna measurements, 12
Multi-carrier transmission, 31
Multi-user MIMO
  fundamentals, 139–170, 173
Multicarrier direct sequence, 219
Multicast, 317
Multilevel coded modulation, 133
Multiple access schemes, 438
Multiuser detection, 130

Network coding, 317
Network state information, 205
Nonlinear precoding, 239

OFDM, 29–61, 229
  advantages, 42
  analytical model, 35–42
  co-channel interference in cellular, 46
  disadvantages, 45–46
  history and development, 30
  SCFDE analogies and differences, 54
  SCFDE and interoperability, 56
  spectral efficiency, 43
  system design issues, 46
  transceiver design, 34
OFDM in mobile applications
  Doppler sensitivity, 235
OFDM-CDMA, 50
OFDM-TDMA, 49, 51
OFDMA, 49
  future applications, 58
Optimization, 318

Packet data convergence sublayer, 458
Parity check codes, 307
Partially Observable Markov Decision Process
    (POMDP), 276
Peak-to-average power ratio(PAPR), 45
Periodic correlation, 72
Physical control channels, 448
Physical layer cooperation, 200
Physical resource blocks, 443
Power distribution law, 100–106
Precoding
  lattice reduction method, 246
Prediction error filter, 229
Primal recovery, 326
Primary exclusive regions, 268
Proportional fairness, 271

Radio channel, 9

Radio resource control, 458
Random access channel, 450

SC-FDMA
    overview, 442
    uplink multiple access, 441
Sequences
    GMW, 65
    low correlation, 63–88
    low correlation QAM, 76
    low correlation zone, 84
Signalling constellations, 312
Single carrier direct sequence, 219
Single-input multiple-output (SIMO), 273
Spectrum map, 278
Spectrum overlay, 254
Spectrum underlay, 253
Sum-product algorithm, 121
Superposition coding, 297

TDE vs FDE, 52
Theory of optimal stopping, 277
Time dispersion, 7
Traversal algorithms, 210
Two-way relaying
    decoding at the relay, 295
    without decoding at the relay, 302

Ultra mobile braodband
    architecture, 352–354
Ultra mobile broadband, 351–386
    MAC layer, 374
    physical layer, 355–356
    power control, 383
    reverse link, 366–369
Ultrawideband wireless
    channel models, 13
UMB, see ultra mobile braodband, 351
Uncoded multiarrier systems
    interference mitigation, 224

Vector metwork analyzer, 12
Vector precoding, 244
Vertical spectrum sharing, 252
Virtual array, 17

Wideband channels
    linear time varying, 8
WiMAX
    channel coding, 407
    forum, 391
    frame structure, 406
    handoff in MAC layer, 423
    link adaptation, 413
    MAC layer, 416
    mobile, 388–428
    mobility management, 397
    multicast broadcast in MAC layer, 422
    network architecture, 393
    performance, 426
    physical layer, 399
    power control, 413
    power saving mode in MAC layer, 421
    QoS in MAC layer, 420
    S-OFDMA frame structure, 401
    security and authentication, 425
    subchannel permutation, 402
Wireless channels
    channel models, MIMO, 16
    characterization, 4
    measurement, 11–13
    measurement and modeling, 1–24
Wireless networks
    coded bidirectional relaying, 291–315
Wireless systems
    joint rate and power adaptation, 108–111
    optimal data rate allocation, 106–108
    Resource allocation, 93–115, 117
    system model, 95–98

XOR denoising, 312

LaVergne, TN USA
09 September 2009

157354LV00001B/34/P